T0299076

A FIRST COURSE IN
PARTIAL
DIFFERENTIAL
EQUATIONS

A FIRST COURSE IN
PARTIAL
DIFFERENTIAL
EQUATIONS

J Robert Buchanan
Zhoude Shao

Millersville University, USA

World Scientific

NEW JERSEY · LONDON · SINGAPORE · BEIJING · SHANGHAI · HONG KONG · TAIPEI · CHENNAI · TOKYO

Published by

World Scientific Publishing Co. Pte. Ltd.

5 Toh Tuck Link, Singapore 596224

USA office: 27 Warren Street, Suite 401-402, Hackensack, NJ 07601

UK office: 57 Shelton Street, Covent Garden, London WC2H 9HE

Library of Congress Cataloging-in-Publication Data
Names: Buchanan, J. Robert, author. | Shao, Zhoude, author.
Title: A first course in partial differential equations / by J. Robert Buchanan
    (Millersville University, USA), Zhoude Shao (Millersville University, USA).
Description: New Jersey : World Scientific, 2017. | Includes bibliographical references and index.
Identifiers: LCCN 2017030405 | ISBN 9789813226432 (hc : alk. paper)
Subjects: LCSH: Differential equations, Partial--Textbooks.
Classification: LCC QA377 .B8287 2017 | DDC 515/.353--dc23
LC record available at https://lccn.loc.gov/2017030405

British Library Cataloguing-in-Publication Data
A catalogue record for this book is available from the British Library.

Desk Editors: V. Vishnu Mohan/Kwong Lai Fun

Typeset by Stallion Press
Email: enquiries@stallionpress.com

Printed in Singapore

Dedication

JRB, for Monika

ZS, for Yanfen

Preface

This textbook can serve as a student's introduction to partial differential equations, a selection of their solution techniques, properties of their solutions, some of their elementary applications, as well as other related topics. The prerequisites for this material are a course in multivariable calculus and a course in ordinary differential equations (particularly skill with solving first-order and second-order linear ODEs). To the extent possible this textbook is self-contained. Students will find the techniques, results, and theoretical material fully justified by material covered in the prerequisites. In some cases for convenience, concepts from complex variables and linear algebra will be employed. Two appendices are provided on these topics. Instructors can assign those appendices as background reading or expand on the topics as necessary. For the few instances when justification of a result requires theoretical material beyond the scope of undergraduate mathematics, references to the justification in the literature are provided.

There is more material in this textbook than can be covered in a typical, one-semester, undergraduate course in partial differential equations. The authors have field tested this material several times at their home university and have found that a one-semester introduction to PDEs covers material in the first six chapters (possibly omitting the chapter on first-order PDEs and/or including the chapter on Sturm-Liouville theory). For students with no prior exposure to PDEs these first six chapters comprise an introduction to the topic. Instructors teaching a second course in partial differential equations or supervising students conducting an independent study in PDEs may choose from the topics in the last six chapters of this textbook.

The table of contents provides a more detailed outline of the topic coverage in the textbook and should be consulted after this brief introduction.

Chapter 1 introduces the terminology and symbolism of partial differential equations, boundary and initial conditions, and the solution approach of separation of variables. This chapter also includes derivations of the heat and wave equations from physical principles. Chapter 2 covers first-order linear and quasi-linear PDEs and the method of characteristics for solving such equations. This chapter can be skipped or postponed until the second-order PDEs have been explored. With the exception of a later chapter on nonlinear partial differential equations, the remainder of the textbook is independent of Chap. 2. Fourier series, their convergence, and properties are explored in Chap. 3. At a minimum, students should be very comfortable with Chaps. 1 and 3 before proceeding with the later chapters. Chapter 4 focuses on the heat equation and its solution, subject to a variety of boundary conditions. This chapter also includes coverage of the fundamental solution to the heat equation and Maximum/Minimum Principles. Chapter 5 explores the solution to the wave equation and its properties. A discussion of d'Alembert's solution to the wave equation is found there as well. Chapter 6 is devoted to boundary value problems of Laplace's equation, their solutions, and Maximum/Minimum Principles. This chapter also introduces the solution of PDEs on two-dimensional domains related to disks and the Poisson kernel.

The second half of the textbook starts with Chap. 7 on Sturm-Liouville boundary value problems, their solutions and properties. This chapter is more theoretical than the previous chapters and given the students' level of interest can be skimmed before exploring the material on special types of functions for solving boundary value problems in Chap. 8. The latter chapter is essential for an understanding of the applications of partial differential equations explored in Chap. 9. An attempt was made in Chap. 8 to put in one place an introduction to families of functions often seen in the physical sciences. Chapter 9 contains a selection of applications of the solution techniques covered in earlier chapters. This chapter also extends solution techniques to higher dimensional domains such as disks and spheres. Chapter 10 introduces Duhamel's Principle and its use in solving nonhomogeneous initial boundary value problems. Chapter 11 explores nonlinear partial differential equations including Burgers' equation and the Korteweg-de Vries equation. Nonlinear PDEs usually require specialized solution techniques and therefore this chapter is independent of topics in the remainder of the textbook except the first-order topics found in Chap. 2. Chapters 10 and 11 can be studied before Chaps. 7, 8, and 9 if the reader chooses. The final chapter introduces the finite differences method for approximating solutions to initial boundary value problems. Chapter 12 relies on many results gathered from complex variables and linear algebra which are summarized in App. A and App. B respectively.

Instructors and students will find a students' solution manual for a subset of the homework exercises available online at the authors' institutional webpage: `www.millersville.edu/math/`. From that page follow the links to the authors' individual pages. A complete instructors' solution manual to the end-of-chapter exercises is available upon request and verification of faculty status. Please email the authors if interested. This manuscript has been read many times for mathematical accuracy and proofread by both humans and computers, but should any errors have persisted, please email a description of the error (with pages and line numbers) to the authors.

J. R. Buchanan
Z. Shao

About the Authors

Dr. J. Robert Buchanan received a BS in Physics from Davidson College (Davidson, NC) and an MS and PhD in Applied Mathematics from North Carolina State University (Raleigh, NC). His mathematical interests include differential equations and financial mathematics. He is currently a Professor of Mathematics at Millersville University of Pennsylvania (Millersville, PA).

Dr. Zhoude Shao received his BS and MS in Mathematics from Shandong University (Jinan, China) and PhD in Mathematics from the University of Minnesota (Minneapolis and Saint Paul, MN). His mathematical interest lies in differential equations and nonlinear dynamical systems. He is currently a Professor of Mathematics at Millersville University of Pennsylvania (Millersville, PA).

Acknowledgments

The publication of this textbook is the culmination of an eight-year writing project conducted primarily during summer recesses until 2016/2017 when the authors were granted sabbaticals to complete the book. Our thanks are extended to Millersville University and the Sabbatical Leave Committee for the sabbatical leaves which enabled us to complete this project. Special thanks also go to the Francine G. McNairy Library of Millersville University and the Fondren Library of Rice University for access to mathematical books, journals, and other resources used in the composition this book. The authors also thank the teachers, instructors, and professors who introduced us to the subject of partial differential equations and the students of Millersville University who have helped field test sections of this textbook in the classroom.

Contents

Chapter 1

Introduction

When modeling physical phenomena, often it is necessary to consider functions that depend on more than one variable. Therefore, if rates of changes are involved, partial derivatives of functions must be used and this leads to mathematical models that are **partial differential equations**. Loosely speaking, a partial differential equation is an equation that involves an unknown function of two or more variables and its partial derivatives. The following are some examples of partial differential equations:

$$u_t + u\,u_x = 0,$$

$$u_{xx} + u_{yy} = f(x, y),$$

$$u_{xx} + xy^3 u_{yy} - e^x u_z + xyu = 0,$$

$$u_t - u_{xx} + (u_x)^2 = -u,$$

$$u_t + c\,uu_x + u_{xxx} = 0.$$

1.1 Preliminaries: Notation, Definitions, and the Principle of Superposition

In each of the equations above, u represents the unknown function and

$$u_t = \frac{\partial u}{\partial t}, \quad u_{xx} = \frac{\partial^2 u}{\partial x^2}, \quad u_{xy} = \frac{\partial^2 u}{\partial y \partial x}, \quad \cdots$$

represent the partial derivatives of u with respect to the indicated variables. Given any partial differential equation, one of the basic problems is to find the function (or functions) that satisfies (or satisfy) the equation. For any partial differential equation, a function is called a **solution** to the equation if the function and its partial derivatives involved in the equation are all defined on a certain domain and satisfy the partial differential equation on that domain. For example, a solution to the third equation above is defined on a domain in the three-dimensional Euclidean space $\mathbb{R}^3$ and solutions of all the other examples are defined on domains in the two-dimensional Euclidean space $\mathbb{R}^2$. In general, the n-dimensional Euclidean space will be denoted by $\mathbb{R}^n$.

Recall that ordinary differential equations are classified based on the concepts of order and linearity. Similar concepts arise in partial differential equations. The **order** of a partial differential equation is defined to be the highest order of all the partial derivatives of the unknown function that appear in the partial differential equation. In the examples above, the first equation is a first-order partial differential equation, the last one is a third-order equation, and the rest are all second-order equations. A partial differential equation is said to be **linear** if all the terms in the equation are linear in the unknown function and its partial derivatives, that is, each term in the equation contains at most one instance of the unknown function or one of its partial derivatives raised to the first power. If an equation is not linear, it is said to be **nonlinear**. For example, the second and third equations in the examples above are linear, while the rest are nonlinear. Later, nonlinear partial differential equations will be further classified as semi-linear, quasilinear, and (truly) nonlinear. **Systems of partial differential equations** can also be considered. This is necessary in many situations. For example, the real part $u(x, y)$ and the imaginary part $v(x, y)$ of any complex analytic function[1] must satisfy the so-called **Cauchy[2]–Riemann[3] equations:**

$$u_x = v_y,$$
$$u_y = -v_x$$

which is a simple system of (first-order, linear) partial differential equations.

A very important class of partial differential equations in physics and other fields of science and engineering consists of linear partial differential equations of the second order. The general form of second-order, linear partial differential equations with two independent variables, t and x (generally

---

[1] A complex-valued function is analytic if it is complex differentiable on an open set in the complex plane [Krantz (1999)]. While this textbook is devoted to partial differential equations, many useful identities, formulas, and relationships will be taken from the field of complex analysis. Much of this background will be developed in place within the text as it is needed. The reader wishing to see all of the complex arithmetic, algebra, and calculus collected in one place, should consult App. A.

[2] Augustin-Louis Cauchy, French mathematician (1789–1857).

[3] Georg Riemann, German mathematician (1826–1866).

interpreted as time and position respectively), is the following equation

$$Au_{tt} + Bu_{tx} + Cu_{xx} + Du_t + Eu_x + Fu = G, \qquad (1.1)$$

where A, B, C, ..., G are functions of t and x only (one or more of them can be constants or even 0). Equation (1.1) is said to be **homogeneous** if $G(x,t) \equiv 0$. In this case, the equation becomes

$$Au_{tt} + Bu_{tx} + Cu_{xx} + Du_t + Eu_x + Fu = 0. \qquad (1.2)$$

The theory of second-order, linear partial differential equations arises frequently in applications and involves a minimal amount of technicality without sacrificing the richness of the theory of partial differential equations. In fact, the discussion of second-order, linear partial differential equations of various types constitutes a large part of this text.

Recall that in the theory of linear ordinary differential equations, the general solution of an equation can be written as a linear combination of a set of specific solutions. This is possible due to the **Principle of Superposition**. This is not the case for nonlinear differential equations in general. However, the Principle of Superposition still holds for linear partial differential equations. The Principle of Superposition in the context of second-order, linear partial differential equations of two independent variables is stated below.

Theorem 1.1 (Principle of Superposition). *If u_i is a solution of the second-order, linear partial differential equation*

$$Au_{tt} + Bu_{tx} + Cu_{xx} + Du_t + Eu_x + Fu = G_i$$

for $i = 1, 2, \ldots, n$ on a domain $\Omega \subset \mathbb{R}^2$, then for any constants $c_1, c_2, \ldots, c_n$ the function $u = c_1 u_1 + c_2 u_2 + \cdots + c_n u_n$ is a solution of

$$Au_{tt} + Bu_{tx} + Cu_{xx} + Du_t + Eu_x + Fu = \sum_{i=1}^{n} c_i G_i$$

on Ω. In particular, if $G_i = 0$ and u_i is a solution to Eq. (1.2) for $i = 1, 2, \ldots, n$ on a domain $\Omega \subset \mathbb{R}^2$, then for any constants $c_1, c_2, \ldots, c_n$ the function $u = c_1 u_1 + c_2 u_2 + \cdots + c_n u_n$ is also a solution of Eq. (1.2) on Ω.

The proof of Theorem 1.1 is routine and left to the reader in Exercise 1. As corollaries, the sum of any finite number of solutions of Eq. (1.2) and a solution to Eq. (1.1) is also a solution of Eq. (1.1) and the difference of any two solutions of Eq. (1.1) is a solution of the corresponding homogeneous equation of Eq. (1.2).

Given any partial differential equation, the first question to ask is whether all of its solutions can be found. This question does not have a simple answer. An ordinary differential equation usually has infinitely many solutions, and initial conditions are used to determine a certain specific solution. For partial differential equations, there are often **boundary condition(s)** necessitated by the underlying physical problem and associated with the domain of definition of the solution to the equation. For example, for a solution of a given partial differential equation with two independent variables defined on a planar region in $\mathbb{R}^2$, the boundary condition could be specified on the curve that forms the boundary of this region. Distinguishing between spatial variables and the time variable is purely a matter of convenience of interpretation. For partial differential equations that depend on time t, a boundary condition at $t = 0$ is often imposed. Such a boundary condition is usually called an **initial condition**. A partial differential equation along with specified initial and boundary conditions is called an **initial boundary value problem** (IBVP). It is called a **boundary value problem** (BVP) if only boundary (spatial variable) conditions are imposed.

One of the fundamental objectives in the study of partial differential equations is the determination of whether or not an initial boundary value problem or a boundary value problem possesses a unique solution. An (initial) boundary value problem is **well-posed** if there exists a unique solution and the solution depends continuously on the initial and boundary conditions. Continuous dependence means that the solution of an initial boundary value problem varies continuously with respect to the initial and boundary conditions, that is, "small" changes in the initial and/or boundary conditions lead to "small" changes in the solution. In order to be more precise, the notion of "small" must be defined, but this is left to standard texts on real analysis and will not be explored in this textbook. Continuous dependence upon the boundary and initial conditions is sometimes called **stability**. An initial boundary value problem is **ill-posed** if it is not well-posed. This book is mainly concerned with the existence and uniqueness of solutions to (initial) boundary value problems. The question of continuous dependence on initial and boundary conditions will only be dealt with in some special cases.

This section concludes with some simple examples demonstrating the concepts introduced above.

Example 1.1. Find the general solution to the partial differential equation

$$u_{xy} = 0 \text{ for } (x,y) \in \mathbb{R}^2. \tag{1.3}$$

Solution. It is not hard to find the solutions of this partial differential equation. Integrating both sides of Eq. (1.3) with respect to y yields

$$u_x(x,y) = f(x),$$

where f is an arbitrary function of $x \in \mathbb{R}$. Integrating the last equation with respect to x produces

$$u(x,y) = F(x) + G(y), \tag{1.4}$$

where G is an arbitrary function of $y \in \mathbb{R}$ and $F(x)$ is any antiderivative of $f(x)$, that is $F'(x) = f(x)$ for $x \in \mathbb{R}$. The reader can verify that $u(x,y)$ given in Eq. (1.4) represents all the solutions of Eq. (1.3).

Example 1.2. Consider the equation

$$u_{xx} + u_{yy} = 0. \tag{1.5}$$

This is an example of one of the most important types of partial differential equations, **Laplace's**[4] **equation**. It is not as easy to find all the solutions to Laplace's equation as it was to find the general solution to Eq. (1.3) in the last example. However, by observation and trial-and-error, the reader can see that functions

$$u_1(x,y) = x + y,$$
$$u_2(x,y) = x^2 - y^2,$$
$$u_3(x,y) = e^x \cos y,$$
$$u_4(x,y) = \ln(x^2 + y^2),$$
$$u_5(x,y) = \frac{x}{x^2 + y^2}$$

are all solutions of Eq. (1.5). More specifically, functions u_1, u_2, and u_3 solve Eq. (1.5) for all $(x,y) \in \mathbb{R}^2$, while functions u_4 and u_5 solve Eq. (1.5) for all $(x,y) \in \mathbb{R}^2 \setminus \{(0,0)\}$. Based on the solutions listed above, many other solutions can be found. In fact, since Eq. (1.5) is linear, any linear combination of finitely many solutions is also a solution by the Principle of Superposition. Also note that any constant or linear function is also a solution of Laplace's equation. If a boundary condition such as

$$u(x,y) = 0 \text{ for } x^2 + y^2 = 1 \tag{1.6}$$

---

[4]Pierre-Simon Laplace, French scholar (1749–1827).

is imposed, then a solution to Eq. (1.5) that also satisfies Eq. (1.6) is needed. Clearly $u_4(x, y)$ is such a solution for $(x, y) \in \mathbb{R}^2 \setminus \{(0,0)\}$.

The boundary condition of Eq. (1.6) can be generalized to that of

$$u(x, y) = f(x, y) \text{ for } x^2 + y^2 = 1, \tag{1.7}$$

where $f(x, y)$ is a given function defined on the unit circle $\{(x, y) \mid x^2 + y^2 = 1\}$. Equation (1.5) and the boundary condition given in Eq. (1.7) (of which Eq. (1.6) is a special case) form a boundary value problem, which will be discussed in detail in Chap. 6.

Example 1.3. Consider the equation

$$u_t - u_{xx} = 0. \tag{1.8}$$

This is an example of a one-dimensional **heat equation**, another class of partial differential equations of great importance. It can be verified directly that

$$u_1(x, t) = t + \frac{1}{2}x^2,$$
$$u_2(x, t) = e^{-t} \sin x$$

are both solutions of Eq. (1.8). Does there exist a solution to the heat equation satisfying the following initial and boundary conditions?

$$u(x, 0) = 100 + \sin x,$$
$$u(0, t) = u(\pi, t) = 100. \tag{1.9}$$

The boundary and initial conditions imply that a solution to the heat equation must be defined for $0 \le x \le \pi$ and $t \ge 0$. Often the domain of the solution of a partial differential equation will be stated explicitly along with the partial differential equation itself. Sometimes (as for example, here) the domain of the solution must be inferred from the boundary and initial conditions. In this textbook, for clarity and ease of interpretation, the domain of a partial differential equation and its associated initial and boundary conditions (if any) will be stated explicitly.

Again, any constant is a solution of Eq. (1.8). Since Eq. (1.8) is linear and homogeneous, then for any constants c and d

$$u(x, t) = c + d e^{-t} \sin x$$

is also a solution. Setting $c = 100$ and $d = 1$ produces a solution of the initial boundary value problem consisting of Eq. (1.8) and Eq. (1.9).

1.2 Classification of Second-Order Partial Differential Equations

In Sec. 1.1, partial differential equations were classified according to their linearity property and the highest order of the partial derivatives that appear in the equation. This introductory chapter continues with a brief discussion of the further classification of second order linear partial differential equations. In this section the classification of equations is limited to those with two independent variables only.

Recall, a general second-order, linear partial differential equation with two independent variables has the form of Eq. (1.1). This equation is further classified based on the coefficients of the second-order derivative terms. Let Ω be a region in $\mathbb{R}^2$, and classify Eq. (1.1) as

- **elliptic** on Ω if $4AC - B^2 > 0$ for all $(x, t) \in \Omega$,
- **parabolic** on Ω if $4AC - B^2 = 0$ for all $(x, t) \in \Omega$,
- **hyperbolic** on Ω if $4AC - B^2 < 0$ for all $(x, t) \in \Omega$.

Based on the definitions above, a second-order, linear partial differential equation could be classified as one type on one region of $\mathbb{R}^2$ and as a different type on another region since the coefficient functions A, B, and C usually depend on the independent variables x and t.

Example 1.4. Let $f(x, t)$ be a function defined on $\mathbb{R}^2$ and classify each of the following second-order equations as elliptic, parabolic, or hyperbolic.

$$\frac{\partial^2 u}{\partial x^2} + \frac{\partial^2 u}{\partial t^2} + \frac{\partial u}{\partial x} - 2\frac{\partial u}{\partial t} = f(x, t), \qquad (1.10)$$

$$\frac{\partial^2 u}{\partial x^2} - \frac{\partial^2 u}{\partial x \partial t} + 2\frac{\partial^2 u}{\partial t^2} + e^t \frac{\partial u}{\partial x} - \sin(x\,t)\frac{\partial u}{\partial t} = f(x, t), \qquad (1.11)$$

$$u_{xx} - 2u_{xt} + u_{tt} + 2u_x - u_t = 0, \qquad (1.12)$$

$$u_{xx} + x\,u_{tt} = 0, \qquad (1.13)$$

$$\frac{\partial^2 u}{\partial x^2} - \beta\frac{\partial u}{\partial t} + \alpha\frac{\partial u}{\partial x} + \gamma\,u + f(x, t) = \frac{\partial u}{\partial t}. \qquad (1.14)$$

Solution. Using the definitions above, each partial differential equation can be classified as outlined at the beginning of this section.

- For Eq. (1.10) the functions $A = 1$, $B = 0$, and $C = 1$. Therefore $4AC - B^2 = 4 > 0$ and the equation is elliptic on $\mathbb{R}^2$.
- For Eq. (1.11) the functions $A = 2$, $B = -1$, and $C = 1$. Therefore $4AC - B^2 = 7 > 0$ and the equation is elliptic on $\mathbb{R}^2$.

- For Eq. (1.12) the functions $A = 1$, $B = -2$, and $C = 1$. Therefore $4AC - B^2 = 0$ and the equation is parabolic on $\mathbb{R}^2$.
- For Eq. (1.13) the functions $A = x$, $B = 0$, and $C = 1$. Therefore $4AC - B^2 = 4x$ and the equation is hyperbolic on $\Omega_1 = \{(x, y) \in \mathbb{R}^2 \mid x < 0\}$ and elliptic on $\Omega_2 = \{(x, y) \in \mathbb{R}^2 \mid x > 0\}$.
- For Eq. (1.14) the functions $A = 0$, $B = 0$, and $C = 1$. Therefore $4AC - B^2 = 0$ and the equation is parabolic on $\mathbb{R}^2$.

In the next two sections, two important second-order, linear partial differential equations – namely the heat and wave equations are derived. Much of the material in this text will support the effort to solve or elicit properties of these equations or special cases of these mathematical models.

1.3   Heat Conduction and the Heat Equation

In many heat conduction problems in physics and engineering, the determination of the temperature distribution in a certain substance is the primary concern. Heat and temperature are not physically the same concept, but as will be seen, temperature is easier to measure and can be used as a surrogate for heat. This leads to an extended discussion of the heat equation. In fact, a simple example of a heat equation was explored in Example 1.3. In the current section the equation that governs the heat conduction process will be derived from physical principles. The reader will see through this process how partial differential equations arise naturally from real world phenomena.

The simplest case to consider is that of heat conduction in a rod of length L, where L can be measured in any convenient unit of length. The rod is located along the positive x-axis with two end faces (or just ends) located at $x = 0$ and $x = L$. Furthermore the following assumptions are made.

- The lateral sides of the rod are insulated completely, that is, no heat exchange takes place with the surrounding media through the lateral surfaces of the rod.
- The cross sectional area and the mass density (mass per unit length) of the rod are functions of x only, denoted as $A(x)$ and $\rho(x)$ respectively. Such a rod is referred to as a **one-dimensional rod**.
- Within each cross-section, the temperature is evenly distributed. This implies that the temperature distribution is a function of location x and time t only.

- The cross sectional area $A(x)$ and the density $\rho(x)$ are both smooth functions of x and the temperature throughout the rod varies smoothly in both x and t (see Fig. 1.1).

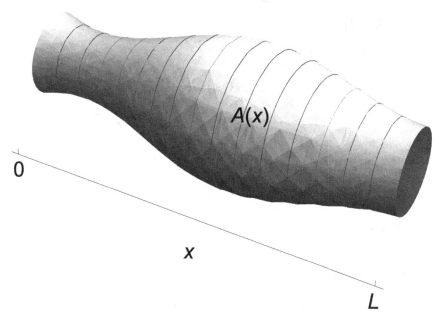

Fig. 1.1   The one-dimensional rod used to illustrate the derivation of the heat equation.

Let $u(x,t)$ denote the temperature of the rod at location $x \in [0, L]$ and at time $t \geq 0$. Let $c(x)$ represent the **specific heat capacity** (or simply specific heat) of the rod (at location $x \in [0, L]$), the amount of heat needed to raise the temperature of the material composing the rod by one degree per unit mass. Specific heat is a physical property of matter: matter with high specific heat capacity needs more heat energy to increase its temperature than matter with low specific heat capacity. Strictly speaking, the specific heat of a material should depend on the temperature u. However, for simplicity, this dependence is ignored. As mentioned earlier, heat and temperature are not the same thing, but they are related through a simple equation. If $q(t)$ denotes the heat energy contained in an interval $[a, b] \subset [0, L]$ of the rod at time t, then the total amount of heat energy present in the rod at time t in the interval $[a, b]$ is the integral of the product of the specific heat, density, volume, and temperature of the

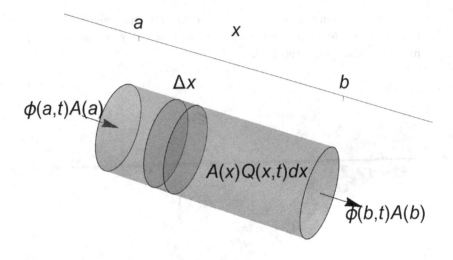

Fig. 1.2   Conservation of heat energy in the portion of the rod occupying the interval $[a, b]$.

rod in the interval, that is

$$q(t) = \int_a^b c(x)\rho(x)A(x)u(x,t)\,dx. \qquad (1.15)$$

How might the heat energy in the interval $[a, b]$ change with time? Two processes will be modeled. First, heat may be generated within the rod itself in the interval $[a, b]$ perhaps by chemical or nuclear reactions. If $Q(x,t)$ denotes the amount of heat generated per unit volume per unit time at location x and time t then the total heat generated per unit time within the interval $[a, b]$ is given by $\gamma(t)$ where

$$\gamma(t) = \int_a^b Q(x,t)A(x)\,dx. \qquad (1.16)$$

The amount of heat energy in $[a, b]$ may also change if heat energy is flowing across the boundaries of the region at $x = a$ and $x = b$. Since the lateral surface of the rod is assumed to be insulated, no heat flow occurs across the lateral surface (see Fig. 1.2). Define the **heat flux** to be the amount of heat energy flowing to the right (the positive x-direction) per unit surface area per unit time. The heat flux at location x and time t is denoted by $\phi(x,t)$.

Since heat energy flowing to the right at $x = a$ is entering the segment of the rod occupying interval $[a, b]$ while heat energy flowing to the right at $x = b$ is leaving, the net balance of heat flux per unit time is

$$\phi(a,t)A(a) - \phi(b,t)A(b) = -\int_a^b \frac{\partial}{\partial x}[\phi(x,t)A(x)]\,dx. \tag{1.17}$$

Thus the change in heat energy in the interval $[a, b]$ per unit time may be described as follows:

$$\begin{pmatrix} \text{Rate of change with} \\ \text{respect to time of the} \\ \text{total heat energy in} \\ \text{the interval } [a,b] \end{pmatrix} = \begin{pmatrix} \text{Rate at which} \\ \text{heat energy} \\ \text{flows in or out} \\ \text{of the two ends} \end{pmatrix} + \begin{pmatrix} \text{Rate at which} \\ \text{heat energy} \\ \text{is generated in} \\ \text{interval } [a,b] \end{pmatrix}.$$

Equations (1.15)–(1.17) express this relation mathematically as

$$\frac{d}{dt}[q(t)] = \phi(a,t)A(a) - \phi(b,t)A(b) + \gamma(t)$$

or

$$\frac{d}{dt}\int_a^b c(x)\rho(x)A(x)u(x,t)\,dx$$

$$= -\int_a^b \frac{\partial}{\partial x}[\phi(x,t)A(x)]\,dx + \int_a^b Q(x,t)A(x)\,dx. \tag{1.18}$$

To further simplify Eq. (1.18) two results from real analysis are employed.

Lemma 1.1 (Leibniz[5] Integral Rule). *If f and $\partial f/\partial t$ are both continuous in some region of the xt-plane, including the region where $a(t) \leq x \leq b(t)$ for $t_1 \leq t \leq t_2$ and if $a(t)$ and $b(t)$ are continuously differentiable for $t_1 \leq t \leq t_2$, then*

$$\frac{d}{dt}\left[\int_{a(t)}^{b(t)} f(x,t)\,dx\right]$$

$$= \int_{a(t)}^{b(t)} \frac{\partial}{\partial t}[f(x,t)]\,dx + f(b(t),t)\frac{db}{dt} - f(a(t),t)\frac{da}{dt}. \tag{1.19}$$

For further generalization of the Leibniz Integral Rule and a proof of Lemma 1.1 see [Flanders (1973)].

Lemma 1.2. *If $f(x)$ is continuous on an interval $[0, L]$, and*

$$\int_a^b f(x)\,dx = 0$$

on any interval $[a, b] \subset [0, L]$, then $f(x) = 0$ for all $x \in [0, L]$.

---

[5]Gottfried Wilhelm Leibniz, German philosopher (1646–1716).

The proof of this lemma can be found in any advanced calculus book, for example [Trench (1978)] or [Gelfand and Fomin (1991), Lemma 1, p. 9]. Combining all three integrals in Eq. (1.18) by using Lemma 1.1, the equation can be re-written as

$$\int_a^b \left[c(x)\rho(x)A(x)\frac{\partial u}{\partial t} + \frac{\partial}{\partial x}\left[\phi(x,t)A(x)\right] - Q(x,t)A(x) \right] dx = 0. \quad (1.20)$$

Since $[a,b] \subset [0,L]$ is arbitrary, then by Lemma 1.2 the integrand in Eq. (1.20) is identically zero on the interval $[0,L]$ and thus the equation below holds.

$$c(x)\rho(x)A(x)\frac{\partial u}{\partial t} = -\frac{\partial}{\partial x}\left[\phi(x,t)A(x)\right] + Q(x,t)A(x), \quad (1.21)$$

where $0 < x < L$ and $t > 0$.

Equation (1.21) actually contains two unknown functions, temperature $u(x,t)$ and heat flux $\phi(x,t)$. Fortunately there is a relationship between temperature and heat flux. This is known as **Fourier's Law of Heat Conduction**. Fourier[6] proposed that heat energy flows from a warmer region toward a cooler region at a rate which is proportional to the difference in the temperatures between the two regions. In terms of heat flux and temperature this may be modeled as

$$\phi(x,t) = -K_0(x)\frac{\partial u}{\partial x}(x,t). \quad (1.22)$$

The nonnegative function $K_0(x)$ is called the **thermal conductivity**, a function that measures the ability of the material to conduct heat. Equation (1.22) may be interpreted as stating that the heat flux (the amount of heat flowing to the right per unit surface area per unit time) is proportional to the rate of change of the temperature with respect to x for any fixed t. For the case of the one-dimensional rod, it is assumed that the thermal conductivity is a function of position only. The negative sign in Eq. (1.22) is due to the fact that heat flows from the hotter part of the rod to the colder part. Substituting Eq. (1.22) into Eq. (1.21) yields

$$c(x)\rho(x)A(x)\frac{\partial u}{\partial t} = -\frac{\partial}{\partial x}\left[-K_0(x)A(x)\frac{\partial u}{\partial x} \right] + Q(x,t)A(x). \quad (1.23)$$

If c, ρ, A, and K_0 are constant, (this is often referred to as the material's **constant thermal property**), Eq. (1.23) can be written as

$$\frac{\partial u}{\partial t} = k\frac{\partial^2 u}{\partial x^2} + \frac{1}{c\rho}Q(x,t), \quad (1.24)$$

Table 1.1 Thermal diffusivities for some common materials.

Material	Diffusivity (cm^2/s)
silver	1.71
copper	1.14
aluminum	0.86
cast iron	0.12
granite	0.011
brick	0.0038
water	0.00144

where $k = K_0/(c\rho) > 0$ is called the **thermal diffusivity** constant. Thermal diffusivity is a material property with units of length^2/time. Table 1.1 lists the thermal diffusivities of some common materials. Equation (1.24) is called the one-dimensional heat equation with heat source. If $Q(x,t) = 0$, that is, there is no heat source inside the rod, the elementary one-dimensional, linear and homogeneous heat equation results,

$$\frac{\partial u}{\partial t} = k\frac{\partial^2 u}{\partial x^2} \text{ for } 0 < x < L \text{ and } t > 0. \tag{1.25}$$

Using the definitions stated in Sec. 1.2 the one-dimensional, homogeneous heat equation is of parabolic type.

1.3.1 *Initial Boundary Value Problems for the Heat Equation*

The heat equation derived above is usually supplemented with initial and boundary conditions. Common initial boundary value problems for the one-dimensional case are discussed in this section.

An **initial condition** represents the temperature distribution at a fixed point in time, usually taken to be $t = 0$. The initial condition will generally be specified using an equation such as

$$u(x,0) = f(x) \text{ for } 0 < x < L, \tag{1.26}$$

where $f(x)$ is a given function defined on $[0, L]$.

There are several types of **boundary conditions** based on the nature of the physical connection between the rod and its environment. Three common boundary conditions are described in the following paragraphs.

Boundary conditions of the first kind, also called **Dirichlet**[7] **boundary conditions**, specify the temperatures at the two ends of the

---

[6] Joseph Fourier, French mathematician and physicist (1768–1830).
[7] Peter Gustav Lejeune Dirichlet, German mathematician (1805–1859).

rod. If the ends are at $x = 0$ and $x = L$, then the general Dirichlet boundary conditions are prescribed as

$$u(0,t) = g_1(t),$$
$$u(L,t) = g_2(t),$$
(1.27)

where $g_1(t)$ and $g_2(t)$ are given functions defined for $t > 0$. A special elementary case of Dirichlet boundary conditions is the following:

$$u(0,t) = u(L,t) = 0 \text{ for } t > 0.$$
(1.28)

The boundary conditions of Eq. (1.28) are homogeneous and may be interpreted as prescribing that the two ends are kept at the constant temperature of zero.

Boundary conditions of the second kind, also called **Neumann**[8] **boundary conditions**, specify the heat fluxes at the two ends of the rod:

$$-K_0 \frac{\partial u}{\partial x}(0,t) = g_1(t),$$
$$K_0 \frac{\partial u}{\partial x}(L,t) = g_2(t),$$
(1.29)

where $g_1(t)$ and $g_2(t)$ are defined for $t > 0$. This is equivalent to specifying the partial derivative with respect to position of the temperature distribution at the two ends. A special case of Neumann boundary conditions is the following requirement for $t > 0$:

$$\frac{\partial u}{\partial x}(0,t) = \frac{\partial u}{\partial x}(L,t) = 0.$$
(1.30)

This homogeneous boundary condition indicates that there is no heat energy flux at the two ends of the rod and may be physically interpreted as implying that the two ends are completely thermally insulated from the surrounding environment.

Boundary conditions of the third kind, also called **Robin**[9] **boundary conditions**, are the following

$$K_0 \frac{\partial u}{\partial x}(0,t) = \alpha(u(0,t) - g_1(t)),$$
$$K_0 \frac{\partial u}{\partial x}(L,t) = -\beta(u(L,t) - g_2(t)),$$
(1.31)

where $g_1(t)$ and $g_2(t)$ are functions defined for $t > 0$. In Eq. (1.31), constants α and β are positive and functions $g_1(t)$ and $g_2(t)$ represent the

---

[8]Carl Gottfried Neumann, German mathematician (1832–1925).
[9]Victor Gustave Robin, French mathematician (1855–1897).

temperature of the media surrounding the two ends of the bar. Therefore, Robin boundary conditions state that the rates of heat loss or gain between the two ends of the rod and surrounding media are proportional to the temperature differences between the ends of the rod and the surrounding media. This is typical in the case when the heat exchange between the rod and the surrounding media is via convection. For example if the rod is hot and the ends are in contact with the air, heat will flow out of the rod. As this is happening, the temperature of the air near the ends will increase and the heat transfer will slow down. From Newton's[10] law of cooling, the rate at which heat leaves the rod is proportional to the temperature difference between the ends of the rod and surrounding media. This leads to the boundary conditions specified by Eq. (1.31). The constants α and β are called the **coefficients of convective heat transfer.**

The one-dimensional heat equation together with an initial condition and a set of boundary conditions described above, forms an initial boundary value problem. The boundary conditions can be one of the three types described above or a combination of the three types, in which case the boundary conditions are described as mixed. In order to avoid confusion and to maintain compatibility between the initial and boundary conditions, note that the initial condition in Eq. (1.26) is specified only for the open interval $(0, L)$ and the boundary conditions of Eqs. (1.27)–(1.31) are imposed only for $t > 0$. Of course, a natural question to ask is whether the initial boundary value problem is well-posed. This question will be explored in Chap. 4.

For higher-dimensional heat equations similar initial boundary value problems can be formulated. This section concludes with several examples illustrating the interconnections between the heat equation and various types of boundary conditions.

Example 1.5. A thin silver bar of length L cm with a completely insulated lateral surface is removed from water boiling at $100°C$. The end of the bar located at $x = 0$ is immersed in a medium that is kept at $0°C$ and the other end is kept at $100°C$. Write down the initial boundary value problem describing the temperature distribution in the silver bar and solve for the steady-state temperature distribution in the bar.

Solution. Since there is no heat source inside the silver bar, the

---

[10]Sir Isaac Newton, English natural philosopher (1642–1727).

temperature distribution $u(x,t)$ should satisfy the heat equation

$$\frac{\partial u}{\partial t} = 1.71\frac{\partial^2 u}{\partial x^2} \text{ for } 0 < x < L \text{ and } t > 0$$

with the initial condition given by

$$u(x,0) = 100 \text{ for } 0 < x < L$$

and the boundary conditions specified as

$$u(0,t) = 0 \text{ and } u(L,t) = 100 \text{ for } t > 0.$$

The value $k = 1.71$ for the thermal diffusivity of silver was obtained from Table 1.1. The steady-state solution will be a time-independent solution to the heat equation. If $U(x)$ is the steady-state temperature distribution in the bar, then

$$1.71U''(x) = 0 \text{ for } 0 < x < L$$

and therefore, $U(x) = c_1 x + c_2$. The boundary conditions imply that $c_2 = 0$, $c_1 = 100/L$, and the steady-state solution is $U(x) = 100x/L$.

Example 1.6. Consider a one-dimensional rod of length L with a completely insulated lateral surface. There is a constant flux of heat energy ϕ_0 into the rod through the end at $x = 0$. The face $x = L$ is kept at constant temperature u_0. Set up the initial boundary value problem for the temperature distribution of the rod with initial temperature distribution given by $f(x)$ and solve for the steady-state temperature distribution of the rod.

Solution. Let $u(x,t)$ represent the temperature distribution in the bar. Then the initial boundary value problem for $u(x,t)$ is given by

$$u_t = k\, u_{xx} \text{ for } 0 < x < L \text{ and } t > 0,$$
$$u(x,0) = f(x) \text{ for } 0 < x < L,$$
$$-K_0 u_x(0,t) = \phi_0 \text{ and } u(L,t) = u_0 \text{ for } t > 0,$$

where k is the thermal diffusivity constant for the rod. For the steady-state temperature distribution, denoted as $U(x)$, it must be the case that

$$k\, U''(x) = 0 \text{ for } 0 < x < L$$

or equivalently $U(x) = c_1 x + c_2$. The boundary conditions can be re-stated as

$$-K_0 U'(0) = \phi_0,$$
$$U(L) = u_0.$$

The boundary condition at $x = 0$ implies

$$U'(0) = c_1 = -\frac{\phi_0}{K_0}$$

which yields the equation

$$U(x) = -\frac{\phi_0}{K_0}x + c_2.$$

Imposing the boundary condition at $x = L$ results in the steady-state solution,

$$U(x) = \frac{\phi_0}{K_0}(L - x) + u_0.$$

Example 1.7. Consider a one-dimensional copper rod of length 5 cm with constant thermal properties. Assume there is a heat source of the form $Q(x,t) = K_0 x^2$ and find the steady-state temperature distribution, if it exists, for the copper rod if the boundary conditions are $u(0,t) = 30C$ and $u_x(5,t) = 0$.

Solution. According to Table 1.1 the thermal diffusivity of copper is $k = K_0/(c\rho) = 1.14$. By Eq. (1.24), the temperature distribution $u(x,t)$ satisfies

$$\frac{\partial u}{\partial t} = 1.14\frac{\partial^2 u}{\partial x^2} + 1.14x^2.$$

Therefore the steady-state temperature distribution, denoted as $U(x)$, satisfies the following equation

$$U''(x) + x^2 = 0.$$

Integrating this equation twice with respect to x produces a steady-state temperature distribution function with two undetermined constants.

$$U(x) = -\frac{1}{12}x^4 + c_1 x + c_2.$$

The boundary condition $u(0,t) = 30$ implies that $c_2 = 30$ while the boundary condition $u_x(5,0) = 0$ implies that $-(5^3)/3 + c_1 = 0$ or $c_1 = 125/3$. Therefore there exists a steady-state solution,

$$U(x) = -\frac{1}{12}x^4 + \frac{125}{3}x + 30.$$

1.3.2   Heat Equations in Higher Dimensions

In much the same way as the one-dimensional heat equation was derived, the higher dimensional versions of the heat equation can be obtained. The details will not be provided here. For example, without a heat source, the homogeneous two-dimensional heat equation takes the form

$$u_t = k(u_{xx} + u_{yy}) \text{ for } (x, y) \in \Omega \subset \mathbb{R}^2 \text{ and } t > 0.$$

The homogeneous three-dimensional heat equation can be written as

$$u_t = k(u_{xx} + u_{yy} + u_{zz}) \text{ for } (x, y, z) \in \Omega \subset \mathbb{R}^3 \text{ and } t > 0.$$

In general, even the n-dimensional case can be considered. The homogeneous n-dimensional heat equation without a heat source is the following

$$u_t = k \triangle u \text{ for } \mathbf{x} \in \Omega \subset \mathbb{R}^n \text{ and } t > 0,$$

where $\mathbf{x} = (x_1, x_2, \ldots, x_n)$. The expression $\triangle u$ is a concise notation for the n-dimensional Laplacian operator defined as

$$\triangle u = \frac{\partial^2 u}{\partial x_1^2} + \frac{\partial^2 u}{\partial x_2^2} + \cdots + \frac{\partial^2 u}{\partial x_n^2}.$$

In each case, if there are internal heat sources, a nonhomogeneous equation is produced. In the two-dimensional case,

$$u_t = k(u_{xx} + u_{yy}) + Q(x, y, t) \text{ for } (x, y) \in \Omega \subset \mathbb{R}^2 \text{ and } t > 0.$$

The source term $Q(x, y, t)$ is derived from the heat energy generated inside the material. In the three-dimensional case,

$$u_t = k(u_{xx} + u_{yy} + u_{zz}) + Q(x, y, z, t) \text{ for } (x, y, z) \in \Omega \subset \mathbb{R}^3 \text{ and } t > 0,$$

where function $Q(x, y, z, t)$ is derived from the heat energy generated inside the material.

1.4   Vibrating Strings and the Wave Equation

In this section the partial differential equation governing the vibration of a one-dimensional string is derived from physical principles and Newton's laws of motion. The equation represents another class of important partial differential equations in mathematical physics – the **wave equation**. The appropriate initial boundary value conditions for vibrating strings in common physical situations will be discussed also. The corresponding higher dimensional versions of the wave equation are taken up in this section as well.

Consider the vibrational movement of a one-dimensional string whose horizontal coordinates lie in the interval $[0, L]$ along the positive portion of the x-axis. The equilibrium position of the string is the straight line in the xu-plane connecting the points with coordinates $(0,0)$ and $(L,0)$. The motion of the string is constrained to be vertical (in the u direction). The string can be set into motion from its equilibrium position by an initial displacement, an initial velocity, an external force, or a combination of these disturbances. The displacement of the point on the string at position $x \in [0, L]$ at time $t \geq 0$ is denoted by $u(x, t)$. Formally the following assumptions and definitions are made:

- The displacement $u(x, t)$ of the string at all points is small and occurs in the xu-plane.
- The string considered is perfectly flexible and offers no resistance to bending.
- The density (mass per unit length) of the string is denoted by $\rho(x)$.
- The string is under tension (a force) whose magnitude is denoted as $T(x, t)$. The tension acts in the direction tangent to the string at point $(x, u(x, t))$. The angle between the tangent line to the string at each point and the positive x-axis will be denoted by $\theta(x, t)$. These angles will be assumed small.
- The vertical component (component parallel to the direction of displacement) of all the external forces per unit length acting on the string is denoted as $f(x, t)$.

The reader may question the assumption that the string at rest assumes the configuration of a straight line. A more realistic assumption would be that, under the influence of gravity, the string assumes the shape of a hanging cable, or **catenary**. However, nothing is gained by this complicating assumption, since $u(x, t)$ could be re-defined as the displacement of the string from the equilibrium position (the catenary shape) and the remaining derivation would proceed as below.

The derivation of the wave equation begins with Newton's second law of motion applied to the string element whose endpoints are at coordinates $(x, u(x, t))$ and $(x + \Delta x, u(x + \Delta x, t))$. The mass of the string element is approximately $\rho(x)\sqrt{(\Delta x)^2 + (\Delta u)^2}$, where $\Delta u = u(x + \Delta x, t) - u(x, t)$. The vertical component of the tension force at the left end of the string element is $-T(x, t) \sin \theta(x, t)$. See Fig. 1.3. Similarly the vertical component of the tension at the right end of the string element is $T(x + \Delta x, t) \sin \theta(x + \Delta x, t)$. Newton's second law states that the mass of the string element multiplied

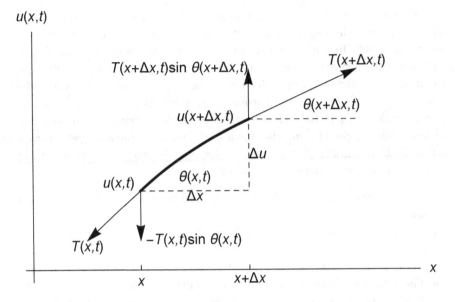

Fig. 1.3   An element of the string and the tension forces on the element.

by the acceleration of the string element (in the vertical direction) is the sum of the forces acting on the element,

$$\rho(x)\sqrt{(\Delta x)^2 + (\Delta u)^2}\frac{\partial^2 u}{\partial t^2} = T(x + \Delta x, t)\sin\theta(x + \Delta x, t)$$
$$- T(x, t)\sin\theta(x, t) + f(x, t)\Delta x. \qquad (1.32)$$

Dividing both sides of Eq. (1.32) by Δx produces

$$\rho(x)\sqrt{1 + \left(\frac{\Delta u}{\Delta x}\right)^2}\frac{\partial^2 u}{\partial t^2} = \frac{T(x + \Delta x, t)\sin\theta(x + \Delta x, t) - T(x, t)\sin\theta(x, t)}{\Delta x}$$
$$+ f(x, t).$$

Taking the limit of both sides of this equation as $\Delta x \to 0$ yields

$$\rho(x)\sqrt{1 + \left(\frac{\partial u}{\partial x}\right)^2}\frac{\partial^2 u}{\partial t^2} = \frac{\partial}{\partial x}\left[T(x, t)\sin\theta(x, t)\right] + f(x, t)$$
$$= \frac{\partial T}{\partial x}(x, t)\sin\theta(x, t)$$
$$+ T(x, t)\frac{\partial\theta}{\partial x}(x, t)\cos\theta(x, t) + f(x, t). \qquad (1.33)$$

The following relationships simplify Eq. (1.33):

$$\tan \theta(x,t) = \frac{\partial u}{\partial x},$$

$$\sin \theta(x,t) = \frac{\frac{\partial u}{\partial x}}{\sqrt{1 + \left(\frac{\partial u}{\partial x}\right)^2}},$$

$$\cos \theta(x,t) = \frac{1}{\sqrt{1 + \left(\frac{\partial u}{\partial x}\right)^2}}.$$

The assumption that angle $\theta(x,t)$ is small implies $\partial u/\partial x$ is small, and justifies the following approximations,

$$\sqrt{1 + \left(\frac{\partial u}{\partial x}\right)^2} \approx 1,$$

$$\sin \theta(x,t) \approx \frac{\partial u}{\partial x},$$

$$\cos \theta(x,t) \approx 1.$$

Substituting these approximations into Eq. (1.33) results in

$$\rho(x)\frac{\partial^2 u}{\partial t^2} = \frac{\partial T}{\partial x}\frac{\partial u}{\partial x} + T(x,t)\frac{\partial \theta}{\partial x} + f(x,t). \tag{1.34}$$

Equation (1.34) contains three unknown functions: u, T, and θ. The third of these can be eliminated by another approximation. Since $\theta(x,t)$ is small, $\partial^2 u/\partial x^2 = \sec^2 \theta(x,t)\partial\theta/\partial x \approx \partial\theta/\partial x$. Therefore

$$\rho(x)\frac{\partial^2 u}{\partial t^2} = \frac{\partial T}{\partial x}\frac{\partial u}{\partial x} + T(x,t)\frac{\partial^2 u}{\partial x^2} + f(x,t). \tag{1.35}$$

Equation (1.35) can be further simplified by considering the horizontal components of the forces on the ends of the string element. The horizontal components of the tension cancel one another since there is no acceleration in the horizontal direction,

$$T(x + \Delta x, t)\cos \theta(x + \Delta x, t) - T(x,t)\cos \theta(x,t) = 0. \tag{1.36}$$

Dividing both sides of Eq. (1.36) by Δx and taking the limit as $\Delta x \to 0$ produce the derivative,

$$\frac{\partial}{\partial x}[T(x,t)\cos \theta(x,t)] = \frac{\partial T}{\partial x}\cos \theta(x,t) - T(x,t)\frac{\partial \theta}{\partial x}\sin \theta(x,t) = 0.$$

Employing the approximations from above, it is shown that

$$\frac{\partial T}{\partial x} \approx T(x,t)\frac{\partial u}{\partial x}\frac{\partial^2 u}{\partial x^2}.$$

Substituting this approximation into Eq. (1.35) produces

$$\rho(x)\frac{\partial^2 u}{\partial t^2} = T(x,t)\left(\frac{\partial u}{\partial x}\right)^2 \frac{\partial^2 u}{\partial x^2} + T(x,t)\frac{\partial^2 u}{\partial x^2} + f(x,t).$$

Finally since $\partial u/\partial x$ is small, the first term on the right-hand side of the previous equation can be ignored leaving

$$\rho(x)\frac{\partial^2 u}{\partial t^2} = T(x,t)\frac{\partial^2 u}{\partial x^2} + f(x,t). \tag{1.37}$$

For a perfectly elastic and uniform string, when only small vibrations are considered, the tension force can be approximated by a constant T_0, that is $T(x,t) \approx T_0$. The uniformity of the string allows the replacement of the density function by a constant, that is $\rho(x) = \rho_0$, and thus

$$\frac{\partial^2 u}{\partial t^2} = c^2\frac{\partial^2 u}{\partial x^2} + \frac{f(x,t)}{\rho_0}, \tag{1.38}$$

where $c^2 = T_0/\rho_0$. Note that since T_0 has units of force (mass-length/time^2) and ρ_0 has units of mass/length, then c^2 has units of length^2/time^2 which implies that constant c has units of length/time or velocity.

If there is no external force acting on the string, Eq. (1.38) simplifies to

$$\frac{\partial^2 u}{\partial t^2} = c^2\frac{\partial^2 u}{\partial x^2}. \tag{1.39}$$

Equations (1.38) and (1.39) are often referred to as the one-dimensional wave equation and the homogeneous one-dimensional wave equation respectively. According to the definitions given in Sec. 1.2, the wave equation is of hyperbolic type.

Two special cases of Eq. (1.38) often occur in applications. If the only external force is the gravitational force Eq. (1.38) becomes

$$\frac{\partial^2 u}{\partial t^2} = c^2\frac{\partial^2 u}{\partial x^2} - g. \tag{1.40}$$

If the only external force is that of a viscous resistance imparted by the medium surrounding the string (for example, air friction), and the resisting force is assumed to be proportional to velocity, the wave equation becomes

$$\frac{\partial^2 u}{\partial t^2} = c^2\frac{\partial^2 u}{\partial x^2} - \gamma\frac{\partial u}{\partial t}, \tag{1.41}$$

where $\gamma > 0$ is the proportionality constant.

1.4.1 *Initial Boundary Value Problems for the Wave Equation*

Similar to the case of the one-dimensional heat equation, the one-dimensional wave equation is supplemented with initial and boundary conditions. The initial conditions prescribe the initial position and the initial velocity of the string. Usually the initial time is taken to be $t = 0$ and the initial conditions take the form

$$u(x,0) = f(x),$$
$$\frac{\partial u}{\partial t}(x,0) = g(x) \tag{1.42}$$

for $x \in (0,L)$. Boundary conditions may be of different types depending on the underlying physical situation.

Boundary conditions of the first kind or **Dirichlet boundary conditions** specify the displacements of the endpoints of the string as functions of time. They may have the form

$$u(0,t) = f_1(t),$$
$$u(L,t) = f_2(t) \tag{1.43}$$

for $t > 0$. Actually, many physical situations call for the position of the endpoints of the string to change dynamically, that is, $f_1(t)$ and $f_2(t)$ may be the solutions of ordinary differential equations. The simplest form of Dirichlet boundary conditions for the wave equation is that of fixed (zero displacement) endpoints for the string, *i.e.*,

$$u(0,t) = u(L,t) = 0 \tag{1.44}$$

for all $t > 0$. The boundary conditions of Eq. (1.44) are homogeneous.

Boundary conditions of the second kind or **Neumann boundary conditions** have the form

$$T_0 \frac{\partial u}{\partial x}(0,t) = g_1(t),$$
$$T_0 \frac{\partial u}{\partial x}(L,t) = g_2(t) \tag{1.45}$$

for $t > 0$ where T_0 is the tension force of the string. A special case is the following

$$\frac{\partial u}{\partial x}(0,t) = \frac{\partial u}{\partial x}(L,t) = 0 \tag{1.46}$$

for all $t > 0$. Boundary conditions as given in Eq. (1.45) specify the slopes of the string at the endpoints. The special case stated in Eq. (1.46) represents the case that the slopes of the string are zero at both ends. This can be achieved when the ends of the string are free to move vertically along a frictionless track. The boundary condition described next is more natural.

Boundary conditions of the third kind or **Robin boundary conditions** take the form

$$T_0 \frac{\partial u}{\partial x}(0,t) = k_1(u(0,t) - d_1(t)),$$
$$T_0 \frac{\partial u}{\partial x}(L,t) = -k_2(u(L,t) - d_2(t)) \tag{1.47}$$

for $t > 0$, where k_1 and k_2 are nonnegative constants and $d_1(t)$ and $d_2(t)$ are given functions. Assuming that the end of the string at $x = 0$ is attached to one end of a spring and the position of the other end of the spring is given by function $d_1(t)$, then the tensile force experienced by the string at the end located at $x = 0$ is given by the first condition in Eq. (1.47). A similar interpretation can be given for the second condition in Eq. (1.47). If $d_1(t) = 0$ and $d_2(t) = 0$ then a special case of homogeneous boundary conditions results,

$$T_0 \frac{\partial u}{\partial x}(0,t) = k_1 u(0,t),$$
$$T_0 \frac{\partial u}{\partial x}(L,t) = -k_2 u(L,t). \tag{1.48}$$

In the limit as $T_0 \to 0$ the boundary conditions of Eq. (1.47) become equivalent to the homogeneous Dirichlet boundary conditions of Eq. (1.44). Similarly in the limit as $k_1 \to 0$ and $k_2 \to 0$ the boundary conditions of Eq. (1.47) become equivalent to the homogeneous Neumann boundary conditions of Eq. (1.46).

An initial boundary value problem for the one-dimensional wave equation will consist of the wave equation (simplest form given in Eq. (1.39)), a set of initial conditions as in Eq. (1.42) and some combination of boundary conditions from Eqs. (1.43)–(1.48).

1.4.2 *Wave Equations in Higher Dimensions*

The motion of a vibrating string can be modeled by an initial boundary value problem involving the one-dimensional wave equation. The two-dimensional analogue of a string is known as a **membrane**. Examples of vibrating membranes are common such as a drum head, a trampoline, or simply a tightly stretched rubber sheet. The equilibrium position of the membrane lies flat in a region Ω of the xy-plane and $u(x,y,t)$ is the vertical displacement of the membrane from the xy-plane at position (x, y) and time t. Under reasonable physical and mathematical assumptions the partial differential equation for $u(x, y, t)$ becomes

$$u_{tt} = c^2(u_{xx} + u_{yy}) \text{ for } (x, y) \in \Omega \subset \mathbb{R}^2 \text{ and } t > 0,$$

where $c = \sqrt{T/\rho}$ with T representing the tension force of the membrane and ρ the mass density of membrane, both of which are assumed to be constant. This is called the wave equation in two dimensions. There is also a three-dimensional version of the wave equation. The three dimensional wave equation is often used to describe the disturbance in a uniform media such as a fluid, an electromagnetic wave, or a sound wave. Let R be a subset of three-dimensional space then the simplest form of the three-dimensional wave equation can be stated as

$$u_{tt} = c^2(u_{xx} + u_{yy} + u_{zz}) \text{ for } (x, y, z) \in R \subset \mathbb{R}^3 \text{ and } t > 0.$$

If external forces are present in the two- and three-dimensional wave equations, then the nonhomogeneous versions of these equations are

$$u_{tt} = c^2(u_{xx} + u_{yy}) + \frac{f}{\rho}, \text{ for } (x, y) \in \Omega \subset \mathbb{R}^2 \text{ and } t > 0$$

in the two-dimensional case, and

$$u_{tt} = c^2(u_{xx} + u_{yy} + u_{zz}) + \frac{f}{\rho} \text{ for } (x, y, z) \in R \subset \mathbb{R}^3 \text{ and } t > 0$$

in the three-dimensional case. Of course, these higher dimensional wave equations are also supplemented by initial and boundary conditions just as in the one-dimensional case.

1.5 Laplace's Equation

In the previous sections, both the heat equation and wave equation were derived from physical principles. These mathematical models share a common feature. Their equilibrium solutions satisfy Laplace's equation or, more generally, Poisson's[11] equation. In the one-dimensional case, if $u(x, t) \equiv U(x)$ is an equilibrium, or steady-state, solution of a one-dimensional heat equation with constant thermal properties and no heat source, or if $U(x)$ is the solution to a wave equation without external forces, then $U(x)$ satisfies the equation $U''(x) = 0$ subject to the appropriate boundary conditions. This is merely a simple second-order linear ordinary differential equation.

---

[11]Siméon Denis Poisson, French mathematician (1781–1840).

The general solution of the equation is $U(x) = cx + d$, where c and d are arbitrary constants. Considering a heat conduction problem and assuming the two ends of the rod are insulated, *i.e.*, there is no heat flow into or out of the rod, physical intuition suggests that the distribution of heat energy within the rod should be nearly uniform after a sufficiently long time. Consequently the temperature should approach a uniform distribution. In this case, $U'(0) = U'(L) = 0$ which implies $c = 0$. From the boundary conditions alone, there is not enough information to determine the constant d. In fact d can be an arbitrary constant.

More generally the one-dimensional Poisson's equation is stated as

$$U''(x) + f(x) = 0 \tag{1.49}$$

on $(0, L)$ along with appropriate boundary conditions at 0 and L. The solution $U(x)$ can be considered as an equilibrium solution of the one-dimensional wave equation with a forcing term or the one-dimensional heat equation with a heat source. Equation (1.49) can be solved easily by elementary integration. For the case in which the nonhomogeneous term depends on $u = U(x)$ and/or $u' = U'(x)$, the equation takes the form of

$$U''(x) + f(x, U, U') = 0$$

which is more complicated. However there are still methods from the study of ordinary differential equations to deal with such problems under various assumptions on the form of function f. See, for example [Coddington and Levinson (1955)] or [Hale (2009)].

In higher dimensional problems, the situation is different. For example, in two dimensions, the steady state solution $u(x, y)$ of the heat or wave equation satisfies an equation of the form

$$u_{xx} + u_{yy} = f(x, y). \tag{1.50}$$

Of course, in the more general situation $f(x, y)$ could depend on u, u_x, u_y. Equation (1.50) is defined on a domain in the xy-plane denoted as Ω and is commonly referred to as the **two-dimensional Poisson's equation**. If $f(x, y) = 0$ then Poisson's equation can be written as

$$u_{xx} + u_{yy} = 0 \text{ for } (x, y) \in \Omega$$

which is known as the **two-dimensional Laplace's equation**. Laplace's equation can be thought of as a time-independent or steady-state solution to the homogeneous two-dimensional heat equation. In the three-dimensional case, Poisson's equation has the form

$$u_{xx} + u_{yy} + u_{zz} = f(x, y, z)$$

and the corresponding three-dimensional Laplace's equation is

$$u_{xx} + u_{yy} + u_{zz} = 0.$$

Laplace's equation, or more generally, Poisson's equation are elliptic equations, another class of important partial differential equations. In addition to the close relationship between equilibrium solutions of the heat and wave equations, Laplace's and Poisson's equations arise naturally in many real world applications such as electrostatics, magnetostatics, and the study of incompressible liquid flow. Since there is no time variable in these equations, they are supplemented by boundary conditions only. For example, for the case of two-dimensional domains, common boundary conditions are Dirichlet boundary conditions,

$$u(x, y) = \phi(x, y) \text{ for } (x, y) \in \partial\Omega$$

and Neumann boundary conditions,

$$\nabla u \cdot \mathbf{n} = \psi(x, y) \text{ for } (x, y) \in \partial\Omega,$$

where $\partial\Omega$ represents the boundary of Ω, $\mathbf{n}$ is the unit outward normal vector of $\partial\Omega$, and $\nabla u \cdot \mathbf{n}$ is the directional derivative of u in the direction of unit vector $\mathbf{n}$. These types of problems will be considered in more detail in Chap. 6.

1.6 Separation of Variables

This section introduces a fundamental technique, the method of **separation of variables**, for solving boundary value problems associated with partial differential equations. The key idea behind separation of variables is to reduce the problem of solving a partial differential equation to that of solving a set of ordinary differential equations. The method will be presented and demonstrated on several examples before applying it to the one-dimensional heat equation. This technique will be used extensively throughout the remainder of this textbook to solve initial boundary value problems.

In using the method of separation of variables, the solution to a partial differential equation is assumed to be the product of two or more functions, each of which depends on only one of the independent variables. This type of solution is sometimes called a **product solution** or **separated solution**. After substituting a product solution into a given partial differential equation, the equation is re-written to see if the variables can be separated. If so, the method of separation of variables applies and a set of ordinary differential equations results. The ordinary differential equations are solved for the functions of a single independent variable. The method of separation of variables is demonstrated in the following examples.

Example 1.8. Apply the method of separation of variables to the equation

$$x^2 u_{xx} - 2y\, u_y = 0,$$

and find the corresponding set of ordinary differential equations.

Solution. Let $u(x,y) = X(x)\, Y(y)$, a product solution. Substituting u into the partial differential equation produces

$$x^2 X''(x)\, Y(y) - 2y\, X(x)\, Y'(y) = 0.$$

Dividing both sides of the equation by $u(x,y) = X(x)\, Y(y)$ and moving the expressions involving variable y to the right-hand side of the equation yield

$$x^2 \frac{X''(x)}{X(x)} = 2y\, \frac{Y'(y)}{Y(y)}.$$

Since the left side of the last equation is a function of x only and the right side is a function y only, the equation can hold only if both sides equal a constant of the same value. The first time a reader encounters this statement, it may seem difficult to accept; however, a change in the independent variable x will leave the right-hand side of the equation (the side dependent on y only) unchanged. Thus the left-hand side must equal a constant for all admissible values of x. Consequently the right-hand side takes on the same constant value. Hence $u(x,y) = X(x)\, Y(y)$ is a solution of the given partial differential equation if and only if

$$x^2 \frac{X''(x)}{X(x)} = 2y\, \frac{Y'(y)}{Y(y)} = c,$$

where c is a constant. A natural question to ask is, what is the value of constant c? The constant is not necessarily arbitrary. In general it may take on values that enable the product solution $u(x,y)$ to satisfy boundary conditions associated with the partial differential equation. In this particular example no boundary conditions were specified, so all that can be concluded is that factors $X(x)$ and $Y(y)$ must satisfy the following ordinary differential equations respectively,

$$x^2 X''(x) - c\, X(x) = 0,$$
$$2y\, Y'(y) - c\, Y(y) = 0.$$

Example 1.9. Apply the method of separation of variables to the following equation and determine the corresponding set of ordinary differential equations.

$$u_{xx} + u_x + 2u_y - u \sin x = 0.$$

Solution. Substituting $u(x, y) = X(x) Y(y)$ into the partial differential equation produces the equation,

$$X''(x) Y(y) + X'(x) Y(y) + 2X(x) Y'(y) - X(x) Y(y) \sin x = 0.$$

Dividing both sides of the equation by $u(x, y)$ yields

$$\frac{X''(x)}{X(x)} + \frac{X'(x)}{X(x)} + 2\frac{Y'(y)}{Y(y)} - \sin x = 0$$

which may be re-written as

$$\frac{X''(x)}{X(x)} + \frac{X'(x)}{X(x)} - \sin x = -2\frac{Y'(y)}{Y(y)}.$$

Since the left-hand side of the equation depends on x alone while the right-hand side depends on y alone, the method of separation of variables has been successful. As in the previous example, $u(x, y) = X(x) Y(y)$ is a solution if and only if the left-hand and right-hand sides of the equation above are equal to the same constant (say c) and hence $X(x)$ and $Y(y)$ satisfy the ordinary differential equations below,

$$X''(x) + X'(x) - X(x) \sin x = c X(x),$$
$$2Y'(y) + c Y(y) = 0.$$

Once again no boundary conditions were specified, so the permissible values of c cannot be determined from these differential equations alone.

Example 1.10. Determine if the method of separation of variables can be applied to the following partial differential equation. If so, determine the resulting ordinary differential equations.

$$u_x + (x + y)u_y = 0.$$

Solution. Set $u(x, y) = X(x) Y(y)$, differentiate and substitute into the given partial differential equation to produce

$$X'(x) Y(y) + (x + y)X(x) Y'(y) = 0.$$

Since it is not possible to rewrite the above equation in a form in which the variables are separated (x on one side and y on the other), the method of separation of variables does not apply to this equation.

Example 1.11. Determine if the method of separation of variables can be applied to the partial differential equation below. If so, determine the ordinary differential equations in each variable which result.

$$u_{rr} + \frac{1}{r}u_r + \frac{1}{r^2}u_{\theta\theta} = 0 \text{ for } r > 0.$$

Solution. Note that variables r and θ are commonly used as the polar coordinate variables, though this has no bearing on the method of separation of variables. Let $u(r, \theta) = R(r)T(\theta)$. Differentiating and substituting the results into the partial differential equation yield

$$R''(r)T(\theta) + \frac{1}{r}R'(r)T(\theta) + \frac{1}{r^2}R(r)T''(\theta) = 0.$$

Dividing both sides of the equation by $u(r, \theta)$ produces

$$\frac{R''(r)}{R(r)} + \frac{1}{r}\frac{R'(r)}{R(r)} + \frac{1}{r^2}\frac{T''(\theta)}{T(\theta)} = 0.$$

Multiplying both sides of this equation by r^2 and moving the terms involving θ to the right-hand side give

$$r^2\frac{R''(r)}{R(r)} + r\frac{R'(r)}{R(r)} = -\frac{T''(\theta)}{T(\theta)}.$$

Yet again, since the left-hand side depends only on r and the right-hand side depends only on θ, both sides must equal the same constant, call it c. Thus the ordinary differential equations implied for R and T are

$$r^2R''(r) + rR'(r) - cR(r) = 0,$$
$$T''(\theta) + cT(\theta) = 0.$$

Before leaving this example, a comment on the role that boundary conditions play in the choice of constant c is appropriate. Since the original partial differential equation employed polar coordinates, a natural expectation is that solutions are periodic in θ with period 2π. This imposes the requirement that $T(\theta + 2\pi) = T(\theta)$. The reader will show in Exercise 19 that the constant $c = n^2$ where n is a nonnegative integer and the function $T(\theta)$ indexed by n has the form,

$$T_n(\theta) = A_n\cos(n\theta) + B_n\sin(n\theta),$$

where A_n and B_n are constants and $n = 0, 1, \ldots$. The function $R(r)$ must satisfy the equation,

$$r^2R''(r) + rR'(r) - n^2R(r) = 0.$$

When $n = 0$, the solution to this differential equation is $R_0(r) = C_0 + D_0 \ln r$. For $n \in \mathbb{N}$, the set of natural numbers, the solution can be written as $R_n(r) = C_n r^n + D_n r^{-n}$. Therefore the product solutions are

$$u_0(r, \theta) = C_0 + D_0 \ln r$$

if $n = 0$, where the constant $T_0(\theta) = A_0$ is absorbed in the arbitrary constants C_0 and D_0, and

$$u_n(r, \theta) = (C_n r^n + D_n r^{-n})(A_n \cos(n\theta) + B_n \sin(n\theta))$$

for $n \in \mathbb{N}$.

The reader should note that the solution above is valid for $r > 0$. However, in many situations, a desirable property of a solution is that it remains bounded as $r \to 0^+$. This boundedness requirement at the origin requires that $D_n = 0$ for $n = 0, 1, \ldots$. This produces $R_n(r) = C_n r^n$ for $r \geq 0$ and $n = 0, 1, \ldots$ and the product solutions take on the form

$$u_n(r, \theta) = r^n [A_n \cos(n\theta) + B_n \sin(n\theta)]$$

for $n = 0, 1, \ldots$. The multiplicative constant C_n has been absorbed into the constants A_n and B_n.

The final example of this section treats an important equation of this text, the initial boundary value problem for the heat equation with Dirichlet boundary conditions. Only the issues associated with separation of variables will be explored since this problem will be studied in more generality in Chap. 4.

Example 1.12. Consider the one-dimensional, homogeneous heat equation with Dirichlet boundary conditions and an initial condition as below:

$$u_t = k\, u_{xx} \text{ for } 0 < x < L \text{ and } t > 0,$$
$$u(0, t) = u(L, t) = 0 \text{ for } t > 0,$$
$$u(x, 0) = f(x) \text{ for } 0 < x < L.$$

Apply the method of separation of variables to this initial boundary value problem and determine the product solutions which satisfy the given boundary conditions.

Solution. Consider a product solution $u(x, t) = X(x)\, T(t)$. Differentiating and substituting this function into the heat equation yield

$$X(x)\, T'(t) = k\, X''(x)\, T(t),$$

which can be rewritten as

$$\frac{X''(x)}{X(x)} = \frac{1}{k}\frac{T'(t)}{T(t)} = -c, \tag{1.51}$$

where c, and consequently $-c$, is a constant. Substituting the product solution into the boundary conditions yields

$$X(0)T(t) = 0,$$
$$X(L)T(t) = 0.$$

To obtain a nontrivial product solution $u(x,t) = X(x)T(t)$, it must be the case that $X(0) = 0$ and $X(L) = 0$, otherwise $T(t) = 0$ for all $t > 0$ which implies $u(x,t) = 0$. Hence the unknown function $X(x)$ must solve the following boundary value problem.

$$X''(x) + cX(x) = 0, \tag{1.52}$$
$$X(0) = 0, \tag{1.53}$$
$$X(L) = 0. \tag{1.54}$$

The constant c may take on any value for which a solution $X(x)$ to the boundary value problem above may be found. There are three cases to consider.

Case $c = 0$: the general solution of Eq. (1.52) is $X(x) = Ax+B$. Applying the boundary conditions results in $X(0) = B = 0$ and $X(L) = AL+B = 0$ which imply $A = B = 0$. Thus the only solution to this case is the trivial solution $X(x) = 0$.

Case $c < 0$: for the sake of notation, let $c = -\lambda^2$ where $\lambda > 0$. In this case, the general solution of Eq. (1.52) is $X(x) = Ae^{-\lambda x} + Be^{\lambda x}$ and by Eqs. (1.53) and (1.54)

$$X(0) = A + B = 0,$$
$$X(L) = Ae^{-\lambda L} + Be^{\lambda L} = 0.$$

This linear system for unknowns A and B yields only the solution $A = B = 0$. Again there is no nontrivial solution for the boundary value problem when $c < 0$.

Case $c > 0$: again for the sake of notation, let $c = \lambda^2$ where $\lambda > 0$. The general solution of Eq. (1.52) is $X(x) = A\cos(\lambda x) + B\sin(\lambda x)$. The boundary conditions imply that

$$X(0) = A = 0,$$
$$X(L) = A\cos(\lambda L) + B\sin(\lambda L) = 0.$$

In order to find a nontrivial product solution, it is required that

$$\sin(\lambda L) = 0,$$

otherwise $B = 0$ which implies $X(x) = 0$ and thus $u(x, t) = 0$. This implies that $\lambda L = n\pi$ for $n \in \mathbb{N}$. Thus the permissible values of c which lead to nontrivial solutions of the boundary value problem stated in Eqs. (1.52)–(1.54) are indexed by the natural numbers as

$$c_n = \lambda_n^2 = \frac{n^2 \pi^2}{L^2} \text{ for } n \in \mathbb{N}.$$

In the following chapters, these constants will be called **eigenvalues** of the boundary value problem expressed in Eqs. (1.52)–(1.54). The corresponding nontrivial solutions,

$$X_n(x) = B_n \sin\left(\frac{n\pi x}{L}\right)$$

will be called **eigenfunctions**.

Now that the boundary value problem has been solved, the eigenvalues are used in Eq. (1.51) to obtain the time-dependent portion of the product solution, $T_n(t)$ for $n \in \mathbb{N}$:

$$T_n'(t) = -\lambda_n^2 k\, T_n(t) = -\frac{k n^2 \pi^2}{L^2} T_n(t). \tag{1.55}$$

This is a first-order, linear ordinary differential equation and its general solution for each $n \in \mathbb{N}$ is

$$T_n(t) = C_n e^{-n^2 \pi^2 k\, t / L^2}.$$

Therefore the product solutions which solve the homogeneous one-dimensional heat equation on $[0, L]$ with Dirichlet boundary conditions can be expressed as

$$u_n(x, t) = B_n e^{-n^2 \pi^2 k\, t / L^2} \sin\left(\frac{n\pi x}{L}\right), \text{ for } n \in \mathbb{N}. \tag{1.56}$$

These solutions are called the **fundamental solutions** of the initial boundary value problem of Eqs. (1.25), (1.26), and (1.28).

Once the boundary conditions are satisfied by the product solution, attention can be given to the initial condition. Since the heat equation is linear and homogeneous, the Principle of Superposition implies that a linear combination of finitely many product solutions as given in Eq. (1.56) will also solve the boundary value problem of Eqs. (1.25) and (1.28).

$$u(x, t) = \sum_{n=1}^{N} B_n e^{-n^2 \pi^2 k\, t / L^2} \sin\left(\frac{n\pi x}{L}\right). \tag{1.57}$$

To find the solution of the initial boundary value problem given in Eqs. (1.25), (1.26), and (1.28), it is necessary that

$$u(x,0) = f(x) = \sum_{n=1}^{N} B_n \sin \left(\frac{n \pi x}{L} \right)$$

for $0 < x < L$. When the initial condition function $f(x)$ involves only a finite sum of sine functions of the same form, a solution to the corresponding initial, boundary value problem can be found. For instance if

$$f(x) = -2 \sin \frac{4 \pi x}{L} + 5 \sin \frac{10 \pi x}{L}$$

then choose $N = 10$, $B_4 = -2$, $B_{10} = 5$, and all other $B_n = 0$. A solution to the initial boundary value problem is then

$$u(x,t) = -2e^{-16\pi^2 k\,t/L^2} \sin \frac{4 \pi x}{L} + 5e^{-100\pi^2 k\,t/L^2} \sin \frac{10 \pi x}{L}.$$

A more difficult (and ultimately more interesting) question is, what if the initial condition $u(x,0) = f(x)$ is not a finite linear combination of sine functions? For example, what can be done for the case in which

$$f(x) = x(1 - x)$$

or some other smooth function? The reader can show that it is impossible to choose $N \in \mathbb{N}$ and $B_n \in \mathbb{R}$ such that

$$f(x) = x(1 - x) = \sum_{n=1}^{N} B_n \sin \frac{n \pi x}{L}.$$

In these cases an infinite series of sine functions is used with coefficients chosen so that

$$f(x) = \sum_{n=1}^{\infty} B_n \sin \frac{n \pi x}{L}.$$

The reader should be familiar with writing functions as power series or Taylor series in calculus and ordinary differential equations. In the study of partial differential equations the representation of smooth functions as infinite series of trigonometric functions is the subject called **Fourier series** which will be discussed in Chap. 3.

In this chapter the basic concepts of partial differential equations have been introduced and the one-dimensional heat and wave equations have been derived from physical principles. The interpretations and imposition of different boundary conditions that may arise in different physical situations have been described. In Sec. 1.1, partial differential equations were classified according to their linearity property and the highest order of the partial derivatives that appear in the equation. The material discussed in Sec. 1.2 gives the reader additional ways to classify second-order linear partial differential equations with two independent variables. The heat, wave, and Laplace's equations are the prototypes of three different classes of second-order, linear partial differential equations. Even though the heat equation, the wave equation, and Laplace's or Poisson's equations are all second-order, linear partial differential equations, the properties of the solutions to the equations are different, as will be demonstrated in later chapters.

1.7 Exercises

(1) Prove the Principle of Superposition as stated in Theorem 1.1.
(2) Prove that if $u_1(x, t)$ and $u_2(x, t)$ both solve Eq. (1.1) then $v(x, t) = u_2(x, t) - u_1(x, t)$ solves Eq. (1.2).
(3) Verify the given function is a solution of each of the following partial differential equations on the indicated domain.

   (a) $u_{xx} + u_{yy} = 0$, $u(x, y) = e^{ax} \sin ay$ for all $(x, y) \in \mathbb{R}^2$, where a is an arbitrary constant.
   (b) $u_{xx} + u_{yy} + u_{zz} = 0$, $u(x, y, z) = (x^2 + y^2 + z^2)^{-1/2}$ for all $(x, y, z) \neq (0, 0, 0)$.
   (c) $u_t = k\, u_{xx}$, $u(x, t) = t^{-1/2} e^{-(x-a)^2/(4kt)}$ for $t > 0$, $-\infty < x < \infty$, where $k > 0$ and a are constants.
   (d) $u_{tt} = u_{xx}$, $u(x, t) = e^{a(t-x)}$ for all $(x, t) \in \mathbb{R}^2$, where a is an arbitrary constant.

(4) Classify each of the following partial differential equations as linear or nonlinear and state the order of each equation.

   (a) $x^2 u_x + y^2 u_y + u_{xy} = 2xy$.
   (b) $(x - y)^2 u_x + u\, u_y = 1$.
   (c) $x^2 u_{xx} + (y^2 + 1)u_{yy} = xyu$.
   (d) $u_{xx} + u_{yy} + 3u_x = \sin(u)$.
   (e) $u_{xx} + u_x u_y + u_{xy} = 1$.
   (f) $u_t + u_{xxxx} + \sqrt{a + u} = 0$.

(5) Find all constants a and b such that $u(x, t) = e^{at} \sin(bx)$ satisfies the partial differential equation,

$$u_t - u_{xx} = 0.$$

(6) Consider the partial differential equation,

$$u_{xx} = 0 \text{ for } 0 < x < 1 \text{ and } t > 0.$$

(a) Find all solutions $u = u(x,t)$ for the equation.
(b) Find a solution of the equation such that $u(0,t) = t^2$ and $u(1,t) = 1$ for $t > 0$.

(7) Consider the partial differential equation with two independent variables

$$u_{xx} + u_{yy} = 1.$$

(a) Verify that $u_c(x,y) = cx^2/2 + (1-c)y^2/2$ is a solution for this equation, where c is an arbitrary constant.
(b) In Exercise 3(a) the reader verified that $u(x,y) = e^{ax} \sin ay$ is a solution of the equation $u_{xx} + u_{yy} = 0$. Without taking the partial derivative, explain why $v(x,y) = e^{ax} \sin ay + u_c(x,y)$ is a solution of the nonhomogeneous equation.
(c) Find a solution of the nonhomogeneous equation that satisfies the condition

$$u(0,y) = \sin 2y - y^2 \text{ for } y \in \mathbb{R}.$$

(8) Classify the types (elliptic, hyperbolic, or parabolic) of the following partial differential equations.

(a) $3u_{tt} + 6u_{tx} + 4u_{xx} - 2u_t - t\,u_x + t^2 x\,u = 0.$
(b) $u_{tt} - 4u_{tx} + 4u_{xx} + t\,u_t - u_x = 0.$
(c) $u_{tt} + 3u_{tx} + u_{xx} + 2u_t - u_x = f(t,x).$
(d) $u_{tt} - 3u_{tx} - 2u_{xx} + x\,u_t - 3u = 0.$

(9) Let $f(x)$ be a continuous function on an interval $[a,b]$ such that $\int_{\alpha}^{\beta} f(x)\,dx = 0$ for any $[\alpha,\beta] \subset [a,b]$. Show that $f(x) = 0$ for all $a \le x \le b$.

(10) At time $t = 0$, an aluminum bar of length L cm with completely insulated lateral surfaces and constant thermal properties is removed from boiling water.

(a) If the two ends of the bar are immersed in a medium with constant temperature $10°C$ immediately, write down the initial boundary value problem for the temperature distribution $u(x,t)$.
(b) If the end at $x = 0$ is immersed in a medium with temperature $0°C$ and the end at $x = L$ is completely insulated, write down the initial boundary value problem for the temperature distribution $u(x,t)$.

(c) If the end at $x = 0$ is immersed in a medium with temperature $0°C$ and the end at $x = L$ is kept in the boiling water, write down the initial boundary value problem for the temperature distribution $u(x, t)$.

(d) Without actually solving for $u(x, t)$, describe the temperature distribution in the bar as $t \to \infty$ for each of situations described in (a), (b), (c) based on physical intuition.

(11) For each of the following boundary value problems, determine a steady state temperature distribution, if one exists.

(a) $u_t = u_{xx} + 1$, $u_x(0, t) = 1$, $u(L, t) = 1 - L$.

(b) $u_t = k\, u_{xx}$, $u(0, t) = T_0$, $-K_0 u_x(L, t) = h(u(L, t) - T_L)$ where $h > 0$, $K_0 > 0$, T_0, T_L are all constants.

(c) $u_t = k\, u_{xx}$, $u_x(0, t) = u(0, t) - T_0$, $u_x(L, t) = 0$ where T_0 is a constant.

(12) Consider a uniform vibrating string of length L. Assume the only external force acting on the string is the gravitational force, that is, the motion is governed by Eq. (1.40). If the homogeneous Dirichlet boundary conditions are imposed at the two ends, find the equilibrium position of the string.

(13) Consider an elastic string of length 30 units, whose ends are held fixed at 0 displacement.

(a) If the string is set into motion from the initial position given by

$$f(x) = \begin{cases} x/15 & \text{if } 0 \le x \le 15, \\ (30 - x)/15 & \text{if } 15 \le x \le 30 \end{cases}$$

with no initial velocity, write down the initial boundary value problem for the displacement $u(x, t)$ of the string.

(b) If the string is set into motion from the equilibrium position with an initial velocity

$$g(x) = \begin{cases} x/15 & \text{if } 0 \le x \le 15, \\ (30 - x)/15 & \text{if } 15 \le x \le 30 \end{cases}$$

write down the initial boundary value problems for the displacement $u(x, t)$ of the string.

(14) Determine if the method of separation of variables applies to the following equations. If so find the corresponding ordinary differential equations for each independent variable.

(a) $u_{xx} + 3u_x + 4u_{yy} = 0$.

(b) $u_{xx} + u_{xt} + 3u_t = 0$.

(c) $x\,u_{yy} + y\,u_{xx} = 0$.

(d) $u_{xx} + (x+t)u_{tx} + u = 0$.

(15) Consider the following initial boundary value problem

$$u_t = k\,u_{xx} \text{ for } 0 < x < 1 \text{ and } t > 0,$$

$$u(0,t) = u(1,t) = 0 \text{ for } t > 0,$$

$$u(x,0) = \sin(\pi x) \text{ for } 0 < x < 1.$$

(a) Give a physical interpretation of the problem.

(b) Explain intuitively that the solution of $u(x,t)$ of the initial boundary value problem tends to zero as $t \to \infty$ for all $x \in [0,1]$.

(c) Find a solution of the initial boundary value problem and verify the result of (b) analytically.

(16) Find a solution of the following initial boundary value problem

$$u_t = \frac{1}{4}u_{xx} \text{ for } 0 < x < 2 \text{ and } t > 0,$$

$$u(0,t) = u(2,t) = 0 \text{ for } t > 0,$$

$$u(x,0) = f(x) \text{ for } 0 < x < 2,$$

for

(a) $f(x) = \sin(\pi x/2) + 4\sin(2\pi x)$,

(b) $f(x) = \sin(\pi x) - \sin(3\pi x/2)$.

(17) Find the solution of the initial boundary value problem,

$$u_t = 4u_{xx} \text{ for } 0 < x < 2 \text{ and } t > 0,$$

$$u(0,t) = u(2,t) = 0 \text{ for } t > 0,$$

$$u(x,0) = \sin(\pi x/2) - \frac{1}{2}\sin(\pi x) + 3\sin(3\pi x) \text{ for } 0 < x < 2.$$

(18) Apply the method of separation of variables to find all the eigenvalues and the corresponding of the eigenfunctions of the boundary value problem

$$u_t = k\,u_{xx} \text{ for } 0 < x < L \text{ and } t > 0,$$

$$u_x(0,t) = u_x(L,t) = 0 \text{ for } t > 0,$$

where k and L are constants.

(19) Determine the values of constant c for which the boundary value problem

$$X''(x) + c\,X(x) = 0 \text{ for } -\infty < x < \infty,$$

$$X(x) = X(x+2\pi) \text{ for } -\infty < x < \infty$$

has nontrivial solutions and find the corresponding nontrivial solutions.

(20) Apply the method of separation of variables to find the eigenvalues and the corresponding eigenfunctions of the boundary value problem,

$$u_t = k\,u_{xx} \text{ for } 0 < x < L \text{ and } t > 0,$$
$$u(0,t) = u_x(L,t) = 0 \text{ for } t > 0,$$

where k and L are positive constants.

(21) Let a, b, and c be constants and consider the partial differential equation

$$u_{xx} - u_{yy} + a\,u_x + b\,u_y = c\,u.$$

Suppose $u(x,y) = e^{\alpha x + \beta y}v(x,y)$. Determine constants α, β, and γ so that

$$v_{xx} - v_{yy} = \gamma\,v.$$

(22) Let a, b, and c be constants and consider the partial differential equation

$$u_t = a\,u_{xx} + b\,u_x + c\,u.$$

Suppose $u(x,t) = e^{\alpha x + \beta t}v(x,t)$. Determine constants α and β so that

$$v_t = a\,v_{xx}.$$

(23) Suppose m and n are positive integers and k is any constant. Show that

$$u(x,y,t) = k\,\sin(n\pi x)\sin(m\pi y)\sin(\sqrt{m^2 + n^2}c\pi t)$$

solves the homogeneous two-dimensional wave equation where $c > 0$ is the constant wave speed.

Chapter 2

First-Order Partial Differential Equations

This chapter is focused on first-order partial differential equations. This foray in first-order partial differential equations and their applications is self contained and independent of the remaining chapters of this text, thus it can be skipped or postponed without disrupting most other topics in the book. However, this chapter should be read before exploring the topics in Chap. 11 pertaining to nonlinear partial differential equations.

First-order partial differential equations can be solved using methods based on geometrical considerations and the theory of ordinary differential equations. Although the method is applicable to more general partial differential equations, to keep the technicalities to a minimum, only first-order partial differential equations with two independent variables will be treated. The general form of a first-order partial differential equation is

$$F(x, y, u, u_x, u_y) = 0$$

where $u(x, y)$ is the unknown function. First-order partial differential equations can be further classified according to their linearity properties. Section 2.1 will define and describe a solution method for first-order **linear** partial differential equations while Sec. 2.2 will treat first-order **quasilinear** partial differential equations. Some applications of first-order linear and quasilinear partial differential equations are discussed in Sec. 2.3.

2.1 First-Order Linear Equations

To introduce the method for solving first-order partial differential equations, consider the following example,

$$u_x + c\, u_y = 0 \tag{2.1}$$

where c is a constant. The labels given to the independent variables, x and y, are arbitrary. In some applications one of the two independent variables represents time and thus the independent variables could be labeled t and x in these cases. However, the solution method explored in this chapter frequently makes use of a parameter and it is convenient to use the symbol t as the parameter, even though the parameter does not necessarily denote time. Equation (2.1) is a special case of an important class of partial differential equations, called **transport equations**. If ∇u and $\nabla \cdot u$ denote the gradient and divergence of u, the general form of scalar transport equations can be written as

$$u_x + \nabla \cdot f(x, \mathbf{y}, u, \nabla u) = g(x, \mathbf{y}, u)$$

where $\mathbf{y} = (y_1, y_2, \ldots, y_n) \in \mathbb{R}^n$ and f and g are functions defined on their appropriate domains. Transport equations can be used to model a variety of phenomena in applications such as heat transfer, air pollution, and traffic flow with u representing the temperature, the density of the pollutant, and the traffic density at position $\mathbf{y}$ and time x, respectively. For the moment, consider the simple equation in Eq. (2.1). The left-hand side of the equation can be viewed as the directional derivative of u in the direction specified by the vector $\langle 1, c \rangle$. Therefore, Eq. (2.1) simply states that the function $u(x, y)$ is constant along each line parallel to the vector $\langle 1, c \rangle$. This family of lines can be described by the equation $y = cx + k$, where k is an arbitrary constant. See Fig. 2.1. This implies that the solution u must be a function of k or equivalently $y - cx$ alone. In other words,

$$u(x, y) = f(y - cx) \tag{2.2}$$

where f is an arbitrary function. The reader may directly verify that the function $u(x, y)$ given in Eq. (2.2) is a solution of Eq. (2.1) for any differentiable function f, see Exercise 1.

Expressions defined by Eq. (2.2) are referred to as the general solution of Eq. (2.1). With the imposition of appropriate boundary conditions, the solution $u(x, y)$ can be further determined. It is natural to specify the values of u on the line $x = 0$, a boundary condition. The boundary condition can be expressed as $u(0, y) = \phi(y)$ for all $y \in (-\infty, \infty)$ where $\phi(y)$ is a given function. This condition will uniquely determine the solution $u(x, y) = \phi(y - cx)$. Note that the graph of $u(x, y)$ is just a shift of the graph of $\phi(y)$ by cx units horizontally. Geometrically, the initial profile or wave front (perhaps modeling the density of a certain substance) is propagated at speed c as x increases. Mathematically speaking the boundary condition can be imposed on any curve in the xy-plane which is not tangent to the vector $\langle 1, c \rangle$ at any point. More will be said about this issue later.

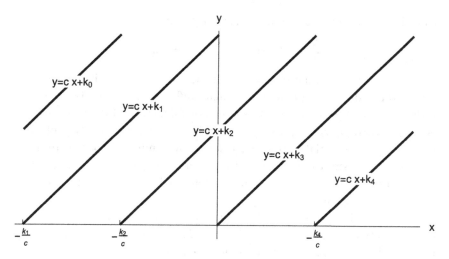

Fig. 2.1 The solution of Eq. (2.1) must be constant along the family of parallel lines of the form $y = cx + k$.

The technique used in the elementary example above can be extended to general linear first-order partial differential equations of the form

$$a(x,y)u_x + b(x,y)u_y = c(x,y)u + d(x,y). \tag{2.3}$$

Following the same line of reasoning as above, if $u(x,y)$ is a solution of Eq. (2.3) on a domain $\Omega \subset \mathbb{R}^2$, then for any $(x,y) \in \Omega$, the directional derivative of u in the direction of the vector $\langle a(x,y), b(x,y) \rangle$ is $c(x,y)u(x,y) + d(x,y)$. The vectors $\langle a(x,y), b(x,y) \rangle$ define a **vector field** on Ω, which determines parametric curves $(x(t), y(t))$ that are tangent to the vector $\langle a(x,y), b(x,y) \rangle$ at every point in $(x,y) \in \Omega$. Therefore $(x(t), y(t))$ is an **integral curve** of the following system of ordinary differential equations:

$$\begin{aligned} \frac{dx}{dt} &= a(x,y), \\ \frac{dy}{dt} &= b(x,y), \end{aligned} \tag{2.4}$$

or equivalently,

$$\frac{dy}{dx} = \frac{b(x,y)}{a(x,y)}. \tag{2.5}$$

Suppose the parametric curve $(x(t), y(t))$ is a solution to Eq. (2.4) and $u(x,y)$ is solution to Eq. (2.3). According to the chain rule for derivatives,

$$\frac{du}{dt} = \frac{d}{dt}\left[u(x(t), y(t))\right] = u_x \frac{dx}{dt} + u_y \frac{dy}{dt} = a(x,y)u_x + b(x,y)u_y$$

and thus along each integral curve $(x(t), y(t))$ of Eq. (2.4),

$$\frac{du}{dt} = c(x(t), y(t))u + d(x(t), y(t)). \qquad (2.6)$$

Thus the problem of finding solutions of the partial differential equation of Eq. (2.3) is reduced to solving the ordinary differential Eqs. (2.4) and (2.6). If the solution $(x(t), y(t))$ to Eq. (2.4) can be found, then Eq. (2.6) is solved on each integral curve $(x(t), y(t))$ to obtain $u \equiv u(t)$. Finally the solution u in terms of the variables x and y is found by eliminating parameter t.

It is often easier to solve the first-order ordinary differential equation in Eq. (2.5) than to solve the system of equations in Eq. (2.4). In this case, the general solution of Eq. (2.5) is often determined in the implicit form $\phi(x, y) = k$, where k is an arbitrary constant. Assume $\phi_y \neq 0$ so that y can be treated as a function of x. For each k, along each curve $\phi(x, y) = k$, Eq. (2.3) becomes an ordinary differential equation

$$\frac{du}{dx} = u_x + u_y \frac{dy}{dx} = u_x + u_y \frac{b(x, y)}{a(x, y)} = \frac{c(x, y)u + d(x, y)}{a(x, y)}. \qquad (2.7)$$

Solving this ordinary differential equation along the curve determined implicitly by $\phi(x, y) = k$, the right-hand side of Eq. (2.7) can be treated as a function of parameter k and variables x and u. Equation (2.7) is a linear ordinary differential equation and its solution u is a function of x and k. After eliminating k using the relation $\phi(x, y) = k$, the solution $u(x, y)$ is obtained.

Equations (2.4) and (2.6) are called the **characteristic equations** of Eq. (2.3). The plane curves $(x(t), y(t))$ are called the **characteristics** of Eq. (2.3). Figure 2.2 illustrates a typical vector field and a sample of characteristics. The corresponding curves $(x(t), y(t), u(x(t), y(t)))$ are called the **characteristic curves**. Figure 2.3 depicts a few characteristic curves corresponding to the characteristics pictured in Fig. 2.2. Note that the characteristics are just the projections of the characteristic curves onto the xy-plane. The solution method described above is usually referred to as the **method of characteristics**.

Since Eq. (2.6) or Eq. (2.7) is solved on each characteristic which depends on an arbitrary constant, denoted as k, its general solution depends on an arbitrary function of k. Therefore, after eliminating k, the solution $u(x, y)$ contains an arbitrary function. This form of the solution is often called the general solution of Eq. (2.3). The arbitrary function can be determined if appropriate boundary conditions are given. What constitutes "appropriate" boundary conditions will be discussed in the next section.

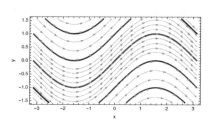

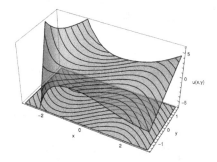

Fig. 2.2 The vector field (arrows) and characteristics (solid curves) taken from Example 2.2.

Fig. 2.3 The characteristic curves (drawn on surface) and characteristics (at bottom of box) taken from Example 2.2.

The remainder of this section contains examples illustrating the solution procedure for first-order linear partial differential equations.

Example 2.1. Use the method of characteristics to find the general solution of the partial differential equation

$$\frac{\partial u}{\partial x} + 3\frac{\partial u}{\partial y} = e^y u. \tag{2.8}$$

Solution. The characteristics for this first-order linear partial differential equation are the straight lines described by the ordinary differential equation

$$\frac{dy}{dx} = 3.$$

Thus the characteristics are the family of straight lines $y - 3x = k$ and on each of these lines the solution $u(x, y) = u(x, y(x))$ satisfies

$$\frac{du}{dx} = u_x + u_y\frac{dy}{dx} = e^y u = e^{3x+k}u.$$

The general solution to this first-order linear ordinary differential equation is the following:

$$u = f(k)e^{\frac{1}{3}e^{3x+k}}$$

where f is an arbitrary function. Eliminating k using the relation $k = y - 3x$, the general solution of Eq. (2.8) becomes

$$u(x, y) = f(y - 3x)e^{\frac{1}{3}e^y}.$$

Since no boundary condition was given for this example, the arbitrary function f cannot be determined.

Example 2.2. Find the general solution of the partial differential equation

$$u_x + (\cos x)u_y = x\,y - u. \tag{2.9}$$

Solution. Solving the ordinary differential equation

$$\frac{dy}{dx} = \cos x$$

yields the characteristics $y - \sin x = k$ or $y = \sin x + k$. For each fixed k, along the curve $y = \sin x + k$,

$$\frac{du}{dx} = u_x + u_y\frac{dy}{dx} = x\,y - u = x(\sin x + k) - u$$

or equivalently

$$\frac{du}{dx} + u = x(\sin x + k).$$

The ordinary differential equation above is linear and nonhomogeneous. Its solution can be found using the integrating factor $\mu(x) = e^x$.

$$u = e^{-x}\left(f(k) + \int xe^x \sin x\, dx + k\int xe^x\, dx \right)$$

$$= e^{-x}\left(f(k) + \frac{1}{2}e^x \cos x - \frac{x}{2}e^x \cos x + \frac{x}{2}e^x \sin x + k(x-1)e^x \right)$$

$$= f(k)e^{-x} + \frac{1}{2}\left[(1-x)\cos x + x\sin x \right] + k(x-1).$$

Substituting $k = y - \sin x$ in the expression above yields the general solution of Eq. (2.9),

$$u(x,y) = f(y - \sin x)e^{-x} + \frac{1}{2}\left[(1-x)\cos x + x\sin x \right] + (x-1)(y - \sin x).$$

Once again, without an appropriate boundary condition the arbitrary function f cannot be determined.

Example 2.3. Consider the partial differential equation

$$3u_x + 4u_y = 8u \text{ for } (x,y) \in \mathbb{R}^2. \tag{2.10}$$

(1) Find the general solution of the equation using the method of characteristics.
(2) Find a solution of the equation that satisfies the boundary condition $u(x,0) = x\sin x$.

Solution. The first order of business is determining the characteristics by solving an ordinary differential equation of the form given in Eq. (2.5). In this case $b(x, y) = 4$ and $a(x, y) = 3$ and thus

$$\frac{dy}{dx} = \frac{4}{3}$$

which has the general solution $y = 4x/3 + k$ or equivalently $3y - 4x = k$. For this example the characteristics are straight lines. Along each of these straight lines, the unknown function $u(x, y(x))$ must satisfy an ordinary differential equation of the form shown in Eq. (2.7) which in this example takes the form

$$\frac{du}{dx} = u_x + u_y \frac{dy}{dx} = u_x + \frac{4}{3} u_y = \frac{3u_x + 4u_y}{3} = \frac{8u}{3}. \tag{2.11}$$

The general solution to Eq. (2.11) is

$$u = f(k)e^{8x/3}$$

where f is an arbitrary function of k, the level set of the characteristic $3y - 4x = k$. Therefore, the general solution of Eq. (2.10) is

$$u(x, y) = f(3y - 4x)e^{8x/3}.$$

To find a solution that satisfies the condition $u(x, 0) = x \sin x$ the function f must satisfy the equation

$$u(x, 0) = f(-4x)e^{8x/3} = x \sin x \iff f(-4x) = xe^{-8x/3} \sin x.$$

To find an expression for f, replace x in the expression above with $-x/4$ which results in

$$f(x) = \frac{x}{4} e^{2x/3} \sin \left(\frac{x}{4} \right).$$

This produces the following expression for the solution of Eq. (2.10) satisfying the condition $u(x, 0) = x \sin x$.

$$u(x, y) = \frac{1}{4} e^{8x/3} e^{2(3y - 4x)/3}(3y - 4x) \sin \left(\frac{3y - 4x}{4} \right)$$

$$= \frac{3y - 4x}{4} e^{2y} \sin \left(\frac{3y - 4x}{4} \right).$$

Figure 2.4 illustrates the integral (solution) surface and several characteristic curves. The characteristic curves differ in their values of the constant $k = 3y - 4x$.

Example 2.4. Find the general solution using the method of characteristics to the linear partial differential equation

$$u_x + y\,u_y = 0. \tag{2.12}$$

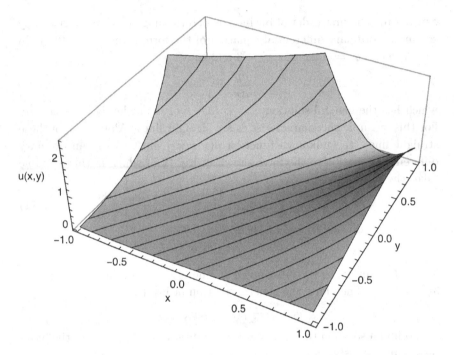

Fig. 2.4   The solution $u(x, y)$ of Eq. (2.10) found in Example 2.3 and several of its characteristic curves.

Solution. Solving the ordinary differential equation

$$\frac{dy}{dx} = y,$$

yields the characteristics

$$y = k\, e^x \text{ or } y\, e^{-x} = k$$

where k is an arbitrary constant. Thus along each curve $y\, e^{-x} = k$,

$$\frac{du}{dx} = u_x + u_y \frac{dy}{dx} = 0$$

which implies $u(x, y) = u(x, ke^x)$ is constant. Denoting the constant as $f(k)$, then the solution is given by $u(x, y) = f(k) = f(y\, e^{-x})$, where f is an arbitrary function.

Though no boundary condition was given in the last example, if it were now chosen to be $u(0, y) = y^2$ (note the condition is specified along the y-axis) then $f(k) = f(y) = y^2$ and hence $u(x, y) = y^2 e^{-2x}$. In fact, the boundary condition imposed on $u(x, y)$ could be any differentiable function defined on the y-axis. However, in the example above arbitrary boundary conditions cannot be specified on the line $y = 0$ (the x-axis) since $y = 0$ is a characteristic corresponding to $k = 0$. In fact, since $u(x, 0) = f(0)$, unless the function specified on the line $y = 0$ is a constant, there will be no solution. Section 2.2 will reveal that a sufficient condition to guarantee that the boundary value problem is well-posed is that the curve, on which the boundary condition is specified, is not tangent to any of the characteristics at any point.

2.2 First-Order Quasilinear Equations

Applications of partial differential equations often involve equations which are only linear in the highest order partial derivatives of the unknown function and are nonlinear in the unknown function or its lower order partial derivatives. A first-order **quasilinear** partial differential equation with two independent variables has the following form,

$$a(x, y, u)u_x + b(x, y, u)u_y = c(x, y, u). \tag{2.13}$$

As a special case, Eq. (2.13) is called **semilinear** if the functions a and b are both independent of the unknown function u. The method of characteristics can be extended to solve Eq. (2.13).

Let $\Omega \subset \mathbb{R}^2$ and consider Eq. (2.13) on the domain Ω. If function $u(x, y)$ is a solution of Eq. (2.13) defined on Ω, then the graph of $u(x, y)$ over its domain is called an **integral surface** of Eq. (2.13). To apply the method of characteristics, rewrite Eq. (2.13) in the form

$$a(x, y, u)u_x + b(x, y, u)u_y - c(x, y, u) = 0. \tag{2.14}$$

For any $(x, y) \in \Omega$, since $\langle u_x, u_y, -1 \rangle$ is a normal vector to the integral surface, Eq. (2.14) simply states, geometrically, that the vector $\langle a(x, y, u), b(x, y, u), c(x, y, u) \rangle$ is perpendicular to this normal vector, and thus is tangent to the integral surface $u(x, y)$ at any (x, y). The vector $\langle a(x, y, u), b(x, y, u), c(x, y, u) \rangle$ defines a vector field on $\Omega \times \mathbb{R}$. It is natural to consider curves determined by this vector field, that is, the integral curves $(x(t), y(t), u(t))$ in the xyu-space of the system of ordinary differential equations

$$\frac{dx}{dt} = a(x, y, u),$$

$$\frac{dy}{dt} = b(x, y, u), \tag{2.15}$$

$$\frac{du}{dt} = c(x, y, u),$$

or equivalently, in symmetric form

$$\frac{dx}{a(x,y,u)} = \frac{dy}{b(x,y,u)} = \frac{du}{c(x,y,u)}. \tag{2.16}$$

Note that the symbol u is used to denote both the solution to the partial differential equation and the third component of the system of ordinary differential equations. When the potential for confusion exists, the symbol z will be used for the third component of the system of ordinary differential equations in Eq. (2.15) or (2.16). The system of ordinary differential equations in Eq. (2.15) or (2.16) is known as the **characteristic system**, and their integral curves are called the **characteristic curves** of Eq. (2.13). The projections of the characteristics curves on the xy-plane are called the **characteristics**. The vector $\langle a(x,y,u), b(x,y,u), c(x,y,u) \rangle$ are called the **characteristic directions** of Eq. (2.13). Figure 2.5 illustrates the characteristic curves and the characteristics for a first-order quasilinear partial differential equation. The close relation between the solutions of Eq. (2.15) (or Eq. (2.16)) and the solutions of Eq. (2.13) is described in the following theorem.

Theorem 2.1. *Assume that functions a, b, and c are continuously differentiable on $\Omega \times (-\infty, \infty)$. Then $u = u(x,y)$ is a solution of Eq. (2.13) on Ω if and only if, for any $(x_0, y_0) \in \Omega$, the solution $(x(t), y(t), z(t))$ of Eq. (2.15) through the point $(x_0, y_0, u(x_0, y_0))$ satisfies $z(t) = u(x(t), y(t))$ for all t in its domain of definition, that is, the integral curve of Eq. (2.15) lies entirely on the surface $u(x,y)$ for t in the domain of its definition as long as the initial value is on the surface.*

Theorem 2.1 simply states that the integral surface of $u(x,y)$ of Eq. (2.13) is comprised of the characteristic curves determined by the characteristic system of Eq. (2.15) or (2.16). On the other hand, any smooth surface that is a union of characteristic curves is an integral surface of Eq. (2.13). The reader should note that a first-order linear or semilinear partial differential equation can be considered as a special case of a quasilinear equation, and thus the method of characteristics applies, as was demonstrated in Sec. 2.1. For linear or semilinear equations, the characteristic system of Eq. (2.15) is decoupled and the method is reduced to that described in Sec. 2.1. The proof of Theorem 2.1 is presented below.

Proof. The proof is based on the existence and uniqueness theorem of ordinary differential systems. To prove the necessity of the stated conditions assume that $u(x,y)$ is a solution to Eq. (2.13) and (x_0, y_0) is any fixed point

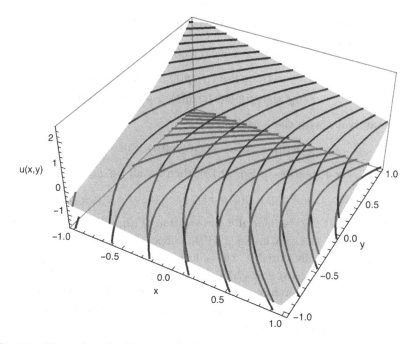

Fig. 2.5   The surface plot illustrates the characteristic curves of a first-order quasilinear partial differential equation while the curves projected onto the plane where $u = -3/2$ (for clarity) are the characteristics.

in its domain Ω. Let $(x(t), y(t), z(t))$ be a solution to the system of ordinary differential equations in Eq. (2.15) with initial values,

$$(x(0), y(0), z(0)) = (x_0, y_0, u(x_0, y_0)) \qquad (2.17)$$

defined in an open interval I containing $t = 0$. Note that z has been used for the third component of the solution of Eq. (2.15). For $t \in I$, define $U(t) = z(t) - u(x(t), y(t))$. Since $(x(t), y(t), z(t))$ is a solution to the system in Eq. (2.15), then for $t \in I$,

$$\frac{dU}{dt} = \frac{dz}{dt} - \frac{d}{dt}[u(x(t), y(t))]$$

$$= c(x(t), y(t), z(t)) - u_x(x(t), y(t))\frac{dx}{dt} - u_y(x(t), y(t))\frac{dy}{dt}$$

$$= c(x, y, z) - u_x(x, y)a(x, y, z) - u_y(x, y)b(x, y, z)$$

$$= c(x, y, U(t) + u(x, y)) - u_x(x, y)a(x, y, U(t) + u(x, y))$$

$$\quad - u_y(x, y)b(x, y, U(t) + u(x, y))$$

where the explicit dependence of x and y on t has been suppressed for brevity. Thus $U(t)$ is a solution of the ordinary differential equation,

$$\frac{dw}{dt} = c(x(t), y(t), w + u(x(t), y(t)))$$
$$- u_x(x(t), y(t))a(x(t), y(t), w + u(x(t), y(t)))$$
$$- u_y(x(t), y(t))b(x(t), y(t), w + u(x(t), y(t))) \qquad (2.18)$$

on the interval I with the initial value,

$$w(0) = z(0) - u(x(0), y(0)) = 0. \qquad (2.19)$$

Since $u(x, y)$ is a solution to Eq. (2.13), then for all $t \in I$,

$$a(x, y, u(x, y))u_x(x, y) - b(x, y, u(x, y))u_y(x, y) = c(x, y, u(x, y)).$$

This implies that $w(t) \equiv 0$ for all t is a solution of Eq. (2.18) with the same initial condition as given in Eq. (2.19). Since all the functions involved are continuously differentiable, the solution of the initial value problem comprised of Eqs. (2.18) and (2.19) is unique (see Sec. 2.8 of [Boyce and DiPrima (2012)]). Therefore, it be must be the case that $U(t) \equiv 0$ for all $t \in I$. This implies that $z(t) = u(x(t), y(t))$, *i.e.*, $(x(t), y(t), z(t))$ lies on the surface $u = u(x, y)$ for all $t \in I$.

Now suppose (x_0, y_0) is any point in the domain of $u(x, y)$ and $(x(t), y(t), z(t))$ be the solution of Eq. (2.15) satisfying the initial condition given in Eq. (2.17). The existence theorem for solutions of ordinary differential equations implies that the solution $(x(t), y(t), z(t))$ exists on an open interval (α, β) containing $t = 0$. By assumption $z(t) = u(x(t), y(t))$ for all $t \in (\alpha, \beta)$. Therefore,

$$c(x(t), y(t), u(x(t), y(t)))$$
$$= \frac{d}{dt}[u(x(t), y(t))] = u_x(x(t), y(t))\frac{dx}{dt} + u_y(x(t), y(t))\frac{dy}{dt}$$
$$= u_x(x(t), y(t))a(x(t), y(t), z(t)) + u_y(x(t), y(t))b(x(t), y(t), z(t))$$
$$= u_x(x(t), y(t))a(x(t), y(t), u(x(t), y(t)))$$
$$+ u_y(x(t), y(t))b(x(t), y(t), u(x(t), y(t))).$$

Setting $t = 0$ produces

$$c(x_0, y_0, u(x_0, y_0))$$
$$= u_x(x_0, y_0)a(x_0, y_0, u(x_0, y_0)) + u_y(x_0, y_0)b(x_0, y_0, u(x_0, y_0))$$

which shows $u(x, y)$ satisfies the partial differential equation in Eq. (2.13) at the point (x_0, y_0). Since (x_0, y_0) is an arbitrary point in the domain of $u(x, y)$ then $u(x, y)$ solves Eq. (2.13) at every point in its domain. $\qquad \square$

Based on Theorem 2.1, an integral surface (or the graph of a solution) of a quasilinear partial differential equation of the form shown in Eq. (2.13) is a union of the characteristic curves determined by the characteristic system of ordinary differential equations given in Eq. (2.15). Of course, there are many integral surfaces for a given quasilinear partial differential equation. A particular solution can be determined if additional conditions are imposed on the general solution. The remainder of this section describes the **Cauchy initial value problem,** which calls for finding a solution of Eq. (2.13) that contains a given curve in xyu-space. Specifically, if Γ is a given space curve described parametrically as

$$x = f(s), \quad y = g(s), \quad z = h(s),$$

where s is the parameter, then a Cauchy initial value problem, or simply a Cauchy problem of Eq. (2.13) is the problem of finding a solution $u(x, y)$ satisfying the condition,

$$h(s) = u(f(s), g(s)). \tag{2.20}$$

The solution is an integral surface passing through the given curve Γ. In the special case that Γ is the x-axis, the solution $u(x, y)$ satisfies the condition that $u(x, 0) = h(x)$ where h is a given function. It can be shown that, under appropriate assumptions, the solution of the Cauchy initial value problem of Eqs. (2.13) and (2.20) exists and is unique as described by the theorem below.

Theorem 2.2. *Let f, g, and h be continuously differentiable functions of s on a closed interval $[\alpha, \beta]$ and assume that a, b, and c are continuously differentiable on a domain containing the curve $\Gamma = \{(f(s), g(s), h(s)) \,|\, \alpha \leq s \leq \beta\}$. Assume further that*

$$g'(s)a(f(s), g(s), h(s)) - f'(s)b(f(s), g(s), h(s)) \neq 0 \tag{2.21}$$

for $s \in [\alpha, \beta]$. Then there exists a unique, continuously differentiable solution of the Cauchy initial value problem of Eqs. (2.13) and (2.20) defined in a neighborhood of the projection of the initial curve Γ.

Note that condition specified in Eq. (2.21) states that the projection of the initial curve Γ on the xy-plane is not parallel to any characteristic anywhere. In this case, through every point $(f(s), g(s), h(s))$, there is a unique integral curve of the system of ordinary differential equations given in Eq. (2.15). These integral curves form an integral surface of Eq. (2.13). A rigorous proof of Theorem 2.2 is omitted here, but can be found in [Myint-U and Debnath (2007)] or [Courant and Hilbert (1989b)].

The solution of the Cauchy problem given by Eqs. (2.13) and (2.20) can be found by solving Eq. (2.15) with the initial conditions

$$x(0) = f(s), \quad y(0) = g(s), \quad z(0) = h(s).$$

In general, the solution (x, y, z) will depend on parameters s and t which parameterize the integral surface determined by the solution u. Often it is possible to eliminate t and s and obtain an expression $u(x, y)$. This section concludes with several examples illustrating the procedure.

Example 2.5. Find a solution to the Cauchy problem

$$u\, u_x + u_y = 1,$$
$$x = 2s, \quad y = 2s, \quad u = s.$$

Solution. The first step is to solve the characteristic system,

$$\frac{dx}{dt} = a(x, y, u) = u, \quad \frac{dy}{dt} = b(x, y, u) = 1, \quad \frac{du}{dt} = c(x, y, u) = 1$$

with the initial conditions,

$$x(0) = 2s, \quad y(0) = 2s, \quad u(0) = s.$$

The second and third equations are readily solved to yield $y(t) = t + 2s$ and $u(t) = t + s$. Substituting these in the ordinary differential equation for x and integrating the resulting equation produce

$$x = 2s + \frac{t^2}{2} + s\, t.$$

Solving the simultaneous equations $y = t + 2s$ and $u = t + s$ for s and t and substituting these results in the equation for variable x immediately above eliminate parameters t and s to give the following expression for u as a function of x and y.

$$x = -\frac{y^2}{2} + y\, u + 2y - 2u \iff u = \frac{2x + y^2 - 4y}{2(y - 2)}.$$

Example 2.6. Find the solution to the Cauchy problem:

$$x\, u_x + u_y = e^u,$$
$$u = 0 \text{ when } y = x - 1.$$

Solution. The curve for which $y = x - 1$ and $u = 0$ can be represented in parameterized form as

$$x = s, \quad y = s - 1, \quad u = 0.$$

The next step is to solve the system of ordinary differential equations for the characteristic curves. The characteristic system can be expressed as

$$\frac{dx}{dt} = x, \quad \frac{dy}{dt} = 1, \quad \frac{du}{dt} = e^u$$

with initial conditions,

$$x(0) = s, \quad y(0) = s - 1, \quad u(0) = 0.$$

The ordinary differential equations are decoupled and the solutions to the initial value problems are

$$x = s\,e^t, \quad y = t + s - 1, \quad e^{-u} = t + 1.$$

Solving the last two expressions for s and t and substituting the results into the solution for x produce the following implicitly defined solution to the Cauchy problem,

$$x = (y - e^{-u} + 2)e^{(e^{-u} - 1)}.$$

Example 2.7. Solve the initial value problem

$$u_x + 3u_y = u^2,$$
$$u(x, 0) = f(x)$$

where $f(x)$ is a given smooth function.

Solution. Solving this boundary value problem is equivalent to finding a solution which passes through the curve Γ parameterized as

$$x = s, \quad y = 0, \quad u = f(s).$$

The first step is to solve the system of ordinary differential equations for the characteristic curves:

$$\frac{dx}{dt} = 1, \quad \frac{dy}{dt} = 3, \quad \frac{du}{dt} = u^2$$

with the initial values,

$$x(0) = s, \quad y(0) = 0, \quad u(0) = f(s).$$

As in the previous example the ordinary differential equations are decoupled and the solution to the initial value problem is

$$x = t + s, \quad y = 3t, \quad u = \frac{f(s)}{1 - t\,f(s)}. \tag{2.22}$$

Solving for parameters s and t from the first two equations in Eq. (2.22) and substituting for s and t in the third equation produce

$$u(x, y) = \frac{f(x - y/3)}{1 - (y/3)f(x - y/3)}.$$

The reader should see from the examples above, solutions of a quasi-linear partial differential equation are not necessary defined on the domain where the partial differential equation is defined, as seen in the study of ordinary differential equations. For instance in Example 2.5, the partial differential equation is defined everywhere even though the solution is un-defined along the line $y = 2$.

The method of characteristics has been applied to first-order linear, semilinear, or quasilinear partial differential equations. Actually, it can also be used to solve first-order nonlinear partial differential equations and higher order partial differential equations in some cases. Consider a general first-order partial differential equation with two independent variables having the following form,

$$F(x, y, u, u_x, u_y) = 0. \tag{2.23}$$

Replace u_x with variable p and u_y with q. Differentiating Eq. (2.23) with respect to x and y respectively yields the following system of equations.

$$F_x + F_u p + F_p p_x + F_q q_x = 0, \tag{2.24}$$

$$F_y + F_u q + F_p p_y + F_q q_y = 0. \tag{2.25}$$

Assuming that $u(x, y)$ has continuous second partial derivatives, by Schwarz's[1] theorem, $q_x = u_{yx} = u_{xy} = p_y$ [Trench (1978)]. Therefore q_x can be replaced by p_y in Eq. (2.24) and then p satisfies

$$F_x + F_u p + F_p p_x + F_q p_y = 0. \tag{2.26}$$

Equation (2.26) is a quasilinear first-order partial differential equation in p. Applying the method of characteristics to Eq. (2.26) produces the following system of characteristic equations:

$$\frac{dx}{F_p} = \frac{dy}{F_q} = \frac{dp}{-F_x - F_u p}. \tag{2.27}$$

Similarly if p_y is replaced by q_x in Eq. (2.25) another system of characteristic equations results with the form,

$$\frac{dx}{F_p} = \frac{dy}{F_q} = \frac{dq}{-F_y - F_u q}. \tag{2.28}$$

The differential of u can be expressed as

$$du = p\, dx + q\, dy = (p\, F_p + q\, F_q)\frac{dx}{F_p}. \tag{2.29}$$

---

[1]Hermann Schwarz, German mathematician (1843–1921).

Combining Eqs. (2.27), (2.28), and (2.29) results in

$$\frac{dx}{F_p} = \frac{dy}{F_q} = \frac{dp}{-F_x - F_u p} = \frac{dq}{-F_y - F_u q} = \frac{du}{p F_p + q F_q}. \tag{2.30}$$

This system of ordinary differential equations with unknowns x, y, p, q, and u is the characteristic system for Eq. (2.23). Solving Eq. (2.30) produces the characteristic curves in $xypqu$-space. The reader interested in a more complete discussion of the method of characteristics should consult the books [McOwen (2002)] or [Debnath (2012)] and the references therein.

2.3 Applications

The method of characteristics is a general procedure which can be used for solving linear and quasilinear partial differential equations of first order. This section presents some applications of first-order linear partial differential equations. This section can be skipped without hindering the reader's understanding the rest of this text until Chap. 11. The first application is to a model of population growth for which the age of individuals in the population is significant.

2.3.1 *Population Growth with Age Structure*

In population dynamics, if only the size of the population (the number or density of individuals in the population) as a function of time is considered, the population can be modeled using ordinary differential equations. For example, the reader may be familiar with exponential growth/decay models and logistic models described in most elementary textbooks on mathematical modeling or ordinary differential equations. A good introduction may be found in [Boyce and DiPrima (2012)]. However, if the extra dimension of the demographics of age structure within the population is to be modeled then partial differential equations are required. Age structure could be of interest if the species under study exhibits different stages of a life cycle (for example egg, larva, juvenile, and adult). In the following discussion, a population in a stable ecosystem is considered. When modeling populations, the density of the population (individuals per unit area or volume) is often modeled rather than the absolute number of individuals, since numbers of individuals are discrete variables while the density is more appropriately thought of as a continuous variable. Let $p \equiv p(x, t)$ denote the age-density of the population, where x represents the age of individuals within the population and t represents time. In this subsection the symbol t is interpreted as an independent variable and not as a parameter used to describe characteristic curves. Note that, if $N(t)$ denotes the total population (absolute number of individuals) of all ages at time t, then

$$N(t) = \int_0^\infty p(x,t)\,dx.$$

Convergence of the improper integral is not an issue since in all realistic populations $p(x,t) = 0$ for large values of x (while some species may have long lifespans, no species has an *arbitrarily* long life). To derive a differential equation involving $p(x,t)$, let Δt and Δx represent the increments of time and age, respectively, and consider the change of the population with ages between x and $x + \Delta x$ when the time is changed from t to $t + \Delta t$. To simplify the derivation, assume that x represents the "true" (or calendar) age of an individual. In this case $\Delta x = \Delta t$. Another way of thinking about this is that individuals in the population age at the same rate that time passes. However, in a more general model, this relationship might not be true since the variable x might represent some non-time related physiological measure (for example body mass). Differentials will be used to approximate the changes in the demographics of population under consideration. First, the population with ages between x and $x + \Delta x$ at time $t + \Delta t$ can be approximated by $p(x, t + \Delta t)\,\Delta x$. Since $\Delta x = \Delta t$ then at time $t + \Delta t$ (ignoring for the moment any deaths which may occur) the population with ages between x and $x + \Delta x$ should be the same as the population with ages between $x - \Delta x$ and x was at time t. The latter can be approximated by $p(x - \Delta x, t)\,\Delta x = p(x - \Delta t, t)\,\Delta x$. Finally, suppose $\mu(x)$ represents the rate (or probability) of death of individuals in the population at age x. The incidence of deaths in the population with ages between $x - \Delta x$ and x between times t and $t + \Delta t$ can be approximated by $\mu(x - \Delta x)p(x - \Delta t, t)\,\Delta x\,\Delta t = \mu(x - \Delta t)p(x - \Delta t, t)\,\Delta x\,\Delta t$. Therefore, the following relationship holds,

$$p(x, t + \Delta t)\,\Delta x \approx p(x - \Delta t, t)\,\Delta x - \mu(x - \Delta t)p(x - \Delta t, t)\,\Delta x\,\Delta t.$$

Dividing both sides of this approximation by Δx and rearranging terms produce

$$p(x, t + \Delta t) - p(x - \Delta t, t) \approx -\mu(x - \Delta t)p(x - \Delta t, t)\,\Delta t.$$

By adding and subtracting $p(x,t)$ on the left-hand side this approximation can be rewritten as

$$p(x, t + \Delta t) - p(x,t) - [p(x - \Delta t, t) - p(x,t)] \approx -\mu(x - \Delta t)p(x - \Delta t, t)\,\Delta t.$$

Dividing both sides of this approximation by Δt and taking the limit as $\Delta t \to 0$ yield the equation

$$p_t(x,t) + p_x(x,t) = -\mu(x)p(x,t). \tag{2.31}$$

Equation (2.31) is sometimes called the **von Foerster**[2] **equation** and is a linear partial differential equation of first order. The solutions to this partial differential equation will of course depend on the boundary conditions imposed on $p(x,t)$. A natural boundary condition is an initial condition describing the population age density at time $t = 0$,

$$p(x,0) = f(x). \tag{2.32}$$

Additionally there should be a boundary condition for $p(x,t)$ specified at $x = 0$, which represents the new births (population members of age $x = 0$) at time t. A simple condition is that $p(0,t)$ is given by a function of t only, for example

$$p(0,t) = \phi(t). \tag{2.33}$$

The boundary conditions must be compatible in the sense that $f(0) = p(0,0) = \phi(0)$ or otherwise a discontinuity in the population density results. Figure 2.6 illustrates the domain of $p(x,t)$ along with the boundary conditions of Eqs. (2.32) and (2.33).

The method of characteristics can be used to find the general solution of Eq. (2.31) subject to the boundary conditions of Eqs. (2.32) and (2.33). The projection of the characteristic curves on the xt-plane is given by $x = t + k$, where k is an arbitrary constant. For each fixed k, along the line $x = t + k$ solution $p(x,t) = p(t+k,t) = \hat{p}(t)$ satisfies the ordinary differential equation

$$\frac{d\hat{p}}{dt} = -\mu(t+k)\hat{p}(t). \tag{2.34}$$

Equation (2.34) is valid for $t > 0$ if $k \geq 0$ and for $t > -k$ if $k < 0$. The general solution is $\hat{p}(t) = C(k)e^{-\int_{t_0}^t \mu(s+k)\,ds}$ where $C(k)$ is an arbitrary function of k and $t_0 = 0$ if $k \geq 0$ while $t_0 = -k$ if $k < 0$. From the initial and boundary conditions of Eqs. (2.32) and (2.33), the initial value for $\hat{p}(t)$ is $\hat{p}(0) = p(k,0) = f(k)$ if $k = x - t \geq 0$ or $x \geq t$ and $\hat{p}(-k) = p(-k+k, -k) = \phi(-k)$ if $k = x - t < 0$ or $x < t$. The solution satisfying the initial values is

$$\hat{p}(t) = \begin{cases} f(k)e^{-\int_0^t \mu(s+k)\,ds} & \text{if } t \leq x, \\ \phi(-k)e^{-\int_{-k}^t \mu(s+k)\,ds} & \text{if } t > x. \end{cases}$$

---

[2]Heinz von Foerster, Austrian American scientist (1911–2002).

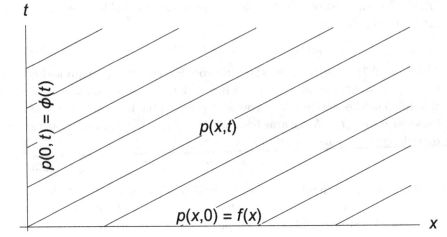

Fig. 2.6 The variable p represents the density of individuals in a population of age x and time t. The boundary condition $p(0,t) = \phi(t)$ represents new births at time t while $p(x,0) = f(x)$ models the initial age distribution of the population. The sloping lines represent characteristics.

Recalling that $k = x - t$ then the age-structured population density can be written as

$$p(x,t) = \begin{cases} f(x-t)e^{\int_0^t \mu(s+x-t)\,ds} & \text{if } t \le x, \\ \phi(t-x)e^{-\int_{-k}^t \mu(s+x-t)\,ds} & \text{if } t > x, \end{cases}$$

$$= \begin{cases} f(x-t)e^{-\int_{x-t}^x \mu(s)\,ds} & \text{if } t \le x, \\ \phi(t-x)e^{-\int_0^x \mu(s)\,dx} & \text{if } t > x. \end{cases} \tag{2.35}$$

The first line of Eq. (2.35) can be interpreted as describing the evolution of the portion of the population that was already alive when $t = 0$ while the second line describes the fraction of the population born since $t = 0$. Figure 2.7 illustrates the integral surface for an example of the von Foerster equation for which the initial age distribution is normal with a mean of $5/2$ and a standard deviation of 1, the mortality rate is $\mu(x) = 1 - e^{-x/5}$, and the time-dependent birth rate is $\phi(t) = (1 - e^{-t})/2$.

The reader might immediately question the validity of the boundary condition given in Eq. (2.33). It is certainly reasonable to expect that the new births, at a given time t, depend on the distribution of ages in the population or even the total population at the same time t. For example, assuming the birth rate with respect to age x is given as $\beta(x)$, then the

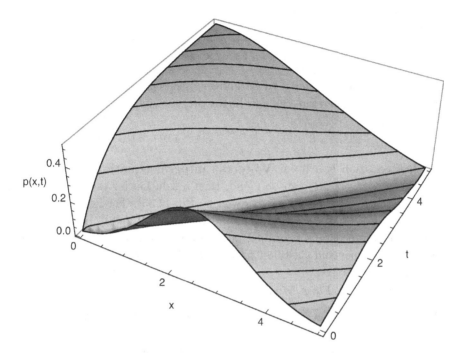

Fig. 2.7 A surface plot of an example of the von Foerster model of age-structured population dynamics given in Eq. (2.31). The mesh lines represent the characteristic curves for which $x - t = k$.

new births should be given by the integral over the time

$$p(0, t) = \int_0^\infty \beta(x)p(x, t)\, dx. \tag{2.36}$$

Such a boundary condition is called a **nonlocal boundary condition**, which is given by an integral that depends on the unknown function. Solving Eq. (2.31) with the initial value of Eq. (2.32) and the new boundary condition of Eq. (2.36) results in a solution similar to that given by Eq. (2.35) except that now $\phi(t)$ must be specified as

$$\phi(t) = p(0, t) = \int_0^\infty \beta(x)p(x, t)\, dx. \tag{2.37}$$

Using Eq. (2.35), then Eq. (2.37) becomes

$$\begin{aligned}
\phi(t) &= \int_0^t \beta(x)p(x, t)\, dx + \int_t^\infty \beta(x)p(x, t)\, dx \\
&= \int_0^t \beta(x)\phi(t - x)e^{-\int_0^x \mu(s)\, ds}\, dx \\
&\quad + \int_t^\infty \beta(x)f(x - t)e^{-\int_{x-t}^x \mu(s)\, ds}\, dx.
\end{aligned} \tag{2.38}$$

Setting

$$\psi(t) = \int_t^\infty \beta(x) f(x-t) e^{-\int_{x-t}^x \mu(s)\,ds}\,dx$$

then $\psi(t)$ can be considered to be known since $f(x)$, $\beta(x)$ and $\mu(x)$ are given functions. Therefore the new births $\phi(t)$ satisfy

$$\phi(t) = \int_0^t \beta(x) e^{-\int_0^x \mu(s)\,ds} \phi(t-x)\,dx + \psi(t). \qquad (2.39)$$

The last equation is a linear **Volterra**[3] **integral equation** with kernel $\beta(x) e^{-\int_0^x \mu(s)\,ds}$. If $\phi(t)$ can be found, then a solution for the population density with respect to the age is given by Eq. (2.35). For details on the solution of Eq. (2.39), see [Polyanin and Manzhirov (2008)]. Readers interested in more details regarding population models with age structure should see [Brauer and Castillo-Chavez (2014)] and [Cushing (1994)].

2.3.2 A Traffic Flow Model

This section explores a simple model for traffic flow on a stretch of a congested highway. The approach taken follows closely that of [Haberman (1998)] which the reader should consult for additional background and extensions of the model. The following assumptions are used in the development of this model of traffic flow.

- Vehicles are not allowed to pass each other. The reader may wish to imagine the road is a single lane with all traffic traveling in the same direction.
- The road has no exits or entrances. All vehicles on the road are already present. No additional vehicles will be added and none may leave.
- Function $\rho(x,t)$ denotes the density of vehicles, the number of vehicles per unit length of road (for example, cars/mile) at location x at time t.
- Function $q(x,t)$ denotes the flow of traffic, the number of vehicles per unit time passing location x at time t.

To derive an equation governing the traffic flow under the assumptions above, consider an arbitrary section of the highway between locations $x = a$ and $x = b$. The total number of vehicles in interval $[a, b]$ given by

$$N(t) = \int_a^b \rho(x,t)\,dx.$$

---

[3]Vito Volterra, Italian mathematician (1860–1940).

The rate of change of $N(t)$ is

$$\frac{dN}{dt} = \frac{d}{dt}\left[\int_a^b \rho(x,t)\,dx\right] = q(a,t) - q(b,t). \tag{2.40}$$

Intuitively this may be thought of as stating that the rate of change of the number of vehicles in interval $[a,b]$ is the rate at which vehicles enter interval $[a,b]$ (the traffic flow $q(a,t)$) minus the rate at which vehicles leave interval $[a,b]$ (the traffic flow $q(b,t)$). Recall the assumption that there are no entrances or exits on the highway. Equation (2.40) is the integral form of "conservation of vehicles". On the other hand, according to the Fundamental Theorem of Calculus

$$q(a,t) - q(b,t) = -\int_a^b \frac{\partial}{\partial x}[q(x,t)]\,dx.$$

Combining these two equations yields

$$\frac{d}{dt}\left[\int_a^b \rho(x,t)\,dx\right] = -\int_a^b \frac{\partial}{\partial x}[q(x,t)]\,dx.$$

Assuming $\rho(x,t)$ is continuously differentiable, the derivative with respect to time t can be taken inside the integral (Lemma 1.1). This produces

$$\int_a^b \frac{\partial}{\partial t}[\rho(x,t)]\,dx = -\int_a^b \frac{\partial}{\partial x}[q(x,t)]\,dx$$

or by moving both integrals to the left-hand side of the equation,

$$\int_a^b \left(\frac{\partial}{\partial t}[\rho(x,t)] + \frac{\partial}{\partial x}[q(x,t)]\right)dx = 0.$$

Since a and b are arbitrary, then by Lemma 1.2,

$$\frac{\partial}{\partial t}[\rho(x,t)] + \frac{\partial}{\partial x}[q(x,t)] = 0. \tag{2.41}$$

Equation (2.41) is underdetermined in the sense that it contains two unknown functions, $q(x,t)$ and $\rho(x,t)$. An additional assumption is needed, namely that q depends on the function ρ, that is $q \equiv q(\rho)$. Suppose that $u(x,t)$ is the velocity of the vehicle located at position x at time t. Observations of real traffic situations suggest that when traffic is light (corresponding to low values of ρ) vehicles' velocities are at or near their maximum legal values. Conversely when traffic density increases, vehicle velocities decrease and in very high density cases, vehicle traffic on an entire stretch of the highway can come to a stop. Using this insight u should be a function of the traffic density $\rho(x, t)$. Assuming a simple linear relationship between vehicle velocity and traffic density leads to

$$u(\rho(x, t)) = u_{\max}\left(1 - \frac{\rho(x, t)}{\rho_{\max}}\right), \tag{2.42}$$

where $u_{\max}$ is the maximum velocity of vehicles and $\rho_{\max}$ is the maximum traffic density. Therefore the traffic flux $q(x, t)$ takes the form

$$q(x, t) = u_{\max}\rho(x, t)\left(1 - \frac{\rho(x, t)}{\rho_{\max}}\right). \tag{2.43}$$

Therefore, Eq. (2.41) becomes

$$\frac{\partial}{\partial t}[\rho(x, t)] + \frac{\partial}{\partial x}[\rho(x, t)u(\rho(x, t))] = 0$$

or equivalently

$$\frac{\partial}{\partial t}[\rho(x, t)] + c(\rho)\frac{\partial}{\partial x}[\rho(x, t)] = 0 \tag{2.44}$$

where $c(\rho) = q'(\rho) = \frac{d}{d\rho}[\rho u(\rho)]$ is considered a known function of ρ. Equation (2.44) is a first-order quasilinear partial differential equation. Assuming the initial traffic density is given by

$$\rho(x, 0) = f(x) \tag{2.45}$$

then Eqs. (2.44) and (2.45) form a Cauchy problem. The characteristic curve of Eq. (2.44) is determined by

$$\frac{dt}{ds} = 1, \quad \frac{dx}{ds} = c(\rho), \quad \frac{d\rho}{ds} = 0.$$

The system of ordinary differential equations above is decoupled. The third equation says that the traffic density $\rho(x, t)$ is a constant along the characteristics. In turn if ρ is constant $c(\rho)$ is constant as well and this implies that the characteristics are all straight lines according to the first two equations of the characteristic system. More specifically, for any fixed position x_0, from the initial condition given in Eq. (2.45), it is known that $\rho(x_0, t) = \rho(x_0, 0) = f(x_0)$ on the characteristic passing through the point $(x_0, 0)$. This implies that the characteristic must be the straight line whose equation is $x = c(f(x_0))t + x_0$. Note that the lines $x = c(f(x_0))t + x_0$ corresponding different values x_0 are not parallel (if $c(\rho)$ is not constant) since they have different slopes.

The remainder of this section will explore some of the interesting dynamics of this traffic model and some of the consequences of the solution. The reader is encouraged to think about the results of the mathematical model in the context of real traffic situations experienced or observed. To further the discussion, assume $q(x, t)$ has the form shown in Eq. (2.43). With this choice Eq. (2.44) takes the form

$$\rho_t + u_{\max} \left(1 - \frac{2\rho}{\rho_{\max}} \right) \rho_x = 0.$$

If the initial distribution of traffic density is $f(x)$ then the characteristics are straight lines parametrically expressed as

$$x = u_{\max} \left(1 - \frac{2f(s)}{\rho_{\max}} \right) t + s$$

(note that x and t are the variables and s is the parameter) and the density of traffic along the characteristic is $\rho = f(s)$.

If $f(s) = f_0$ a constant, then the traffic density is uniform and all vehicles move along the road at a constant speed maintaining the constant density. It is rare that traffic maintains a constant density. If f is a decreasing function of s, the slope of the characteristic in the xt-plane is also a decreasing function of s and thus the characteristics "fan out" as illustrated in Fig. 2.8. Vehicles in the front of the cluster of cars experience lower traffic density than vehicles further back in the pack. Thus those vehicles at the front can accelerate away from the group, further lowering the density and ultimately allowing more vehicles to travel faster.

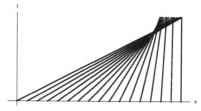

Fig. 2.8 If the initial density distribution of vehicles is a decreasing function then characteristics fan out. In terms of vehicle behavior, cars at the front of the initial configuration may drive faster and pull away from the cars behind. In this way the density of vehicles becomes more rarified.

Fig. 2.9   If the initial density distribution of vehicles is an increasing function then characteristics will intersect at some point where $t > 0$. This phenomenon is called a shock and results in the solution being multi-valued and hence not well-defined.

On the other hand, if $f(s)$ is an increasing function of s, then the slopes of the characteristic lines are increasing as well. Under these conditions the characteristics will intersect at some positive value of t (see Fig. 2.9). At the intersections of these lines, the solution is not well-defined (since the density will have two or more different values) or the solution has a jump discontinuity at the line. This introduces the phenomena called **shock waves**. The surface plot of traffic density in Fig. 2.10 illustrates that for some $t > 0$ the density folds over on itself and becomes triple-valued. The reader should note that the shock forms at a density lower than the maximum density, thus shocks are not necessarily the result of high traffic density. Several of the exercises will explore the specifics of shock formation in this model. Chapter 11 will take up the formation of shocks in partial differential equations and methods for determining which (if any) of the multiple values of the traffic density is the "correct" value.

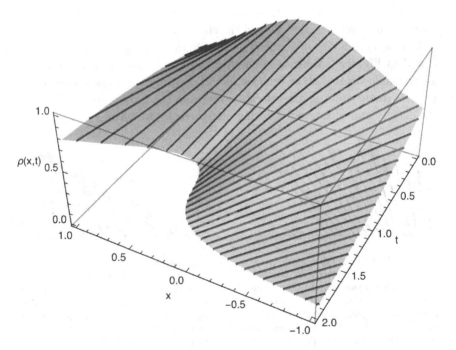

Fig. 2.10 If the initial density of traffic is an increasing function of x the solution becomes multi-valued for some $t > 0$ and hence the solution is not well-defined.

2.4   Exercises

(1) Assume $f(z)$ is any differentiable function defined on $(-\infty, \infty)$. Show that $u(x, y) = f(y - cx)$ is a solution of Eq. (2.1).

(2) Consider the following first-order linear constant coefficient partial differential equation,

$$au_x + bu_y + cu = 0.$$

The symbols a, b, and c represent constants with $ab \neq 0$ and $u \equiv u(x, y)$. Alice says the general solution is

$$u(x, y) = e^{-cx/a} f(bx - ay)$$

for an arbitrary continuously differentiable function f while Bob says the general solution is

$$u(x, y) = e^{-cy/b} g(bx - ay)$$

for an arbitrary continuously differentiable function g. Who, if anyone, is correct?

(3) Consider the initial value problem:

$$u_x + c\, u_y = \phi(x, y) \text{ for } x > 0 \text{ and } -\infty < y < \infty,$$
$$u(0, y) = \psi(y) \text{ for } -\infty < y < \infty,$$

where $\phi(x, y)$ and $\psi(y)$ are given functions.

(a) Explain why $u(x, y) = v(x, y) + w(x, y)$ is a solution of the initial value problem above if $v(x, y)$ and $w(x, y)$ are solutions of

$$v_x + c\, v_y = 0 \text{ for } x > 0 \text{ and } -\infty < y < \infty,$$
$$v(0, y) = \psi(y) \text{ for } -\infty < y < \infty$$

and

$$w_x + c\, w_y = \phi(x, y) \text{ for } x > 0 \text{ and } -\infty < y < \infty,$$
$$w(0, y) = 0 \text{ for } -\infty < y < \infty.$$

(b) Solve for $v(x, y)$ and $w(x, y)$. Note that this problem indicates in general how to deal with nonhomogeneous equations and nonhomogeneous boundary conditions.

(4) Consider the initial value problem,

$$x\, u_x + u_y = y \text{ for } x > 0 \text{ and } -\infty < y < \infty,$$
$$u(x, 0) = x^2 \text{ for } x > 0.$$

The partial differential equation has characteristic equations

$$\frac{dx}{dt} = x \text{ and } \frac{dy}{dt} = 1$$

which can be rewritten as

$$\frac{dy}{dx} = \frac{1}{x} \text{ or } \frac{dx}{dy} = x.$$

(a) Find the general solution of the partial differential equation by solving

$$\frac{dx}{dy} = x$$

for the characteristic curves, then find the solution of the initial value problem.

(b) Find the general solution of the partial differential equation by solving

$$\frac{dy}{dx} = \frac{1}{x}$$

for the characteristic curves, then find a solution of the initial value problem.

(5) Find the general solution for each of the following partial differential equations.

(a) $2u_x + 3u_y + 8u = 0$.

(b) $x\,u_x - y\,u_y = u$.

(c) $x^2 u_x + y\,u_y + x\,y\,u = 0$.

(6) Show there is no solution to the first-order linear partial differential equation

$$2u_x + 3u_y + 8u = 0$$

satisfying the boundary condition

$$u(x, (3x - 1)/2) = e^x.$$

(7) For each of the following partial differential equations, (i) solve the characteristic equations and sketch the graph of a few typical characteristic curves, (ii) find the general solution of the given partial differential equation, and (iii) find a solution of the given partial differential equation that satisfies the given condition.

(a) $2u_x + 5u_y + 6u = 0$, $u(x,0) = x \cos x$ for $x \in \mathbb{R}$.

(b) $2u_x + 5u_y + 6u = 0$, $u(0,y) = e^y \cos y$ for $y \in \mathbb{R}$.

(c) $x^2 u_x - 2u_y - x\,u = x^2$, $u(x,0) = e^x$ for $x \in \mathbb{R}$.

(d) $x\,u_x + y\,u_y + 4 = 0$, $u(x,0) = x^2$ for $x \in \mathbb{R}$.

(8) Consider the first-order partial differential equation

$$u_x + e^x u_y + e^z u_z = (2x - e^x)e^u$$

with three independent variables. Use the method of characteristics to find the general solution to this equation.

(9) Consider the partial differential equation $u\,u_x + u_y = 0$. Let $u_1(x,y)$ and $u_2(x,y)$ be any two nontrivial nonidentical solutions of the equation. Show the following:

  (a) $k\,u_1(x,y)$ is not a solution of the equation unless $k = 0$ or $k = 1$.

  (b) $u_1(x,y) + u_2(x,y)$ is not a solution of the equation.

  (c) Explain why the Principle of Superposition does not hold for this equation.

(10) Solve the initial value problem in Exercise 3 using the method for solving initial value problems of quasilinear partial differential equations described in Sec. 2.2.

(11) For each of the following partial differential equations, find a solution that satisfies the following conditions.

  (a) $u_x - u_y = u^2$, $u = 4$ on $y = 2x - 1$.

  (b) $x\,u_x + y\,u_y = \sec u$, $u = 0$ on $y = x^3$.

  (c) $u_x + y^2 u_y = \cos u$, $u(x,y) = 1$ if $x - y^2 = 0$.

  (d) $(y+u)u_x + y\,u_y = x - y$ for $x \in \mathbb{R}$ and $y > 0$ where $u(x,1) = x+1$.

(12) The partial differential equation

$$u_t + c\,u_x = 0$$

with $c \neq 0$ is often called the first order **wave equation**.

  (a) Introduce the new variables $\xi = x + ct$ and $\eta = x - ct$ and show that the wave equation can be rewritten as

  $$u_\xi = 0.$$

  (b) Use the change of variables to show the general solution to the first-order wave equation is $u(x,t) = F(x - ct)$ where F is an arbitrary differentiable function.

(13) The first-order partial differential equation

$$u_t + u\,x^2 t\,u_x = 0$$

is an example of a homogeneous **advection equation**. Suppose the initial condition is $u(x,0) = F(x)$ for some suitable function F. Find the general solution to this advection equation.

(14) The first-order partial differential equation

$$u_x + u\, u_y = 0$$

is known as the **inviscid Burgers' equation**. Show that the solution to this equation with boundary condition $u(0, y) = f(y)$, where f is a differentiable function, can be expressed implicitly as

$$u(x, y) = f(y - u(x, y)x).$$

(15) Suppose that a typical car is 5 meters in length and that all traffic comes to a halt when there is only a half car length between cars on the highway. What is the maximum traffic density in units of cars per kilometer?

(16) Suppose that a typical car is 5 meters in length and that when there is one car length between cars, the traffic can move at 10 kilometers per hour. What is the traffic flux in this case? Use the result of Exercise 15 and Eq. (2.42).

(17) The formation of a shock in the traffic flow model is initiated by the occurrence of a vertical tangent to the density surface $\rho(x, t)$. Use the chain rule for derivatives to show that

$$\frac{d\rho}{dx} = \frac{f'(s)}{1 - \frac{2u_{\max}t}{\rho_{\max}}f'(s)}.$$

(18) Suppose the shock in the traffic flow model occurs at the earliest time for which a vertical tangent to the density surface exists. Use Exercise 17 to show that the shock forms at

$$t_0 = \frac{\rho_{\max}}{2u_{\max}f'(s)}.$$

(19) Suppose that $f(s) = \rho_{\max}/(3 + 3s^2)$ is the initial distribution of traffic density in the traffic flow model. Find the location (x-coordinate) of the maximum density at time $t > 0$.

(20) Using the initial density mentioned in Exercise 19 find the time and location of the first shock. Show that the density at the point where the shock forms is below the maximum density at the time the first shock forms.

Chapter 3

Fourier Series

The heat equation was used to illustrate the method of separation of variables in Sec. 1.6. Separation of variables produced product solutions to the partial differential equation describing heat conduction in a uniform, one-dimensional rod. The product solutions were chosen to satisfy the Dirichlet boundary conditions selected for Example 1.12. To find a solution to a chosen initial condition the Principle of Superposition is employed. This necessitates expanding the initial condition function as a series of eigenfunctions (in the example, as a series of trigonometric sine functions). This produces a formal solution of an initial boundary value problem for the heat equation. As seen in the examples of Sec. 1.6, if the initial condition function is a finite linear combination of sine functions, then by equating coefficients of the sine functions, the solution to the initial boundary value problem of the one-dimensional heat equation is found in closed form. French mathematician and physicist/engineer, Jean Baptiste Joseph Fourier is known for initiating the investigation of the representation of functions as trigonometric series (now called **Fourier series**) and the application of this method to problems of heat transfer. In fact, Fourier essentially developed the procedure of separation of variables and presented many concrete examples of trigonometric expansions in connection with boundary value problems related to heat conduction problems. German mathematician Peter Dirichlet is believed to be the first person to describe sufficient conditions that guarantee the convergence of a Fourier series. The work of Fourier and Dirichlet leads to the important branch of mathematics known as **Fourier analysis**, or more generally, **harmonic analysis** today.

Roughly speaking, a Fourier series expansion of a function is a representation of a given function on an interval $[-L, L]$ as the sum of sine and cosine functions of the form

$$\frac{a_0}{2} + \sum_{n=1}^{\infty} \left(a_n \cos \frac{n\pi x}{L} + b_n \sin \frac{n\pi x}{L} \right), \tag{3.1}$$

where a_n and b_n are constants. Series of the form given in Eq. (3.1) are called **Fourier series**. This is somewhat analogous to what is done with Taylor series in elementary calculus. Under appropriate assumptions, a given function $f(x)$ defined on an interval containing x_0 can be represented by its Taylor series, that is

$$f(x) = \sum_{n=0}^{\infty} \frac{f^{(n)}(x_0)}{n!} (x - x_0)^n.$$

Such a series will converge for $x = x_0$ and may converge for $|x - x_0| < \rho$ where $\rho > 0$. The reader should note that the equality of $f(x)$ and the Taylor series occurs only within the interval of convergence for the Taylor series and only for functions whose Taylor remainder converges to zero.[1]

The representation of functions by Fourier series raises several questions similar to those which must be addressed when representing a function by a Taylor series.

- What kind of functions $f(x)$ can be written as a Fourier series?
- If a function can be represented in the form of a Fourier Series, what are the constants a_n and b_n?
- Will the Fourier series converge?
- Provided the Fourier series converges, does it converge to $f(x)$ at all points in the interval $[-L, L]$?

This chapter answers such questions and discusses some fundamental properties of Fourier series. Section 3.1 introduces the basic properties of periodic functions, in particular, sine and cosine functions. Derivation of the coefficients of Fourier series is discussed in Secs. 3.3–3.5. Convergence issues of Fourier series, including the Gibbs[2] phenomenon, are discussed in Secs. 3.6 and 3.7. Properties of Fourier series, such as differentiation and integration of Fourier series and topics related to approximation are covered in Secs. 3.8 and 3.9. This is followed by the complex form of Fourier series in Sec. 3.10. The chapter concludes with a proof of the Dirichlet Convergence Theorem and some other theoretical results in Sec. 3.11, as well as some

---

[1] While not required for this text, a reader wishing to review Taylor series should consult a textbook on real analysis such as [Trench (1978)].

[2] Josiah Willard Gibbs, American scientist (1839–1903).

supplemental remarks and comments in Sec. 3.12. [Bhatia (2005)] provides a focused introduction to Fourier series, the history of their development, and their properties and applications.

3.1 Periodic Functions

This section discusses some basic properties of periodic functions and the trigonometric systems. Most readers should be familiar with these properties.

A function $f(x)$ is said to be **periodic** if there exists a constant $T > 0$ such that, for any x in the domain of f, the element $x + T$ is in its domain and $f(x + T) = f(x)$ holds for all such x. In this case, T is called a **period** of $f(x)$ and $f(x)$ is said to be T-**periodic** or **periodic with period** T. For example, for any constant $a > 0$, $\sin(ax)$ and $\cos(ax)$ are periodic with period $2\pi/a$. As another example, the function $f(x) = x - [\![x]\!]$, where $[\![x]\!]$ represents the greatest integer less than or equal to x, is a periodic function with period 1 (see Fig. 3.1).

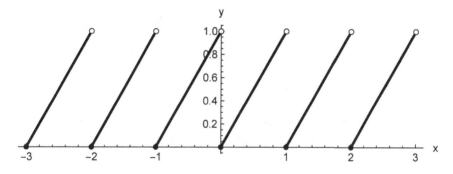

Fig. 3.1   The graph of $f(x) = x - [\![x]\!]$.

The following is a list of some elementary properties of periodic functions.

- Any constant function is trivially periodic and any positive number T is a period.
- If T is a period of a periodic function $f(x)$, then kT is also a period of f for any positive integer k.
- If $f(x)$ and $g(x)$ are two periodic functions with a common period T, then for any constant c, $cf(x)$, $f(x) \pm g(x)$, $f(x) \cdot g(x)$, and $f(x)/g(x)$ are all periodic with period T on their respective domains.

- If $f(x)$ is a periodic function with period T, then $f'(x)$ is also T-periodic on its domain. In other words, the derivative of a periodic function is periodic with the same period wherever the function is differentiable.
- If $f(x)$ is T-periodic, integrable, and $\int_0^T f(x)\,dx = 0$, then $\int_0^x f(t)\,dt$ is T-periodic.
- If $f(x)$ is an integrable, periodic function with period T defined on $(-\infty, \infty)$, then for any constant $a \in \mathbb{R}$,

$$\int_a^{a+T} f(x)\,dx = \int_0^T f(x)\,dx. \tag{3.2}$$

Most of the properties mentioned above can be verified directly from the definition of periodic functions, and only the proof for the property expressed in Eq. (3.2) for the case that $f(x)$ is continuous is given here. Assuming continuity of $f(x)$ makes the proof a simple application of integration by substitution. Writing the integral on the left-hand side of Eq. (3.2) as

$$\int_a^{a+T} f(x)\,dx = \int_a^0 f(x)\,dx + \int_0^T f(x)\,dx + \int_T^{a+T} f(x)\,dx$$

$$= -\int_0^a f(x)\,dx + \int_0^T f(x)\,dx + \int_T^{a+T} f(x)\,dx. \tag{3.3}$$

In the third integral of Eq. (3.3), making the substitution $x = u + T$ yields

$$\int_T^{a+T} f(x)\,dx = \int_0^a f(u+T)\,du = \int_0^a f(u)\,du$$

where the T–periodicity of f is used in the last equality. Substituting this into Eq. (3.3) produces Eq. (3.2) and completes the proof.

In applications of Fourier series to partial differential equations, the **periodic extension** of a function is often used. Suppose function $f(x)$ is defined on a finite interval $[-L, L]$ for some $L > 0$. The $2L$-periodic extension of $f(x)$ denoted as $F(x)$ can be defined on $(-\infty, \infty)$ in the following way.

- If $x \in (-L, L]$, then $F(x) = f(x)$.
- If $x \notin (-L, L]$ and k is an integer such that $x + k(2L) \in (-L, L]$, then $F(x) = f(x + 2kL)$.

Clearly, the function $F(x)$ is periodic with period $2L$. When no confusion can result $f(x)$ is often used to denote its own periodic extension. Note that the function $F(x)$ defined above is not a "true" extension of $f(x)$ defined on the closed interval $[-L, L]$ unless $f(-L) = f(L)$. Later it will be shown if a function $f(x)$ is not defined at finitely many values in $[-L, L]$ or its values are changed at finitely many values in $[-L, L]$, the Fourier series representation and its properties will not be changed. Therefore there is no harm by allowing such an ambiguity.

3.2 The Trigonometric System and Orthogonality

The concept of orthogonality is familiar to readers who have studied vector calculus or linear algebra. In this section the orthogonality of functions, in particular the trigonometric functions sine and cosine, is introduced. If $u(x)$ and $v(x)$ are two integrable functions on an interval $[a, b]$, the **inner product** of u and v on $[a, b]$, is denoted as $\langle u, v \rangle$, and is defined as

$$\langle u, v \rangle = \int_a^b u(x)v(x)\, dx.$$

The functions u and v are said to be **orthogonal** on $[a, b]$ if

$$\langle u, v \rangle = \int_a^b u(x)v(x)\, dx = 0.$$

This is a generalization of the concept of orthogonality for vectors in $\mathbb{R}^n$ studied in linear algebra. The quantity $\langle u, v \rangle$ defined above represents the inner product of any two functions in an appropriate function space. A set of integrable functions on $[a, b]$ is said to be a **mutually orthogonal set** if each pair of distinct functions in the set is orthogonal.

Let $L > 0$ be a constant, then the **trigonometric system** is defined to be the set of functions

$$\left\{ 1, \cos\frac{\pi x}{L}, \sin\frac{\pi x}{L}, \cos\frac{2\pi x}{L}, \sin\frac{2\pi x}{L}, \ldots, \cos\frac{n\pi x}{L}, \sin\frac{n\pi x}{L}, \ldots \right\}.$$

Note that all the functions in this trigonometric system have a common period of $2L$. Most importantly, this trigonometric system is a mutually orthogonal set on the interval $[-L, L]$. In fact, it can be directly verified that for any $m = 0, 1, 2, \ldots$ and $n \in \mathbb{N}$,

$$\int_{-L}^{L} \cos\frac{m\pi x}{L} \cos\frac{n\pi x}{L}\, dx = \begin{cases} 0 & \text{if } m \neq n, \\ L & \text{if } m = n, \end{cases} \tag{3.4}$$

$$\int_{-L}^{L} \cos\frac{m\pi x}{L} \sin\frac{n\pi x}{L}\, dx = 0, \tag{3.5}$$

$$\int_{-L}^{L} \sin\frac{m\pi x}{L} \sin\frac{n\pi x}{L}\, dx = \begin{cases} 0 & \text{if } m \neq n, \\ L & \text{if } m = n \end{cases} \tag{3.6}$$

by using the trigonometric product-to-sum formulas:

$$\cos \alpha \cos \beta = \frac{1}{2}(\cos(\alpha + \beta) + \cos(\alpha - \beta)), \qquad (3.7)$$

$$\cos \alpha \sin \beta = \frac{1}{2}(\sin(\alpha + \beta) - \sin(\alpha - \beta)), \qquad (3.8)$$

$$\sin \alpha \sin \beta = \frac{1}{2}(\cos(\alpha - \beta) - \cos(\alpha + \beta)). \qquad (3.9)$$

Note that if $m = 0$, then $\cos(m\pi x/L) = 1$. The orthogonality of the trigonometric system will be essential to the determination of the coefficients of a Fourier series representation for a function.

3.3   Euler-Fourier Formulas and Fourier Series

In this section a method for obtaining the coefficients a_n and b_n in Eq. (3.1) is presented. For any fixed $L > 0$, function $f(x)$ is assumed to be an integrable function on $[-L, L]$ and has been extended periodically with period $2L$ as described by the method in Sec. 3.1.

Assuming that $f(x)$ can be written as a series of trigonometric functions as in Eq. (3.1), then

$$f(x) = \frac{a_0}{2} + \sum_{n=1}^{\infty} \left(a_n \cos \frac{n\pi x}{L} + b_n \sin \frac{n\pi x}{L} \right). \qquad (3.10)$$

For the moment the reader should assume equality holds for all $x \in [-L, L]$ except possibly at finitely many isolated points. It will be shown that a_n and b_n must be given by

$$a_0 = \frac{1}{L} \int_{-L}^{L} f(x)\,dx, \qquad (3.11)$$

$$a_n = \frac{1}{L} \int_{-L}^{L} f(x) \cos \frac{n\pi x}{L}\,dx, \qquad (3.12)$$

$$b_n = \frac{1}{L} \int_{-L}^{L} f(x) \sin \frac{n\pi x}{L}\,dx, \qquad (3.13)$$

for $n \in \mathbb{N}$. The formulas in Eqs. (3.11)–(3.13) are commonly referred to as the **Euler-Fourier formulas** and the constants a_n and b_n are referred to as the **Fourier coefficients** of the function $f(x)$.

In fact, assuming the series on the right-hand side of Eq. (3.10) can be integrated term by term over $[-L, L]$, then

$$\int_{-L}^{L} f(x)\,dx = \int_{-L}^{L} \frac{a_0}{2}\,dx = a_0 L,$$

which yields Eq. (3.11). Note that the orthogonality of $\cos(n\pi x/L)$ and $\sin(n\pi x/L)$ was used. For any integer $m > 0$, multiplying both sides of Eq. (3.10) by $\cos m\pi x/L$ and integrating over $[-L, L]$ yield the following

$$\int_{-L}^{L} f(x) \cos \frac{m\pi x}{L}\, dx$$

$$= \sum_{n=1}^{\infty} \int_{-L}^{L} a_n \cos \frac{n\pi x}{L} \cos \frac{m\pi x}{L}\, dx + \int_{-L}^{L} b_n \sin \frac{n\pi x}{L} \cos \frac{m\pi x}{L}\, dx$$

$$= \begin{cases} a_m L & \text{if } m = n, \\ 0 & \text{if } m \neq n, \end{cases}$$

where Eqs. (3.4) and (3.5) have been used. This produces the formula in Eq. (3.12). In a similar manner, Eq. (3.13) can be established. Note that the orthogonality relations in Eqs. (3.4)–(3.6) of the trigonometric system play a crucial role in deriving the Euler-Fourier formulas.

For any given integrable function on $[-L, L]$ the Euler-Fourier formulas can be used to find the Fourier coefficients of $f(x)$ and produce a Fourier series of the form in Eq. (3.1). The relationship between $f(x)$ and its Fourier series will be denoted as follows

$$f(x) \sim \frac{a_0}{2} + \sum_{n=1}^{\infty} \left(a_n \cos \frac{n\pi x}{L} + b_n \sin \frac{n\pi x}{L} \right) \tag{3.14}$$

where a_n and b_n are given by Eqs. (3.11)–(3.13). The symbol "$\sim$" rather than "$=$" is employed since it has not yet been established whether the series on the right-hand side converges. Even if the series on the right-hand side converges, at this point in the development of the theory it cannot be claimed that the sum is $f(x)$. It is a convention that the constant term in the Fourier series is denoted as $a_0/2$, which the reader will note is the average of the function on $[-L, L]$. Adopting this convention allows Eq. (3.11) to be subsumed as a special case of Eq. (3.12).

Please note that to use the Euler-Fourier formula, the only assumption made about $f(x)$ is that it be integrable. In some situations function $f(x)$ might even not be defined everywhere on $[-L, L]$. Furthermore, in applications, $f(x)$ is thought of as a function defined on $(-\infty, \infty)$, if necessary by extending $f(x)$ periodically with period $2L$ as described in Sec. 3.1.

Before exploring some specific examples, it is worthwhile to point out that the Fourier coefficients for a linear combination of two functions of $f(x)$ and $g(x)$, say $c\, f(x) + d\, g(x)$, are just the linear combination of the Fourier coefficients of $f(x)$ and $g(x)$. This can be used to simplify the process of finding Fourier coefficients and to find the Fourier series of new functions from known Fourier series.

Example 3.1. Consider the function

$$f(x) = \begin{cases} 1 & \text{if } -1 < x < 0, \\ 2 & \text{if } 0 < x < 1. \end{cases}$$

Find the Fourier series of $f(x)$.

Solution. This is a function with a discontinuity at 0 and the periodic extension of $f(x)$ has discontinuities at integer values of x. The graph of its periodic extension is shown in Fig. 3.2.

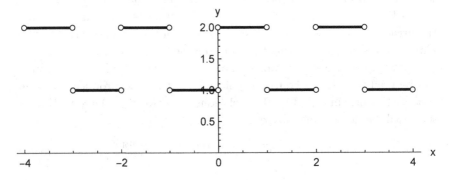

Fig. 3.2   Graph of the periodic extension of $f(x)$.

Applying the Euler-Fourier formulas in Eqs. (3.11)–(3.13) produces:

$$a_0 = \frac{1}{1} \int_{-1}^{1} f(x)\,dx = \int_{-1}^{0} 1\,dx + \int_{0}^{1} 2\,dx = 3,$$

$$a_n = \int_{-1}^{1} f(x)\cos n\pi x\,dx = \int_{-1}^{0} \cos n\pi x\,dx + \int_{0}^{1} 2\cos n\pi x\,dx = 0,$$

$$b_n = \int_{-1}^{1} f(x)\sin n\pi x\,dx = \int_{-1}^{0} \sin n\pi x\,dx + \int_{0}^{1} 2\sin n\pi x\,dx$$

$$= \frac{1-(-1)^n}{n\pi} = \begin{cases} 0 & \text{if } n = 2k \text{ is even,} \\ \frac{2}{(2k-1)\pi} & \text{if } n = 2k-1 \text{ is odd.} \end{cases}$$

The Fourier series of $f(x)$ can be written as

$$f(x) \sim \frac{3}{2} + \frac{2}{\pi}\sum_{k=1}^{\infty} \frac{1}{2k-1}\sin(2k-1)\pi x. \tag{3.15}$$

Example 3.2. Consider the piecewise-defined function
$$f(x) = \begin{cases} x + 2 & \text{if } -2 < x < 0, \\ 0 & \text{if } 0 \le x \le 2. \end{cases}$$
Find the Fourier series representation of $f(x)$.

Solution. This is a function with a discontinuity at 0 and the graph of its periodic extension is shown in Fig. 3.3.

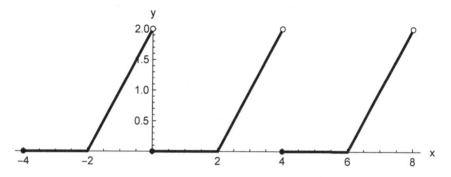

Fig. 3.3   Graph of the periodic extension of $f(x)$.

Again, the Euler-Fourier formulas yield
$$a_0 = \frac{1}{2} \int_{-2}^{2} f(x)\, dx = \frac{1}{2} \int_{-2}^{0} (x + 2)\, dx = 1,$$

$$a_n = \frac{1}{2} \int_{-2}^{2} f(x) \cos \frac{n\pi x}{2}\, dx = \frac{1}{2} \int_{-2}^{0} (x + 2) \cos \frac{n\pi x}{2}\, dx$$

$$= \frac{1}{2} \frac{2}{n\pi} \left(\left[(x + 2) \sin \frac{n\pi x}{2} \right]_{-2}^{0} - \int_{-2}^{0} \sin \frac{n\pi x}{2}\, dx \right)$$

$$= -\frac{1}{n\pi} \left[-\frac{2}{n\pi} \cos \frac{n\pi x}{2} \right]_{-2}^{0} = \frac{2}{n^2 \pi^2} (1 - \cos(-n\pi))$$

$$= \frac{2}{n^2 \pi^2} (1 - (-1)^n),$$

$$b_n = \frac{1}{2} \int_{-2}^{2} f(x) \sin \frac{n\pi x}{2}\, dx = \frac{1}{2} \int_{-2}^{0} (x + 2) \sin \frac{n\pi x}{2}\, dx$$

$$= -\frac{1}{2} \frac{2}{n\pi} \left(\left[(x + 2) \cos \frac{n\pi x}{2} \right]_{-2}^{0} - \int_{-2}^{0} \cos \frac{n\pi x}{2}\, dx \right) = -\frac{2}{n\pi}.$$

Therefore, the Fourier series representation of $f(x)$ is
$$f(x) \sim \frac{1}{2} + \frac{2}{\pi} \sum_{n=1}^{\infty} \left(\frac{(1 - (-1)^n)}{n^2 \pi} \cos \frac{n\pi x}{2} - \frac{1}{n} \sin \frac{n\pi x}{2} \right).$$

Example 3.3. Consider the function

$$f(x) = \begin{cases} 0 & \text{if } -\pi < x \le 0, \\ \sin x & \text{if } 0 < x \le \pi. \end{cases}$$

Find the Fourier series of $f(x)$.

Solution. Using the Euler-Fourier formulas produces

$$a_0 = \frac{1}{\pi} \int_{-\pi}^{\pi} f(x)\, dx = \frac{1}{\pi} \int_0^{\pi} \sin x\, dx = \frac{2}{\pi}$$

and for $n = 2, 3, \ldots$,

$$
\begin{aligned}
a_n &= \frac{1}{\pi} \int_{-\pi}^{\pi} f(x) \cos nx\, dx = \frac{1}{\pi} \int_0^{\pi} \sin x \cos nx\, dx \\
&= \frac{1}{2\pi} \int_0^{\pi} [\sin(n+1)x + \sin(1-n)x]\, dx \\
&= -\frac{1}{2\pi} \left[\frac{1}{n+1} \cos(n+1)x + \frac{1}{1-n} \cos(1-n)x \right]_0^{\pi} \\
&= -\frac{1 + (-1)^n}{\pi(n^2 - 1)} = \begin{cases} 0 & \text{if } n = 2k+1 \text{ and } n \ne 1, \\ -\frac{2}{\pi(4k^2 - 1)} & \text{if } n = 2k, \end{cases} \\
b_n &= \frac{1}{\pi} \int_{-\pi}^{\pi} f(x) \sin nx\, dx = \frac{1}{\pi} \int_0^{\pi} \sin x \sin nx\, dx \\
&= \frac{1}{2\pi} \int_0^{\pi} [\cos(1-n)x - \cos(1+n)x]\, dx \\
&= \frac{1}{2\pi} \left[\frac{1}{1-n} \sin(1-n)x - \frac{1}{1+n} \sin(1+n)x \right]_0^{\pi} = 0.
\end{aligned}
$$

Recall that $\sin x$ and $\cos nx$ are orthogonal on $[-\pi, \pi]$; however, this does not imply that $\sin x$ and $\cos nx$ are orthogonal on $[0, \pi]$. Hence not all the Fourier coefficients a_n vanish, in fact those for which n is an even integer are nonzero. There remain two Fourier coefficients to calculate.

$$a_1 = \frac{1}{\pi} \int_{-\pi}^{\pi} f(x) \cos x\, dx = \frac{1}{\pi} \int_0^{\pi} \sin x \cos x\, dx = \frac{1}{2\pi} \int_0^{\pi} \sin 2x\, dx = 0,$$

$$b_1 = \frac{1}{\pi} \int_{-\pi}^{\pi} f(x) \sin x\, dx = \frac{1}{\pi} \int_0^{\pi} \sin^2 x\, dx = \frac{1}{2\pi} \int_0^{\pi} (1 - \cos 2x)\, dx = \frac{1}{2}.$$

Therefore the Fourier series of $f(x)$ can be expressed as

$$f(x) \sim \frac{1}{\pi} + \frac{1}{2} \sin x - \frac{2}{\pi} \sum_{k=1}^{\infty} \frac{1}{4k^2 - 1} \cos(2k\,x).$$

Example 3.4. Find the Fourier series representation of $g(x) = |\sin x|$ on $[-\pi, \pi]$.

Solution. Let $f(x)$ be the function defined in Example 3.3, then the reader can verify that

$$g(x) = 2f(x) - \sin x.$$

Since the Fourier series representation of $\sin x$ is just $\sin x$ itself, the Fourier series of $g(x)$ is

$$g(x) \sim 2 \left(\frac{1}{\pi} + \frac{1}{2} \sin x - \frac{2}{\pi} \sum_{k=1}^{\infty} \frac{1}{4k^2 - 1} \cos(2k\,x) \right) - \sin x$$

$$= \frac{2}{\pi} - \frac{4}{\pi} \sum_{k=1}^{\infty} \frac{1}{4k^2 - 1} \cos(2k\,x).$$

3.4 Even and Odd Functions

The Fourier series for functions possessing certain symmetry properties can be simplified. In this section, the Fourier series of even and odd functions are discussed.

The reader should first recall that a function $f(x)$ is said to be **even** if, for any x in its domain, $-x$ is in its domain and $f(-x) = f(x)$ for all such x. A function $f(x)$ is **odd**, if $f(-x) = -f(x)$ for all real numbers x such that x and $-x$ are in the domain of f. Note that, for any given function $f(x)$ defined on $(-\infty, \infty)$, $f(x)$ can be written as the sum of an even function and an odd function. In fact,

$$f(x) = \frac{f(x) + f(-x)}{2} + \frac{f(x) - f(-x)}{2}$$

where $(f(x) + f(-x))/2$ is even (sometimes called the even part of f) and $(f(x) - f(-x))/2$ is odd (likewise called the odd part of f). The following are some basic properties of even and odd functions.

- The graph of an even function is symmetric about the y-axis and the graph of an odd function is symmetric about the origin.
- If $f(x)$ is odd and $f(0)$ is defined, then $f(0) = 0$.
- If $f(x)$ and $g(x)$ are both even functions, then so are $f(x) \pm g(x)$, $f(x) \cdot g(x)$, and $f(x)/g(x)$.
- If $f(x)$ and $g(x)$ are both odd functions, then $f(x) \pm g(x)$ is odd, and $f(x) \cdot g(x)$ and $f(x)/g(x)$ are even.

- If $f(x)$ is even (odd) and $g(x)$ is odd (even), then $f(x)\cdot g(x)$ and $f(x)/g(x)$ (if well defined) are both odd.
- If $f(x)$ is an even integrable function on $[-L, L]$, then

$$\int_{-L}^{L} f(x)\, dx = 2 \int_{0}^{L} f(x)\, dx.$$

- If $f(x)$ is an odd integrable function on $[-L, L]$, then

$$\int_{-L}^{L} f(x)\, dx = 0.$$

All the properties except the last two can be verified directly from the definitions of even and odd functions. The last two can be proved using integration by substitution.

The symmetry properties of even and odd functions have implications for the calculation of Fourier coefficients. If $f(x)$ is an even, integrable function on $[-L, L]$, then $f(x)\sin(n\pi x/L)$ is an odd function and $f(x)\cos(n\pi x/L)$ is an even function on $[-L, L]$. Hence for $n \in \mathbb{N}$,

$$b_n = \frac{1}{L} \int_{-L}^{L} f(x) \sin \frac{n\pi x}{L}\, dx = 0$$

and for $n = 0, 1, 2, \ldots$,

$$a_n = \frac{1}{L} \int_{-L}^{L} f(x) \cos \frac{n\pi x}{L}\, dx = \frac{2}{L} \int_{0}^{L} f(x) \cos \frac{n\pi x}{L}\, dx. \qquad (3.16)$$

Therefore the Fourier series of $f(x)$ consists of only cosine terms, and the Fourier series of an even function will have the form

$$f(x) \sim \frac{a_0}{2} + \sum_{n=1}^{\infty} a_n \cos \frac{n\pi x}{L},$$

where a_n is given by Eq. (3.16). Such a Fourier series is called a **Fourier cosine series**. Similarly, if $f(x)$ is an odd, integrable function on $[-L, L]$, the Fourier series of $f(x)$ consists of only the sine terms and has the form

$$f(x) \sim \sum_{n=1}^{\infty} b_n \sin \frac{n\pi x}{L}$$

which is called a **Fourier sine series**, where b_n is given by

$$b_n = \frac{2}{L} \int_{0}^{L} f(x) \sin \frac{n\pi x}{L}\, dx$$

for $n \in \mathbb{N}$.

Example 3.5. Find the Fourier series representation of

$$f(x) = \begin{cases} -1 & \text{if } -\pi < x < 0, \\ 0 & \text{if } x = 0, \\ 1 & \text{if } 0 < x < \pi. \end{cases}$$

Solution. This function is odd and its periodic extension is commonly referred to as the **square wave function** whose graph is shown in Fig. 3.4.

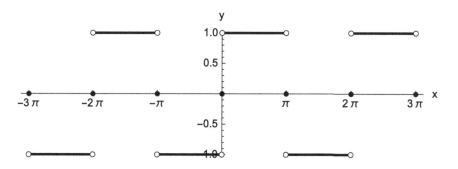

Fig. 3.4   The graph of the square wave function.

Since $f(x)$ is odd then

$$a_n = 0 \quad \text{for } n = 0, 1, 2, \dots,$$

$$b_n = \frac{2}{\pi} \int_0^{\pi} f(x) \sin nx \, dx = \frac{2}{\pi} \int_0^{\pi} \sin nx \, dx$$

$$= \left[-\frac{2}{n\pi} \cos nx \right]_0^{\pi} = \frac{2}{\pi} \left(\frac{1 - (-1)^n}{n} \right) = \begin{cases} \frac{4}{n\pi} & \text{if } n \text{ is odd}, \\ 0 & \text{if } n \text{ is even.} \end{cases}$$

Therefore the Fourier (sine) series representation of $f(x)$ is

$$f(x) \sim \frac{4}{\pi} \sum_{k=1}^{\infty} \frac{1}{2k-1} \sin(2k-1)x.$$

Example 3.6. Find the Fourier series representation of the 2π-periodic extension of $f(x) = x^2$ for $-\pi \le x \le \pi$.

Solution. The function $f(x)$ is an even function and the graph of its periodic extension is shown in Fig. 3.5.

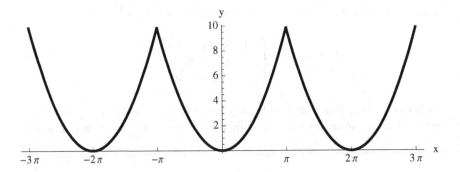

Fig. 3.5   Graph of the 2π-periodic extension of $f(x) = x^2$ on $[-\pi, \pi]$.

Since the function $f(x)$ is even, $b_n = 0$ and

$$a_0 = \frac{2}{\pi} \int_0^\pi f(x) \, dx = \frac{2}{\pi} \int_0^\pi x^2 \, dx = \frac{2}{3}\pi^2,$$

$$a_n = \frac{2}{\pi} \int_0^\pi f(x) \cos \frac{n\pi x}{\pi} \, dx = \frac{2}{\pi} \int_0^\pi x^2 \cos nx \, dx$$

$$= \frac{2}{n\pi} \left([x^2 \sin nx]_0^\pi - \int_0^\pi 2x \sin nx \, dx \right)$$

$$= \frac{4}{n^2\pi} \left([x \cos nx]_0^\pi - \int_0^\pi \cos nx \, dx \right) = \frac{4}{n^2} \cos n\pi = \frac{4(-1)^n}{n^2}.$$

The Fourier series representation is expressed as

$$f(x) \sim \frac{\pi^2}{3} + 4 \sum_{n=1}^\infty \frac{(-1)^n}{n^2} \cos(nx). \tag{3.17}$$

Example 3.7. Let $f(x) = x^3$, for $-\pi < x < \pi$ and find the Fourier series representation of the 2π–periodic extension of $f(x)$.

Solution. This is an odd function and the graph of its 2π-periodic extension is shown in Fig. 3.6.

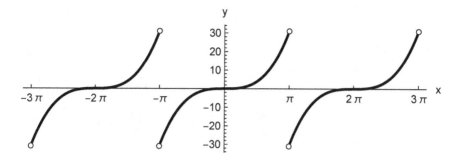

Fig. 3.6 Graph of the 2π-periodic extension of $f(x) = x^3$ on $(-\pi, \pi)$.

The function $f(x)$ is odd and thus $a_n = 0$ for $n = 0, 1, 2, \ldots$ and

$$
\begin{aligned}
b_n &= \frac{2}{\pi} \int_0^\pi f(x) \sin nx \, dx = \frac{2}{\pi} \int_0^\pi x^3 \, d\left(-\frac{1}{n}\cos nx\right) \\
&= -\frac{2}{n\pi}\left([x^3 \cos nx]_0^\pi - \int_0^\pi 3x^2 \cos nx \, dx\right) \\
&= -\frac{2}{n\pi}\left(\pi^3(-1)^n - 3\int_0^\pi x^2 \, d\left(\frac{1}{n}\sin nx\right)\right) \\
&= -\frac{2}{n\pi}\left(\pi^3(-1)^n + \frac{3}{n}\int_0^\pi 2x \sin nx \, dx\right) \\
&= -\frac{2}{n\pi}\left(\pi^3(-1)^n + \frac{6}{n}\left(\left[-\frac{1}{n}x \cos nx\right]_0^\pi + \frac{1}{n}\int_0^\pi \cos nx \, dx\right)\right) \\
&= -\frac{2}{n}\left(\pi^2(-1)^n - \frac{6}{n^2}(-1)^n\right) = \frac{2(-1)^n(6 - n^2\pi^2)}{n^3}.
\end{aligned}
$$

This produces the Fourier (sine) series representation of $f(x)$ as

$$
f(x) \sim 2\sum_{n=1}^\infty \frac{(-1)^n(6 - n^2\pi^2)}{n^3} \sin(nx). \tag{3.18}
$$

Example 3.8. Let $f(x) = 2x^3 - x^2$, for $x \in (-\pi, \pi]$, and find the Fourier series representation of $f(x)$.

Solution. Since the Fourier series for x^2 and x^3 were found in Eq. (3.17) and Eq. (3.18) respectively, it is readily seen that the Fourier series for $f(x)$ is just a combination of the previous results using the linearity property of

Fourier coefficients.

$$f(x) = 2x^3 - x^2$$

$$\sim 4 \sum_{n=1}^{\infty} \frac{(-1)^n(6 - n^2\pi^2)}{n^3} \sin(n\,x) - \left(\frac{\pi^2}{3} + 4\sum_{n=1}^{\infty} \frac{(-1)^n}{n^2} \cos(n\,x)\right)$$

$$= -\frac{\pi^2}{3} + 4\sum_{n=1}^{\infty} \left[\frac{(-1)^n(6 - n^2\pi^2)}{n^3} \sin(n\,x) - \frac{(-1)^n}{n^2} \cos(n\,x)\right].$$

3.5   Even or Odd Extension of Functions

To this point in the discussion, the Fourier series for the $2L$-periodic extension of any function defined on $[-L, L]$ can be found by using the Euler-Fourier formulas given in Eqs. (3.11)–(3.13). Special cases of these formulas exist for even and odd functions. In applications, functions are often defined on the interval $[0, L]$ as opposed to being defined on $[-L, L]$ and require expansion as Fourier series. For example, in the heat and wave equations, the initial conditions are typically given on the interval $[0, L]$. In order to represent functions defined on $[0, L]$ as Fourier series, the definition of $f(x)$ may be extended to $[-L, L]$ and then extended $2L$-periodically to $(-\infty, \infty)$. Theoretically, the extension of $f(x)$ from domain $[0, L]$ to $[-L, L]$ may be performed in any way convenient as long as the resulting function is integrable. Three common ways used to extend functions are described below.

- Extend $f(x)$ from $[0, L]$ as an even, periodic function $F(x)$ with period $2L$ by defining $F(x) = f(x)$ for any $x \in [0, L]$ and $F(x) = f(-x)$ for any $x \in [-L, 0)$. Subsequently, the resulting function $F(x)$ is then extended $2L$-periodically to $(-\infty, \infty)$. The $F(x)$ obtained is called the **even periodic extension** of $f(x)$ and will result in a Fourier cosine series expansion of $f(x)$ on $[0, L]$.

- Extend $f(x)$ from $[0, L]$ as an odd function $F(x)$ with period $2L$ by defining $F(x) = f(x)$ for $x \in (0, L)$ and $F(x) = -f(-x)$ for $x \in (-L, 0)$. The resulting function $F(x)$ is then extended $2L$-periodically onto $(-\infty, \infty)$. The $F(x)$ obtained is called the **odd periodic extension** of $f(x)$ and will result in a Fourier sine series expansion of $f(x)$ on $[0, L]$. In this case for each integer value of k, define $F(k\,L) = 0$ for convenience.

- Extend the definition of $f(x)$ to $(-L, L)$ so that the extended function is defined on $(-L, 0)$ in any way such that the resulting function satisfies the conditions of the Fourier convergence theorem (to be covered in Sec. 3.6).

Then extend the function to $(-\infty, \infty)$ with period $2L$. For example, define $f(x) = 0$ on $(-L, 0)$ for simplicity. Such an extension will result in a Fourier series representation of $f(x)$ on $[0, L]$ containing both sine and cosine terms.

Example 3.9. Let $f(x) = x$ on $[0, 2]$, sketch the even, 4-periodic extension of $f(x)$ and find its Fourier cosine series.

Solution. The graph of the even periodic extension of $f(x)$ is shown in Fig. 3.7.

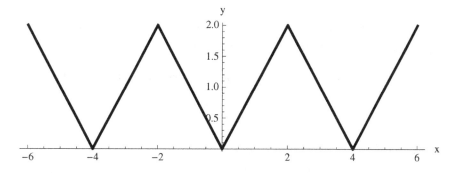

Fig. 3.7   The even, 4-periodic extension of $f(x) = x$ defined on $[0, 2]$.

For such an extension, $b_n = 0$ and

$$
a_0 = \frac{2}{2} \int_0^2 f(x)\, dx = \int_0^2 x\, dx = 2
$$

$$
a_n = \frac{2}{2} \int_0^2 x \cos \frac{n\pi x}{2}\, dx = \left[\frac{2x}{n\pi} \sin \frac{n\pi x}{2} \right]_0^2 - \frac{2}{n\pi} \int_0^2 \sin \frac{n\pi x}{2}\, dx
$$

$$
= \frac{4}{n^2 \pi^2}((-1)^n - 1) = \begin{cases} -\frac{8}{(2k-1)^2 \pi^2} & \text{if } n = 2k - 1 \text{ is odd,} \\ 0 & \text{if } n = 2k \text{ is even.} \end{cases}
$$

Therefore the Fourier cosine series of the even 4-periodic extension of $f(x)$ is expressed as

$$
f(x) \sim 1 - \frac{8}{\pi^2} \sum_{k=1}^{\infty} \frac{1}{(2k-1)^2} \cos \frac{(2k-1)\pi x}{2}. \tag{3.19}
$$

Example 3.10. Find the Fourier cosine series representation of $g(x) = 2 - 3x$ for $0 \le x \le 2$.

Solution. Using the linearity of the Fourier series expansion, the Fourier cosine series for $g(x) = 2 - 3x$ is readily found. Note the Fourier cosine series of 2 is simply 2 and the Fourier cosine series of the even, 4-periodic extension of $f(x) = x$ from $[0,2]$ to $(-\infty, \infty)$ is expressed in Eq. (3.19). Therefore the Fourier cosine series of $g(x) = 2 - 3x$ on $[0,2]$ is

$$g(x) \sim 2 - 3 \left(1 - \frac{8}{\pi^2} \sum_{k=1}^{\infty} \frac{1}{(2k-1)^2} \cos \frac{(2k-1)\pi x}{2} \right)$$

$$= -1 + \frac{24}{\pi^2} \sum_{k=1}^{\infty} \frac{1}{(2k-1)^2} \cos \frac{(2k-1)\pi x}{2}.$$

Example 3.11. Let $f(x) = e^x$ on $(0, \pi)$. Sketch the odd 2π-periodic extension of $f(x)$ and find its Fourier sine series.

Solution. The graph of the odd 2π-periodic extension of $f(x)$ is shown in Fig. 3.8.

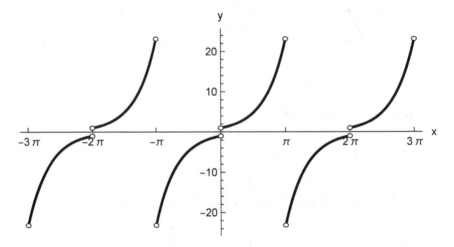

Fig. 3.8   Graph of the odd, periodic extension of $f(x) = e^x$ defined for $x \in (0, \pi)$.

Since the extension is odd, $a_n = 0$ and

$$b_n = \frac{2}{\pi} \int_0^{\pi} e^x \sin nx \, dx.$$

The integral on the right-hand side can be evaluated by integration by parts.

$$\frac{b_n \pi}{2} = \int_0^\pi e^x \sin nx \, dx = [e^x \sin nx]_0^\pi - n \int_0^\pi e^x \cos nx \, dx$$

$$= -n \left([e^x \cos nx]_0^\pi + n \int_0^\pi e^x \sin nx \, dx \right) = n \left(1 - (-1)^n e^\pi \right) - n^2 \frac{b_n \pi}{2}.$$

This equation implies

$$b_n = \frac{2n \left(1 - (-1)^n e^\pi \right)}{\pi(n^2 + 1)}.$$

Therefore the Fourier sine series of the odd, 2π-periodic extension of $f(x)$ is

$$f(x) \sim \frac{2}{\pi} \sum_{n=1}^\infty \frac{n \left(1 - (-1)^n e^\pi \right)}{n^2 + 1} \sin(n \, x).$$

Example 3.12. Let $f(x) = \sin x$ defined for $x \in [0, \pi]$. Sketch the even, 2π-periodic extension of $f(x)$ and find the Fourier cosine series of this extension.

Solution. The graph of the even, 2π-periodic extension of $f(x)$ is shown in Fig. 3.9.

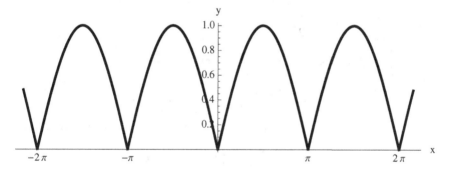

Fig. 3.9   The graph of the even, 2π-periodic extension of $f(x) = \sin x$ defined on $[0, \pi]$.

The extension of the function $f(x)$ as an even, 2π-periodic function, is equivalent to the function $|\sin x|$ whose Fourier series representation was found earlier in Example 3.4. The reader should review the details of the calculations of Fourier coefficients in that example. Note that the extended function is continuous everywhere.

Example 3.13. Consider the piecewise-defined function

$$f(x) = \begin{cases} 1 - x & \text{if } 0 < x \leq 1, \\ 0 & \text{if } 1 < x < 2. \end{cases}$$

A First Course in Partial Differential Equations

Sketch the even and odd 4-periodic extensions of $f(x)$ and find the Fourier cosine and Fourier sine series respectively.

Solution. The graph of the even, 4-periodic extension of $f(x)$ is shown in Fig. 3.10.

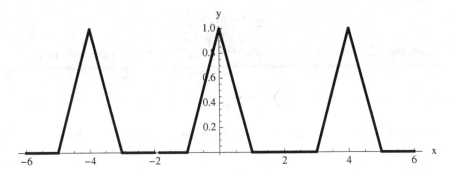

Fig. 3.10   Graph of the even, 4-periodic extension of $f(x)$ described in Example 3.13.

For the Fourier cosine series of this even extension,

$$a_0 = \frac{2}{2} \int_0^2 f(x)\, dx = \frac{1}{2},$$

$$a_n = \frac{2}{2} \int_0^2 f(x) \cos \frac{n\pi x}{2}\, dx = \int_0^1 (1-x) \cos \frac{n\pi x}{2}\, dx$$

$$= \frac{2}{n\pi} \left(\left[(1-x) \sin \frac{n\pi x}{2} \right]_0^1 + \int_0^1 \sin \frac{n\pi x}{2}\, dx \right) = \frac{4}{n^2\pi^2} \left(1 - \cos \frac{n\pi}{2} \right),$$

and thus the Fourier cosine series for the even, 4–periodic extension is

$$f(x) \sim \frac{1}{4} + \frac{4}{\pi^2} \sum_{n=1}^{\infty} \frac{1}{n^2} \left(1 - \cos \frac{n\pi}{2} \right) \cos \frac{n\pi x}{2}. \tag{3.20}$$

The graph of the odd, 4-periodic extension of $f(x)$ is shown in Fig. 3.11. The Fourier sine series coefficients are calculated as

$$b_n = \frac{2}{2} \int_0^2 f(x) \sin \frac{n\pi x}{2}\, dx = \int_0^1 (1-x) \sin \frac{n\pi x}{2}\, dx$$

$$= -\frac{2}{n\pi} \left(\left[(1-x) \cos \frac{n\pi x}{2} \right]_0^1 + \int_0^1 \cos \frac{n\pi x}{2}\, dx \right) = \frac{2}{n\pi} \left(1 - \frac{2}{n\pi} \sin \frac{n\pi}{2} \right)$$

and thus the Fourier sine series can be expressed as

$$f(x) \sim \frac{2}{\pi} \sum_{n=1}^{\infty} \frac{1}{n} \left(1 - \frac{2}{n\pi} \sin \frac{n\pi}{2} \right) \sin \frac{n\pi x}{2}. \tag{3.21}$$

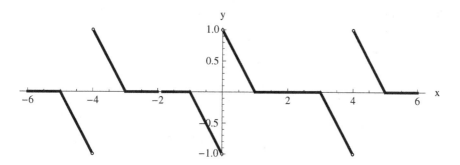

Fig. 3.11   Graph of the odd, 4-periodic extension of $f(x)$ given in Example 3.13.

In some applications of partial differential equations, functions must be expressed as Fourier series that only involve certain, particular trigonometric functions. For example, there may be a reason to include only even- or odd-indexed trigonometric terms. The remainder of this section describes how can this be done in one case, and the reader will be asked to explore other cases in the exercises.

Assume that $f(x)$ is any function defined on the interval $[0, L]$ and a Fourier series representation for $f(x)$ of the following form is sought

$$f(x) \sim \sum_{k=1}^{\infty} c_k \sin \frac{(2k-1)\pi x}{2L}. \tag{3.22}$$

Note that the series is a Fourier sine series containing only odd-indexed sine terms. The function $f(x)$ can be expressed in this form if it is extended appropriately. The appropriate extension takes place in two steps. First $f(x)$ is extended from $[0, L]$ to $[0, 2L]$ as function $\hat{f}(x)$ by defining for any $x \in [L, 2L]$,

$$\hat{f}(x) = f(2L - x).$$

Thus $\hat{f}(x)$ has even symmetry across the line $x = L$. Next function $\hat{f}(x)$, which is now defined for $x \in [0, 2L]$, is extended as an odd, periodic function on $(-\infty, \infty)$ with a period of $4L$. The Fourier expansion of this extension will have the form shown in Eq. (3.22). In fact, since the extension is odd, the Fourier series expansion must be a Fourier sine series and

$$b_n = \frac{2}{2L} \int_0^{2L} \hat{f}(x) \sin \frac{n\pi x}{2L} \, dx.$$

The manner in which $f(x)$ was extended implies that

$$
\begin{aligned}
b_{2k} &= \frac{2}{2L} \int_0^{2L} \hat{f}(x) \sin \frac{2k\pi x}{2L} \, dx \\
&= \frac{1}{L} \left(\int_0^L f(x) \sin \frac{k\pi x}{L} \, dx + \int_L^{2L} f(2L - x) \sin \frac{k\pi x}{L} \, dx \right) \\
&= \frac{1}{L} \left(\int_0^L f(x) \sin \frac{k\pi x}{L} \, dx - \int_L^0 f(s) \sin \frac{k\pi(2L - s)}{L} \, ds \right) \\
&= \frac{1}{L} \left(\int_0^L f(x) \sin \frac{k\pi x}{L} \, dx - \int_0^L f(s) \sin \frac{k\pi s}{L} \, ds \right) = 0.
\end{aligned}
$$

Calculating the odd-indexed Fourier sine coefficients results in

$$
\begin{aligned}
b_{2k-1} &= \frac{2}{2L} \int_0^{2L} \hat{f}(x) \sin \frac{(2k - 1)\pi x}{2L} \, dx \\
&= \frac{1}{L} \left(\int_0^L f(x) \sin \frac{(2k - 1)\pi x}{2L} \, dx + \int_L^{2L} f(2L - x) \sin \frac{(2k - 1)\pi x}{2L} \, dx \right) \\
&= \frac{1}{L} \left(\int_0^L f(x) \sin \frac{(2k - 1)\pi x}{2L} \, dx - \int_L^0 f(s) \sin \frac{(2k - 1)\pi(2L - s)}{2L} \, ds \right) \\
&= \frac{2}{L} \int_0^L f(x) \sin \frac{(2k - 1)\pi x}{2L} \, dx.
\end{aligned}
$$

Therefore a Fourier sine series for $f(x)$ in the form of Eq. (3.22) results, with coefficients

$$
c_k = b_{2k-1} = \frac{2}{L} \int_0^L f(x) \sin \frac{(2k - 1)\pi x}{2L} \, dx.
$$

3.6   Convergence Theorem

The formal process for expanding a given integrable function $f(x)$ as a Fourier series has been illustrated with several examples. This section addresses the theoretical issues related to the convergence of Fourier series. The following questions are of paramount importance.

- What are the conditions that guarantee the Fourier series of a given function $f(x)$ converges?
- If the Fourier series of a given function $f(x)$ converges, does it converge to the value of $f(x)$ at a given x?

The answers to these questions can be found for a large class of functions. However in this text attention will be restricted to the class of piecewise continuous functions to be defined below. This family of functions is broad enough to encompass most of the functions needed in applications of partial differential equations.

A function $f(x)$ is **piecewise continuous** on an interval $[a, b]$ (or (a, b)) if there are finitely many points $a = x_0 < x_1 < x_2 < x_3 < \cdots < x_n = b$, such that

(1) function $f(x)$ is continuous on (x_{i-1}, x_i) for all $i = 1, 2, \ldots, n$,

(2) the one-sided limits

$$\lim_{x \to x_i^-} f(x) \quad \text{and} \quad \lim_{x \to x_i^+} f(x)$$

exist for all $i = 1, 2, \ldots, n - 1$, and

(3) the one-sided limits

$$\lim_{x \to a^+} f(x) \quad \text{and} \quad \lim_{x \to b^-} f(x)$$

both exist.

By definition, a function f is piecewise continuous on $[a, b]$ if f continuous on $[a, b]$ except at finitely many values of x where f has removable or jump discontinuities (including the two endpoints, a and b). A function $f(x)$ is piecewise continuous on $(-\infty, \infty)$ if it is piecewise continuous on every finite interval $[a, b]$. A function $f(x)$ is **piecewise smooth** on $[a, b]$ if $f'(x)$ is piecewise continuous on $[a, b]$. See Exercise 14. Note that any piecewise continuous function on $[a, b]$ is integrable on $[a, b]$. Convergence issues aside, this implies that the Fourier series of any piecewise continuous function exists in the sense that the Fourier coefficients are well-defined.

For the sake of brevity of notation, for any $c \in [a, b]$, define

$$\lim_{x \to c^+} f(x) = f(c+) \quad \text{and} \quad \lim_{x \to c^-} f(x) = f(c-).$$

Clearly, $f(x)$ is continuous at c if and only if $f(c+) = f(c-) = f(c)$.

As examples, suppose $L > 0$ is a constant, then the functions

$$f(x) = |x| \quad \text{and} \quad g(x) = \begin{cases} x & \text{if } -L < x < 0, \\ x^2 + 2 & \text{if } 0 \leq x < L \end{cases}$$

are both piecewise smooth on the interval $[-L, L]$. However, the functions

$$p(x) = \frac{1}{x^2} \quad \text{and} \quad q(x) = \sin \frac{1}{x}$$

are not piecewise continuous on $[-L, L]$ since 0 is not the location of a removable or jump discontinuity of these two functions. It can also be shown from the definitions of piecewise continuity and piecewise smoothness that the functions

$$r(x) = x^{2/3} \quad \text{and} \quad s(x) = x^2 \sin \frac{1}{x}$$

are not piecewise smooth, though they are piecewise continuous.

To consider the convergence of Fourier series in Eq. (3.1), it is natural to consider the partial sum

$$S_N(x) = \frac{a_0}{2} + \sum_{n=1}^{N} \left(a_n \cos \frac{n\pi x}{L} + b_n \sin \frac{n\pi x}{L} \right),$$

where N is a positive integer. By definition, for each fixed x, the Fourier series converges to $f(x)$ if and only if $\lim_{N \to \infty} S_N(x)$ exists and $\lim_{N \to \infty} S_N(x) = f(x)$. This is made formal in the following theorem.

Theorem 3.1 (Dirichlet Convergence Theorem). *Assume that $f(x)$ is a piecewise smooth function on the interval $[-L, L]$ extended to $(-\infty, \infty)$ periodically with period $2L$. Then the Fourier series of $f(x)$ converges for all x to the value,*

$$\frac{1}{2} \left(f(x+) + f(x-) \right).$$

The theorem above is commonly referred to as the **Dirichlet Convergence Theorem** for Fourier series. Under the assumptions of the convergence theorem, if f is continuous at z, that is $f(z+) = f(z-) = f(z)$, then the Fourier series representation of $f(x)$ converges for $x = z$ to $f(z)$. If f has a jump discontinuity at $x = z$, then the Fourier series of $f(x)$ converges to the average of the one-sided limit values $f(z+)$ and $f(z-)$. The reader should note in the latter case, the sum of the Fourier series is independent of the value of function f at z, if it is defined. The proof of the theorem can be established with only elementary calculus and is postponed until Sec. 3.11.

When a function $f(x)$ defined on $[-L, L]$ is periodically extended to $(-\infty, \infty)$, often the extended function is not well-defined at values $x = \pm L, \pm 3L, \dots$. However, if the function $f(x)$ is piecewise continuous on the interval, the two one-sided limits of the extended function exist at those points and the theorem still applies. As an example, consider again the function presented in Example 3.1:

$$f(x) = \begin{cases} 1 & \text{if } -1 < x < 0, \\ 2 & \text{if } 0 < x < 1. \end{cases}$$

Clearly $f(x)$ is piecewise smooth (in fact, $f'(x) = 0$ for all non-integer value x). Therefore, the Fourier series of $f(x)$ given in Eq. (3.15) converges everywhere to the periodic extension of $f(x)$ on $(-\infty, \infty)$ except when x is an integer. In the cases when x is an integer, the series converges to $3/2$, the average of the limits from the left and the right when x is an integer. This is certainly true for the Fourier series expansion of $f(x)$ given in Eq. (3.15). Any integer value of x results in the vanishing of the infinite series portion of the expansion of $f(x)$ in Eq. (3.15).

The Dirichlet Convergence Theorem often can be used to find the sums of some infinite series which are difficult to find otherwise.

Example 3.14. Use the Fourier series for $f(x) = x^2$ found in Example 3.6 to find the sums of the following two infinite series,

$$\sum_{n=1}^{\infty} \frac{(-1)^n}{n^2} \text{ and } \sum_{n=1}^{\infty} \frac{1}{n^2}.$$

Solution. Both of these series are routinely studied in elementary calculus and both are easily shown to converge. However, confirmation of convergence does not reveal the numerical sum of the series. The Dirichlet Convergence Theorem can be used to find their sums.

The even extension of $f(x) = x^2$ is continuous on $[-\pi, \pi]$. Since both limits $\lim_{x \to \pi^-} f'(x)$ and $\lim_{x \to \pi^+} f'(x)$ exist, $f'(x)$ is piecewise continuous on $[-\pi, \pi]$. The Dirichlet Convergence Theorem implies that the Fourier series of $f(x)$ converges everywhere to the periodic extension of $f(x)$, which is still denoted as $f(x)$ on $(-\infty, \infty)$, that is

$$f(x) = \frac{\pi^2}{3} + 4 \sum_{n=1}^{\infty} \frac{(-1)^n}{n^2} \cos(n x) \text{ for } x \in (-\infty, \infty). \qquad (3.23)$$

In particular, setting $x = 0$ yields

$$0 = \frac{\pi^2}{3} + 4 \sum_{n=1}^{\infty} \frac{(-1)^n}{n^2}$$

which implies that

$$\sum_{n=1}^{\infty} \frac{(-1)^n}{n^2} = -\frac{\pi^2}{12}.$$

Similarly, setting $x = \pi$ in Eq. (3.23) produces

$$\pi^2 = \frac{\pi^2}{3} + 4 \sum_{n=1}^{\infty} \frac{(-1)^n}{n^2} \cos(n\pi) = \frac{\pi^2}{3} + 4 \sum_{n=1}^{\infty} \frac{1}{n^2}$$

which implies

$$\sum_{n=1}^{\infty} \frac{1}{n^2} = \frac{\pi^2}{6}.$$

Example 3.15. Find the Fourier series representation of the function

$$f(x) = \begin{cases} 0 & \text{for } -\pi \le x \le 0, \\ x & \text{for } 0 < x < \pi \end{cases}$$

discuss its convergence, and then use the Fourier series for $f(x)$ and the Dirichlet Convergence Theorem to find the sums of the following two series:

$$\sum_{k=1}^{\infty} \frac{1}{(2k-1)^2} \quad \text{and} \quad \sum_{k=1}^{\infty} \frac{(-1)^{k-1}}{2k-1}.$$

Solution. By the Euler-Fourier coefficient formulas, the Fourier coefficients for $f(x)$ are

$$a_0 = \frac{1}{\pi} \int_{-\pi}^{\pi} f(x)\,dx = \frac{1}{\pi} \int_{0}^{\pi} x\,dx = \frac{\pi}{2},$$

$$a_n = \frac{1}{\pi} \int_{-\pi}^{\pi} f(x) \cos nx\,dx = \frac{1}{\pi} \int_{0}^{\pi} x \cos nx\,dx = \left[\frac{1}{n^2 \pi} \cos nx \right]_{0}^{\pi}$$

$$= \frac{1}{n^2 \pi} \left((-1)^n - 1 \right),$$

$$b_n = \frac{1}{\pi} \int_{0}^{\pi} x \sin nx\,dx = -\left[\frac{1}{n\pi} (x \cos nx) \right]_{0}^{\pi} + \frac{1}{n\pi} \int_{0}^{\pi} \cos nx\,dx = -\frac{(-1)^n}{n}.$$

Therefore the Fourier series for $f(x)$ is expressed as

$$f(x) \sim \frac{\pi}{4} + \sum_{n=1}^{\infty} \left(\frac{1}{n^2 \pi} \left((-1)^n - 1 \right) \cos nx - \frac{(-1)^n}{n} \sin nx \right)$$

$$= \frac{\pi}{4} - \sum_{k=1}^{\infty} \frac{2}{(2k-1)^2 \pi} \cos(2k-1)x - \sum_{n=1}^{\infty} \frac{(-1)^n}{n} \sin nx.$$

Since the only discontinuities in the 2π–periodic extension of $f(x)$ occur at odd integer multiples of π, then for x in the open interval $(-\pi, \pi)$ the Dirichlet Convergence Theorem reveals that

$$f(x) = \frac{\pi}{4} - \sum_{k=1}^{\infty} \frac{2}{(2k-1)^2 \pi} \cos([2k-1]x) - \sum_{n=1}^{\infty} \frac{(-1)^n}{n} \sin(nx). \quad (3.24)$$

In particular, the Fourier series converges at $x = 0$ to $f(0) = 0$, that is,

$$0 = f(0) = \frac{\pi}{4} - \sum_{k=1}^{\infty} \frac{2}{(2k-1)^2 \pi}$$

which implies

$$\sum_{k=1}^{\infty} \frac{1}{(2k-1)^2} = \frac{\pi^2}{8}.$$

Similarly, setting $x = \pi/2$ in Eq. (3.24) yields

$$\frac{\pi}{2} = \frac{\pi}{4} - \sum_{k=1}^{\infty} \frac{2}{(2k-1)^2 \pi} \cos \frac{(2k-1)\pi}{2} - \sum_{n=1}^{\infty} \frac{(-1)^n}{n} \sin \frac{n\pi}{2}$$

$$= \frac{\pi}{4} - \sum_{k=1}^{\infty} \frac{(-1)^{2k-1}}{2k-1} \sin \frac{(2k-1)\pi}{2} = \frac{\pi}{4} + \sum_{k=1}^{\infty} \frac{(-1)^{k-1}}{2k-1}$$

which implies

$$\sum_{k=1}^{\infty} \frac{(-1)^{k-1}}{2k-1} = \frac{\pi}{4}.$$

Many other interesting sums can be obtained in this way with little more effort than a simple application of the Dirichlet Convergence Theorem.

3.7 The Gibbs Phenomenon and Uniform Convergence

According to the Dirichlet Convergence Theorem (Theorem 3.1) the Fourier series of any piecewise smooth function f converges to the function at every point where the function is continuous and to the average of the left-hand and right-hand limits of the function at any point where the function has a jump or removable discontinuity. In this section the nature of the convergence is explored. The following example will motivate the discussion.

Example 3.16. Consider the function

$$f(x) = \begin{cases} -1 & \text{if } -\pi < x < 0, \\ 0 & \text{if } x = 0, \\ 1 & \text{if } 0 < x < \pi. \end{cases} \tag{3.25}$$

In Example 3.5, the Fourier series representation of $f(x)$ is found as

$$f(x) \sim \frac{4}{\pi} \sum_{k=1}^{\infty} \frac{\sin(2k-1)x}{2k-1}. \tag{3.26}$$

By the Dirichlet Convergence Theorem, the series converges to $f(x)$ at any fixed $x \in (-\pi, \pi)$, in other words

$$\lim_{n \to \infty} S_n(x) = f(x) \tag{3.27}$$

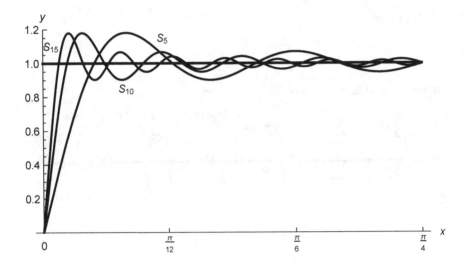

Fig. 3.12   The Gibbs phenomenon illustrated by graphing $S_n(x)$ for $n = 5, 10, 15$.

for any $x \in (-\pi, \pi)$, where $S_n(x)$ is defined by

$$S_n(x) = \frac{4}{\pi} \left(\sin x + \frac{\sin 3x}{3} + \cdots + \frac{\sin(2n-1)x}{2n-1} \right).$$

The jump discontinuity in f occurring at $x = 0$ does not affect the convergence of the Fourier series to $f(0)$ since the average of the limits from the left and the right of $f(x)$ at $x = 0$ is $0 = f(0)$. The graphs of $f(x)$ and $S_n(x)$ on interval $[0, \pi/4]$ for several values of n are shown in Fig. 3.12. The figure indicates that there is a "bump" in the graph of $S_n(x)$ that overshoots the graph of $f(x)$ by a fixed amount and the bump occurs at x values near 0. In fact, it will be shown below that

$$\lim_{n \to \infty} \max_{0 \leq x \leq \pi/4} S_n(x) \approx 1.179.$$

Elementary calculus determines the location of the extrema. The bumps are located at values of x for which $S_n(x)$ reaches a maximum. Note that $S_n(x)$ is continuously differentiable on the interval $[-\pi, \pi]$ and

$$S_n'(x) = \frac{4}{\pi} \left(\cos x + \cos 3x + \cdots + \cos(2n-1)x \right).$$

Multiplying $S_n'(x)$ by $\sin x$ and applying the product-to-sum formula in

Eq. (3.8) yield

$$(\sin x)S_n'(x) = \frac{4}{\pi}(\sin x)(\cos x + \cos 3x + \cdots + \cos(2n-1)x)$$

$$= \frac{2}{\pi}\left(\sin 2x + \sum_{k=1}^{n-1}[\sin 2(k+1)x - \sin 2kx]\right)$$

$$= \frac{2}{\pi}\sin(2nx).$$

This implies the extrema of $S_n(x)$ on $(0, \pi)$ occur at

$$x = \frac{m\pi}{2n} \quad \text{for } m = 1, 2, \ldots, (2n-1).$$

Consider the maximum in the graph of $S_n(x)$ occurring for positive x nearest to 0. The first derivative test reveals that at $x = \pi/(2n)$, the function $S_n(x)$ reaches a maximum value of

$$S_n\left(\frac{\pi}{2n}\right) = \frac{4}{\pi}\left(\sin\frac{\pi}{2n} + \frac{\sin\frac{3\pi}{2n}}{3} + \cdots + \frac{\sin\frac{(2n-1)\pi}{2n}}{2n-1}\right)$$

$$= \frac{2}{\pi}\left[\left(\frac{\sin\frac{\pi}{2n}}{\pi/(2n)} + \frac{\sin\frac{3\pi}{2n}}{3\pi/(2n)} + \cdots + \frac{\sin\frac{(2n-1)\pi}{2n}}{(2n-1)\pi/(2n)}\right)\frac{\pi}{n}\right].$$

The terms inside the brackets on the right-hand side can be regarded as the Riemann sum of the function $(\sin x)/x$ on the interval $[0, \pi]$ corresponding to a regular partition of $[0, \pi]$ into subintervals of size $\Delta x = \pi/n$ and evaluation points at the midpoint of each subinterval. Therefore

$$\lim_{n\to\infty} S_n\left(\frac{\pi}{2n}\right) = \frac{2}{\pi}\int_0^\pi \frac{\sin x}{x}\,dx \approx 1.179,$$

where the definite integral has been evaluated numerically. This implies that, for any fixed $x \in (0, \pi)$, the Fourier series converges to $f(x)$, that is, $\lim_{n\to\infty} S_n(x) = f(x)$. However, if $x = \pi/(2n)$, then $S_n(\pi/(2n))$ will remain a fixed distance away from $f(\pi/(2n))$ no matter how large n is.

The behavior exhibited in this example is actually a common phenomenon which is referred to as the **Gibbs phenomenon**. The Gibbs phenomenon describes the fact that the partial sum of the Fourier series of a function $f(x)$ does not uniformly approximate $f(x)$. Rather surprisingly, the graph of the partial sum $S_n(x)$ actually oscillates "wildly" about the graph of $f(x)$ in a neighborhood of any discontinuity of $f(x)$. In general, it can be shown that the amount of the deviation from the graph of $f(x)$ on either side of a discontinuity due to the Gibbs phenomenon is about 9% of the jump size at the discontinuity, that is, a quantity of magnitude roughly $0.09|f(c+) - f(c-)|$, for large n. For details, see [Walker (1988)] or [Korner (1988)]. Note that this does not contradict the Dirichlet Convergence Theorem which states that if $f(x)$ is a piecewise smooth function on $[-L, L]$ and x is any fixed point at which $f(x)$ is continuous, then the Fourier series of $f(x)$ converges to $f(x)$. Such convergence is usually referred to as **pointwise convergence**. To clarify the conclusion of the Dirichlet Convergence Theorem, the Fourier series representation of $f(x)$ converges pointwise to $f(x)$ at every x where $f(x)$ is continuous. A stronger type of convergence is called uniform convergence and is discussed next.

Mathematically, the Gibbs phenomenon shows that the Fourier series of a function does not converge to $f(x)$ uniformly near a discontinuity of $f(x)$. The topic of uniform convergence will not be treated in great detail in this text, rather merely the essentials are presented. A sequence of functions $\{f_n(x)\}_{n=1}^{\infty}$ defined on $[a, b]$ **converges uniformly** to a limit function $f(x)$ defined on $[a, b]$ if there exists a sequence of positive real numbers $\{\epsilon_n\}_{n=1}^{\infty}$ such that $\epsilon_n \to 0$ as $n \to \infty$ and

$$|f_n(x) - f(x)| \leq \epsilon_n \text{ for all } x \in [a, b] \text{ and } n \in \mathbb{N}.$$

An infinite series $\sum_{k=1}^{\infty} u_n(x)$ **converges uniformly** if its sequence of partial sums converges uniformly. Some simple examples are given in the following list.

- The sequence $\{x^n\}_{n=1}^{\infty}$ does not converge uniformly on $[0, 1]$. Note that the pointwise limit function is

$$f(x) = \begin{cases} 0 & \text{if } 0 \leq x < 1, \\ 1 & \text{if } x = 1. \end{cases}$$

- The sequence $\{x^n\}_{n=1}^{\infty}$ converges uniformly to $f(x) = 0$ on $[0, b]$ for any fixed b such that $0 < b \leq 1$.
- The sequence $\left\{ \frac{1-x^n}{1-x} \right\}_{n=1}^{\infty}$ does not converge uniformly on $(-1, 1)$ to its pointwise limit function $f(x) = 1/(1 - x)$.
- The infinite series $\sum_{n=0}^{\infty} x^n$ does not converge uniformly on $(-1, 1)$ to its pointwise sum $1/(1 - x)$.

The following theorem summarizes the properties of a uniformly convergent series. The proofs of these results can be found in almost any advanced calculus book, for example [Rudin (1976)].

Theorem 3.2. *Consider the series $\sum_{n=1}^{\infty} u_n(x)$.*

- *Assume that $\sum_{n=1}^{\infty} u_n(x)$ converges uniformly to its sum $u(x)$ on $[a,b]$, and, for each n, $u_n(x)$ is continuous on $[a,b]$. The sum $u(x)$ is continuous on $[a,b]$. That is, the sum of a uniformly convergent series of continuous functions is a continuous function.*
- *Assume that $\sum_{n=1}^{\infty} u_n(x)$ converges uniformly to its sum $u(x)$, and for each n, $u_n(x)$ is integrable on $[a,b]$, then the sum $u(x)$ is integrable on $[a,b]$, and*

$$\int_a^b u(x)\,dx = \int_a^b \left(\sum_{n=1}^{\infty} u_n(x) \right) dx = \sum_{n=1}^{\infty} \int_a^b u_n(x)\,dx.$$

That is, the sum of a uniformly convergent series of integrable functions is integrable and the series may be integrated term by term.

- *Assume $\sum_{n=1}^{\infty} u_n(x)$ converges to a sum $u(x)$ on $[a,b]$, and for each n, $u_n'(x)$ exists and $\sum_{n=1}^{\infty} u_n'(x)$ converges uniformly on $[a,b]$. The sum $u(x)$ is differentiable on $[a,b]$ and the derivative can be obtained by differentiating the series term by term,*

$$u'(x) = \left(\sum_{n=1}^{\infty} u_n(x) \right)' = \sum_{n=1}^{\infty} u_n'(x) \quad \text{for all } x \in [a,b].$$

These properties of uniform convergence will be useful for studying the calculus of Fourier series. The following theorem gives a convenient and powerful test for uniform convergence of an infinite series of functions.

Theorem 3.3 (Weierstrass[3] M-test). *Let $\sum_{n=1}^{\infty} u_n(x)$ be a series of functions defined on an interval $[a,b]$ and suppose that for each n there is a non-negative number M_n such that $|u_n(x)| \le M_n$ for all $x \in [a,b]$ and $\sum_{n=1}^{\infty} M_n$ converges, then $\sum_{n=1}^{\infty} u_n(x)$ converges uniformly on $[a,b]$.*

The proof of this theorem and other criteria for uniform convergence can be found in any advanced calculus book, for example [Trench (1978)]. As a simple application of Theorem 3.3, the reader can verify that $\sum_{n=0}^{\infty} \frac{\sin nx}{n^2}$ converges uniformly on $(-\infty, \infty)$ since $\left| \frac{\sin nx}{n^2} \right| \le 1/n^2$ and $\sum_{n=0}^{\infty} 1/n^2$ converges.

Returning to the topic of Fourier series, the following theorem on uniform convergence is presented.

---

[3]Karl Weierstrass, German mathematician (1815–1897).

Theorem 3.4. *Assume that $f(x)$ is continuous on $[-L, L]$ with $f(-L) = f(L)$ and $f'(x)$ is piecewise continuous on $[-L, L]$, then the Fourier series of $f(x)$ converges uniformly to $f(x)$ on $[-L, L]$.*

The proof of this theorem will be presented later in Sec. 3.11.

3.8   Differentiation and Integration of Fourier Series

When Fourier series are used to solve initial boundary value problems, the series must be differentiated and/or integrated. Often these calculus operations are carried out on Fourier series in a term-by-term manner. However, this must be done under an appropriate set of assumptions. This process involves the interchange of the operations of differentiation (or integration) and infinite summation which is not always valid. In this section, some criteria for differentiating or integrating Fourier series are presented.

To experience the pitfalls of invalidly interchanging the order of summation and differentiation, consider the function $f(x) = x$ defined on $(-\pi, \pi)$. The function $f(x)$ is an odd function and its Fourier sine series is

$$f(x) = 2 \sum_{n=1}^{\infty} \frac{(-1)^{n+1}}{n} \sin(n\,x) \text{ for } -\pi < x < \pi,$$

see Exercise 7(c). The derivative of $f(x)$ is 1 on $(-\pi, \pi)$. However differentiating the series on the right-hand side term by term yields

$$2 \sum_{n=1}^{\infty} (-1)^{n+1} \cos(n\,x)$$

which diverges for all $x \in (-\pi, \pi)$ by the nth term test [Smith and Minton (2012)]. In this case the derivative $f'(x)$ cannot be found using term-by-term differentiation of the Fourier series of $f(x)$. For comparison, consider the function $g(x) = x^2$ on $[-\pi, \pi]$ whose Fourier series is given by

$$g(x) = \frac{\pi^2}{3} + 4 \sum_{n=1}^{\infty} \frac{(-1)^n}{n^2} \cos(n\,x)$$

for all $x \in [-\pi, \pi]$, see Example 3.6. Differentiating term-by-term produces

$$4 \sum_{n=1}^{\infty} \frac{(-1)^{n+1}}{n} \sin(n\,x)$$

which is the Fourier series of $g'(x) = 2x$. Therefore the Fourier series of $g(x) = x^2$ on $(-\pi, \pi)$ can be differentiated term by term and the result converges to the Fourier series of $g'(x)$. In general, Fourier series may be differentiated term by term under the circumstances laid out in the next theorem.

Theorem 3.5 (Differentiation of Fourier Series). *Assume that $f(x)$ is continuous on $(-L, L)$ with $f(-L+) = f(L-)$, and $f'(x)$ is piecewise smooth on $(-L, L)$. Then the Fourier series of $f(x)$ can be differentiated term-by-term.*

Before proving Theorem 3.5 an implication of the theorem will be explored. Since the assumptions on $f(x)$ imply that the $2L$-periodic extension of $f(x)$ is continuous on $(-\infty, \infty)$, then by the Dirichlet Convergence Theorem (Theorem 3.1), $f(x)$ can be expressed as the Fourier series

$$f(x) = \frac{a_0}{2} + \sum_{n=1}^{\infty} \left(a_n \cos \frac{n\pi x}{L} + b_n \sin \frac{n\pi x}{L} \right)$$

for all $x \in (-\infty, \infty)$. Theorem 3.5 simply states that the Fourier series of $f'(x)$ is given by

$$f'(x) \sim \sum_{n=1}^{\infty} \frac{n\pi}{L} \left(b_n \cos \frac{n\pi x}{L} - a_n \sin \frac{n\pi x}{L} \right).$$

Since $f'(x)$ is assumed to be piecewise smooth, then by the Dirichlet Convergence Theorem, for any $x \in (-\infty, \infty)$,

$$\frac{f'(x+) + f'(x-)}{2} = \sum_{n=1}^{\infty} \frac{n\pi}{L} \left(b_n \cos \frac{n\pi x}{L} - a_n \sin \frac{n\pi x}{L} \right) \qquad (3.28)$$

where a_n and b_n are the Fourier coefficients of $f(x)$ defined in Eqs. (3.11)–(3.13).

The following is an outline of the proof of Theorem 3.5.

Proof. Since $f(x)$ is continuous and $f(-L+) = f(L-)$, if $f(L)$ is defined as $f(L) = f(L-) = f(L+)$, then the periodic extension of $f(x)$ is continuous on $(-\infty, \infty)$ and the Fourier series for $f(x)$ is

$$f(x) = \frac{a_0}{2} + \sum_{n=1}^{\infty} \left(a_n \cos \frac{n\pi x}{L} + b_n \sin \frac{n\pi x}{L} \right)$$

where a_n and b_n are given by Eqs. (3.11)–(3.13). Since $f'(x)$ is piecewise smooth, $f'(x)$ can be expressed as

$$f'(x) = \frac{\alpha_0}{2} + \sum_{n=1}^{\infty} \left(\alpha_n \cos \frac{n\pi x}{L} + \beta_n \sin \frac{n\pi x}{L} \right)$$

for every x in $[-L, L]$ except possibly at finitely many points with α_n and β_n given by Eqs. (3.11)–(3.13) with $f(x)$ replaced by $f'(x)$. The reader should note that

$$\alpha_0 = \frac{1}{L} \int_{-L}^{L} f'(x)\, dx = \frac{1}{L}(f(L) - f(-L)) = 0.$$

Integration by parts yields for all $n \in \mathbb{N}$,

$$
\begin{aligned}
\alpha_n &= \frac{1}{L} \int_{-L}^{L} f'(x) \cos \frac{n\pi x}{L} \, dx \\
&= \left[\frac{1}{L} f(x) \cos \left(\frac{n\pi x}{L} \right) \right]_{-L}^{L} - \frac{1}{L} \int_{-L}^{L} f(x) \left(-\frac{n\pi}{L} \sin \frac{n\pi x}{L} \right) dx = \frac{n\pi}{L} b_n, \\
\beta_n &= \frac{1}{L} \int_{-L}^{L} f'(x) \sin \frac{n\pi x}{L} \, dx \\
&= \left[\frac{1}{L} f(x) \sin \left(\frac{n\pi x}{L} \right) \right]_{-L}^{L} - \frac{1}{L} \int_{-L}^{L} f(x) \left(\frac{n\pi}{L} \cos \frac{n\pi x}{L} \right) dx = -\frac{n\pi}{L} a_n.
\end{aligned}
$$

These are the same Fourier coefficients that were found using term-by-term differentiation in Eq. (3.28). Thus the Fourier series may be differentiated term by term. $\qquad\qquad\qquad\qquad\qquad\qquad\qquad\qquad\qquad\qquad\qquad\qquad\quad\square$

Please note that the function $f(x) = x$ defined for $-\pi < x < \pi$, does not satisfy the assumptions of Theorem 3.5 since $f(-\pi+) \neq f(\pi-)$. There is no continuous 2π-periodic extension of f to $(-\infty, \infty)$ which agrees with f on the interval $(-\pi, \pi)$. Theorem 3.5 also applies to Fourier sine and Fourier cosine series for functions defined on $[0, L]$ if the even or odd extension of the function satisfy the conditions of the theorem. Note that the even extension of any continuous function $f(x)$ on $[0, L]$ always satisfies the condition $f(-L) = f(L)$. Therefore, a Fourier cosine series can always be differentiated term by term as long as $f(x)$ is continuous and $f'(x)$ is piecewise smooth on $[0, L]$.

The next result states that a Fourier series can always be integrated term by term.

Theorem 3.6. *Let $f(x)$ be piecewise continuous on $[-L, L]$. The Fourier series of $f(x)$ can be integrated term by term and the resulting series always converges to the integral of $f(x)$ on $[-L, L]$. That is, if*

$$
f(x) \sim \frac{a_0}{2} + \sum_{n=1}^{\infty} \left(a_n \cos \frac{n\pi x}{L} + b_n \sin \frac{n\pi x}{L} \right)
$$

then

$$
\begin{aligned}
\int_0^x f(s) \, ds &= \int_0^x \frac{a_0}{2} \, ds + \sum_{n=1}^{\infty} \int_0^x \left(a_n \cos \frac{n\pi s}{L} + b_n \sin \frac{n\pi s}{L} \right) ds \\
&= \frac{a_0}{2} x + \frac{L}{\pi} \sum_{n=1}^{\infty} \left[\frac{a_n}{n} \sin \frac{n\pi x}{L} + \frac{b_n}{n} \left(1 - \cos \frac{n\pi x}{L} \right) \right].
\end{aligned} \qquad (3.29)
$$

Note that Eq. (3.29) implies for any $[a, b] \subset [-L, L]$,

$$\int_a^b f(x)\, dx = \frac{a_0}{2}(b - a) + \sum_{n=1}^{\infty} \int_a^b \left(a_n \cos \frac{n\pi x}{L} + b_n \sin \frac{n\pi x}{L} \right) dx.$$

Proof. Define $F(x) = \int_0^x f(s)\, ds$ and $G(x) = F(x) - a_0 x/2$, then both $F(x)$ and $G(x)$ are continuous on $(-L, L)$ and

$$G(-L) = \int_0^{-L} f(s)\, ds + \frac{a_0 L}{2} = -\frac{a_0 L}{2} + \int_0^L f(s)\, ds = G(L).$$

The Dirichlet Convergence Theorem implies that

$$G(x) = \frac{A_0}{2} + \sum_{n=1}^{\infty} \left(A_n \cos \frac{n\pi x}{L} + B_n \sin \frac{n\pi x}{L} \right), \qquad (3.30)$$

where A_n and B_n are the Fourier coefficients of the function G. For $n \in \mathbb{N}$,

$$
\begin{aligned}
A_n &= \frac{1}{L} \int_{-L}^{L} G(x) \cos \frac{n\pi x}{L}\, dx \\
&= \frac{1}{n\pi} \left(\left[G(x) \sin \frac{n\pi x}{L} \right]_{-L}^{L} - \int_{-L}^{L} \left(f(x) - \frac{a_0}{2} \right) \sin \frac{n\pi x}{L}\, dx \right) \\
&= -\frac{1}{n\pi} \int_{-L}^{L} f(x) \sin \frac{n\pi x}{L}\, dx + \frac{a_0}{2n\pi} \int_{-L}^{L} \sin \frac{n\pi x}{L}\, dx = -\frac{L}{n\pi} b_n,
\end{aligned}
$$

$$
\begin{aligned}
B_n &= \frac{1}{L} \int_{-L}^{L} G(x) \sin \frac{n\pi x}{L}\, dx \\
&= -\frac{1}{n\pi} \left(\left[G(x) \cos \frac{n\pi x}{L} \right]_{-L}^{L} - \int_{-L}^{L} \left(f(x) - \frac{a_0}{2} \right) \cos \frac{n\pi x}{L}\, dx \right) \\
&= \frac{1}{n\pi} \int_{-L}^{L} f(x) \cos \frac{n\pi x}{L}\, dx - \int_{-L}^{L} \frac{a_0}{2n\pi} \cos \frac{n\pi x}{L}\, dx = \frac{L}{n\pi} a_n,
\end{aligned}
$$

where a_n and b_n are the Fourier coefficients of $f(x)$. Therefore, Eq. (3.30) can be written as

$$G(x) = \frac{A_0}{2} + \sum_{n=1}^{\infty} \left(-\frac{L}{n\pi} b_n \cos \frac{n\pi x}{L} + \frac{L}{n\pi} a_n \sin \frac{n\pi x}{L} \right).$$

To determine A_0, let $x = 0$ in the equation above and obtain

$$0 = G(0) = \frac{A_0}{2} - \sum_{n=1}^{\infty} \frac{L}{n\pi} b_n \iff \frac{A_0}{2} = \sum_{n=1}^{\infty} \frac{L}{n\pi} b_n.$$

The last equation implies

$$G(x) = \sum_{n=1}^{\infty} \frac{L}{n\pi} b_n + \sum_{n=1}^{\infty} \left(-\frac{L}{n\pi} b_n \cos \frac{n\pi x}{L} + \frac{L}{n\pi} a_n \sin \frac{n\pi x}{L} \right)$$

$$= \sum_{n=1}^{\infty} \frac{L}{n\pi} \left(b_n \left(1 - \cos \frac{n\pi x}{L} \right) + \frac{L}{n\pi} a_n \sin \frac{n\pi x}{L} \right).$$

This in turn implies that

$$F(x) = \frac{a_0}{2} x + \frac{L}{\pi} \sum_{n=1}^{\infty} \left(\frac{b_n}{n} \left(1 - \cos \frac{n\pi x}{L} \right) + \frac{a_n}{n} \sin \frac{n\pi x}{L} \right)$$

which is Eq. (3.29) and completes the proof.   □

3.9   Mean Square Approximation and Parseval's Identity

Under fairly reasonable assumptions a function $f(x)$ can be expressed as a series of trigonometric functions, namely the Fourier series. This section addresses the question of approximating the function $f(x)$ by the partial sum

$$S_N(x) = \frac{a_0}{2} + \sum_{n=1}^{N} \left(a_n \cos \frac{n\pi x}{L} + b_n \sin \frac{n\pi x}{L} \right) \tag{3.31}$$

where a_n and b_n are given by Eqs. (3.11)–(3.13). How accurate is this approximation? From the discussion in Sec. 3.7, often the convergence is only of the pointwise variety. This prompts consideration of a kind of average measurement of the error and leads to the concept of **mean square approximation** of functions.

An interesting property of the Fourier coefficients can be obtained directly from the Euler-Fourier formulas. Let $S_N(x)$ be the partial sum of the Fourier series as defined in Eq. (3.31). Multiplying the partial sum $S_N(x)$ by $f(x)$ and integrating over the interval $[-L, L]$ produces

$$\int_{-L}^{L} f(x) S_N(x) \, dx$$

$$= \frac{1}{2} a_0 \int_{-L}^{L} f(x) \, dx + \sum_{n=1}^{N} \int_{-L}^{L} \left(a_n f(x) \cos \frac{n\pi x}{L} + b_n f(x) \sin \frac{n\pi x}{L} \right) dx$$

$$= \frac{L}{2} a_0^2 + L \sum_{n=1}^{N} \left(a_n^2 + b_n^2 \right). \tag{3.32}$$

Multiplying the partial sum $S_N(x)$ by itself and integrating over the interval $[-L, L]$ produce

$$\int_{-L}^{L} S_N(x) S_N(x)\, dx$$

$$= \frac{1}{2} a_0 \int_{-L}^{L} S_N(x)\, dx + \sum_{n=1}^{N} \int_{-L}^{L} \left(a_n S_N(x) \cos \frac{n\pi x}{L} + b_n S_N(x) \sin \frac{n\pi x}{L} \right) dx$$

$$= \frac{L}{2} a_0^2 + L \sum_{n=1}^{N} \left(a_n^2 + b_n^2 \right), \tag{3.33}$$

where the orthogonality of the trigonometric functions summarized in Eqs. (3.4)–(3.6) is used. This implies that

$$\int_{-L}^{L} (f(x) - S_N(x))^2\, dx$$

$$= \int_{-L}^{L} (f(x))^2\, dx - 2 \int_{-L}^{L} f(x) S_N(x)\, dx + \int_{-L}^{L} (S_N(x))^2\, dx$$

$$= \int_{-L}^{L} (f(x))^2\, dx - \left(\frac{L}{2} a_0^2 + L \sum_{n=1}^{N} \left(a_n^2 + b_n^2 \right) \right).$$

Since the integral on the left-hand side of the equation is non-negative,

$$\frac{1}{2} a_0^2 + \sum_{n=1}^{N} \left(a_n^2 + b_n^2 \right) \leq \frac{1}{L} \int_{-L}^{L} (f(x))^2\, dx. \tag{3.34}$$

This inequality is called **Bessel's inequality** for Fourier coefficients (after Friedrich Bessel[4]). According to the derivation above this inequality holds for any function defined on $[-L, L]$ such that its Fourier coefficients are defined. Since the right-hand side of the inequality is independent of N, the inequality holds as $N \to \infty$. Note the presence of the square of function f in the integrand of Eq. (3.34). A real-valued function f for which $\int_{-L}^{L} (f(x))^2\, dx < \infty$ is said to be square integrable on $[-L, L]$. In particular, Bessel's inequality implies that, for any square integrable function $f(x)$, the infinite series

$$\frac{1}{2} a_0^2 + \sum_{n=1}^{\infty} \left(a_n^2 + b_n^2 \right)$$

converges, that is, the sum of squares of the Fourier coefficients always converges and its sum is less than or equal to the twice the average value

---

[4] Friedrich Bessel, German astronomer and mathematician (1784–1846).

of $(f(x))^2$ on the interval $[-L, L]$. As a consequence of Bessel's inequality and the nth term test, for any square integrable function $f(x)$, the Fourier coefficients satisfy the limits,

$$\lim_{n \to \infty} a_n = 0 = \lim_{n \to \infty} b_n. \tag{3.35}$$

In fact, if $f(x)$ is piecewise smooth, then a stronger result than the one in Eq. (3.35) holds. See Exercise 23.

For a large class of functions, Bessel's inequality can be strengthened to an equality as shown below. For any positive integer N, define the **mean** (or **total**) **square error** of the partial sum relative to f by

$$E_N = \frac{1}{2L} \int_{-L}^{L} (f(x) - S_N(x))^2 \, dx.$$

As the name suggests, E_N is the average of the function $(f(x) - S_N(x))^2$ on the interval $[-L, L]$ and measures how well the function is approximated by $S_N(x)$ in this mean square sense. The following theorem holds.

Theorem 3.7. *Assume that $f(x)$ is a square integrable function on $[-L, L]$, then $\lim_{N \to \infty} E_N = 0$.*

The theorem implies that the sequence of partial sums $\{S_N(x)\}_{n=1}^{\infty}$ of the Fourier series of a square integrable function $f(x)$ on $[-L, L]$ approximates $f(x)$ in the sense that the mean square error approaches zero as $N \to \infty$. In this case $S_N(x)$ is said to **approximate** $f(x)$ **in the mean**.

The proof of this theorem in the general case is beyond the scope of this text. Interested readers may consult [Walker (1988)] or [Folland (1992)]. A proof for the case that $f(x)$ is continuous and piecewise smooth with $f(-L+) = f(L-)$ is presented here. Under these assumptions, the proof becomes elementary. By Theorem 3.4, $S_N(x)$ converges uniformly to $f(x)$ on $[-L, L]$. Therefore, by definition, there exists a sequence $\{\epsilon_n\}_{n=1}^{\infty}$ such that $\lim_{n \to \infty} \epsilon_n = 0$ and

$$|f(x) - S_N(x)| \le \epsilon_n$$

holds for all $x \in [-L, L]$. This implies that

$$|E_n| = \frac{1}{2L} \int_{-L}^{L} (f(x) - S_N(x))^2 \, dx \le \frac{1}{2L} (\epsilon_n^2 2L) \to 0$$

as $n \to \infty$.

The following important result is a corollary to Theorem 3.7.

Corollary 3.1. *If $f(x)$ is a square integrable function on $[-L, L]$, then*

$$\frac{1}{2L} \int_{-L}^{L} (f(x))^2 \, dx = \frac{1}{4} a_0^2 + \frac{1}{2} \sum_{n=1}^{\infty} \left(a_n^2 + b_n^2 \right) \qquad (3.36)$$

where a_n and b_n are the Fourier coefficients of $f(x)$.

Proof. From the definition of the mean squared error,

$$E_N = \frac{1}{2L} \int_{-L}^{L} (f(x))^2 \, dx - \frac{1}{L} \int_{-L}^{L} f(x) S_N(x) \, dx$$
$$+ \frac{1}{2L} \int_{-L}^{L} (S_N(x))^2 \, dx. \qquad (3.37)$$

From Eqs. (3.32) and (3.33),

$$\frac{1}{2L} \int_{-L}^{L} (S_N(x))^2 \, dx = \frac{1}{4} a_0^2 + \frac{1}{2} \sum_{n=1}^{N} (a_n^2 + b_n^2) \qquad (3.38)$$

and

$$\frac{1}{L} \int_{-L}^{L} f(x) S_N(x) \, dx = \frac{1}{2} a_0^2 + \sum_{n=1}^{N} (a_n^2 + b_n^2). \qquad (3.39)$$

Using Eqs. (3.38) and (3.39) on the right-hand side of Eq. (3.37) produces

$$E_N = \frac{1}{2L} \int_{-L}^{L} (f(x))^2 \, dx - \frac{1}{4} a_0^2 - \frac{1}{2} \sum_{n=1}^{N} (a_n^2 + b_n^2).$$

By Theorem 3.7, $\lim_{N \to \infty} E_N = 0$ completing the proof of the corollary. $\square$

Equation (3.36) in the corollary is called **Parseval's identity**[5], which is a form of Pythagorean theorem valid in the context of Fourier series. On a cautionary note, though the series $a_0^2/2 + \sum_{n=1}^{\infty} \left(a_n^2 + b_n^2 \right)$ converges, in general this does not imply $\sum_{n=1}^{\infty} a_n$ or $\sum_{n=1}^{\infty} b_n$ converge. For example consider the square wave function and its associated Fourier series from Example 3.5:

$$\sum_{n=1}^{\infty} b_n = \frac{4}{\pi} \sum_{k=1}^{\infty} \frac{1}{2k-1},$$

which diverges while

$$\sum_{n=1}^{\infty} b_n^2 = \frac{16}{\pi^2} \sum_{k=1}^{\infty} \frac{1}{(2k-1)^2}$$

converges to 2.

---

[5]Marc-Antoine Parseval, French mathematician (1755–1836).

Another interesting observation is that the approximation of $f(x)$ by a function of the form $S_N(x)$ is the best (*i.e.*, with the smallest mean squared error) when a_n and b_n are given by Eqs. (3.11)–(3.13). In fact, considering E_N as a function of the coefficients a_n and b_n then

$$\frac{\partial E_N}{\partial a_n} = \frac{\partial}{\partial a_n} \left[\int_{-L}^{L} (f(x) - S_N(x))^2 \, dx \right]$$

$$= 2 \int_{-L}^{L} (f(x) - S_N(x)) \frac{\partial}{\partial a_n} [S_N(x)] \, dx$$

$$= 2 \int_{-L}^{L} (f(x) - S_N(x)) \cos \frac{n\pi x}{L} \, dx$$

$$= 2 \left(\int_{-L}^{L} f(x) \cos \frac{n\pi x}{L} \, dx - a_n L \right)$$

where the orthogonality relations in Eqs. (3.4)–(3.6) have been employed. When a_n is chosen to be the Fourier coefficient, then $\partial E_N / \partial a_n = 0$ and similarly, it can be shown that $\partial E_N / \partial b_n = 0$ when b_n is given by Eq. (3.13). From these observations and the fact that E_N is quadratic in $(a_0, a_1, \ldots, a_N)$ and $(b_1, b_2, \ldots, b_N)$, it can be seen that E_N reaches a minimum when a_n and b_n are chosen to be the Fourier coefficients.

3.10　Complex Form of the Fourier Series

Fourier series are often conveniently expressed in terms of complex exponential functions instead of sine and cosine functions. The **Euler**[6] **identity**:

$$e^{i\theta} = \cos \theta + i \sin \theta \tag{3.40}$$

connects the trigonometric functions and the exponential functions. The reader is asked to verify this identity in Exercise 28. Readers needing some additional background in complex functions should consult App. A. From Eq. (3.40) the usual cosine and sine functions can be written as

$$\cos \theta = \frac{e^{i\theta} + e^{-i\theta}}{2}, \tag{3.41}$$

$$\sin \theta = \frac{e^{i\theta} - e^{-i\theta}}{2i}. \tag{3.42}$$

---

[6]Leonhard Euler, Swiss mathematician and physicist (1707–1783).

This produces the complex form of the Fourier series of any integrable function $f(x)$ on $[-L, L]$ as

$$f(x) \sim \frac{a_0}{2} + \sum_{n=1}^{\infty} \left[\frac{a_n}{2}(e^{in\pi x/L} + e^{-in\pi x/L}) + \frac{ib_n}{2}(e^{-in\pi x/L} - e^{in\pi x/L}) \right]$$

$$= \frac{a_0}{2} + \sum_{n=1}^{\infty} \left(\frac{a_n - ib_n}{2} e^{in\pi x/L} + \frac{a_n + ib_n}{2} e^{-in\pi x/L} \right).$$

If $c_0 = a_0/2$, $c_n = (a_n - ib_n)/2$, and $c_{-n} = (a_n + ib_n)/2$ for $n \in \mathbb{N}$, the Fourier series for $f(x)$ can be expressed as

$$f(x) \sim c_0 + \sum_{n=1}^{\infty} (c_n e^{in\pi x/L} + c_{-n} e^{-in\pi x/L}) = \sum_{n=-\infty}^{\infty} c_n e^{in\pi x/L}. \quad (3.43)$$

This is called the Fourier series in complex form. Moreover, introducing $b_0 = 0$ for convenience, then for $n = 0, 1, \ldots$,

$$c_n = \frac{a_n - ib_n}{2} = \frac{1}{2} \left(\frac{1}{L} \int_{-L}^{L} f(x) \cos \frac{n\pi x}{L} \, dx - \frac{i}{L} \int_{-L}^{L} f(x) \sin \frac{n\pi x}{L} \, dx \right)$$

$$= \frac{1}{2L} \int_{-L}^{L} f(x) \left(\cos \frac{n\pi x}{L} - i \sin \frac{n\pi x}{L} \right) dx$$

$$= \frac{1}{2L} \int_{-L}^{L} f(x) e^{-in\pi x/L} \, dx.$$

Likewise for $n \in \mathbb{N}$,

$$c_{-n} = \frac{a_n + ib_n}{2} = \frac{1}{2} \left(\frac{1}{L} \int_{-L}^{L} f(x) \cos \frac{n\pi x}{L} \, dx + \frac{i}{L} \int_{-L}^{L} f(x) \sin \frac{n\pi x}{L} \, dx \right)$$

$$= \frac{1}{2} \left(\frac{1}{L} \int_{-L}^{L} f(x) \cos \frac{-n\pi x}{L} \, dx - \frac{i}{L} \int_{-L}^{L} f(x) \sin \frac{-n\pi x}{L} \, dx \right)$$

$$= \frac{1}{2L} \int_{-L}^{L} f(x) e^{in\pi x/L} \, dx.$$

Therefore, for $n = 0, \pm 1, \pm 2, \ldots$,

$$c_n = \frac{1}{2L} \int_{-L}^{L} f(x) e^{-in\pi x/L} \, dx. \quad (3.44)$$

Note that if $f(x)$ is a real-valued function, then c_n and c_{-n} are complex conjugates for $n \in \mathbb{N}$. In particular, if $L = \pi$, then

$$f(x) \sim \sum_{n=-\infty}^{\infty} c_n e^{inx} \quad \text{with } c_n = \frac{1}{2\pi} \int_{-\pi}^{\pi} f(x) e^{-inx} \, dx.$$

For any given function $f(x)$, the complex form of its Fourier series given in Eq. (3.43) is usually conveniently found using the coefficient formula given in Eq. (3.44). This is particularly true when $f(x)$ is an exponential function.

Example 3.17. Suppose a is a constant and $a \notin \{ki \,|\, k = 0, 1, 2, \ldots\}$. Find the complex form of the Fourier series of $f(x) = e^{ax}$ on $(-\pi, \pi)$.

Solution. Using Eq. (3.44),

$$
\begin{aligned}
c_n &= \frac{1}{2\pi} \int_{-\pi}^{\pi} e^{ax} e^{-inx} \, dx = \frac{1}{2(a - in)\pi} \left(e^{(a-in)\pi} - e^{-(a-in)\pi} \right) \\
&= \frac{1}{(a - in)\pi} \left(\frac{e^{(a-in)\pi} - e^{-(a-in)\pi}}{2} \right) = \frac{\sinh(a - in)\pi}{(a - in)\pi}.
\end{aligned}
$$

3.11   Proofs of Two Theoretical Results

The proofs of the Dirichlet Convergence Theorem (Theorem 3.1) and the theorem on uniform convergence of Fourier series (Theorem 3.4) have been postponed to this section. Even though the results of these two theorems are quite profound, the proof is surprisingly accessible with just a background in elementary analysis. The Dirichlet Convergence Theorem is taken up first and its proof is facilitated by two lemmas presented next. The first lemma is a special case of the Riemann-Lesbegue[7] Lemma (see [Gradshteyn and Ryzhik (2007)]).

Lemma 3.1. *For any fixed $L > 0$ and piecewise continuous function $g(x)$ on interval $[0, L]$,*

$$
\lim_{N \to \infty} \int_0^L g(x) \sin \frac{(2N + 1)\pi x}{2L} \, dx = 0.
$$

Proof. Using the sum of angles formula for the sine function,

$$
\begin{aligned}
&\int_0^L g(x) \sin \frac{(2N + 1)\pi x}{2L} \, dx \\
&= \int_0^L \left[g(x) \sin \frac{\pi x}{2L} \right] \cos \frac{N \pi x}{L} \, dx + \int_0^L \left[g(x) \cos \frac{\pi x}{2L} \right] \sin \frac{N \pi x}{L} \, dx.
\end{aligned}
$$

The first integral on the right side of the equation is a constant multiple of the Nth Fourier cosine coefficient of the function $g(x) \sin \pi x/(2L)$ on the interval $[0, L]$ which approaches zero as $N \to \infty$ by Eq. (3.35). Similarly, the second term vanishes as $N \to \infty$. □

---

[7] Henri Léon Lesbegue, French mathematician (1875–1941).

It is necessary to show that $S_N(x)$ as defined by Eq. (3.31) converges to the value claimed in the Dirichlet Convergence Theorem. Function $S_N(x)$ can be written in an integral form using Eqs. (3.11)–(3.13) and the difference of angles formula for the cosine:

$$
\begin{aligned}
S_N(x) &= \frac{a_0}{2} + \sum_{n=1}^{N} \left(a_n \cos \frac{n\pi x}{L} + b_n \sin \frac{n\pi x}{L} \right) \\
&= \frac{1}{2L} \int_{-L}^{L} f(t)\, dt \\
&\quad + \frac{1}{L} \sum_{n=1}^{N} \int_{-L}^{L} f(t) \left(\cos \frac{n\pi t}{L} \cos \frac{n\pi x}{L} + \sin \frac{n\pi t}{L} \sin \frac{n\pi x}{L} \right) dt \\
&= \frac{1}{L} \int_{-L}^{L} f(t) \left(\frac{1}{2} + \sum_{n=1}^{N} \cos \frac{n\pi(t-x)}{L} \right) dt \\
&= \frac{1}{L} \int_{-L}^{L} f(t) D_N(t-x)\, dt
\end{aligned}
$$

where the function $D_N(x)$ is defined as

$$
D_N(x) = \frac{1}{2} + \sum_{n=1}^{N} \cos \frac{n\pi x}{L}, \tag{3.45}
$$

and is called the **Dirichlet Kernel**. The reader should note that $D_N(x)$ is an even function and for all N,

$$
\int_{0}^{L} D_N(x)\, dx = \frac{L}{2}. \tag{3.46}
$$

Next $D_N(x)$ is re-written in a form to which Lemma 3.1 can be applied. Consider the product,

$$
\begin{aligned}
D_N(x) \sin \frac{\pi x}{L} &= \left(\frac{1}{2} + \sum_{n=1}^{N} \cos \frac{n\pi x}{L} \right) \sin \frac{\pi x}{L} \\
&\quad - \frac{1}{2} \sin \frac{\pi x}{L} + \frac{1}{2} \sum_{n=1}^{N} \left(\sin \frac{(n+1)\pi x}{L} \quad \sin \frac{(n-1)\pi x}{L} \right) \\
&= \frac{1}{2} \left(\sin \frac{N\pi x}{L} + \sin \frac{(N+1)\pi x}{L} \right) \\
&= \frac{1}{2} \left(\sin \frac{2N\pi x}{2L} + \sin \frac{2(N+1)\pi x}{2L} \right) \\
&= \frac{1}{2} \left[\sin \left(\frac{(2N+1)\pi x}{2L} + \frac{\pi x}{2L} \right) + \sin \left(\frac{(2N+1)\pi x}{2L} - \frac{\pi x}{2L} \right) \right] \\
&= \sin \left(\frac{(2N+1)\pi x}{2L} \right) \cos \frac{\pi x}{2L}.
\end{aligned}
$$

Therefore,

$$D_N(x) = \frac{\sin\left(\frac{(2N+1)\pi x}{2L}\right)\cos\frac{\pi x}{2L}}{\sin\frac{\pi x}{L}} = \frac{\sin\left(\frac{(2N+1)\pi x}{2L}\right)\cos\frac{\pi x}{2L}}{2\sin\frac{\pi x}{2L}\cos\frac{\pi x}{2L}}$$

$$= \frac{\sin\left(\frac{(2N+1)\pi x}{2L}\right)}{2\sin\frac{\pi x}{2L}}. \tag{3.47}$$

Moreover, Eq. (3.47) holds for any x that is not an integer multiple of $2L$. When x is an integer multiple of $2L$, for instance $2kL$, the limit of $D_N(x)$ as $x \to 2kL$ exists and, as a result, $D_N(2kL)$ can be considered to be continuous on $(-\infty, \infty)$. From the expression in Eq. (3.47) an important property of the Dirichlet kernel can be derived and is stated in following lemma.

Lemma 3.2. *If $g(x)$ is a piecewise smooth function on $[0, L]$, then*

$$\lim_{N\to\infty} \int_0^L g(x)D_N(x)\,dx = \frac{L}{2}g(0+). \tag{3.48}$$

Proof. From Eq. (3.46),

$$\int_0^L g(x)D_N(x)\,dx = \int_0^L \left(g(x) - g(0+) + g(0+)\right)D_N(x)\,dx$$

$$= \int_0^L \left(g(x) - g(0+)\right)D_N(x)\,dx + g(0+)\int_0^L D_N(x)\,dx$$

$$= \int_0^L \left(g(x) - g(0+)\right)D_N(x)\,dx + \frac{L}{2}g(0+).$$

Thus it suffices to show that

$$\lim_{N\to\infty} \int_0^L \left(g(x) - g(0+)\right)D_N(x)\,dx = 0. \tag{3.49}$$

By Eq. (3.47),

$$\int_0^L \left(g(x) - g(0+)\right)D_N(x)\,dx = \int_0^L \frac{g(x) - g(0+)}{2\sin\frac{\pi x}{2L}}\sin\left(\frac{(2N+1)\pi x}{2L}\right)dx.$$

Define $G(x) = (g(x) - g(0+))/(2\sin\frac{\pi x}{2L})$. Since $g'(x)$ is piecewise continuous on $[0, L]$, there exists a $\delta > 0$ such that $g(x)$ and $g'(x)$ are both continuous on $(0, \delta]$. Consequently the following two limits are valid:

$$\lim_{x\to 0^+} g(x) = g(0+),$$

$$\lim_{x\to 0^+} g'(x) = g'(0+).$$

Defining $g(0) = g(0+)$ if necessary, then $g(x)$ is a continuous function on $[0, \delta]$ and the Mean Value Theorem can be applied to obtain

$$\lim_{x \to 0^+} \frac{g(x) - g(0+)}{x} = \lim_{x \to 0^+} g'(c) = g'(0+).$$

This implies

$$\lim_{x \to 0^+} G(x) = \lim_{x \to 0^+} \frac{g(x) - g(0+)}{2 \sin \frac{\pi x}{2L}} = \lim_{x \to 0^+} \frac{g(x) - g(0+)}{x} \frac{x}{2 \sin \frac{\pi x}{2L}} = \frac{L}{\pi} g'(0+).$$

Therefore $G(x)$ is piecewise continuous on $[0, L]$. By Lemma 3.1, Eq. (3.49) holds and the lemma is proved. $\qquad\qquad\square$

All the groundwork is complete for proving the Dirichlet Convergence Theorem.

Proof of Theorem 3.1. As derived earlier,

$$S_N(x) = \frac{1}{L} \int_{-L}^{L} f(t) D_N(t - x) \, dt = \frac{1}{L} \int_{x-L}^{x+L} f(t) D_N(t - x) \, dt,$$

since $f(t) D_N(t-x)$ is $2L$-periodic. The substitution $s = t - x$ in the definite integral yields:

$$
\begin{aligned}
S_N(x) &= \frac{1}{L} \int_{-L}^{L} f(x + s) D_N(s) \, ds \\
&= \frac{1}{L} \int_{-L}^{0} f(x + s) D_N(s) \, ds + \frac{1}{L} \int_{0}^{L} f(x + s) D_N(s) \, ds \\
&= \frac{1}{L} \int_{0}^{L} f(x - s) D_N(s) \, ds + \frac{1}{L} \int_{0}^{L} f(x + s) D_N(s) \, ds \\
&= I_N(x) + J_N(x),
\end{aligned}
$$

where

$$I_N(x) = \frac{1}{L} \int_{0}^{L} f(x - s) D_N(s) \, ds,$$

$$J_N(x) = \frac{1}{L} \int_{0}^{L} f(x + s) D_N(s) \, ds.$$

Since f is piecewise smooth and $2L$–periodic on $(-\infty, \infty)$, for any fixed x, functions $f(x - s)$ and $f(x + s)$ are both piecewise smooth and $2L$–periodic on $(-\infty, \infty)$. Consequently, by Lemma 3.2,

$$\lim_{N \to \infty} I_N(x) = \lim_{N \to \infty} \frac{1}{L} \int_{0}^{L} f(x - s) D_N(s) \, ds = \frac{1}{2} f(x-),$$

and

$$\lim_{N\to\infty} J_N(x) = \lim_{N\to\infty} \frac{1}{L}\int_0^L f(x+s)D_N(s)\,ds = \frac{1}{2}f(x+).$$

Therefore

$$\lim_{N\to\infty} S_N(x) = \frac{1}{2}\left(f(x-) + f(x+)\right)$$

and the proof is completed.   □

Now attention can be directed to the proof of Theorem 3.4 regarding the uniform convergence of Fourier series. In this proof the theorem regarding the differentiation of Fourier series (Theorem 3.5) and Bessel's inequality found in Eq. (3.34) will be used. Note that the proofs of these results do not depend on Theorem 3.4.

Proof of Theorem 3.4. Under the assumptions of Theorem 3.5, the Fourier series of $f(x)$ can be differentiated term by term and the Fourier coefficients of $f'(x)$ are given by

$$\alpha_n = \frac{n\pi b_n}{L} \text{ and } \beta_n = -\frac{n\pi a_n}{L} \text{ for } n \in \mathbb{N},$$

where a_n and b_n are the Fourier coefficients of $f(x)$. By Bessel's inequality given in Eq. (3.34),

$$\sum_{n=1}^{N} \left(\alpha_n^2 + \beta_n^2\right) \leq \frac{1}{L}\int_{-L}^{L} \left(f'(x)\right)^2 dx. \qquad (3.50)$$

For any $x \in [-L, L]$ and $k \in \mathbb{N}$, the reader will show in Exercise 29 that

$$\left| a_k \cos\frac{k\pi x}{L} + b_k \sin\frac{k\pi x}{L} \right| \leq \sqrt{a_k^2 + b_k^2}. \qquad (3.51)$$

The next step in the proof requires a fundamental result known as the Cauchy-Schwarz inequality (see [Steele (2004)]). For real numbers A_n and B_n,

$$\sum_{n=1}^{\infty}(A_n B_n) \leq \left(\sum_{n=1}^{\infty} A_n^2\right)^{1/2}\left(\sum_{n=1}^{\infty} B_n^2\right)^{1/2}. \qquad (3.52)$$

For any positive integer N, let $S_N(x)$ be the partial sum of the Fourier series given by Eq. (3.31). By Eq. (3.51) and the Cauchy-Schwarz inequality the following inequality holds:

$$|f(x) - S_N(x)| = \left| \sum_{k=N+1}^{\infty} \left(a_k \cos \frac{k\pi x}{L} + b_k \sin \frac{k\pi x}{L} \right) \right|$$

$$\leq \sum_{k=N+1}^{\infty} \sqrt{a_k^2 + b_k^2} = \sum_{k=N+1}^{\infty} \frac{L}{k\pi} \sqrt{\alpha_k^2 + \beta_k^2}$$

$$\leq \left(\sum_{k=N+1}^{\infty} \frac{L^2}{k^2 \pi^2} \right)^{1/2} \left(\sum_{k=N+1}^{\infty} (\alpha_k^2 + \beta_k^2) \right)^{1/2}$$

$$= \frac{L}{\pi} \left(\sum_{k=N+1}^{\infty} \frac{1}{k^2} \right)^{1/2} \left(\sum_{k=N+1}^{\infty} (\alpha_k^2 + \beta_k^2) \right)^{1/2}. \quad (3.53)$$

The sum $\sum_{k=n+1}^{\infty} \frac{1}{k^2}$ can be estimated as follows:

$$\sum_{k=N+1}^{\infty} \frac{1}{k^2} \leq \int_N^{\infty} \frac{1}{x^2} \, dx = \frac{1}{N}.$$

Therefore, by Eq. (3.53),

$$|f(x) - S_N(x)| \leq \frac{\sqrt{L}}{\pi \sqrt{N}} \left(\int_{-L}^{L} (f'(x))^2 \, dx \right)^{1/2}.$$

The last inequality implies that $S_N(x)$ converges to $f(x)$ uniformly as $N \to \infty$, in other words, the Fourier series for $f(x)$ converges uniformly. $\square$

3.12 Historical and Supplemental Remarks

The theory of Fourier series began with the work of Fourier on heat conduction in solid bodies in the early 1800s. However, the notion of trigonometric series appeared much earlier. For example, in the works of Leonhard Euler and Daniel Bernoulli[8] on vibrating strings, Bernoulli had claimed that every solution of the wave equation can be represented as a trigonometric series and Euler claimed that not all curves (a notion equivalent to functions at that time) can be represented as a trigonometric series. Fourier first made extensive use of such trigonometric series in his work on heat conduction as published in his prize-winning work *Theórie Analytique de la Cheleur*. In fact, Fourier claimed that any graphs (or functions) on a closed interval with zero boundary conditions can be written as a series of sine

---

[8]Daniel Bernoulli, Swiss mathematician and physicist (1700–1782).

functions. Although Fourier implied in his work that the series converge, he did not provide any rigorous proof. Fourier's work on heat conduction and trigonometric series had profound implications on the future of mathematical physics. However, Fourier's work, not supported by rigorous mathematical proof, was criticized by many of his peers, Laplace and Lagrange, among many others. In fact it was not until 1828, when Dirichlet published his work on the convergence of trigonometric series and proved that Fourier was almost correct when he (Fourier) claimed all functions can be represented by trigonometric series, see [Dirichlet (1829)].

Many other mathematicians contributed to the theory of Fourier series, such as the work of Fejér[9] on $(C, 1)$ summability and uniform convergence of Fourier series for continuous functions. In 1922, Kolmogorov[10] published an example of a Lebesgue[11] integrable function on $[-\pi, \pi]$ with a nowhere convergent Fourier series [Kolmogorov (1927)]. Carleson[12] published an example of a continuous function with an everywhere divergent Fourier series in 1966 [Carleson (1996)]. The theory of Fourier series along with the Fourier integral not only find applications in heat conduction, wave propagation, and many other problems of mathematical physics, but also in many modern problems, such as optics and signal processing.

The theory is also a prelude to the more abstract branch of modern analysis called harmonic analysis.

3.13   Exercises

(1) Let $f(x)$ be a T-periodic function.

   (a) Show that $f'(x)$ is T-periodic.
   (b) Show that, if $f(x)$ is integrable on $[0, T]$ and $\int_0^T f(x)\,dx = 0$ then $G(x) = \int_a^x f(t)\,dt$ is T-periodic.

(2) Prove the relations stated in Eqs. (3.4)–(3.6).

(3) Sketch the graph of the periodic extension and find the Fourier series of each of the following functions:

   (a) $f(x) = \begin{cases} 1 & \text{if } -2 < x < 0, \\ 2 & \text{if } 0 < x \le 2. \end{cases}$

---

[9]Lipót Fejér, Hungarian mathematician (1880–1959).
[10]Andrey Nikolaevich Kolmogorov, Soviet mathematician (1903–1987).
[11]Henri Léon Lebesgue, French mathematician (1875–1941).
[12]Lennart Axel Edvard Carleson, Swedish mathematician (1928–).

(b) $f(x) = \begin{cases} 2 + x & \text{if } -2 < x < 0, \\ 1 & \text{if } 0 < x < 2. \end{cases}$

(c) $f(x) = \begin{cases} x & \text{if } -\pi < x < 0, \\ -1 & \text{if } 0 < x < \pi. \end{cases}$

(d) $f(x) = \begin{cases} 0 & \text{if } -\pi < x < 0, \\ \cos x & \text{if } 0 < x < \pi. \end{cases}$

(e) $f(x) = \begin{cases} 0 & \text{if } -2 < x < 0, \\ 2 & \text{if } 0 < x < 1, \\ 1 & \text{if } 1 < x < 2. \end{cases}$

(f) $f(x) = e^{ax}$, for $-\pi < x < \pi$, where $a \neq 0$ is a constant.

(4) Without using the Euler-Fourier formulas, find the Fourier series of the following functions for $-\pi < x < \pi$:

(a) $f(x) = \sin x + \cos 2x$.
(b) $f(x) = -2\sin(3x) + 2\sin(5x)$.
(c) $f(x) = \cos^2 x$.
(d) $f(x) = \sin x \cos 2x$. (*Hint*: use an appropriate trigonometric identity.)
(e) $f(x) = \sin x + \frac{1}{4}|\sin x|$. (*Hint*: use Example 3.4.)

(5) Find the Fourier series of $\cos(ax)$ on $(-\pi, \pi)$, where a is a constant.

(6) Prove the following properties:

(a) If $f(x)$ is an even, integrable function on $[-L, L]$, show that $\int_{-L}^{L} f(x)\, dx = 2\int_0^L f(x)\, dx$.
(b) If $f(x)$ is an odd, integrable function on $[-L, L]$, show that $\int_{-L}^{L} f(x)\, dx = 0$.

(7) Plot the graph of the even and odd periodic extensions and find the Fourier cosine and sine series of each of the following functions defined on $(0, L)$:

(a) $f(x) = \begin{cases} 1 & \text{if } 0 < x < 1, \\ 2 & \text{if } 1 < x < 2. \end{cases}$

(b) $f(x) = \begin{cases} 0 & \text{if } 0 < x < 1, \\ 1 - x & \text{if } 1 < x < 2. \end{cases}$

(c) $f(x) = x$ for $0 < x < \pi$.
(d) $f(x) = 1 - x$ for $0 < x < 2$.
(e) $f(x) = x^3$ for $0 < x < \pi$.
(f) $f(x) = e^{2x}$ for $0 < x < \pi$.

(8) Find the Fourier cosine series of $f(x) = x(2 - x^2)$ for $0 < x < \pi$.

(9) Find the Fourier cosine and sine series of $f(x) = e^{ax}$ for $0 < x < \pi$, where $a \neq 0$ is a constant.

(10) Given $y = f(x)$ defined on $[0, L]$ and extending the function appropriately, write $f(x)$ as a Fourier series in the following form and determine the coefficients of the series in terms of an integral involving the function $f(x)$ on the interval $[0, L]$.

$$f(x) \sim \sum_{k=1}^{\infty} c_k \cos \frac{(2k-1)\pi x}{2L}.$$

(11) Determine if the following functions are piecewise continuous, and/or piecewise smooth, or neither on the given interval:

(a) $f(x) = x - [\![x]\!]$ on $(-\infty, \infty)$, where $[\![x]\!]$ is the greatest integer that is less than or equal to x.

(b) $f(x) = \frac{x}{x^2-1}$ on $[-2, 2]$.

(c) $f(x) = \sin \frac{1}{x}$ on $[-\pi, \pi]$.

(d) $f(x) = x^2 \sin \frac{1}{x}$ on $[-\pi, \pi]$.

(e) $f(x) = x^4 \sin \frac{1}{x}$ on $[-\pi, \pi]$.

(f) $f(x) = \sqrt{9 - x^2}$ on $[-3, 3]$.

(g) $f(x) = e^{-1/x^2}$ on $(-\infty, \infty)$.

(h) $f(x) = x^{3/2}$ on $[-1, 1]$.

(12) Show that each of the functions in Exercise 3 is piecewise smooth and determine the sum of its Fourier series.

(13) Find the Fourier sine and cosine series of

$$f(x) = \begin{cases} x & \text{for } 0 \leq x \leq 1, \\ -1 & \text{for } 1 < x \leq 2, \\ 2 & \text{for } 2 < x \leq \pi, \end{cases}$$

and determine the sums of both series for each x value on $[-\pi, \pi]$.

(14) Show that if $f'(x)$ is piecewise continuous on $[a, b]$, then so is $f(x)$.

(15) Prove that if f is piecewise smooth on $[-L, L]$ and all of its Fourier coefficients are 0, then $f(x) = 0$ at all points where f is continuous.

(16) Suppose functions f and g are continuous and piecewise smooth on $[-L, L]$ and the Fourier coefficients of f are the same as the Fourier coefficients of g. Show that $f(x) = g(x)$ for $-L < x < L$.

(17) Find the Fourier series of

$$f(x) = \begin{cases} 1 & \text{if } -1 \leq x < 0, \\ 2 & \text{if } 0 \leq x \leq 1 \end{cases}$$

and use the result to prove $\sum_{k=1}^{\infty} \frac{(-1)^{k-1}}{2k-1} = \frac{\pi}{4}$.

(18) Using the Fourier cosine series of $\sin x$ for $0 < x < \pi$ (see Example 3.12) to find the sums

$$\sum_{n=1}^{\infty} \frac{1}{4n^2 - 1} \quad \text{and} \quad \sum_{n=1}^{\infty} \frac{(-1)^n}{4n^2 - 1}.$$

(19) Use the Fourier series for $\cos(a\,x)$ for $-\pi < x < \pi$, to establish the identity

$$\frac{a\pi}{\sin a\pi} = 1 + 2a^2 \sum_{n=1}^{\infty} \frac{(-1)^{n+1}}{n^2 - a^2}$$

for $a \neq 0, \pm 1, \pm 2, \ldots$.

(20) In Example 3.9, it was found that the Fourier cosine series of $f(x) = x$ for $0 \leq x \leq 2$ is

$$1 - \frac{8}{\pi^2} \sum_{k=1}^{\infty} \frac{1}{(2k-1)^2} \cos \frac{(2k-1)\pi x}{2}.$$

(a) Verify that the even extension of $f(x)$ satisfies the conditions of Theorem 3.4 and therefore, the Fourier cosine series converges uniformly to $f(x)$ on $[0, 2]$.

(b) Prove directly using the Weierstrass M-test that the Fourier cosine series converges uniformly to $f(x)$ on $[0, 2]$.

(21) Show that the Fourier cosine series of $f(x) = x$ for $0 < x < \pi$ can be differentiated term by term.

(22) Consider the 2π-periodic extension of the function

$$f(x) = \begin{cases} -1 & \text{for } -\pi < x \leq -\pi/2, \\ 1 & \text{for } -\pi/2 < x \leq \pi/2, \\ -1 & \text{for } \pi/2 < x < \pi. \end{cases}$$

(a) Show that $F(x) = \int_0^x f(s)\,ds$ is 2π-periodic.

(b) Plot both $f(x)$ and $F(x)$ on an interval of length at least three periods.

(c) Find the Fourier series of $f(x)$.

(d) Find the Fourier series of $F(x)$ by integrating the Fourier series of $f(x)$ term by term.

(e) Determine the sums of the Fourier series for $f(x)$ and $F(x)$ for each $x \in [-\pi, \pi]$.

(23) Assume that $f(x)$ is a piecewise smooth function on $[-L, L]$ and $f(-L+) = f(L-)$. Show that the Fourier coefficients a_n and b_n of $f(x)$ satisfy

$$\lim_{n \to \infty} n a_n = 0 = \lim_{n \to \infty} n b_n.$$

(24) Integrate the following Fourier series term by term and find the function represented by the new series.

(a) $2x = \sum_{k=1}^{\infty} \frac{4(-1)^{k+1}}{k} \sin(k\,x)$.

(b) $f(x) = \frac{3}{2} + \frac{1}{\pi} \sum_{n=1}^{\infty} \frac{1 - (-1)^n}{n} \sin(n\,x)$, where

$$f(x) = \begin{cases} 1 & \text{if } -\pi < x < 0, \\ 2 & \text{if } 0 < x < \pi. \end{cases}$$

(25) Consider the Fourier series

$$x = 2 \sum_{n=1}^{\infty} \frac{(-1)^{n+1}}{n} \sin(n\,x)$$

for $-\pi < x < \pi$.

(a) Use Parseval's identity to show

$$\sum_{n=1}^{\infty} \frac{1}{n^2} = \frac{\pi^2}{6}.$$

(b) Use the result of Exercise (25a) to show that

$$\sum_{n=1}^{\infty} \frac{1}{(2n-1)^2} = \frac{\pi^2}{8}.$$

(26) Use the Fourier cosine series of $y = x^2$ on $(-\pi, \pi)$ and Parseval's identity to show that

$$\sum_{k=1}^{\infty} \frac{1}{k^4} = \frac{\pi^4}{90}.$$

(27) With the help of Parseval's inequality, show that

$$\sum_{n=1}^{\infty} \frac{1}{(2k-1)^6} = \frac{\pi^6}{960}.$$

(*Hint:* use the Fourier series for x^3 on $(-\pi, \pi)$.)

(28) Expand $e^{i\theta}$ as a Taylor's series about $x_0 = 0$ to establish Euler's Identity.

(29) Prove the inequality shown in Eq. (3.51).

Chapter 4

The Heat Equation

In this chapter solutions to one of the basic equations of mathematical physics will be presented. The previous material on separation of variables (Sec. 1.6) and Fourier series (Chap. 3) will be used extensively. The heat equation also arises in situations where a quantity (whether it is heat energy, a chemical, or something more abstract such as "information") diffuses through a continuous medium, thus an understanding of its properties and solution is important to any applied mathematician or engineer. The sections of this chapter are organized according to the spatial domains and the boundary conditions imposed on the heat equation. The first section will treat the heat equation on a closed, bounded (i.e., compact) interval. After three different types of homogeneous boundary conditions on the closed, bounded interval have been studied, nonhomogeneous boundary conditions, such as boundary conditions which depend on time will be studied. One of the most important theoretical results related to the heat equation is that under relatively mild assumptions the maximum and minimum properties of the solution can be described and the uniqueness of the solution to the initial boundary value problem can be proved. Section 4.3 covering the Maximum Principle and the uniqueness of solutions can be skipped on a first reading. In Sec. 4.4 the boundedness of the spatial domain will be relaxed and solutions on infinite and semi-infinite intervals will be explored. The chapter will close with an extension of the one-dimensional results to solutions of the heat equation on a two-dimensional rectangular spatial domain and suggestions for further explorations of the heat equation.

4.1   Homogeneous Boundary Value Problems on Bounded Intervals

Consider the heat equation on a one-dimensional spatial domain comprised of the interval $[0, L]$ where $L > 0$ is constant. The temperature at location $0 \leq x \leq L$ at time $t \geq 0$ will be denoted as $u(x,t)$. Generically the partial differential equation and initial condition for the heat equation on this domain can be written as

$$u_t = k\, u_{xx} \text{ for } 0 < x < L \text{ and } t > 0, \qquad (4.1)$$

$$u(x,0) = f(x) \text{ for } 0 < x < L. \qquad (4.2)$$

The constant k represents the coefficient of diffusion or thermal diffusivity of the material occupying the spatial domain. It would be premature to solve the heat equation without imposing a set of boundary conditions. In this section three standard types of homogeneous boundary conditions will be studied: Dirichlet type, Neumann type, and boundary conditions of the third kind, sometimes called Robin boundary conditions. See Sec. 1.3.1 on p. 13 for background information and assumptions on these types of boundary conditions.

4.1.1   *Dirichlet Boundary Conditions*

The Dirichlet boundary condition specifies the temperature at the boundary of the domain. The simplest case would be to fix the temperature at $x = 0$ and $x = L$ to be zero on some temperature scale. Physically this could be achieved by a homogeneous rod that is insulated along its length and is in perfect thermal contact with a block of ice at each end of the rod (if temperature is measured on the Celsius[1] scale). Mathematically this can be specified as

$$u(0,t) = u(L,t) = 0 \text{ for } t > 0. \qquad (4.3)$$

Equations (4.1), (4.2), and (4.3) constitute an initial boundary value problem sometimes called a **Fourier problem**. The first step in solving this initial boundary value problem is to assume that it possesses a **product solution** $u(x,t) = X(x)T(t)$ and to use the method of separation of variables to solve the boundary value problem for $X(x)$. In Sec. 1.6 separation of variables was used and it was found that $X(x)$ and $T(t)$ must satisfy Eq. (1.51). This yields product solutions of the form

$$u_n(x,t) = a_n e^{-kn^2\pi^2 t/L^2} \sin\left(\frac{n\pi x}{L}\right), \qquad (4.4)$$

---

[1] Anders Celsius, Swedish astronomer (1701–1744).

where a_n is an arbitrary constant and $n \in \mathbb{N}$. The reader should note that the solution $u_n(x,t)$ given in Eq. (4.4) satisfies the heat equation of Eq. (4.1) and the Dirichlet boundary condition of Eq. (4.3). The Principle of Superposition (Theorem 1.1) and the homogeneity of the boundary conditions imply that any finite linear combination of solutions of the form given in Eq. (4.4) will also solve the heat equation with the same boundary conditions. Thus a function of the form

$$u(x,t) = \sum_{n=1}^{N} a_n e^{-kn^2\pi^2 t/L^2} \sin\left(\frac{n\pi x}{L}\right)$$

also solves the heat equation with the Dirichlet boundary conditions.

This observation enables the solution to the initial boundary value problem of Eqs. (4.1), (4.2), and (4.3) to be found when the initial function $f(x)$ is in the form of

$$f(x) = \sum_{n=1}^{N} a_n \sin\left(\frac{n\pi x}{L}\right).$$

In this case the solution to the Fourier problem would be

$$u(x,t) = \sum_{n=1}^{N} a_n e^{-kn^2\pi^2 t/L^2} \sin\left(\frac{n\pi x}{L}\right).$$

For the more general case in which $u(x,0) = f(x)$, a solution of the form

$$u(x,t) = \sum_{n=1}^{\infty} a_n e^{-kn^2\pi^2 t/L^2} \sin\left(\frac{n\pi x}{L}\right)$$

is sought. This calls for the initial function to be represented as

$$f(x) = \sum_{n=1}^{\infty} a_n \sin\left(\frac{n\pi x}{L}\right).$$

For this situation as long as $f(x)$ is piecewise smooth, Fourier series (Chap. 3) provide exactly what is needed. Function $f(x)$ can be extended as an odd $2L$-periodic function on $(-\infty, \infty)$ and while a_n for $n \in \mathbb{N}$ are chosen to be the Fourier sine coefficients, that is

$$a_n = \frac{2}{L} \int_0^L f(x) \sin\left(\frac{n\pi x}{L}\right) dx. \tag{4.5}$$

Thus formally, a solution to this Fourier problem is

$$u(x,t) = \sum_{n=1}^{\infty} a_n e^{-kn^2\pi^2 t/L^2} \sin\left(\frac{n\pi x}{L}\right) \tag{4.6}$$

where a_n is given by Eq. (4.5).

It is worth noting that the formal solution defined in Eq. (4.6) is well-defined for all $(x, t) \in [0, L] \times [0, \infty)$. In fact, for any $t > 0$, since $\{a_n\}_{n=1}^{\infty}$ is bounded (see Exercise 23 of Chap. 3), there exists a constant $M > 0$ such that $|a_n| \leq M$ and

$$\left| a_n e^{-kn^2\pi^2 t/L^2} \sin\left(\frac{n\pi x}{L}\right) \right| \leq M \left(e^{-k\pi^2 t/L^2} \right)^n.$$

Since the infinite series $\sum_{n=1}^{\infty} M (e^{-k\pi^2 t/L^2})^n$ converges (as a geometric series), it can be shown that the series in Eq. (4.6) converges uniformly with respect to $x \in [0, L]$. In much that same way, it can be shown that the "derivative series" corresponding to $\partial u/\partial t$, $\partial u/\partial x$, and $\partial^2 u/\partial x^2$ converge uniformly for $x \in [0, L]$. Therefore when the initial condition $f(x)$ is sufficiently smooth on $[0, L]$ the function given in Eq. (4.6) is a formal solution to the heat equation with homogeneous Dirichlet boundary conditions. The presence of the decaying exponential term in Eq. (4.6) suggests that $u(x, t) \to 0$ on $[0, L]$ as $t \to \infty$. This asymptotic behavior agrees with the physical intuition that if the ends of the rod are kept at zero temperature, then for large t the entire rod should be near the zero temperature.

Example 4.1. Solve the following Fourier problem:

$$u_t = u_{xx} \text{ for } 0 < x < 1 \text{ and } t > 0,$$
$$u(0, t) = u(1, t) = 0 \text{ for } t > 0,$$
$$u(x, 0) = 100 \text{ for } 0 < x < 1.$$

Solution. From Eq. (4.5) the coefficients of the Fourier series solution to this Fourier problem are

$$a_n = 2 \int_0^1 100 \sin(n\pi x) \, dx = \left[\frac{-200}{n\pi} \cos(n\pi x) \right]_0^1 = \begin{cases} \frac{400}{n\pi} & \text{if } n \text{ is odd,} \\ 0 & \text{if } n \text{ is even.} \end{cases}$$

Thus the Fourier solution can be written as

$$u(x, t) = \frac{400}{\pi} \sum_{n=1}^{\infty} \frac{1}{2n-1} e^{-(2n-1)^2\pi^2 t} \sin((2n-1)\pi x).$$

Figure 4.1 shows a contour plot of the solution.

4.1.2 *Neumann Boundary Conditions*

Much of the work in this subsection will resemble the work of the previous subsection, since only the type of the boundary condition changes. While

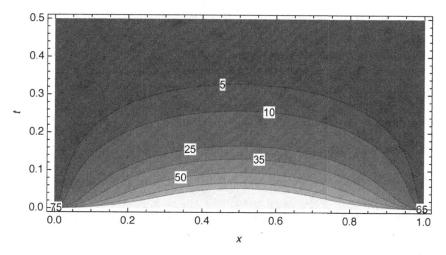

Fig. 4.1   A contour plot of the solution to the Fourier problem of Example 4.1. Warmer areas are in lighter shades while cooler areas are darker.

this change is important and leads to a solution with different behavior, much of the analysis is the same. The Neumann boundary condition corresponds physically to the situation in which the ends of the domain are perfectly insulated from the environment. There is no heat flux at the boundary. This can be stated mathematically as

$$u_x(0,t) = u_x(L,t) = 0 \text{ for } t > 0. \tag{4.7}$$

Equations (4.1), (4.2), and (4.7) form a new initial boundary value problem. Once again, the method of separation of variables is used to obtain a product solution $u(x,t) = X(x)T(t)$. In Exercise 18 of Chap. 1 the reader was asked to show that the boundary value problem for $X(x)$ has eigenvalues and eigenfunctions of the form

$$\lambda_n = \frac{n\pi}{L}, \tag{4.8}$$

$$X_n(x) = \cos\left(\frac{n\pi x}{L}\right) \tag{4.9}$$

for $n = 0, 1, 2, \ldots$.

Substituting λ_n into Eq. (1.55) and solving the resulting equation produce

$$T_n(t) = e^{-kn^2\pi^2 t/L^2}.$$

Note that $T_0(t) = 1$. This yields a product solution with the form

$$u_n(x,t) = e^{-k n^2 \pi^2 t/L^2} \cos\left(\frac{n\pi x}{L}\right), \tag{4.10}$$

for $n = 0, 1, 2, \ldots$. As before the reader should note the product solution $u_n(x, t)$ given in Eq. (4.10) satisfies the heat equation of Eq. (4.1) and the Neumann boundary condition of Eq. (4.7). Due to the linearity and homogeneity properties of the heat equation and the Neumann boundary conditions, any finite linear combination of solutions of the form given in Eq. (4.10) will also solve the heat equation and the boundary conditions. Thus for any positive integer N and constants a_n,

$$u(x, t) = \frac{a_0}{2} + \sum_{n=1}^{N} a_n e^{-kn^2\pi^2 t/L^2} \cos\left(\frac{n\pi x}{L}\right)$$

also solves the heat equation and satisfies the Neumann boundary conditions.

The only remaining detail of the solution to this initial boundary value problem is that of satisfying the initial condition of Eq. (4.2). A formal solution is sought of the form,

$$u(x, t) = \frac{a_0}{2} + \sum_{n=1}^{\infty} a_n e^{-kn^2\pi^2 t/L^2} \cos\left(\frac{n\pi x}{L}\right).$$

To satisfy $u(x, 0) = f(x)$, it is necessary that

$$f(x) = \frac{a_0}{2} + \sum_{n=1}^{\infty} a_n \cos\left(\frac{n\pi x}{L}\right).$$

Making use of the results on Fourier series, $f(x)$ can be extended as an even $2L$-periodic function on $(-\infty, \infty)$ and a_n for $n = 0, 1, \ldots$ can be the Fourier cosine coefficients of this extension. That is,

$$a_n = \frac{2}{L} \int_0^L f(x) \cos\left(\frac{n\pi x}{L}\right) dx \qquad (4.11)$$

for $n = 0, 1, \ldots$. This produces a formal solution that satisfies the initial condition stated in Eq. (4.2) when a_n is given by Eq. (4.11).

Example 4.2. Solve the following initial boundary value problem:

$$u_t = u_{xx} \text{ for } 0 < x < 1 \text{ and } t > 0,$$
$$u_x(0, t) = u_x(1, t) = 0 \text{ for } t > 0,$$
$$u(x, 0) = 1 + \sin(2\pi x) \text{ for } 0 < x < 1.$$

Solution. From Eq. (4.11) the coefficients of the Fourier series solution to this initial boundary value problem are

$$a_0 = 2 \int_0^1 (1 + \sin(2\pi x)) \, dx = 2 \quad \text{and}$$

$$a_n = 2 \int_0^1 (1 + \sin(2\pi x)) \cos(n\pi x) \, dx = \begin{cases} \frac{-8}{(n^2-4)\pi} & \text{if } n \text{ is odd,} \\ 0 & \text{if } n \text{ is even.} \end{cases}$$

Thus the Fourier solution can be written as

$$u(x,t) = 1 - \frac{8}{\pi}\sum_{k=1}^{\infty}\frac{e^{-(2k-1)^2\pi^2 t}}{(2k-1)^2-4}\cos((2k-1)\pi x).$$

Figure 4.2 shows a contour plot of the solution.

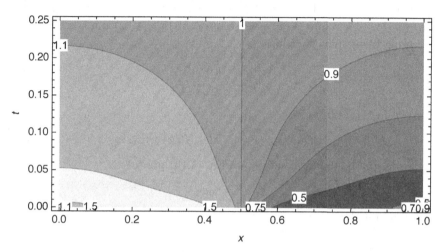

Fig. 4.2   A contour plot of the Fourier series solution to the initial boundary value problem of Example 4.2. Warmer areas are in lighter shades while cooler areas are darker.

4.1.3   *Boundary Conditions of the Third Kind*

The conditions of constant temperature at the ends of the domain (Dirichlet boundary conditions) or perfect thermal insulation (Neumann boundary conditions) are nearly impossible to achieve in the physical world. The remaining class of boundary conditions to be studied occupies a middle ground between perfect thermal contact and perfect thermal insulation at the boundaries. Let γ_0 and γ_L be positive constants and consider the boundary conditions of the third kind stated as

$$u_x(0,t) - \gamma_0 u(0,t) = 0 \quad \text{and} \tag{4.12}$$
$$u_x(L,t) + \gamma_L u(L,t) = 0. \tag{4.13}$$

The Dirichlet and Neumann boundary conditions can be thought of as limiting cases of the boundary conditions above. As $\gamma_0 \to 0^+$ and $\gamma_L \to 0^+$, Eq. (4.12) and Eq. (4.13) become Eq. (4.7), the Neumann boundary conditions. On the other hand as $\gamma_0 \to \infty$ and $\gamma_L \to \infty$ then Eq. (4.12) and Eq. (4.13) become Eq. (4.3), the Dirichlet boundary conditions.

Finding a product solution which satisfies the boundary conditions of Eq. (4.12) and Eq. (4.13) is more challenging than in the Dirichlet and Neumann cases. Assuming a product solution of the form $u(x,t) = X(x)T(t)$, differentiating it, and then substituting it into the heat equation produce Eq. (1.51). Therefore $X(x)$ is a solution of Eq. (1.52) and satisfies the following boundary conditions:

$$X'(0) = \gamma_0 X(0), \tag{4.14}$$

$$X'(L) = -\gamma_L X(L). \tag{4.15}$$

If $c = 0$ in Eq. (1.52) then

$$X(x) = a_1 x + a_2 \text{ and } X'(x) = a_1.$$

Imposing the boundary condition of Eq. (4.14) produces

$$a_1 = \gamma_0 a_2 \tag{4.16}$$

and the boundary condition of Eq. (4.15) implies

$$a_1 = -\frac{\gamma_L a_2}{1 + \gamma_L L}. \tag{4.17}$$

Combining Eq. (4.16) and Eq. (4.17) while assuming $a_2 \neq 0$ (else $a_1 = a_2 = 0$ and the solution is trivial) yields the equation

$$\gamma_L = -\gamma_0(1 + \gamma_L L).$$

If $\gamma_L > 0$ as assumed then $\gamma_0 < 0$ contradicting the assumption that $\gamma_0 > 0$. Therefore $c = 0$ is not an eigenvalue.

If $c = \lambda^2 > 0$ in Eq. (1.52) then

$$X(x) = a_1 \cosh(\lambda x) + a_2 \sinh(\lambda x),$$

$$X'(x) = \lambda(a_1 \sinh(\lambda x) + a_2 \cosh(\lambda x)).$$

Proceeding as before, the boundary conditions of Eqs. (4.14) and (4.15) imply that

$$a_1 \gamma_0 = a_2 \lambda, \tag{4.18}$$

$$(a_1 \gamma_L + a_2 \lambda) \cosh(\lambda L) + (a_2 \gamma_L + a_1 \lambda) \sinh(\lambda L) = 0. \tag{4.19}$$

Combining Eq. (4.18) and Eq. (4.19) while assuming $a_2 \neq 0$ and $\gamma_0 > 0$ yields the equation

$$(\gamma_0 + \gamma_L)\lambda \cosh(\lambda L) + (\gamma_0 \gamma_L + \lambda^2) \sinh(\lambda L) = 0.$$

Since $\cosh x \geq 1$ for all x then the previous equation is equivalent to

$$(\gamma_0 + \gamma_L)\lambda = -(\gamma_0\gamma_L + \lambda^2)\tanh(\lambda L).$$

Since the left-hand side of this equation is positive the expression $\tanh(\lambda L)$ must be negative which implies $\lambda L < 0$ and contradicts the assumptions that both of these factors are positive. Consequently any $c > 0$ is not an eigenvalue.

The last case to consider is that for which $c = -\lambda^2 < 0$ in Eq. (1.51) which implies

$$X(x) = a_1 \cos(\lambda x) + a_2 \sin(\lambda x),$$
$$X'(x) = \lambda(a_2 \cos(\lambda x) - a_1 \sin(\lambda x)).$$

Following the same line of reasoning as in the preceding case and assuming $a_2 \neq 0$, the parameter λ must satisfy the equation

$$(\gamma_0 + \gamma_L)\lambda \cos(\lambda L) + (\gamma_0\gamma_L - \lambda^2)\sin(\lambda L) = 0, \tag{4.20}$$

and $a_2 = a_1\gamma_0/\lambda$. Note that $\cos(\lambda L) = 0$ if and only if $\lambda = (2n-1)\pi/(2L)$ for some $n \in \mathbb{N}$. This condition in Eq. (4.20) is in turn equivalent to

$$\gamma_0\gamma_L = \left(\frac{(2n-1)\pi}{2L}\right)^2. \tag{4.21}$$

Therefore, if there exists no $n \in \mathbb{N}$ such that Eq. (4.21) holds, then $\cos(\lambda L) \neq 0$ and Eq. (4.20) may be re-written as

$$\frac{(\gamma_0 + \gamma_L)\lambda}{\lambda^2 - \gamma_0\gamma_L} = \tan(\lambda L). \tag{4.22}$$

Although the solution λ of the equation above cannot be found in closed form, there exist infinitely many solutions denoted as λ_n to the equation which can be approximated numerically (see Fig. 4.3). For example when $\gamma_0 = \gamma_L = 1$ and $L = 1$ the first few positive eigenvalues are approximately

$$\lambda_1 \approx 1.30654, \quad \lambda_2 \approx 3.67319, \quad \lambda_3 \approx 6.58462, \quad \lambda_4 \approx 9.63168.$$

Since

$$\lim_{\lambda \to \infty} \frac{(\gamma_0 + \gamma_L)\lambda}{\lambda^2 - \gamma_0\gamma_L} = 0$$

the eigenvalues of the boundary value problem for the boundary conditions of the third kind asymptotically approach the positive solutions of $\tan(\lambda L) = 0$. Thus in general $\lambda_n \approx (n-1)\pi/L$ for large n.

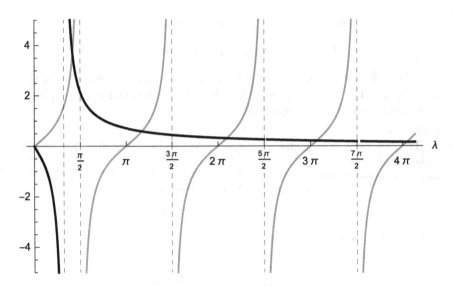

Fig. 4.3 There are infinitely many positive solutions to Eq. (4.22) in the case where $\gamma_0 = \gamma_L = 1$ and $L = 1$. The thick hyperbolic curve is the graph of $(\gamma_0 + \gamma_L)\lambda/(\lambda^2 - \gamma_0\gamma_L)$ while the lighter curve is the graph of $\tan(\lambda L)$.

Once the eigenvalues λ_n have been described, the eigenfunctions can be expressed as

$$X_n(x) = \cos(\lambda_n x) + \frac{\gamma_0}{\lambda_n} \sin(\lambda_n x), \qquad (4.23)$$

and the time-dependent portion of the product solution can be found. This leads to product solutions of the form

$$u_n(x,t) = a_n e^{-\lambda_n^2 kt} \left(\cos(\lambda_n x) + \frac{\gamma_0}{\lambda_n} \sin(\lambda_n x) \right). \qquad (4.24)$$

For the case in which Eq. (4.21) is satisfied for some $N \in \mathbb{N}$ the product solution has the same form, but in this case there is only a single product solution corresponding to $\lambda = (2N - 1)\pi/(2L)$. Initial boundary value problems falling into this category arise rarely in applications. For the case in which $\lambda \neq (2n - 1)\pi/(2L)$ for all $n \in \mathbb{N}$, solving the corresponding initial boundary value problem requires writing the initial condition as a series of the eigenfunctions defined in Eq. (4.23). The boundary value problem given by the ordinary differential equation of Eq. (1.52) and the boundary conditions in Eqs. (4.14) and (4.15) is a special case of a more general class of boundary value problems which will be explored in Chap. 7.

Example 4.3. Solve the following initial boundary value problem with boundary conditions of the third kind.

$$u_t = u_{xx} \text{ for } 0 < x < 1 \text{ and } t > 0,$$

$$u_x(0,t) - 9\pi u(0,t) = 0 \text{ for } t > 0,$$

$$u_x(1,t) + \frac{9\pi}{4}u(1,t) = 0 \text{ for } t > 0,$$

$$u(x,0) = \frac{1}{2}\cos\left(\frac{9\pi x}{2}\right) + \sin\left(\frac{9\pi x}{2}\right) \text{ for } 0 < x < 1.$$

Solution. In this case $\gamma_0 = 9\pi$ and $\gamma_L = \frac{9\pi}{4}$ and therefore

$$\sqrt{\gamma_0\gamma_L} = \frac{9\pi}{2} \quad \text{and} \quad \cos(\sqrt{\gamma_0\gamma_L}) = 0.$$

The solution to the heat equation satisfying the boundary conditions is, according to Eq. (4.24),

$$u(x,t) = Ae^{-81\pi^2 t/4}\left[\cos\left(\frac{9\pi x}{2}\right) + 2\sin\left(\frac{9\pi x}{2}\right)\right].$$

Using the initial condition, by inspection it can be determined that the constant $A = 1/2$. Figure 4.4 shows a contour plot of the solution. If the initial condition is any function other than a linear combination of $\cos(9\pi x/2)$ and $\sin(9\pi x/2)$, it would be more difficult to determine the Fourier coefficients in the solution.

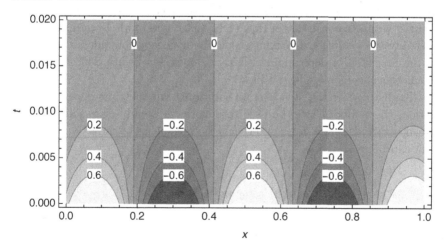

Fig. 4.4 A contour plot of the solution to the initial boundary value problem of Example 4.3. Warmer areas are in lighter shades while cooler areas are darker.

4.2   Nonhomogeneous Boundary Value Problems

In Sec. 4.1 the steps for solving many types of initial boundary value problems with homogeneous boundary conditions were outlined. In the present section, some of the common types of nonhomogeneous boundary conditions will be introduced and techniques for solving these types of problems will be presented. Since the techniques used for the three types of boundary conditions are similar, the solution methods will be explored through examples in some cases. Later in Chap. 10, methods for solving nonhomogeneous versions of partial differential equations along with nonhomogeneous boundary conditions will be discussed in more generality.

The first case to be considered is that of boundary conditions corresponding to fixed, but nonzero, temperatures at the ends of the spatial domain. A general case of this Dirichlet-type boundary condition is presented below.

$$u_t = ku_{xx} \text{ for } 0 < x < L \text{ and } t > 0, \qquad (4.25)$$

$$u(0,t) = u_0 \text{ and } u(L,t) = u_L \text{ for } t > 0, \qquad (4.26)$$

$$u(x,0) = f(x) \text{ for } 0 < x < L. \qquad (4.27)$$

During the earlier study of the solution to the heat equation with homogeneous Dirichlet boundary conditions (given in Eq. (4.3)) it was observed that as $t \to \infty$, the solution $u(x,t) \to 0$ for $0 \leq x \leq L$. The trivial solution $u(x,t) = 0$ could be thought of as the steady-state solution for the Fourier problem consisting of Eqs. (4.1), (4.2), and (4.3). Determination of a steady-state solution satisfying the nonhomogeneous boundary conditions is the first step in determining the time-dependent solution in the case of nonhomogeneous boundary values.

To distinguish the steady-state solution from the time-dependent solution, the symbol U will be used to denote the steady-state solution. At steady-state, $u(x,t) = U(x)$ and the nonhomogeneous boundary value problem can be re-stated as

$$U''(x) = 0,$$

$$U(0) = u_0 \text{ and } U(L) = u_L.$$

For any choice of constants c_1 and c_2, the function $U(x) = c_1 x + c_2$ solves the differential equation and the reader may confirm that

$$U(x) = \frac{u_L - u_0}{L}x + u_0 \qquad (4.28)$$

satisfies not only the differential equation, but also the boundary conditions.

The complete solution $u(x,t)$ to the initial boundary value problem can be written as $u(x,t) = v(x,t) + U(x)$ where the "transient" portion of the solution $v(x,t)$ is yet to be determined. Since $v_t = u_t$ and $v_{xx} = u_{xx}$, then $v(x,t)$ must solve the heat equation, Eq. (4.25). At $x = 0$,

$$u(0,t) = u_0 = v(0,t) + U(0) \implies v(0,t) = 0,$$

while at $x = L$,

$$u(L,t) = u_L = v(L,t) + U(L) \implies v(L,t) = 0.$$

Therefore $v(x,t)$ satisfies the homogeneous Dirichlet boundary conditions of Eq. (4.3). Lastly when $t = 0$,

$$u(x,0) = f(x) = v(x,0) + U(x) \implies v(x,0) = f(x) - U(x).$$

This initial condition for the transient solution $v(x,t)$ is merely the steady-state solution removed from the original initial condition for solution $u(x,t)$. The function $v(x,t)$ is a solution to a Fourier problem of the type studied in Sec. 4.1.1, therefore

$$v(x,t) = \sum_{n=1}^{\infty} a_n e^{-kn^2\pi^2 t/L^2} \sin\left(\frac{n\pi x}{L}\right),$$

where

$$a_n = \frac{2}{L} \int_0^L (f(x) - U(x)) \sin\left(\frac{n\pi x}{L}\right) dx,$$

for $n \in \mathbb{N}$. Finally the solution to the nonhomogeneous boundary value problem may be written as

$$u(x,t) = u_0 + \frac{u_L - u_0}{L}x + \sum_{n=1}^{\infty} a_n e^{-kn^2\pi^2 t/L^2} \sin\left(\frac{n\pi x}{L}\right).$$

Example 4.4. A cylindrical rod of ice $L = 1/2$ meters in length has a thermal conductivity of $k = 2$ Watts per meter per Kelvin.[2] The rod is insulated along its length and is initially at a temperature of 100K (about $-173°C$). For $t > 0$ the ends of the rod are kept at 60K and 120K. Find the temperature of the ice for $0 \le x \le 1/2$ and $t > 0$.

Solution. The temperature $u(x,t)$ of the rod may be summarized mathematically in the following initial boundary value problem:

$$u_t = 2u_{xx} \text{ for } 0 < x < 1/2 \text{ and } t > 0,$$
$$u(0,t) = 60 \text{ and } u(1/2,t) = 120 \text{ for } t > 0,$$
$$u(x,0) = 100 \text{ for } 0 < x < 1/2.$$

---

[2]William Thomson, 1st Baron Kelvin, British mathematical physicist (1824–1907).

The steady-state solution is

$$U(x) = \frac{120 - 60}{1/2}x + 60 = 120x + 60.$$

The transient portion of the solution $v(x,t)$ solves the following Fourier problem.

$$v_t = 2v_{xx} \text{ for } 0 < x < 1/2 \text{ and } t > 0,$$
$$v(0,t) = 0 \text{ and } v(1/2,t) = 0 \text{ for } t > 0,$$
$$v(x,0) = 40 - 120x \text{ for } 0 < x < 1/2.$$

The coefficients of the Fourier series solution for $n = 1, 2, \ldots$ are

$$a_n = \frac{2}{1/2}\int_0^{1/2}(40 - 120x)\sin\left(\frac{n\pi x}{1/2}\right)\,dx = \frac{40(2 + (-1)^n)}{n\pi}.$$

Consequently the solution to the nonhomogeneous boundary value problem is

$$u(x,t) = 60 + 120x + \frac{40}{\pi}\sum_{n=1}^{\infty}\frac{(2 + (-1)^n)e^{-8n^2\pi^2 t}}{n}\sin(2n\pi x).$$

Figure 4.5 shows a contour plot of the solution.

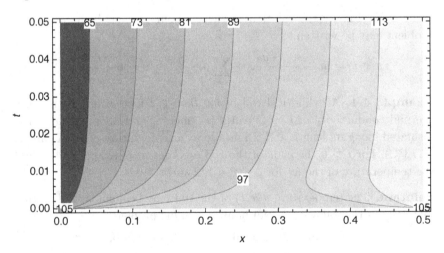

Fig. 4.5   A contour plot of the Fourier series solution to the nonhomogeneous boundary value problem of Example 4.4. Warmer areas are in lighter shades while cooler areas are darker.

For the case in which the boundary conditions are time dependent, the approach is similar, but slightly more complicated. Consider the following initial boundary value problem with time varying boundary conditions:

$$u_t = k u_{xx} \text{ for } 0 < x < L \text{ and } t > 0,$$

$$u(0, t) = a(t) \text{ and } u(L, t) = b(t) \text{ for } t > 0,$$

$$u(x, 0) = f(x) \text{ for } 0 < x < L.$$

The function

$$r(x, t) = \frac{b(t) - a(t)}{L} x + a(t) \tag{4.29}$$

which satisfies the boundary conditions will be called a **reference function**. If the solution to the nonhomogeneous boundary value problem is written as $u(x, t) = v(x, t) + r(x, t)$, the unknown function $v(x, t)$ has the following properties:

$$u(x, 0) = v(x, 0) + r(x, 0) = f(x),$$

which implies

$$v(x, 0) = f(x) - \frac{b(0) - a(0)}{L} x - a(0).$$

When $x = 0$,

$$u(0, t) = v(0, t) + r(0, t) = a(t) \implies v(0, t) = 0$$

and likewise when $x = L$, $v(L, t) = 0$. Thus the unknown function $v(x, t)$ satisfies homogeneous boundary conditions and a relatively familiar initial condition. Attention now turns to the partial differential equation that $v(x, t)$ satisfies. Since

$$u_t = v_t + r_t = v_t + \frac{b'(t) - a'(t)}{L} x + a'(t),$$

$$u_{xx} = v_{xx} + r_{xx} = v_{xx},$$

then substituting into the heat equation,

$$v_t = k v_{xx} - a'(t) + \frac{a'(t) - b'(t)}{L} x$$

which yields a nonhomogeneous version of the heat equation. To summarize, the unknown function $v(x, t)$ must satisfy the following initial boundary value problem with homogeneous, Dirichlet-type boundary conditions.

$$v_t = k v_{xx} - a'(t) + \frac{a'(t) - b'(t)}{L} x \text{ for } 0 < x < L \text{ and } t > 0,$$

$$v(0, t) = 0 \text{ and } v(L, t) = 0 \text{ for } t > 0,$$

$$v(x, 0) = f(x) - \frac{b(0) - a(0)}{L} x - a(0) \text{ for } 0 < x < L.$$

Creating homogeneous boundary conditions for $v(x,t)$ is a step in the right direction. Further leverage may be gained by engineering a homogeneous initial condition. Suppose functions $v_1(x,t)$ and $v_2(x,t)$ solve the following pair of initial boundary value problems respectively:

$$(v_1)_t = k(v_1)_{xx} \text{ for } 0 < x < L \text{ and } t > 0,$$
$$v_1(0,t) = v_1(L,t) = 0 \text{ for } t > 0, \qquad\qquad (4.30)$$
$$v_1(x,0) = f(x) - \frac{b(0) - a(0)}{L}x - a(0) \text{ for } 0 < x < L,$$

and

$$(v_2)_t = k(v_2)_{xx} - a'(t) + \frac{a'(t) - b'(t)}{L}x \text{ for } 0 < x < L \text{ and } t > 0,$$
$$v_2(0,t) = v_2(L,t) = 0 \text{ for } t > 0, \qquad\qquad (4.31)$$
$$v_2(x,0) = 0 \text{ for } 0 < x < L.$$

The sum $v_1(x,t) + v_2(x,t)$ solves the earlier nonhomogeneous initial boundary value problem. Thus $v(x,t) = v_1(x,t) + v_2(x,t)$. The task of finding $v(x,t)$ has been split into finding the solutions to the pair of initial boundary value problems above. The initial boundary value problem for $v_1(x,t)$ is an example of a Fourier problem and thus the techniques of Sec. 4.1.1 will produce its solution. The boundary value problem for $v_2(x,t)$ is nonhomogeneous in the partial differential equation, but does have homogeneous boundary and initial conditions. The Principle of Superposition of solutions which underlies the Fourier series approach to the solution no longer holds in the nonhomogeneous case. However, the reader may recall solving nonhomogeneous ordinary differential equations using the method known as variation of parameters. If the partial differential equation satisfied by $v_2(x,t)$ were homogeneous, then the solution would resemble a Fourier series of the form,

$$\sum_{n=1}^{\infty} a_n e^{-kn^2\pi^2 t/L^2} \sin\left(\frac{n\pi x}{L}\right).$$

For the nonhomogeneous case, rather than assuming the Fourier coefficients a_n are constants, assume they are functions of t and write the solution as

$$\sum_{n=1}^{\infty} a_n(t) e^{-kn^2\pi^2 t/L^2} \sin\left(\frac{n\pi x}{L}\right).$$

Differentiating this expression and substituting it into the nonhomogeneous heat equation produce

$$-a'(t) + \frac{a'(t) - b'(t)}{L}x = \sum_{n=1}^{\infty} a_n'(t) e^{-kn^2\pi^2 t/L^2} \sin\left(\frac{n\pi x}{L}\right).$$

Using the orthogonality of the functions $\{\sin(n\pi x/L)\}_{n=1}^{\infty}$ on $[0, L]$ yields

$$a_n'(t)e^{-kn^2\pi^2 t/L^2} = \frac{2}{L}\int_0^L \left(-a'(t) + \frac{a'(t) - b'(t)}{L}x\right) \sin\left(\frac{n\pi x}{L}\right) dx$$

or integrating and solving for $a_n'(t)$,

$$a_n'(t) = \frac{2}{n\pi}e^{kn^2\pi^2 t/L^2}\left((-1)^n b'(t) - a'(t)\right).$$

Since $v_2(x, 0) = 0$ then $a_n(0) = 0$ for $n \in \mathbb{N}$. Thus the function $a_n(t)$ may be written as

$$a_n(t) = \frac{2}{n\pi}\int_0^t e^{kn^2\pi^2 s/L^2}\left((-1)^n b'(s) - a'(s)\right) ds. \tag{4.32}$$

This implies the solution to the nonhomogeneous portion of the partial differential equation may be expressed as

$$v_2(x, t) = \frac{2}{\pi}\sum_{n=1}^{\infty}\frac{1}{n}\sin\left(\frac{n\pi x}{L}\right)\int_0^t e^{-kn^2\pi^2(t-s)/L^2}\left((-1)^n b'(s) - a'(s)\right) ds.$$

By carefully retracing the dissection of the solution to the original non-homogeneous boundary value problem, the solution can be re-assembled as

$$u(x, t) = r(x, t) + v_1(x, t) + v_2(x, t)$$

$$= a(t) + \frac{b(t) - a(t)}{L}x + \sum_{n=1}^{\infty}b_n e^{-kn^2\pi^2 t/L^2}\sin\left(\frac{n\pi x}{L}\right)$$

$$+ \frac{2}{\pi}\sum_{n=1}^{\infty}\frac{1}{n}\sin\left(\frac{n\pi x}{L}\right)\int_0^t e^{-kn^2\pi^2(t-s)/L^2}\left((-1)^n b'(s) - a'(s)\right) ds,$$

where for $n \in \mathbb{N}$,

$$b_n = \frac{2}{L}\int_0^L \left(f(x) - \frac{b(0) - a(0)}{L}x - a(0)\right)\sin\left(\frac{n\pi x}{L}\right) dx$$

$$= \frac{2((-1)^n b(0) - a(0))}{n\pi} + \frac{2}{L}\int_0^L f(x)\sin\left(\frac{n\pi x}{L}\right) dx.$$

This procedure can be generalized and will be re-visited during a discussion of nonhomogeneous initial boundary value problems in Chap. 10.

Example 4.5. Find a solution to the following initial boundary value problem:

$$u_t = u_{xx} \text{ for } 0 < x < 1 \text{ and } t > 0,$$

$$u(0, t) = 1 \text{ and } u(1, t) = t \text{ for } t > 0,$$

$$u(x, 0) = 1 + \sin(\pi x) - x \text{ for } 0 \le x < 1.$$

Solution. Using the explanation given above as a guide, three components of the solution are needed: (1) a reference function, (2) a solution to the homogeneous heat equation (for $v_1(x,t)$ in Eq. (4.30)), and (3) a solution to a nonhomogeneous heat equation (for $v_2(x,t)$ in Eq. (4.31)). A suitable reference function is

$$r(x,t) = 1 + (t-1)x.$$

Next, since

$$1 + \sin(\pi x) - x - r(x,0) = \sin(\pi x)$$

finding $v_1(x,t)$ is trivial,

$$v_1(x,t) = e^{-\pi^2 t} \sin(\pi x).$$

Lastly using the formula for $a_n(t)$ given in Eq. (4.32) gives

$$a_n(t) = \frac{2(-1)^n}{n\pi} \int_0^t e^{n^2 \pi^2 s} \, ds = \frac{2(-1)^n}{n^3 \pi^3} (e^{n^2 \pi^2 t} - 1)$$

which implies

$$v_2(x,t) = \frac{2}{\pi^3} \sum_{n=1}^{\infty} \frac{(-1)^n}{n^3} (1 - e^{-n^2 \pi^2 t}) \sin(n\pi x).$$

Finally,

$$u(x,t) = 1 + (t-1)x + e^{-\pi^2 t} \sin(\pi x)$$
$$+ \frac{2}{\pi^3} \sum_{n=1}^{\infty} \frac{(-1)^n}{n^3} (1 - e^{-n^2 \pi^2 t}) \sin(n\pi x).$$

Figure 4.6 shows a contour plot of the solution.

Example 4.6. Find a solution to the following initial boundary value problem:

$$u_t = u_{xx} \text{ for } 0 < x < 1 \text{ and } t > 0,$$
$$u_x(0,t) = -t \text{ and } u_x(1,t) = t^2 \text{ for } t > 0,$$
$$u(x,0) = \cos(\pi x) \text{ for } 0 < x < 1.$$

Solution. The approach taken in the last example will be used again to determine the solution to the present problem. As before, three components of the solution are needed: (1) a reference function, (2) a solution to the homogeneous heat equation, and (3) a solution to a nonhomogeneous heat equation. The reference function $r(x,t)$ should be chosen so that the

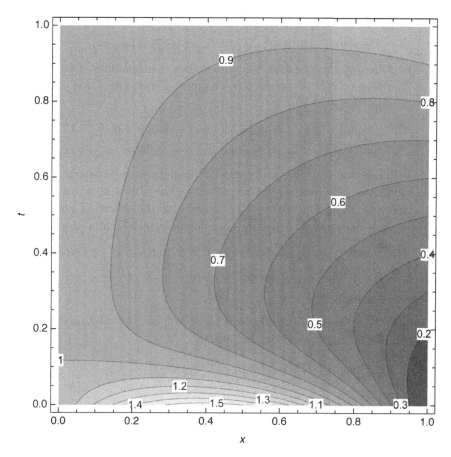

Fig. 4.6   A contour plot of the Fourier series solution to the nonhomogeneous boundary value problem of Example 4.5. Warmer areas are in lighter shades while cooler areas are darker.

function $v(x,t) = u(x,t) - r(x,t)$ satisfies homogeneous Neumann type boundary conditions at $x = 0$ and $x = 1$. A suitable reference function is

$$r_x(x,t) = -t + (t^2 + t)x \implies r(x,t) = \frac{(t^2 + t)x^2}{2} - x\,t.$$

Since $u(x,t) = v(x,t) + r(x,t)$, differentiating $u(x,t)$ and substituting into the heat equation yield the following differential equation for $v(x,t)$,

$$u_t = v_t + r_t = v_{xx} + r_{xx} = u_{xx},$$
$$v_t = v_{xx} + t^2 + t + x - \frac{1}{2}(2t+1)x^2.$$

The initial condition for $v(x,t)$ is

$$v(x,0) + r(x,0) = \cos(\pi x) \implies v(x,0) = \cos(\pi x).$$

The function $v(x,t)$ can be separated into two functions $v_1(x,t)$ and $v_2(x,t)$, where $v_1(x,t)$ solves the following initial boundary value problem,

$$(v_1)_t = (v_1)_{xx} \text{ for } 0 < x < 1 \text{ and } t > 0,$$

$$(v_1)_x(0,t) = 0 \text{ and } (v_1)_x(1,t) = 0 \text{ for } t > 0,$$

$$v_1(x,0) = \cos(\pi x) \text{ for } 0 < x < 1.$$

This initial boundary value problem with homogeneous Neumann boundary conditions was studied in Sec. 4.1.2. The solution is $v_1(x,t) = e^{-\pi^2 t}\cos(\pi x)$. The function $v_2(x,t)$ must solve the following nonhomogeneous heat equation with homogeneous boundary and initial conditions.

$$(v_2)_t = (v_2)_{xx} + t^2 + t + x - \frac{1}{2}(1+2t)x^2 \text{ for } 0 < x < 1 \text{ and } t > 0,$$

$$(v_2)_x(0,t) = 0 \text{ and } (v_2)_x(1,t) = 0 \text{ for } t > 0,$$

$$v_2(x,0) = 0 \text{ for } 0 < x < 1.$$

If the partial differential equation was homogeneous, the solution could be expressed as a Fourier series in the form of

$$\frac{a_0}{2} + \sum_{n=1}^{\infty} a_n e^{-n^2\pi^2 t} \cos(n\pi x)$$

where the coefficients a_n for $n = 0, 1, 2, \ldots$ are constants determined by the initial condition. To accommodate the non-homogeneity, the Fourier coefficients are assumed to be functions of t. Thus a solution of the form

$$v_2(x,t) = \frac{a_0(t)}{2} + \sum_{n=1}^{\infty} a_n(t) e^{-n^2\pi^2 t} \cos(n\pi x)$$

is sought. Since $v_2(x,0) = 0$ then $a_n(0) = 0$ for $n = 0, 1, 2, \ldots$. Differentiating this solution and substituting into the nonhomogeneous partial differential equation yield,

$$(v_2)_t - (v_2)_{xx} = x + t + t^2 - \frac{1}{2}(1+2t)x^2$$

$$= \frac{1}{2}a_0'(t) + \sum_{n=1}^{\infty} a_n'(t) e^{-n^2\pi^2 t} \cos(n\pi x).$$

Integrating both sides of this equation with respect to x over the interval $[0,1]$ produces the following,

$$a_0'(t) = \frac{2}{3} + \frac{4t}{3} + 2t^2$$

which implies

$$a_0(t) = \frac{2}{3}(t + t^2 + t^3),$$

using the previously established initial condition for $a_0(t)$. Multiplying both sides of the partial differential equation by $\cos(n\pi x)$ and integrating with respect to x over the interval $[0, 1]$ produce the following:

$$a'_n(t) = \frac{-(1 + 2(-1)^n t)}{n^2 \pi^2}$$

and consequently

$$a_n(t) = \frac{-(t + (-1)^n t^2)}{n^2 \pi^2},$$

using the initial condition for $a_n(t)$. Therefore $v_2(x, t)$ can be expressed as

$$v_2(x, t) = \frac{1}{3}(t + t^2 + t^3) - \frac{1}{\pi^2} \sum_{n=1}^{\infty} \frac{t + (-1)^n t^2}{n^2} e^{-n^2 \pi^2 t} \cos(n\pi x),$$

and finally the solution to the original nonhomogeneous initial boundary value problem is

$$u(x, t) = r(x, t) + v_1(x, t) + v_2(x, t)$$
$$= \frac{(t^2 + t)x^2}{2} - x t + e^{-\pi^2 t} \cos(\pi x)$$
$$+ \frac{1}{3}(t + t^2 + t^3) - \frac{1}{\pi^2} \sum_{n=1}^{\infty} \frac{(t + (-1)^n t^2)}{n^2} e^{-n^2 \pi^2 t} \cos(n\pi x).$$

Figure 4.7 shows a contour plot of the solution.

4.3 A Maximum Principle and Uniqueness of Solutions

In applications of the heat equation to the physical world, the exact values of the boundary conditions and the initial condition are almost never known. They may be estimated very closely, but there will always be some "fuzziness" in their exact values. In this section the relationship between errors in the estimation of the boundary and initial conditions and the spread of solutions to the initial boundary value problem of the heat equation will be studied. Fortunately for mathematicians and engineers small changes in the initial and boundary conditions translate into small changes in the solution under a mild set of assumptions. Mathematically speaking this implies the solution to the heat equation depends continuously on the boundary and initial conditions. A further (also comforting) consequence is

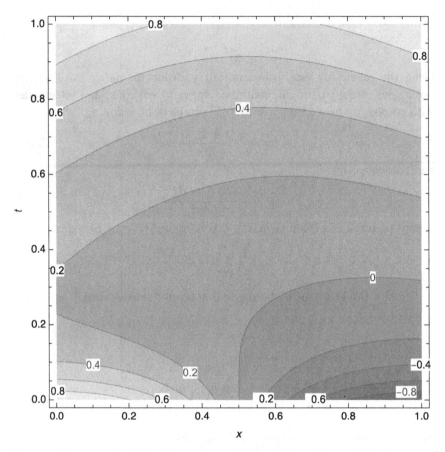

Fig. 4.7   A contour plot of the Fourier series solution to the nonhomogeneous boundary value problem of Example 4.6. Warmer areas are in lighter shades while cooler areas are darker.

that there is only one solution (under this same set of assumptions) to the heat equation for a given set of boundary and initial conditions. There are two main results proved in this section: (1) the Maximum Principle (and its companion Minimum Principle) and (2) the continuous dependence of the solution on the boundary and initial conditions.

Before stating and proving the Maximum Principle the reader may wish to conduct a mental experiment by thinking about cooking some food (perhaps a Thanksgiving turkey) in a closed oven and asking the question, "where is the greatest temperature in the food located?" In terms of the one-dimensional rod with no internal sources of heat, the physical principle that heat energy flows from a warmer region to a cooler region implies the temperature cannot be the highest at a point (x_0, t_0) with $0 < x_0 < L$ and $t_0 > 0$ unless the temperature is the same at all points of the rod. This is mathematically stated as the Maximum Principle.

Theorem 4.1 (Maximum Principle). *Consider the following sets in the xt-plane:*

$$R = \{(x, t) \mid 0 < x < L, \, 0 < t < T\},$$
$$S = \{(x, 0) \mid 0 \le x \le L\} \cup \{(0, t) \mid 0 \le t \le T\} \cup \{(L, t) \mid 0 \le t \le T\},$$
$$\overline{R} = R \cup S \cup \{(x, t) \mid 0 < x < L\}$$

and suppose $u(x, t) \in \mathcal{C}^2(R) \cap \mathcal{C}(\overline{R})$ and solves the heat equation $u_t = k \, u_{xx}$ for $0 < x < L$ and $t > 0$, then

$$\max_{(x,t) \in \overline{R}} u(x, t) = \max_{(x,t) \in S} u(x, t).$$

Sets R and S are illustrated in Fig. 4.8. Set $\overline{R}$ merely includes the top edge of rectangular region $R \cup S$.

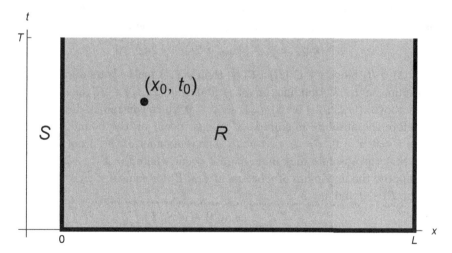

Fig. 4.8   The sets R and S used in the proof of the Maximum Principle for the heat equation.

Before proving the Maximum Principle several remarks are in order. First, the conclusion of the Maximum Principle implies the greatest temperature is found at the boundaries ($x = 0$ or $x = L$) or is initially present ($t = 0$) and no interior temperature in the spatial domain can exceed this greatest temperature (though an interior temperature may equal it). Considering there are no heat sources, this result has an intuitive appeal as mentioned earlier. Second, no particular style of boundary condition is mentioned, so any type may be present and the result still holds. Lastly, there is a stronger version of the Maximum Principle which states the value of $u(x,t)$ at an interior point must be less than the maximum on the boundary unless u is constant with respect to $x \in [0, L]$.

The proof the Maximum Principle depends on several results well-known in elementary calculus (see, for instance [Smith and Minton (2012)]). Since $u \in \mathcal{C}(\overline{R})$ then by the Extreme Value Theorem $u(x,t)$ attains an absolute maximum on $\overline{R}$. Since $u \in \mathcal{C}^2(R)$ if the maximum of u occurs at a point $(x_0, t_0) \in R$, the point (x_0, t_0) must be a critical point of $u(x,t)$. Recall that this implies $u_t(x_0, t_0) = 0$. The assumption that a local maximum is found at this critical point implies $u_{xx}(x_0, t_0) \leq 0$. However, if $u_{xx}(x_0, t_0) < 0$ then $u_t(x_0, t_0) - ku_{xx}(x_0, t_0) > 0$ which contradicts the assumption that the heat equation is satisfied at (x_0, t_0). This observation motivates the following proof.

Proof. Define $v(x,t) = u(x,t) + \epsilon x^2$, where $\epsilon > 0$. Then $v(x,t) \geq u(x,t)$ for all $(x,t) \in \overline{R}$ and $v_t = u_t$ and $v_{xx} = u_{xx} + 2\epsilon$. Therefore

$$v_t - kv_{xx} = u_t - k(u_{xx} + 2\epsilon) = -2k\epsilon < 0 \qquad (4.33)$$

for $(x,t) \in R$. Since $v \in \mathcal{C}^2(R) \cap \mathcal{C}(\overline{R})$ then $v(x,t)$ must attain an absolute maximum on $\overline{R}$. If that maximum is found at $(x_0, t_0) \in R$ then as reasoned above $v_t(x_0, t_0) = 0$ and $v_{xx}(x_0, t_0) \leq 0$ which contradicts Eq. (4.33). Therefore the absolute maximum of v must occur on the boundary of R where $t = 0$, $x = 0$, $x = L$, or $t = T$. The remainder of the proof aims to show that the absolute maximum cannot occur where $t = T$.

Suppose the maximum of v occurs at (x_0, T) for some $0 < x_0 < L$, then $v_{xx}(x_0, T) \leq 0$ and

$$v_t(x_0, T) = \lim_{t \to T^-} \frac{v(x_0, t) - v(x_0, T)}{t - T} \geq 0.$$

Combining the two derivatives, $v_t(x_0, T) - kv_{xx}(x_0, T) \geq 0$ which again contradicts Eq. (4.33). Thus it has been established that

$$\max_{(x,t) \in \overline{R}} v(x,t) = \max_{(x,t) \in S} v(x,t).$$

Using the definition of v then

$$\max_{(x,t) \in S} v(x,t) \leq \max_{(x,t) \in S} u(x,t) + \max_{(x,t) \in S} \epsilon x^2 = \max_{(x,t) \in S} u(x,t) + \epsilon L^2.$$

Likewise by the definition of v,

$$\max_{(x,t)\in\overline{R}} u(x,t) \leq \max_{(x,t)\in\overline{R}} v(x,t).$$

Consequently

$$\max_{(x,t)\in\overline{R}} u(x,t) \leq \max_{(x,t)\in S} v(x,t) \leq \max_{(x,t)\in S} u(x,t) + \epsilon L^2.$$

Taking the limit as $\epsilon \to 0^+$ implies that

$$\max_{(x,t)\in\overline{R}} u(x,t) \leq \max_{(x,t)\in S} u(x,t).$$

Since the last inequality cannot be strict (due to the set containment $S \subset \overline{R}$), then $\max_{(x,t)\in\overline{R}} u(x,t) = \max_{(x,t)\in S} u(x,t)$. $\qquad\square$

The Maximum Principle has a corollary, stated as the Minimum Principle below, which can be derived directly from the Maximum Principle or proved using techniques found in the proof above. The reader will be asked to consider the proof of the Minimum Principle in Exercise 20.

Corollary 4.1 (Minimum Principle). *Let R, S, and $\overline{R}$ be defined as in Theorem 4.1 and suppose $u(x,t) \in C^2(R) \cap C(\overline{R})$ solves the heat equation $u_t = k\,u_{xx}$ for $0 < x < L$ and $t > 0$, then*

$$\min_{(x,t)\in\overline{R}} u(x,t) = \min_{(x,t)\in S} u(x,t).$$

The next example illustrates a simple application of the Maximum and Minimum Principles.

Example 4.7. Find a bound for the solution to the following initial boundary value problem:

$$u_t = ku_{xx} \text{ for } 0 < x < 10 \text{ and } t > 0,$$
$$u(0,t) = 0 \text{ and } u(10,t) = \tan^{-1} t \text{ for } t > 0,$$
$$u(x,0) = \sin\left(\frac{\pi x}{5}\right) \text{ for } 0 < x < 10.$$

Solution. For any $T > 0$, $\tan^{-1} T < \pi/2$. On the interval $[0,10]$ function $\sin \pi x/5 \geq -1$. The Maximum and Minimum Principles imply that

$$-1 \leq u(x,t) \leq \tan^{-1} T < \frac{\pi}{2}$$

for $0 \leq x \leq 10$ and $0 \leq t \leq T$. Since $T > 0$ is arbitrary and the bound on $\tan^{-1} T$ holds for all $T > 0$, then the conclusion can be extended to declare that $-1 \leq u(x,t) < \pi/2$ for $0 \leq x \leq 10$ and $t \geq 0$.

Now the dependence of the solution on the boundary and initial conditions can be explored. The proof of the next result is again made possible by an application of the Maximum and Minimum Principles. The next result will be used later to establish the uniqueness of the solutions to the initial boundary value problem of the heat equation.

Theorem 4.2. *Consider the two initial boundary value problems:*

$$u_t = ku_{xx}, \qquad v_t = kv_{xx},$$
$$u(0,t) = a_1(t), \quad v(0,t) = a_2(t),$$
$$u(L,t) = b_1(t), \quad v(L,t) = b_2(t),$$
$$u(x,0) = f_1(x), \quad v(x,0) = f_2(x)$$

defined for $0 < x < L$ and $t > 0$. Let $T > 0$ and R, S, and $\overline{R}$ be defined as in Theorem 4.1. Suppose there exists $\epsilon \geq 0$ such that

$$|f_1(x) - f_2(x)| \leq \epsilon \text{ for } 0 < x < L,$$
$$|a_1(t) - a_2(t)| \leq \epsilon \text{ for } 0 < t \leq T, \text{ and}$$
$$|b_1(t) - b_2(t)| \leq \epsilon \text{ for } 0 < t \leq T.$$

If $u(x,t)$ and $v(x,t)$ are $C^2(R) \cap C(\overline{R})$ solutions respectively to the two initial boundary value problems, then for all $0 < x < L$ and $0 < t \leq T$,

$$|u(x,t) - v(x,t)| \leq \epsilon.$$

Proof. This result is a straightforward consequence of the Maximum and Minimum Principles. Let $U(x,t) = u(x,t) - v(x,t)$, then

$$U_t = u_t - v_t = ku_{xx} - kv_{xx} = kU_{xx},$$

in other words $U(x,t)$ solves the heat equation. For $x = 0$, $|U(0,t)| = |a_1(t) - a_2(t)|$ and by assumption $-\epsilon \leq U(0,t) \leq \epsilon$. Likewise $-\epsilon \leq U(L,t) \leq \epsilon$ for $0 \leq t \leq T$. In a similar fashion it is shown that for $t = 0$, $-\epsilon \leq U(x,0) \leq \epsilon$ for $0 \leq x \leq L$. Applying the Maximum and Minimum Principles yields the conclusion that $-\epsilon \leq U(x,t) \leq \epsilon$ for $(x,t) \in \overline{R}$ or equivalently that $|u(x,t) - v(x,t)| \leq \epsilon$ for $0 \leq x \leq L$ and $0 \leq t \leq T$. $\square$

This section concludes with a corollary to Theorem 4.2 establishing the uniqueness of solutions to the heat equation.

Corollary 4.2 (Uniqueness). *Consider the initial boundary value problem,*

$$u_t = ku_{xx} + g(x,t) \text{ for } 0 < x < L \text{ and } t > 0,$$
$$u(0,t) = a(t) \text{ and } u(L,t) = b(t) \text{ for } t > 0,$$
$$u(x,0) = f(x) \text{ for } 0 < x < L.$$

Suppose that u and v are both solutions to the initial boundary value problem that are continuous for $0 \leq x \leq L$ and $t \geq 0$ and are C^2 for $0 < x < L$ and $t > 0$, then $u(x,t) = v(x,t)$ for all $(x,t) \in [0, L] \times [0, \infty)$.

Proof. Suppose there are two solutions $u(x,t)$ and $v(x,t)$. Applying Theorem 4.2 with $\epsilon = 0$ yields the conclusion that $u(x,t) = v(x,t)$ for all $0 \leq x \leq L$ and $0 \leq t \leq T$. Since $T > 0$ is arbitrary, the equality holds for all $t > 0$ and the solution is unique. □

4.4   The Heat Equation on Unbounded Intervals

Solutions to the heat equation on two types of unbounded intervals will be explored in this section. The approach to finding a solution and the form of the solution differ from those in the section on solutions for bounded intervals (Sec. 4.1), and thus this section may be safely skipped without hindering the reader's understanding of the remainder of this chapter.

The two canonical unbounded intervals of interest are the real line, $-\infty < x < \infty$ and the semi-infinite interval, $0 \leq x < \infty$. Define the function U as

$$U(x,t) = \frac{1}{\sqrt{4\pi k t}} e^{-x^2/(4kt)}, \tag{4.34}$$

which is defined on any region in the xt-plane away from the line $t = 0$. Differentiating this function for $t > 0$ and $-\infty < x < \infty$ yields

$$U_t(x,t) = \frac{1}{4\sqrt{k\pi t}} e^{-x^2/(4kt)} \left(\frac{x^2}{2kt^2} - \frac{1}{t} \right), \tag{4.35}$$

$$U_{xx}(x,t) = \frac{1}{4k\sqrt{k\pi t}} e^{-x^2/(4kt)} \left(\frac{x^2}{2kt^2} - \frac{1}{t} \right) \tag{4.36}$$

from which it is clear that $U_t = k\,U_{xx}$ and hence the function in Eq. (4.34) is a solution to the heat equation. However, this is not a product solution. The function $U(x,t)$ defined in Eq. (4.34) is known as the **fundamental solution** to the heat equation. Two issues are of interest about this fundamental solution:

- the fundamental solution makes no reference to any type of boundary condition, and
- the fundamental solution is defined only for $t > 0$.

While evaluation of the fundamental solution at $t = 0$ is meaningless, it is possible to determine the limit as t approaches 0 from the right. Assuming

$x \neq 0$ then by l'Hôpital's rule,

$$\lim_{t \to 0^+} U(x,t) = \lim_{t \to 0^+} \frac{1}{\sqrt{4\pi kt}} e^{-x^2/(4kt)} = 0. \tag{4.37}$$

When $x = 0$ then

$$\lim_{t \to 0^+} U(0,t) = \lim_{t \to 0^+} \frac{1}{\sqrt{4\pi kt}} = \infty.$$

While the fundamental solution "blows up" at $x = 0$ the exponential decay of the solution due to the presence of the $e^{-x^2/(4kt)}$ term may lead the reader to suspect the function $U(x,t)$ is integrable over the real line for $t > 0$. Indeed for fixed (but arbitrary) $t > 0$,

$$\int_{-\infty}^{\infty} U(x,t)\,dx = 2\int_0^{\infty} U(x,t)\,dx = 2\left(\int_0^1 U(x,t)\,dx + \int_1^{\infty} U(x,t)\,dx\right).$$

Since integration is taking place with respect to x and $t > 0$ the first integral on the right-hand side of the equation is finite. According to the Comparison Test [Smith and Minton (2012), pp. 491] since

$$0 < U(x,t) \leq \frac{1}{\sqrt{4\pi kt}} e^{-x/(4kt)}$$

for $x \geq 1$ and

$$\int_1^{\infty} \frac{1}{\sqrt{4\pi kt}} e^{-x/(4kt)}\,dx = 2\sqrt{\frac{kt}{\pi}} e^{-1/(4kt)} < \infty$$

for $t > 0$ then,

$$\int_{-\infty}^{\infty} U(x,t)\,dx$$

converges for every $t > 0$. A perhaps surprising result is that the improper integral converges to the same value for every $t > 0$, call this value S. The value of S can be found via a clever, indirect method.

$$S^2 = \left(\int_{-\infty}^{\infty} U(x,t)\,dx\right)^2 = \left(\int_{-\infty}^{\infty} U(x,t)\,dx\right)\left(\int_{-\infty}^{\infty} U(y,t)\,dy\right)$$

$$= \frac{1}{4k\pi t}\int_{-\infty}^{\infty} e^{-y^2/(4kt)}\left[\int_{-\infty}^{\infty} e^{-x^2/(4kt)}\,dx\right]\,dy$$

$$= \frac{1}{4k\pi t}\int_{-\infty}^{\infty}\int_{-\infty}^{\infty} e^{-(x^2+y^2)/(4kt)}\,dx\,dy.$$

Converting the last double integral to polar coordinates yields

$$S^2 = \frac{1}{4\pi kt}\int_0^{2\pi}\int_0^{\infty} re^{-r^2/(4kt)}\,dr\,d\theta = \frac{1}{2kt}\int_0^{\infty} re^{-r^2/(4kt)}\,dr = 1.$$

Consequently,

$$\int_{-\infty}^{\infty} \frac{1}{\sqrt{4\pi kt}} e^{-x^2/(4kt)} \, dx = 1 \qquad (4.38)$$

for any $t > 0$. Taking the limit of this integral as $t \to 0^+$ produces a result of 1 as well. Thus the reader familiar with generalized functions may recognize that

$$\lim_{t \to 0^+} \frac{1}{\sqrt{4\pi kt}} e^{-x^2/(4kt)} = \delta(x)$$

where $\delta(x)$ is the Dirac[3] delta function [Boyce and DiPrima (2012)]. Figure 4.9 shows a surface plot of the fundamental solution of Eq. (4.34). For fixed $t > 0$ the graphs are the familiar Gaussian "bell-shaped" curve well known in statistics. As $t \to 0^+$ the curve becomes more peaked near $x = 0$ while as t becomes larger the curve is flatter. Physically the fundamental solution can be thought of as representing a situation in which a unit amount of heat energy is initially at the origin of an infinite (in both directions) rod. As time starts to move forward, this concentration of heat instantly smooths out. Some heat is transferred to portions of the rod very far from the origin in an arbitrarily short length of time. This may not seem physically possible, but is nonetheless a property of this solution.

Define $r = 1/(4kt)$ and the function $\gamma_r(x) = \sqrt{r/\pi} e^{-rx^2}$. The fundamental solution is then $U(x,t) = \gamma_r(x)$. Clearly $\gamma_r(x) \geq 0$ for all $x \in \mathbb{R}$. It has just been proved that $\int_{-\infty}^{\infty} \gamma_r(x) \, dx = 1$ for all $r > 0$. These last two statements hold in the limit as r approaches positive infinity (or equivalently as $t \to 0^+$).

Lemma 4.1. *For all $\epsilon > 0$ and $c > 0$ there exists $r_0 > 0$ such that if $r > r_0$ then,*

$$\int_{-\infty}^{-c} \gamma_r(x) \, dx + \int_{c}^{\infty} \gamma_r(x) \, dx < \epsilon. \qquad (4.39)$$

Proof. Let $\epsilon > 0$ and $c > 0$ be given. Since $\gamma_r(x)$ is an even function, it will suffice to show that

$$\int_{c}^{\infty} \gamma_r(x) \, dx < \frac{\epsilon}{2}.$$

Making the change of variables $z = x/c$ produces

$$\int_{c}^{\infty} \gamma_r(x) \, dx = \int_{1}^{\infty} \gamma_{c^2 r}(z) \, dz < \int_{1}^{\infty} \sqrt{\frac{c^2 r}{\pi}} e^{-c^2 r z} \, dz = \frac{1}{c\sqrt{\pi r}} e^{-c^2 r}.$$

---

[3]Paul A.M. Dirac, English theoretical physicist (1902–1984).

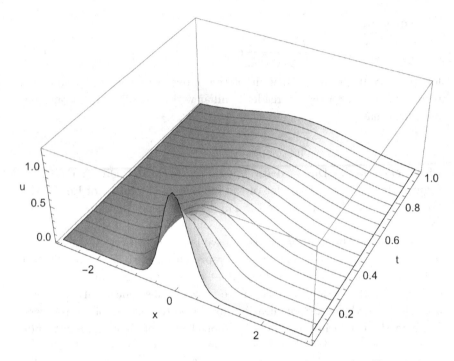

Fig. 4.9   The graph of the fundamental solution (formulated in Eq. (4.34)) to the heat equation. For any $t > 0$ the graph of $U(x,t)$ is a Gaussian exponential curve familiar from the statistics of normally distributed random variables. For $t = 0$ the fundamental solution is the Dirac delta function.

The reader will be asked to confirm in Exercise 16 that the last expression is a monotone decreasing function of r and that its limit as $r \to \infty$ is zero. Hence there exists $r_0 > 0$ such that for all $r > r_0$ the inequality in Eq. (4.39) holds.   □

The family of functions $\gamma_r(x)$ for $r > 0$ is called a **delta sequence**. The following lemma establishes another property of this delta sequence.

Lemma 4.2. *Suppose $g(x)$ is a bounded function and the one-sided limits*

$$\lim_{x \to 0^-} g(x) = g(0-) \quad and \quad \lim_{x \to 0^+} g(x) = g(0+)$$

both exist, then

$$\lim_{r \to \infty} \int_{-\infty}^{\infty} \gamma_r(x) g(x)\, dx = \frac{1}{2}(g(0-) + g(0+)). \tag{4.40}$$

Proof. Using the facts that $\int_{-\infty}^{\infty} \gamma_r(x)\,dx = 1$ and that $\gamma_r(x)$ is an even function,

$$
\int_{-\infty}^{\infty} \gamma_r(x)g(x)\,dx - \frac{1}{2}(g(0-) + g(0+))
$$

$$
= \int_{-\infty}^{\infty} \gamma_r(x)g(x)\,dx - \frac{1}{2}(g(0-) + g(0+)) \int_{-\infty}^{\infty} \gamma_r(x)\,dx
$$

$$
= \int_{-\infty}^{\infty} \gamma_r(x)g(x)\,dx - (g(0-) + g(0+)) \int_{0}^{\infty} \gamma_r(x)\,dx
$$

$$
= \int_{-\infty}^{0} \gamma_r(x)(g(x) - g(0-))\,dx + \int_{0}^{\infty} \gamma_r(x)(g(x) - g(0+))\,dx.
$$

Since g is bounded there exists a finite number $M > 0$ such that $|g(x)| \le M$ for all $x \in \mathbb{R}$. This implies $|g(x) - g(0\pm)| \le 2M$ for all $x \in \mathbb{R}$. Let $\epsilon > 0$, then there exists $c > 0$ such that $|g(x) - g(0-)| \le \epsilon/2$ for $-c < x < 0$ and $|g(x) - g(0+)| \le \epsilon/2$ for $0 < x < c$. With the given ϵ and c and again appealing to the symmetry of $\gamma_r(x)$, Lemma 4.1 suggests there exists $r_0 > 0$ such that

$$
\int_{-\infty}^{-c} \gamma_r(x)\,dx < \frac{\epsilon}{8M} \quad \text{and} \quad \int_{c}^{\infty} \gamma_r(x)\,dx < \frac{\epsilon}{8M} \quad \text{for all } r > r_0.
$$

Therefore if $r > r_0$,

$$
\left| \int_{-\infty}^{0} \gamma_r(x)(g(x) - g(0-))\,dx \right|
$$

$$
\le \int_{-\infty}^{-c} \gamma_r(x)|g(x) - g(0-)|\,dx + \int_{-c}^{0} \gamma_r(x)|g(x) - g(0-)|\,dx
$$

$$
\le 2M \int_{-\infty}^{-c} \gamma_r(x)\,dx + \frac{\epsilon}{2} \int_{-c}^{0} \gamma_r(x)\,dx < 2M \frac{\epsilon}{8M} + \frac{\epsilon}{2} \int_{-\infty}^{0} \gamma_r(x)\,dx = \frac{\epsilon}{2}.
$$

Similarly it can be shown that

$$
\left| \int_{0}^{\infty} \gamma_r(x)(g(x) - g(0+))\,dx \right| < \frac{\epsilon}{2}.
$$

Using the triangle inequality, then

$$
\left| \int_{-\infty}^{\infty} \gamma_r(x)g(x)\,dx - \frac{1}{2}(g(0-) + g(0+)) \right| < \epsilon.
$$

Since $\epsilon > 0$ is arbitrary then Eq. (4.40) holds. $\qquad\square$

As a consequence of Lemma 4.2,

$$\lim_{r \to \infty} \int_{-\infty}^{\infty} \gamma_r(x)g(x)\,dx = \int_{-\infty}^{\infty} \lim_{r \to \infty} \gamma_r(x)g(x)\,dx$$

$$= \int_{-\infty}^{\infty} \delta(x)g(x)\,dx = \frac{1}{2}(g(0-) + g(0+)),$$

for any piecewise continuous function g. Using a substitution this statement can be generalized as, for any $a \in \mathbb{R}$,

$$\int_{-\infty}^{\infty} \delta(x - a)g(x)\,dx = \frac{1}{2}(g(a-) + g(a+)).$$

The fundamental solution can be used to solve the heat equation on the real line with a bounded initial condition. The proposed solution can be expressed as

$$u(x,t) = \begin{cases} \displaystyle\int_{-\infty}^{\infty} U(x - y, t)f(y)\,dy & \text{if } t > 0, \\ f(x) & \text{if } t = 0. \end{cases} \tag{4.41}$$

Two questions must be addressed:

(1) Does $u(x,t)$ given in Eq. (4.41) solve the heat equation for $-\infty < x < \infty$ and $t > 0$?
(2) For a fixed x_0 does $\lim_{t \to 0^+} u(x_0, t) = f(x_0)$?

These issues are taken up in the proof of the following theorem.

Theorem 4.3. *Consider the initial value problem:*

$$u_t = k\,u_{xx} \text{ for } -\infty < x < \infty \text{ and } t > 0,$$
$$u(x,0) = f(x) \text{ for } -\infty < x < \infty,$$

where $f(x)$ is bounded and piecewise continuous on the real number line. The function given in Eq. (4.41) is a bounded solution to the heat equation satisfying the initial condition in the sense that

$$\lim_{t \to 0^+} u(x_0, t) = \frac{1}{2}(f(x_0-) + f(x_0+)). \tag{4.42}$$

Proof. Since f is bounded and piecewise continuous, it can be shown that differentiation inside the integral can be carried out with respect to t and x when $t > 0$. Equations (4.35) and (4.36) imply

$$u_t(x,t) = \int_{-\infty}^{\infty} \frac{1}{4\sqrt{k\pi t}} e^{-(x-y)^2/(4kt)} \left(\frac{(x-y)^2}{2kt^2} - \frac{1}{t} \right) f(y)\,dy$$

and

$$u_{xx}(x,t) = \int_{-\infty}^{\infty} \frac{1}{4k\sqrt{k\pi t}} e^{-(x-y)^2/(4kt)} \left(\frac{(x-y)^2}{2kt^2} - \frac{1}{t} \right) f(y)\, dy$$

and thus $u_t = k\, u_{xx}$, which establishes that the function given in Eq. (4.41) solves the heat equation for $t > 0$.

To establish the initial condition is satisfied as described in Eq. (4.42), fix $x_0 \in \mathbb{R}$, then

$$\lim_{t \to 0^+} u(x_0, t) = \lim_{t \to 0^+} \int_{-\infty}^{\infty} \gamma_r(x_0 - y) f(y)\, dy$$

$$= \lim_{r \to \infty} \int_{-\infty}^{\infty} \gamma_r(y - x_0) f(y)\, dy = \frac{1}{2}(f(x_0-) + f(x_0+)),$$

by Lemma 4.2.

Since the initial condition is bounded there exists $M > 0$ such that $|f(x)| \leq M$ for all $x \in \mathbb{R}$. To show the solution is bounded for $t > 0$ consider

$$|u(x,t)| = \left| \int_{-\infty}^{\infty} U(x - y, t) f(y)\, dy \right| \leq M \int_{-\infty}^{\infty} U(x - y, t)\, dy = M.$$

Hence the solution given in Eq. (4.41) is bounded by the same bound restricting the initial condition. $\qquad\square$

In fact, a slightly stronger result can be shown. If $f(x)$ is continuous at $x = x_0$, then

$$\lim_{(x,t) \to (x_0, 0^+)} u(x,t) = f(x_0).$$

To establish this limit consider,

$$u(x,t) - f(x_0) = \int_{-\infty}^{\infty} \gamma_r(x - y) f(y)\, dy - f(x_0)$$

$$= \int_{-\infty}^{\infty} \gamma_r(x - y)(f(y) - f(x_0))\, dy$$

$$= \int_{x_0-\delta}^{x_0+\delta} \gamma_r(x - y)(f(y) - f(x_0))\, dy$$

$$+ \int_{-\infty}^{x_0-\delta} \gamma_r(x - y)(f(y) - f(x_0))\, dy$$

$$+ \int_{x_0+\delta}^{\infty} \gamma_r(x - y)(f(y) - f(x_0))\, dy.$$

Let $\epsilon > 0$ and since f is continuous at x_0, there exists $\delta > 0$ such that if $|x - x_0| < \delta$ then $|f(x) - f(x_0)| < \epsilon/2$. This implies

$$\left| \int_{x_0-\delta}^{x_0+\delta} \gamma_r(x - y)(f(y) - f(x_0)) \, dy \right| < \frac{\epsilon}{2}.$$

If $|x - x_0| < \delta/2$ then

$$\left| \int_{-\infty}^{x_0-\delta} \gamma_r(x - y)(f(y) - f(x_0)) \, dy \right| = \left| \int_{x-x_0+\delta}^{\infty} \gamma_r(z)(f(x - z) - f(x_0)) \, dz \right|$$

$$\leq 2M \int_{\delta/2}^{\infty} \gamma_r(z) \, dz \to 0$$

as $r \to \infty$ or equivalently as $t \to 0^+$ according to Lemma 4.1. Similarly it can be shown that

$$\left| \int_{x_0+\delta}^{\infty} \gamma_r(x - y)(f(y) - f(x_0)) \, dy \right| \to 0.$$

Thus choose $\delta > 0$ small enough that when $|x - x_0| < \delta/2$ and $0 < t < \delta$, then $|u(x,t) - f(x_0)| < \epsilon$.

To prove the solution to the heat equation on the real number line is unique under the assumptions of Theorem 4.3, a version of the Maximum Principle is needed.

Theorem 4.4 (Maximum Principle for Unbounded Domains). *Suppose $u(x,t)$ is a bounded solution to the heat equation,*

$$u_t = k \, u_{xx} \text{ for } -\infty < x < \infty \text{ and } t > 0$$

and $u(x,t)$ is C^2 on $(-\infty, \infty) \times (0, \infty)$ and continuous on $(-\infty, \infty) \times [0, \infty)$. If $u(x,0) \leq M$ for $x \in \mathbb{R}$ with M constant, then $u(x,t) \leq M$ for all $x \in \mathbb{R}$ and $t > 0$.

Proof. Let $\epsilon > 0$ and fix $x_0 \in \mathbb{R}$ and define

$$v(x,t) = u(x,t) - \epsilon(2kt + (x - x_0)^2).$$

The reader can verify that $v_t = k \, v_{xx}$ so that v is also a solution to the heat equation on the real number line. Let $T > 0$ and consider the regions in the xt-plane defined as

$$R_N = \{(x,t) \mid x_0 - N < x < x_0 + N, \, 0 < t < T\},$$

$$S_N = \{(x,0) \mid x_0 - N \leq x \leq x_0 + N\} \cup \{(x_0 - N, t) \mid 0 \leq t \leq T\}$$

$$\cup \{(x_0 + N, t) \mid 0 \leq t \leq T\},$$

$$\overline{R}_N = R_N \cup S_N \cup \{(x,t) \mid x_0 - N \leq x \leq x_0 + N\}.$$

By the version of the Maximum Principle for finite, bounded domains given in Theorem 4.1,

$$\max_{(x,t)\in \overline{R}_N} v(x,t) = \max_{(x,t)\in S_N} v(x,t).$$

On the set $\{(x,0)\,|\,x_0 - N \le x \le x_0 + N\} \subset S_N$, since $u(x,0) \le M$, then $v(x,0) \le M - \epsilon(x-x_0)^2 \le M$. Since $u(x,t)$ is bounded there exists $\widehat{M} > 0$ such that $|u(x,t)| \le \widehat{M}$ for all $-\infty < x < \infty$ and $t > 0$. On the sets $\{(x_0 \pm N, t)\,|\,0 \le t \le T\} \subset S_N$, the function $v(x_0 \pm N, t) \le \widehat{M} - \epsilon N^2 < M$ for sufficiently large N. Consequently $v(x,t) \le M$ for all $(x,t) \in \overline{R}_N$ when N is sufficiently large. This implies that $v(x_0,t) = u(x_0,t) - 2kt\epsilon \le M$ for all $0 \le t \le T$. Taking the limits as $\epsilon \to 0^+$ results in $u(x_0,t) \le M$ for all $0 \le t \le T$. Since x_0 and T were arbitrarily chosen, then $u(x,t) \le M$ for all $x \in \mathbb{R}$ and $t \ge 0$. $\square$

A similar argument can be used to show that $u(x,t) \ge m$ for all $x \in \mathbb{R}$ and $t > 0$, where m is a constant such that $u(x,0) \ge m$ for all $x \in \mathbb{R}$. This is stated as a corollary below.

Corollary 4.3 (Minimum Principle for Unbounded Domains). *Suppose $u(x,t)$ is a bounded solution to the heat equation,*

$$u_t = k\,u_{xx} \text{ for } -\infty < x < \infty \text{ and } t > 0$$

and $u(x,t)$ is C^2 on $(-\infty,\infty)\times(0,\infty)$ and continuous on $(-\infty,\infty)\times[0,\infty)$. If $u(x,0) \ge m$ for $x \in \mathbb{R}$ with m constant, then $u(x,t) \ge m$ for all $x \in \mathbb{R}$ and $t > 0$.

Note that bounds for the solution to the heat equation derived using the formula in Eq. (4.41) were established in the proof of Theorem 4.3. In fact, in both cases the bounds for the solution turned out to be the minimum and maximum of the piecewise continuous initial condition $f(x)$. However, the bounds established in Theorem 4.4 and Corollary 4.3 do not depend on any particular form of the solution. As a consequence, the uniqueness of the bounded solution can be stated.

Theorem 4.5 (Uniqueness of Solution on Unbounded Domains). *Consider the initial value problem,*

$$u_t = k\,u_{xx} \text{ for } -\infty < x < \infty \text{ and } t > 0,$$

$$u(x,0) = f(x) \text{ for } -\infty < x < \infty,$$

where $f(x)$ is bounded and piecewise continuous on the real number line. The function given in Eq. (4.41) is the unique, bounded solution to the initial value problem.

Proof. Suppose there exist two bounded solutions $u(x,t)$ and $v(x,t)$ which satisfy the same initial condition $f(x)$ for all $x \in \mathbb{R}$. Define the function $w(x,t) = u(x,t) - v(x,t)$. Since the heat equation is linear, then $w(x,t)$ also solves the heat equation with initial condition $w(x,0) = 0$ for all $x \in \mathbb{R}$. Since $|w(x,t)|$ is bounded by zero according to Theorem 4.4 and Corollary 4.3, $u(x,t) = v(x,t)$ for all $x \in \mathbb{R}$ and all $t > 0$. □

If the boundedness assumption on the solution is dropped there can exist infinitely many nontrivial solutions to the heat equation on the real number line which satisfy the initial condition $u(x,0) = 0$. See [Evans (2010), page 59]. While the solution to the initial value of the heat equation is expressed in Eq. (4.41), the evaluation of the integral involved can be challenging. The examples below demonstrate this.

Example 4.8. Find the unique solution to the following initial boundary value problem:

$$u_t = k\,u_{xx} \text{ for } -\infty < x < \infty \text{ and } t > 0,$$

$$u(x,0) = f(x) = \begin{cases} a & \text{if } x < 0, \\ b & \text{if } x \geq 0. \end{cases}$$

Solution. For $t > 0$ the solution may be found from the integral in Eq. (4.41),

$$u(x,t) = \int_{-\infty}^{\infty} U(x-y,t)f(y)\,dy$$

$$= \frac{a}{\sqrt{4\pi k\,t}} \int_{-\infty}^{0} e^{-(x-y)^2/(4kt)}\,dy + \frac{b}{\sqrt{4\pi k\,t}} \int_{0}^{\infty} e^{-(x-y)^2/(4kt)}\,dy$$

$$= -\frac{a\sqrt{4k\,t}}{\sqrt{4\pi k\,t}} \int_{\infty}^{x/\sqrt{4kt}} e^{-z^2}\,dz - \frac{b\sqrt{4k\,t}}{\sqrt{4\pi k\,t}} \int_{x/\sqrt{4kt}}^{-\infty} e^{-z^2}\,dz$$

$$= \frac{a}{\sqrt{\pi}} \int_{x/\sqrt{4kt}}^{\infty} e^{-z^2}\,dz + \frac{b}{\sqrt{\pi}} \int_{-\infty}^{x/\sqrt{4kt}} e^{-z^2}\,dz.$$

While it is mathematically correct to leave the solution in this form, this example presents an opportunity to introduce a useful function defined in terms of an integral, the **error function**, denoted as

$$\text{erf}(x) = \frac{2}{\sqrt{\pi}} \int_{0}^{x} e^{-z^2}\,dz. \tag{4.43}$$

Making use of the error function,

$$
\begin{aligned}
u(x,t) &= \frac{a}{2} \left(\frac{2}{\sqrt{\pi}} \int_0^\infty e^{-z^2}\,dz - \frac{2}{\sqrt{\pi}} \int_0^{x/\sqrt{4kt}} e^{-z^2}\,dz \right) \\
&\quad + \frac{b}{2} \left(\frac{2}{\sqrt{\pi}} \int_{-\infty}^0 e^{-z^2}\,dz + \frac{2}{\sqrt{\pi}} \int_0^{x/\sqrt{4kt}} e^{-z^2}\,dz \right) \\
&= \frac{a}{2} \left(1 - \operatorname{erf}\left(\frac{x}{\sqrt{4kt}} \right) \right) + \frac{b}{2} \left(1 + \operatorname{erf}\left(\frac{x}{\sqrt{4kt}} \right) \right) \\
&= \frac{a+b}{2} + \frac{b-a}{2} \operatorname{erf}\left(\frac{x}{\sqrt{4kt}} \right).
\end{aligned}
$$

In particular, if in Example 4.8, $a = 0$ and $b = 1$, the initial condition becomes the Heaviside[4] function and the corresponding solution is

$$
u(x,t) = \frac{1}{2} + \frac{1}{2} \operatorname{erf}\left(\frac{x}{\sqrt{4kt}} \right).
$$

Note that the partial derivative of $u(x,t)$ with respect to x is

$$
u_x(x,t) = \frac{1}{2} \left(\frac{2}{\sqrt{\pi}} \frac{1}{\sqrt{4kt}} e^{-x^2/(4kt)} \right) = U(x,t),
$$

which is the fundamental solution to the heat equation with initial condition $u(x,0) = \delta(x)$.

Example 4.9. Find the unique solution to the following initial boundary value problem:

$$
u_t = u_{xx} \text{ for } -\infty < x < \infty \text{ and } t > 0,
$$
$$
u(x,0) = e^{-x^2} \sin x \text{ for } -\infty < x < \infty.
$$

Solution. For $t > 0$ the solution may be found from the integral in Eq. (4.41),

$$
u(x,t) = \int_{-\infty}^\infty U(x-y,t) e^{-y^2} \sin y\,dy
$$

where in this example $k = 1$. To evaluate the integral, some techniques from complex analysis are used. Using the Euler identity of Eq. (3.40) the

---

[4]Oliver Heaviside, English electrical engineer, (1850–1925).

improper integral is re-written as

$$u(x,t) = \int_{-\infty}^{\infty} U(x-y,t)e^{-y^2}\,\mathrm{Im}\left(e^{iy}\right)\,dy$$

$$= \frac{e^{-\frac{x^2}{4t}}}{\sqrt{4\pi t}}\,\mathrm{Im}\left(\int_{-\infty}^{\infty} e^{-\frac{(1+4t)y^2 - 2(x+2it)y}{4t}}\,dy\right)$$

$$= \frac{e^{-\frac{x^2}{4t}}}{\sqrt{4\pi t}}\,\mathrm{Im}\left(e^{\frac{(x+2it)^2}{4t(1+4t)}}\int_{-\infty}^{\infty} e^{-\frac{1+4t}{4t}(y-\frac{x+2it}{1+4t})^2}\,dy\right)$$

$$= \frac{e^{-\frac{x^2+t}{1+4t}}}{\sqrt{4\pi t}}\,\mathrm{Im}\left(e^{\frac{ix}{1+4t}}\int_{-\infty}^{\infty} e^{-\frac{1+4t}{4t}(y-\frac{x+2it}{1+4t})^2}\,dy\right). \quad (4.44)$$

In one sense the improper integral in Eq. (4.44) could be evaluated like any other, treating the term involving $i = \sqrt{-1}$ in the exponent like any other constant. However a more rigorous and careful approach will be taken here. Define the new variable z as

$$z = \sqrt{\frac{1+4t}{4t}}\left(y - \frac{x}{1+4t}\right) \iff y = \sqrt{\frac{4t}{1+4t}}z + \frac{x}{1+4t}.$$

Using this change of variable in Eq. (4.44) implies

$$\int_{-\infty}^{\infty} e^{-\frac{1+4t}{4t}(y-\frac{x+2it}{1+4t})^2}\,dy = \sqrt{\frac{4t}{1+4t}}\int_{-\infty}^{\infty} e^{-(z-i\sqrt{\frac{t}{1+4t}})^2}\,dz.$$

Consider the function $f(\zeta) = e^{-\zeta^2}$. Function $f(\zeta)$ is analytic on the complex plane and thus if C is any piecewise smooth simple closed curve in the complex plane, the integral $\oint_C f(\zeta)\,d\zeta = 0$ by the Cauchy-Goursat theorem (Theorem A.6). Consider the piecewise smooth simple closed curve pictured below.

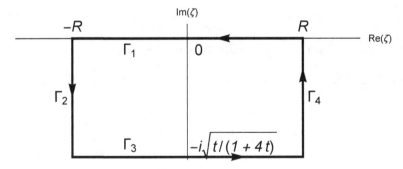

Since $C = \Gamma_1 \cup \Gamma_2 \cup \Gamma_3 \cup \Gamma_4$ and using properties of the contour integral,

$$0 = \int_{\Gamma_1} f(\zeta)\,d\zeta + \int_{\Gamma_2} f(\zeta)\,d\zeta + \int_{\Gamma_3} f(\zeta)\,d\zeta + \int_{\Gamma_4} f(\zeta)\,d\zeta. \quad (4.45)$$

Along the path Γ_2, $\zeta = -R - s\,i\sqrt{t/(1+4t)}$ for $0 \le s \le 1$, and

$$\left| \int_{\Gamma_2} f(\zeta)\,d\zeta \right| \le \int_0^1 |f(\zeta)|\,ds \le e^{-R^2} \to 0 \text{ as } R \to \infty.$$

Similarly, the integral along Γ_4 also vanishes as $R \to \infty$. Along the path Γ_1, $\zeta = -z$ for $-R \le z \le R$. Therefore,

$$\int_{\Gamma_1} f(\zeta)\,d\zeta = -\int_{-R}^R f(-z)\,dz = -\int_{-R}^R e^{-z^2}\,dz,$$

$$\lim_{R \to \infty} \int_{\Gamma_1} f(\zeta)\,d\zeta = -\int_{-\infty}^{\infty} e^{-z^2}\,dz = -\sqrt{\pi}.$$

The integral along path Γ_3 as $R \to \infty$ is the improper integral in Eq. (4.44) needed to determine the solution to the initial value problem for the heat equation. Taking the limit as $R \to \infty$ of both sides of Eq. (4.45) produces

$$\int_{-\infty}^{\infty} e^{-\left(z - i\sqrt{\frac{t}{1+4t}}\right)^2}\,dz = \sqrt{\pi}.$$

Consequently the solution to the initial value problem becomes,

$$u(x,t) = \frac{e^{-(x^2+t)/(1+4t)}}{\sqrt{4\pi t}} \operatorname{Im} \left(e^{ix/(1+4t)} \sqrt{\frac{4\pi t}{1+4t}} \right)$$

$$= \frac{e^{-(x^2+t)/(1+4t)}}{\sqrt{1+4t}} \sin\left(\frac{x}{1+4t} \right).$$

Figure 4.10 shows a contour plot of the solution.

Having done all the heavy lifting to find the solution to the heat equation on the whole real line, finding the solution on the semi-infinite domain is straightforward. The semi-infinite spatial domain consists of the open interval $(0, \infty)$. The origin $x = 0$ is considered a "boundary" for the domain, though the reader will see below that solutions are defined at the origin using limits. The first case considered is that of the homogeneous Dirichlet type boundary condition at the origin. For the sake of brevity the notation $u(x, 0+)$ is used to represent the $\lim_{t \to 0+} u(x, t)$.

Theorem 4.6. *Suppose $f(x)$ is bounded and piecewise continuous on $[0, \infty)$. The initial boundary value problem,*

$$u_t = k\,u_{xx} \text{ for } 0 < x < \infty \text{ and } t > 0,$$

$$u(0,t) = 0 \text{ for } t > 0,$$

$$u(x, 0+) = f(x), \text{ for } 0 < x < \infty$$

has a unique, bounded solution defined for $t > 0$,

$$u(x,t) = \int_0^{\infty} \left(U(x - y, t) - U(x + y, t) \right) f(y)\,dy. \tag{4.46}$$

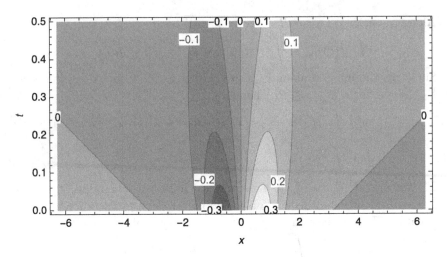

Fig. 4.10   A contour plot of the solution to the heat equation on the real line in Example 4.9. Warmer areas are in lighter shades while cooler areas are darker.

Proof. Extend the initial condition $f(x)$ to the whole real line by defining the odd extension,

$$f_o(x) = \begin{cases} -f(-x) & \text{if } x < 0, \\ f(x) & \text{if } x \geq 0. \end{cases}$$

Furthermore $f_o(x)$ is piecewise continuous. By Theorems 4.3 and 4.5 with $f_o(x)$ used as the initial condition, the unique, bounded solution defined on the whole real line is

$$u(x,t) = \int_{-\infty}^{\infty} U(x-y,t) f_o(y)\, dy.$$

For $x > 0$ this solution satisfies the heat equation and the initial condition of the semi-infinite Dirichlet boundary value problem. Note that

$$u(0,t) = \frac{1}{\sqrt{4\pi k t}} \int_{-\infty}^{\infty} e^{-y^2/(4kt)} f_o(y)\, dy = 0$$

since the integrand is an odd function. Thus $u(x,t)$ satisfies the Dirichlet boundary condition at $x = 0$ as well. Using the odd symmetry of the integrand the solution may be written as

$$u(x,t) = -\int_{-\infty}^{0} U(x-y,t) f(-y)\, dy + \int_{0}^{\infty} U(x-y,t) f(y)\, dy$$

$$= -\int_{0}^{\infty} U(x+z,t) f(z)\, dz + \int_{0}^{\infty} U(x-y,t) f(y)\, dy$$

which is Eq. (4.46).   □

Example 4.10. Find the unique solution to the following initial boundary value problem:

$$u_t = u_{xx} \text{ for } 0 < x < \infty \text{ and } t > 0,$$
$$u(0, t) = 0 \text{ for } t > 0,$$
$$u(x, 0+) = e^{-x^2} \sin x \text{ for } 0 < x < \infty.$$

Solution. For $t > 0$ the solution may be found from the integral in Eq. (4.46) where $k = 1$.

$$u(x, t) = \int_0^\infty \left(U(x - y, t) - U(x + y, t) \right) e^{-y^2} \sin y \, dy$$
$$= \int_0^\infty U(x - y, t) e^{-y^2} \sin y \, dy - \int_0^\infty U(x + y, t) e^{-y^2} \sin y \, dy.$$

Using the substitution $z = -y$ in the second integral above produces

$$u(x, t) = \int_0^\infty U(x - y, t) e^{-y^2} \sin y \, dy - \int_0^{-\infty} U(x - z, t) e^{-z^2} \sin z \, dz$$
$$= \int_{-\infty}^\infty U(x - y, t) e^{-y^2} \sin y \, dy = \frac{e^{-(x^2+t)/(1+4t)}}{\sqrt{1 + 4t}} \sin\left(\frac{x}{1 + 4t} \right)$$

according to Example 4.9. The reader may have recognized that since the partial differential equation and initial condition of this example are the same as in Example 4.9 and the solution of Example 4.9 is an odd function, this example should have the same solution.

If the boundary condition at $x = 0$ is of the zero-flux, Neumann type the solution is found in a similar manner.

Theorem 4.7. *Suppose $f(x)$ is piecewise continuous and bounded on $[0, \infty)$, then the initial boundary value problem*

$$u_t = k\, u_{xx} \text{ for } 0 < x < \infty \text{ and } t > 0,$$
$$u_x(0, t) = 0 \text{ for } t > 0,$$
$$u(x, 0+) = f(x) \text{ for } 0 < x < \infty$$

has a unique, bounded solution defined for $t > 0$,

$$u(x, t) = \int_0^\infty \left(U(x - y, t) + U(x + y, t) \right) f(y) \, dy. \tag{4.47}$$

Proof. Extend the initial condition $f(x)$ continuously to the whole real line by defining the even extension

$$f_e(x) = \begin{cases} f(-x) & \text{if } x < 0, \\ f(x) & \text{if } x \geq 0. \end{cases}$$

By Theorems 4.3 and 4.5 with $f_e(x)$ used as the initial condition, the unique solution defined on the whole real line is

$$u(x,t) = \int_{-\infty}^{\infty} U(x-y,t)f_e(y)\,dy. \qquad (4.48)$$

For $x > 0$ this solution satisfies the heat equation and the initial condition of the semi-infinite Neumann boundary value problem. It can be shown that for $t > 0$ it is permissible to differentiate and taking the limit inside the resulting integral as $x \to 0^+$,

$$u_x(x,t) = -\frac{1}{\sqrt{4\pi kt}} \int_{-\infty}^{\infty} \frac{x-y}{2kt} e^{-(x-y)^2/(4kt)} f_e(y)\,dy,$$

$$\lim_{x\to 0^+} u_x(x,t) = -\frac{1}{\sqrt{4\pi kt}} \int_{-\infty}^{\infty} \left(\lim_{x\to 0^+} \frac{x-y}{2kt} e^{-(x-y)^2/(4kt)} \right) f_e(y)\,dy$$

$$= \frac{1}{\sqrt{4\pi kt}} \int_{-\infty}^{\infty} \frac{y}{2kt} e^{-y^2/(4kt)} f_e(y)\,dy = 0.$$

The last equation holds since the integrand is an odd function of y. Thus $u(x,t)$ satisfies the Neumann boundary condition as $x \to 0^+$. Using the definition of the even function $f_e(y)$ in Eq. (4.48), the solution may be written as

$$u(x,t) = \int_{-\infty}^{0} U(x-y,t)f(-y)\,dy + \int_{0}^{\infty} U(x-y,t)f(y)\,dy$$

$$= \int_{0}^{\infty} U(x+z,t)f(z)\,dz + \int_{0}^{\infty} U(x-y,t)f(y)\,dy$$

which is Eq. (4.47).   □

Example 4.11. Find the unique solution to the following initial boundary value problem:

$$u_t = u_{xx} \text{ for } 0 < x < \infty \text{ and } t > 0,$$
$$u_x(0,t) = 0 \text{ for } t > 0,$$
$$u(x,0+) = e^{-x^2}\cos x \text{ for } 0 < x < \infty.$$

Solution. For $t > 0$ the solution may be found from the integral in Eq. (4.47).

$$u(x,t) = \int_{0}^{\infty} \left(U(x-y,t) + U(x+y,t) \right) e^{-y^2}\cos y\,dy$$

$$= \int_{-\infty}^{\infty} U(x-y,t)e^{-y^2}\cos y\,dy = \int_{-\infty}^{\infty} U(x-y,t)e^{-y^2}\,\mathrm{Re}(e^{iy})\,dy$$

where integration by substitution and Euler's identity from Eq. (3.40) have been employed. Using the work done in Example 4.9,

$$u(x,t) = \frac{e^{-(x^2+t)/(1+4t)}}{\sqrt{4\pi t}} \, \mathrm{Re} \left(e^{ix/(1+4t)} \sqrt{\frac{4\pi t}{1+4t}} \right)$$

$$= \frac{e^{-(x^2+t)/(1+4t)}}{\sqrt{1+4t}} \cos \left(\frac{x}{1+4t} \right).$$

Figure 4.11 shows a contour plot of the solution.

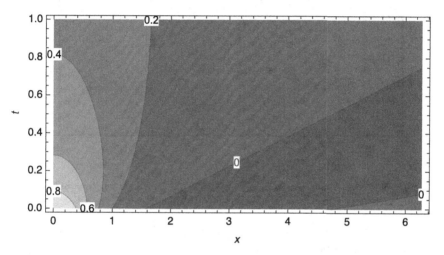

Fig. 4.11   A contour plot of the solution to the heat equation on the real line in Example 4.11. Warmer areas are in lighter shades while cooler areas are darker.

4.5   The Heat Equation on a Rectangular Domain

Later in Chap. 9, partial differential equations such as the heat, wave, and Laplace's equation in higher space dimensions and alternative coordinate systems will be covered in more detail; however, the extension of the heat equation from the one-dimensional spatial domain (the case of the "metal rod") to the two-dimensional "rectangular metal plate" is sufficiently natural that it will be introduced here. Readers interested in covering other classes of partial differential equations now such as the wave equation and Laplace's equation can omit this section for the time being and return to the topic later without loss of continuity.

Consider the rectangular domain $R = \{(x,y) \,|\, 0 \le x \le L, \, 0 \le y \le M\}$ as shown in Fig. 4.12.

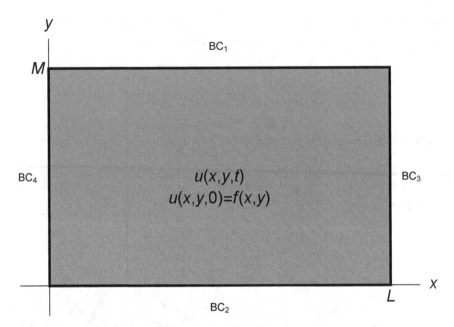

Fig. 4.12   The rectangular spatial domain on which the heat equation, a set of boundary conditions, and an initial condition are assumed to hold.

A corner of the domain does not necessarily have to rest at the origin and the edges do not have to be parallel to the coordinate axes, but this assumption tremendously simplifies the following analysis. The faces of the rectangle (if necessary refer to them as the top and bottom face) are insulated so that the flow of heat energy (if any) into or out of the plate takes place through the four edges. The partial differential equation and initial condition satisfied by heat flow in two dimensions (expressed in Cartesian coordinates) is:

$$u_t = k(u_{xx} + u_{yy}) \text{ for } (x,y) \in (0,L) \times (0,M), \, t > 0, \quad (4.49)$$

$$u(x,y,0) = f(x,y) \text{ for } (x,y) \in [0,L] \times [0,M]. \quad (4.50)$$

The positive constant k is again called the thermal diffusivity constant. Just as was the case for the one-dimensional heat equation, the interesting details of the solution to the two-dimensional heat equation come to light when boundary conditions are assigned to the four edges of the rectangular plate. Fortunately the geometry of rectangles allows the method of separation of variables and Fourier series to figure prominently in the solution when the boundary conditions are of the homogeneous Dirichlet type, Neumann type, or of the third kind.

If the solution to the heat equation on the rectangle factors into a product solution of the form $u(x, y, t) = X(x)Y(y)T(t)$ then by differentiating this function and substituting it into Eq. (4.49) the equation can be rearranged as

$$X(x)Y(y)T'(t) = k(X''(x)Y(y)T(t) + X(x)Y''(y)T(t)) \quad \text{or}$$

$$\frac{T'(t)}{kT(t)} = \frac{X''(x)}{X(x)} + \frac{Y''(y)}{Y(y)} = \alpha.$$

The expression α must be a constant since the left-hand side of the equation is a function of t only while the right-hand side is a function of x and y only. Focusing attention on the right-hand side (the spatially dependent portion) of the equation, this side can be further manipulated to yield

$$\frac{X''(x)}{X(x)} = \alpha - \frac{Y''(y)}{Y(y)} = \beta.$$

Once again β must be a constant. The constants α and β are further restricted by the boundary conditions imposed at the edges of rectangle R. Suppose all the boundary conditions are of Dirichlet type, *i.e.*,

$$u(0, y, t) = 0 \quad \text{and} \quad u(L, y, t) = 0$$

for $0 \le y \le M$ and $t \ge 0$ and also

$$u(x, 0, t) = 0 \quad \text{and} \quad u(x, M, t) = 0$$

for $0 \le x \le L$ and $t \ge 0$, then the reader may recall that nonzero solutions to the boundary value problem,

$$X''(x) - \beta X(x) = 0,$$

$$X(0) = X(L) = 0$$

exist only when $\beta = -\sigma_n^2$ where $\sigma_n = n\pi/L$, $n \in \mathbb{N}$. The nontrivial solutions corresponding to $-\sigma_n^2$ are nonzero constant multiples of

$$X_n(x) = \sin\left(\frac{n\pi x}{L}\right).$$

Having determined the x-dependent portion of the product solution, this partial result may be substituted into the boundary value problem to determine $Y(y)$.

$$\alpha - \frac{Y''(y)}{Y(y)} = \beta = -\left(\frac{n\pi}{L}\right)^2,$$

$$Y''(y) - \left[\alpha + \left(\frac{n\pi}{L}\right)^2\right] Y(y) = 0.$$

Nonzero solutions satisfying the Dirichlet boundary conditions $Y(0) = Y(M) = 0$ of this last ordinary differential equation exist when

$$\alpha = -\pi^2 \left[\left(\frac{m}{M} \right)^2 + \left(\frac{n}{L} \right)^2 \right]$$

and are nonzero constant multiples of

$$Y_m(y) = \sin \left(\frac{m\pi y}{M} \right)$$

for $m = 1, 2, \ldots$ and $n = 1, 2, \ldots$. To draw a stronger connection with the one-dimensional case, the functions

$$X_n(x)Y_m(y) = \sin \left(\frac{n\pi x}{L} \right) \sin \left(\frac{m\pi y}{M} \right)$$

will be called the eigenfunctions of the boundary value problem and the constants

$$\lambda_{n,m} = \pi^2 \left[\left(\frac{m}{M} \right)^2 + \left(\frac{n}{L} \right)^2 \right]$$

for $m, n \in \mathbb{N}$ will be their corresponding eigenvalues. The reader can then verify that the time-dependent factor of the product solution is

$$T_{n,m}(t) = e^{-k\lambda_{n,m}t}$$

and consequently a product solution to the heat equation on the rectangle R with Dirichlet boundary conditions is

$$u_{n,m}(x, y, t) = e^{-k\lambda_{n,m}t} \sin \left(\frac{n\pi x}{L} \right) \sin \left(\frac{m\pi y}{M} \right).$$

Since the boundary conditions and the heat equation on the rectangle are linear and homogeneous then the Principle of Superposition of solutions can be applied and thus any finite linear combination of product solutions will be another solution. In other words

$$u(x, y, t) = \sum_{n=1}^{N} \sum_{m=1}^{K} a_{n,m} e^{-k\lambda_{n,m}t} \sin \left(\frac{n\pi x}{L} \right) \sin \left(\frac{m\pi y}{M} \right)$$

is also a solution to the heat equation and satisfies the Dirichlet boundary conditions imposed on the four sides of the rectangle. The expressions $a_{n,m}$ are constant coefficients. The only remaining condition to be met is the initial condition in Eq. (4.50). If the initial condition is a finite linear combination of eigenfunctions, then by equating coefficients, the solution to the initial boundary value problem can be determined. For more general initial conditions the solution must be expressed formally as a Fourier series,

$$u(x, y, t) = \sum_{n=1}^{\infty} \sum_{m=1}^{\infty} a_{n,m} e^{-k\lambda_{n,m}t} \sin \left(\frac{n\pi x}{L} \right) \sin \left(\frac{m\pi y}{M} \right). \qquad (4.51)$$

The double summation in Eq. (4.51) is sometimes referred to as an infinite double series. It can be shown that the infinite double series in Eq. (4.51) converges and that $u(x, y, t)$ solves the heat equation and satisfies the required Dirichlet boundary conditions. To make sure that $u(x, y, t)$ satisfies the initial condition in Eq. (4.50), it must be the case that

$$u(x, y, 0) = \sum_{n=1}^{\infty} \sum_{m=1}^{\infty} a_{n,m} \sin\left(\frac{n\pi x}{L}\right) \sin\left(\frac{m\pi y}{M}\right) = f(x, y). \qquad (4.52)$$

This can be achieved by choosing the coefficients $a_{n,m}$ to be

$$a_{n,m} = \frac{4}{LM} \int_0^M \int_0^L f(x, y) \sin\left(\frac{m\pi y}{M}\right) \sin\left(\frac{n\pi x}{L}\right) dx\, dy, \qquad (4.53)$$

provided the integrals exist. The infinite double series in Eq. (4.52) is often referred to as a double Fourier sine series.

Example 4.12. Find the solution to the initial boundary value problem:

$$u_t = u_{xx} + u_{yy} \text{ for } (x, y) \in (0, 1) \times (0, 2) \text{ and } t > 0,$$
$$u(0, y, t) = u(1, y, t) = 0 \text{ for } 0 < y < 2 \text{ and } t > 0,$$
$$u(x, 0, t) = u(x, 2, t) = 0 \text{ for } 0 < x < 1 \text{ and } t > 0,$$
$$u(x, y, 0) = 1000x(1 - x)\sin(2\pi y) \text{ for } (x, y) \in [0, 1] \times [0, 2].$$

Solution. Since the boundary conditions are of the Dirichlet type then the formal solution must resemble the infinite double sine series shown in Eq. (4.51) with coefficients $a_{n,m}$ determined by Eq. (4.53),

$$a_{n,m} = \frac{4}{(1)(2)} \int_0^2 \int_0^1 1000x(1 - x)\sin(2\pi y)\sin(n\pi x)\sin\left(\frac{m\pi y}{2}\right) dx\, dy$$

which, by the orthogonality property of the sine functions, vanishes for $m \neq 4$. Thus the nonzero coefficients are

$$a_{n,4} = 2000 \int_0^2 \int_0^1 x(1 - x)\sin^2(2\pi y)\sin(n\pi x)\, dx\, dy$$

$$= 2000 \int_0^1 x(1 - x)\sin(n\pi x)\left[\int_0^2 \sin^2(2\pi y)\, dy\right] dx$$

$$= 2000 \int_0^1 x(1 - x)\sin(n\pi x)\, dx$$

$$= \begin{cases} \frac{8000}{n^3 \pi^3} & \text{for } n \text{ odd}, \\ 0 & \text{for } n \text{ even}. \end{cases}$$

Consequently the formal solution to the initial boundary value problem can be written as

$$u(x, y, t) \sim \frac{8000}{\pi^3} \sin(2\pi y) \sum_{n=1}^{\infty} \frac{e^{-\pi^2((2n-1)^2+4)t}}{(2n-1)^3} \sin((2n-1)\pi x).$$

Figure 4.13 illustrates several contour curves of the solution for $t = 1/10$. This choice of value for t is arbitrary, but the contour plots for other values of t are similar in appearance.

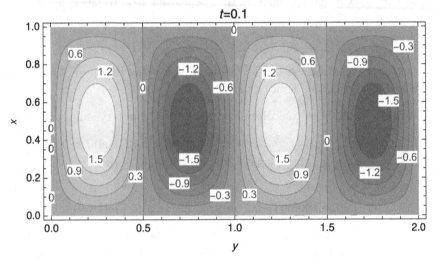

Fig. 4.13   A contour plot of the series solution $u(x, y, t)$ from Example 4.12. The value of t is fixed at $t = 1/10$. Contour plots for other positive values of t are similar in appearance. Note that the graph has been oriented with the y-axis running horizontally to better fit the printed page.

Neumann type and nonhomogeneous boundary conditions on a rectangle are also readily handled by generalizing the one-dimensional results of this chapter.

Example 4.13. Find the solution to the initial boundary value problem:

$$u_t = u_{xx} + u_{yy} \text{ for } (x, y) \in (0, 1) \times (0, 1) \text{ and } t > 0,$$
$$u(0, y, t) = 1 \text{ and } u(1, y, t) = 5 \text{ for } 0 < y < 1 \text{ and } t > 0,$$
$$u_y(x, 0, t) = u_y(x, 1, t) = 0 \text{ for } 0 < x < 1 \text{ and } t > 0,$$
$$u(x, y, 0) = (1 + 4x) + \sin(\pi x)\cos(2\pi y) \text{ for } (x, y) \in [0, 1] \times [0, 1].$$

Solution. Using a bit of physical intuition and the insights gained from the earlier discussions of the nonzero, but constant Dirichlet-type boundary conditions and the Neumann boundary conditions, the steady-state solution to this problem should be a function of x only. The heat in the rectangle should flow so as to "average out" all temperatures differences on lines parallel to the y-axis. Thus the complete solution will be conjectured to be the sum of a steady-state solution of the form $U(x, y) = 1 + 4x$ and a "transient" solution $v(x, y, t)$ which solves the following initial boundary value problem with homogeneous boundary conditions:

$$v_t = v_{xx} + v_{yy} \text{ for } (x, y) \in (0, 1) \times (0, 1) \text{ and } t > 0,$$

$$v(0, y, t) = v(1, y, t) = 0 \text{ for } 0 < y < 1 \text{ and } t > 0,$$

$$v_y(x, 0, t) = v_y(x, 1, t) = 0 \text{ for } 0 < x < 1 \text{ and } t > 0,$$

$$v(x, y, 0) = \sin(\pi x) \cos(2\pi y) \text{ for } (x, y) \in [0, 1] \times [0, 1].$$

The eigenfunctions of this simpler boundary value problem are of the form,

$$v_{n,m}(x, y) = \sin(n\pi x) \cos(m\pi y)$$

while the corresponding eigenvalues are

$$\lambda_{n,m} = \pi^2 (n^2 + m^2),$$

for $m = 0, 1, 2, \ldots$ and $n \in \mathbb{N}$. The reader should note that these eigenfunctions are merely the products of the eigenfunctions of the corresponding one-dimensional homogeneous Dirichlet boundary value problem (for the variable x) and the one-dimensional homogeneous Neumann boundary value problem (for the variable y).

The double series solution for the initial boundary value problem with homogeneous boundary conditions will resemble

$$v(x, y, t) = \sum_{m=0}^{\infty} \sum_{n=1}^{\infty} a_{n,m} e^{-\pi^2 (n^2 + m^2)t} \sin(n\pi x) \cos(m\pi y).$$

The first set of series coefficients can be found for the case in which $m = 0$,

$$a_{n,0} = 2 \int_0^1 \int_0^1 \sin(\pi x) \cos(2\pi y) \sin(n\pi x) \, dy \, dx = 0.$$

When $m \in \mathbb{N}$, then

$$a_{n,m} = 4 \int_0^1 \int_0^1 [\sin(\pi x) \cos(2\pi y)] \cos(m\pi y) \sin(n\pi x) \, dy \, dx = 0,$$

except for the case where $n = 1$ and $m = 2$, in which $a_{1,2} = 1$. Thus the complete solution to the original initial boundary value problem may be written as

$$u(x, y, t) = 1 + 4x + e^{-5\pi^2 t} \sin(\pi x) \cos(2\pi y).$$

The astute reader will notice that the evaluation of the double integrals can be avoided by equating coefficients in the initial conditions for the equation solved by $v(x, y, t)$ since the initial condition is a finite linear combination of eigenfunctions. Figure 4.14 illustrates several of the level surfaces of the series solution.

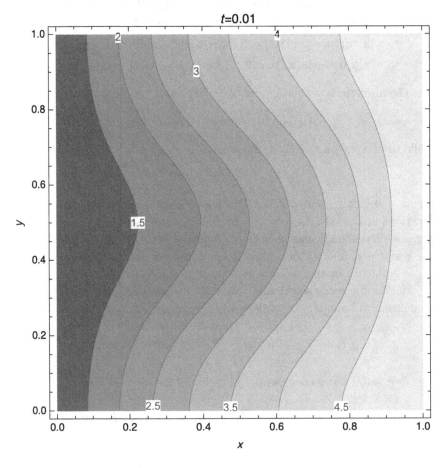

Fig. 4.14   Several level surfaces of $u(x, y, 0.01)$ for the series solution of Example 4.13.

4.6 Supplemental Remarks and Suggestions for Further Study

This chapter has used the method of separation of variables and Fourier series to solve the heat equation under several common and physically important boundary conditions. Students sometimes focus their attention on the format of the final series solution and become overwhelmed trying to memorize all the various product solutions and the boundary conditions to which the solutions correspond. There is no need for memorization of the solutions if the reader remembers a more basic principle. If the partial differential equation and boundary conditions are linear and homogeneous, the solution is likely to be a product solution, linear combination of product solutions, or an infinite series based on product solutions. Students need only to differentiate the product solution, substitute it into the partial differential equation, separate the variables, and determine the spatially dependent portion of the product solution which satisfies the given homogeneous boundary conditions. These steps lead to the eigenvalues and eigenfunctions of the boundary value problem. The time dependent portion of the solution is readily found from the eigenvalues. If the initial condition is not a finite linear combination of eigenfunctions, then the orthogonality property of the eigenfunctions and the definite integral will determine the Fourier series coefficients for the series solution.

Readers may have gathered that much of this chapter (and indeed the remainder of this text) focuses on product solutions and separation of variables. This is due to the fact that the method of separation of variables is one of the fundamental techniques used to solve a homogeneous boundary value problems. However, there are other important techniques. For example, the use of the fundamental solution to the heat equation, is crucial in solving initial value problems of the heat equation on infinite and semi-infinite spatial domains.

This chapter also delved into several theoretical results associated with the heat equation, namely the Maximum Principle and the uniqueness of solutions to the Fourier problem. These results as stated were not the most general formulations of the Maximum Principle and uniqueness of solutions. A Maximum Principle and statement regarding the uniqueness of solutions to heat equation on a rectangle can also be stated and proved. The reader will recall that the Maximum Principle states that the solution $u(x, t)$ to the heat equation achieves its maximum on a portion of the boundary of its domain of definition (the portion where the boundary and initial conditions are specified), but the case that the maximum was matched in the interior of its domain was not ruled out. The behavior cannot be ruled out, since a constant temperature may be the solution to the heat equation. There exists a stricter form of the Maximum Principle known as the Strong Maximum Principle which states that if the solution to the Fourier problem achieves its maximum on the set $\{(x, y) \,|\, 0 < x < L, 0 < t \leq T\}$, then the solution must be a constant. Precise statement of the Strong Maximum Principle and its proof can be found in most intermediate and advanced level books on partial differential equations, for instance [Bleecker and Csordas (1996), pp. 154–156].

4.7 Exercises

(1) Find the solution of the heat conduction problem:

$$u_t = 100u_{xx} \text{ for } 0 < x < 1 \text{ and } t > 0,$$
$$u(0, t) = u(1, t) = 0 \text{ for } t > 0,$$
$$u(x, 0) = \sin 2\pi x - \sin 5\pi x \text{ for } 0 \leq x \leq 1.$$

(2) Find the solution of the heat conduction problem:

$$u_t = 16u_{xx} \text{ for } 0 < x < 1 \text{ and } t > 0,$$
$$u_x(0, t) = u_x(1, t) = 0 \text{ for } t > 0,$$
$$u(x, 0) = 1 + \cos 3\pi x + \cos 8\pi x \text{ for } 0 \leq x \leq 1.$$

(3) Find the solution of the heat conduction problem:

$$u_t = u_{xx} \text{ for } 0 < x < 1 \text{ and } t > 0,$$
$$u(0, t) = u(1, t) = 0 \text{ for } t > 0,$$
$$u(x, 0) = \begin{cases} 0 & \text{if } 0 \leq x < 1/4, \\ 100 & \text{if } 1/4 \leq x \leq 3/4, \\ 0 & \text{if } 3/4 < x \leq 1. \end{cases}$$

(4) Find the solution of the heat conduction problem:

$$u_t = 4u_{xx} \text{ for } 0 < x < 1 \text{ and } t > 0,$$
$$u_x(0, t) = u_x(1, t) = 0 \text{ for } t > 0,$$
$$u(x, 0) = \begin{cases} 100 & \text{if } 0 \leq x \leq 1/2, \\ 0 & \text{if } 1/2 < x \leq 1. \end{cases}$$

(5) A metal rod has a thermal diffusivity constant of $k = 1/4 \text{ m}^2/\text{s}$ and a length of 1 meter. The ends of the rod are kept at $0°\text{C}$. The portion of

the rod from $(0, 1/2)$ is initially $50°$C and the portion of the rod from $[1/2, 1]$ is $0°$C. Find the temperature of the rod at (x, t) for $0 < x < 1$ and $t > 0$.

(6) Consider a uniform rod of length L with constant thermal diffusion constant k. Assume that the initial temperature distribution is given by the function $f(x)$ for $0 \le x \le L$ and the temperature at one end, say $x = 0$, is kept at $0°$C, while the other end is completely insulated.

(a) Write down the initial boundary value problem for the temperature distribution $u(x, t)$ in the rod.

(b) Apply the method of separation of variable to find that the fundamental solutions of the initial boundary value problem. (*Hint*: see Exercise 20 in Chap. 1.)

(c) Determine a formal solution of the initial boundary value problem.

(d) If $L = 2$ and

$$f(x) = \begin{cases} x & \text{if } 0 \le x \le 1, \\ 1 & \text{if } 1 \le x \le 2, \end{cases}$$

find the formal solution of the initial boundary value problem.

(7) Solve the following initial boundary value problem:

$$u_t = u_{xx} \text{ for } 0 < x < 1 \text{ and } t > 0,$$
$$u(0, t) = u_x(1, t) = 0 \text{ for } t > 0,$$
$$u(x, 0) = x(2 - x) \text{ for } 0 \le x \le 1.$$

Hint: First determine the eigenvalues and eigenfunctions of the associated boundary value problem.

(8) A metal rod has a thermal diffusivity constant of $k = 4$ m^2/s and a length of $1/2$ meter. The ends of the rod are insulated. The initial temperature distribution in the rod is given by the piecewise defined function:

$$u(x, 0) = \begin{cases} 100x & \text{if } 0 \le x \le 1/4, \\ 50(1 - 2x) & \text{if } 0 \le x \le 1/2. \end{cases}$$

Find the temperature of the bar at (x, t) for $0 < x < 1/2$ and $t > 0$.

(9) For almost all the examples considered in this chapter, the length of the domain (in most cases described as a rod) has been insulated and if heat could flow into the spatial domain, it did so only at the ends (typically $x = 0$ and $x = L$). Suppose this assumption is relaxed and

the rod now obeys Newton's Law of Cooling along its length. An initial boundary value problem reflecting this situation is the following,

$$u_t = ku_{xx} - cu \text{ for } 0 < x < L \text{ and } t > 0,$$
$$u(0, t) = u(L, t) = 0 \text{ for } t > 0,$$
$$u(x, 0) = f(x) \text{ for } 0 < x < L,$$

with positive constants k and c. Using the method of separation of variables, appropriate eigenfunctions and eigenvalues for the associated boundary value problem, and Fourier series, find the solution $u(x, t)$ to the initial boundary value problem above.

(10) The one-dimensional heat equation has a natural extension from the geometry of the rod to the geometry of a closed circular ring. The initial boundary value problem describing this situation is

$$u_t = ku_{xx} \text{ for } -L < x < L \text{ and } t > 0,$$
$$u(-L, t) = u(L, t) \text{ for } t > 0,$$
$$u_x(-L, t) = u_x(L, t) \text{ for } t > 0,$$
$$u(x, 0) = f(x) \text{ for } -L \leq x \leq L.$$

The ring with circumference $2L$ has been identified with the interval $[-L, L]$.

(a) Interpret and explain the two boundary conditions given above.

(b) Find the solution $u(x, t)$ to the initial boundary value problem.

(11) A metal rod has a thermal diffusivity constant of $k = 1/2$ m^2/s and a length of 1 meter. The end of the rod at $x = 0$ is kept at 20°C. The end of the rod at $x = 1$ is kept at 50°C. The portion of the rod from $(0, 1/2)$ is initially 0°C and the portion of the rod from $[1/2, 1)$ is 80°C. Find the temperature of the rod at (x, t) for $0 < x < 1$ and $t > 0$.

(12) Consider the heat equation with nonhomogeneous boundary conditions:

$$u_t = ku_{xx} \text{ for } 0 < x < L \text{ and } t > 0,$$
$$u(0, t) = A \text{ and } u(L, t) = B \text{ for } t > 0,$$
$$u(x, 0) = \sin \frac{\pi x}{L} \text{ for } 0 < x < L.$$

(a) Find a steady-state temperature distribution, $U(x)$.

(b) Define $v(x, t) = u(x, t) - U(x)$ and determine the initial boundary value problem solved by $v(x, t)$.

(c) Determine $v(x,t)$ formally as a Fourier series.

(d) Express the solution to the original initial boundary value problem as $u(x,t) = v(x,t) + U(x)$.

(13) Consider the heat equation with nonhomogeneous boundary conditions:

$$u_t = k\, u_{xx} \text{ for } 0 < x < L \text{ and } t > 0,$$
$$u_x(0,t) = 0 \text{ and } u_x(L,t) = B \neq 0 \text{ for } t > 0,$$
$$u(x,0) = f(x) \text{ for } 0 < x < L.$$

(a) Find a suitable reference temperature distribution $r(x,t)$.

(b) Define $v(x,t) = u(x,t) - r(x,t)$ and determine the initial boundary value problem solved by $v(x,t)$.

(c) Determine $v(x,t)$ formally as a Fourier series.

(d) Express the solution to the original initial boundary value problem as $u(x,t) = v(x,t) + r(x,t)$.

(14) Consider the nonhomogeneous heat equation with boundary and initial conditions:

$$u_t = k\, u_{xx} + k \text{ for } 0 < x < L \text{ and } t > 0,$$
$$u(0,t) = A \text{ and } u(L,t) = B \text{ for } t > 0,$$
$$u(x,0) = f(x) \text{ for } 0 < x < L.$$

(a) Find an equilibrium temperature distribution, $U(x)$.

(b) Define $v(x,t) = u(x,t) - U(x)$ and determine the initial boundary value problem solved by $v(x,t)$.

(c) Determine $v(x,t)$ formally as a Fourier series.

(d) Express the solution to the original initial boundary value problem as $u(x,t) = v(x,t) + U(x)$.

(15) Use l'Hôpital's Rule [Smith and Minton (2012), pp. 223] to evaluate the limit in Eq. (4.37) for the fundamental solution of the heat equation.

(16) Complete the details to the proof of Lemma 4.1.

(a) Show that $f(r) = e^{-c^2 r}/(c\sqrt{\pi r})$ is a monotone decreasing function of r when $c > 0$.

(b) Show that the limit of $\lim_{r \to \infty} f(r) = 0$.

(17) Solve the initial value problem:

$$u_t = u_{xx} \text{ for } -\infty < x < \infty \text{ and } t > 0,$$
$$u(x,0) = e^{-x^2} \text{ for } -\infty < x < \infty.$$

(18) Solve the initial, boundary value problem:
$$u_t = u_{xx} \text{ for } 0 \le x < \infty \text{ and } t > 0,$$
$$u(0,t) = 0 \text{ for } t > 0,$$
$$u(x,0) = x\,e^{-x^2} \text{ for } 0 \le x < \infty.$$

(19) Solve the initial, boundary value problem:
$$u_t = u_{xx} \text{ for } 0 \le x < \infty \text{ and } t > 0,$$
$$u_x(0,t) = 0 \text{ for } t > 0,$$
$$u(x,0) = e^{-x^2} \text{ for } 0 \le x < \infty.$$

(20) Prove the Minimum Principle stated in Corollary 4.1.

(21) Consider a heat conduction problem with an internal source of heat:
$$u_t = u_{xx} + (t+2)\sin x \text{ for } 0 < x < \pi \text{ and } t > 0,$$
$$u(0,t) = u(\pi,t) = 0, \text{ for } t > 0,$$
$$u(x,0) = \sin x \text{ for } 0 < x < \pi.$$

(a) Verify that $u(x,t) = (t+1)\sin x$ is a solution.

(b) For any $T > 0$, let
$$\overline{R} = \{(x,t)\,|\,0 \le x \le \pi,\ 0 \le t \le T\} \text{ and}$$
$$S = \{(x,t) \in \overline{R}\,|\,x = 0 \text{ or } x = \pi \text{ or } t = 0\}.$$

Show the conclusion of the Maximum Principle does not hold by showing that that
$$\max_{(x,t)\in\overline{R}} u(x,t) > \max_{(x,t)\in S} u(x,t).$$

(c) Explain why this does not contradict the Maximum Principle.

(22) Assume that $u(x,t)$ and $v(x,t)$ are continuous for $0 \le x \le L$ and $t \ge 0$, twice continuously differentiable for $0 < x < L$ and $t > 0$ and are solutions of the heat equation
$$u_t = k\,u_{xx} \text{ for } 0 < x < L \text{ and } t > 0.$$

If it is true that
$$u(0,t) \le v(0,t) \text{ and } u(L,t) \le v(L,t) \text{ for } t \ge 0$$

and
$$u(x,0) \le v(x,0) \text{ for } 0 \le x \le L,$$

show that
$$u(x,t) \le v(x,t) \text{ for } 0 \le x \le L \text{ and } t \ge 0.$$

(23) Without using the Maximum/Minimum Principle, prove that the solution of the initial boundary value problem

$$u_t = k\, u_{xx} \text{ for } 0 < x < L \text{ and } t > 0,$$
$$u(0,t) = \phi(t) \text{ and } u(L,t)v = \psi(t) \text{ for } t > 0,$$
$$u(x,0) = f(x) \text{ for } 0 < x < L$$

is unique. (*Hint*: assume u_1 and u_2 are solutions of the problem and prove that the function $w(t) = \int_0^L (u_1(x,t) - u_2(x,t))^2 \, dx$ must be zero for all t.)

(24) A square metal plate with sides of length 1 has plane faces which are insulated while the edges are kept at $0°C$. If the initial temperature distribution within the plate is

$$u(x,y,0) = \sin(\pi x) \sin(3\pi y),$$

find the temperature at (x,y,t) for $(x,y) \in [0,1] \times [0,1]$ and $t > 0$.

(25) A square metal plate with sides of length 1 has plane faces which are insulated while the edges labeled in counterclockwise order starting with the right are kept a temperatures $10°C$, $20°C$, $30°C$, and $40°C$. Find the steady-state distribution of temperature within the plate. *Hint*: assume the solution $U(x,y) = U_1(x,y) + U_2(x,y) + U_3(x,y) + U_4(x,y)$ where $U_i(x,y)$ is the steady-state solution to the same problem except that the temperature on edge $j \neq i$ is maintained at $0°C$.

(26) Consider the boundary value problem,

$$u_t = 2u_{xx} \text{ for } 0 < x < 1 \text{ and } t > 0,$$
$$u(0,t) = 0 \text{ and } u_x(1,t) + u(1,t) = 0 \text{ for } t > 0.$$

(a) Find the first four positive eigenvalues of the boundary value problem.

(b) Find the product solutions of the partial differential equations corresponding to the eigenvalues just found.

(27) Consider the following Fourier problem:

$$u_t = k\, u_{xx} \text{ for } 0 < x < \pi \text{ and } t > 0,$$
$$u(0,t) = u(\pi,t) = 0 \text{ for } t > 0,$$
$$u(x,0) = 9\sin(2x) - 4\sin(x) \text{ for } 0 \le x \le \pi.$$

Find constants m and M such that $m \le u(x,t) \le M$ for $0 \le x \le \pi$ and $t \ge 0$.

Projects

(1) Generalize the methods of this chapter to solving the heat equation on a rectangular box

$$Q = \{(x, y, z) \,|\, 0 \le x \le L,\, 0 \le y \le M,\, 0 \le z \le N\}.$$

The heat/diffusion equation in three-dimensional Cartesian coordinates is

$$u_t = k(u_{xx} + u_{yy} + u_{zz}) \text{ for } (x, y, z) \in (0, L) \times (0, M) \times (0, N) \text{ and } t > 0.$$

Explore as many of the combinations of fixed temperature and insulated boundary conditions as possible. To get started on a specific set of boundary conditions, the reader may wish to consider

$$u(0, y, z, t) = u(L, y, z, t) = 0,$$
$$u(x, 0, z, t) = u(x, M, z, t) = 0,$$
$$u_z(x, y, 0, t) = u_z(x, y, N, t) = 0,$$

which correspond to the situation of the box being insulated on the top and bottom faces and the remaining four faces kept at temperature 0. The initial condition may be assumed to be $u(x, y, z, 0) = f(x, y, z)$.

(2) The price of a financial instrument known as a European call option satisfies a partial differential equation known as the Black-Scholes equation:

$$F_t = rF - rSF_S - \frac{1}{2}\sigma^2 S^2 F_{SS} \text{ for } S > 0 \text{ and } 0 < t < T.$$

The price of the option is $F(S, t)$, the risk-free interest rate is r, the volatility in the price of the a security S is denoted σ^2, and t represents time. If

$$S = Ke^x,$$
$$t = T - \frac{2\tau}{\sigma^2},$$
$$F(S, t) = Ke^{\alpha x + \beta \tau}u(x, \tau),$$

where $K > 0$ and $T > 0$ are constants, then show that the Black-Scholes equation can be re-written as the one-dimensional heat equation

$$u_\tau = u_{xx} \text{ for } -\infty < x < \infty \text{ and } \tau > 0$$

when constants α and β are chosen so that

$$\alpha = \frac{1}{2}\left(1 - \frac{2r}{\sigma^2}\right) \text{ and } \beta = -\frac{1}{4}\left(1 + \frac{2r}{\sigma^2}\right)^2.$$

Chapter 5

The Wave Equation

In Chap. 1, the one-dimensional wave equation and the corresponding initial boundary value problems arising from various physical situations were described. This chapter will concentrate on the basic techniques of finding and interpreting the solutions of these initial boundary value problems. First, the method of separation of variables is used to solve the initial boundary value problem of Dirichlet type in Sec. 5.1. Next, Sec. 5.2 introduces d'Alembert's solution as well as topics related to wave propagation. The rest of the chapter is devoted to nonhomogeneous initial boundary value problems and some theoretical results including the energy integral and uniqueness of solutions. Solutions of the wave equation possess very different properties compared to those of the heat equation. While the heat equation serves as an example from the family of parabolic equations, the wave equation is a member of the class of hyperbolic equations.

5.1 Wave Equation with Homogeneous Boundary Conditions

For a perfectly elastic and flexible string with fixed ends and no external forces, the one-dimensional wave equation with the following initial and boundary conditions serves as a mathematical model:

$$u_{tt} = c^2 u_{xx} \text{ for } 0 < x < L \text{ and } t > 0,$$
$$u(0,t) = u(L,t) = 0 \text{ for } t > 0, \tag{5.1}$$
$$u(x,0) = f(x) \text{ and } u_t(x,0) = g(x) \text{ for } 0 < x < L.$$

For the initial boundary value problem of Eq. (5.1), the partial differential equation is linear and both the equation and the boundary conditions are homogeneous. Therefore, the method of separation of variables and the

181

Principle of Superposition is expected to apply. In fact, the procedure for solving this initial boundary value problem is similar to the procedure used for the case of the heat equation. The steps will be as follows:

(1) Find the product solutions of the corresponding boundary value problem by the method of separation of variables.
(2) Use the Principle of Superposition to create a formal solution of the initial boundary value problem as a linear combination of the product solutions found in the previous step.
(3) Verify mathematically that the function formed in the previous step is indeed a solution of the wave equation and satisfies the required boundary and initial conditions. The theory of Fourier series will play an essential role in determining the coefficients of the infinite series.

5.1.1 *Separation of Variables*

Similar to the case of the one-dimensional heat equation, the first step in finding the solution to the wave equation is to find the product solutions of the boundary value problem.

$$u_{tt} = c^2 u_{xx} \text{ for } 0 < x < L \text{ and } t > 0,$$
$$u(0,t) = u(L,t) = 0 \text{ for } t > 0. \tag{5.2}$$

Assuming a product solution of the form $u(x,t) = X(x)T(t)$ and substituting it in the wave equation produce

$$X(x)T''(t) = c^2 X''(x)T(t) \text{ which implies } \frac{X''}{X} = \frac{T''}{c^2 T}.$$

The last equality holds only when both X''/X and $T''/(c^2 T)$ are equal to the same constant, say $-\sigma$. Therefore, $u(x,t) = X(x)T(t)$ is a solution of the wave equation if and only if

$$X'' + \sigma X = 0 \text{ and } T'' + \sigma c^2 T = 0,$$

where σ is referred to as the separation constant in Sec. 1.6. The boundary conditions of Eq. (5.2) imply that

$$X(0)T(t) = X(L)T(t) = 0.$$

Therefore $u(x,t) = X(x)T(t)$ is a nonzero (or nontrivial) solution of Eq. (5.2) if and only if

$$X''(x) + \sigma X(x) = 0,$$
$$X(0) = X(L) = 0, \tag{5.3}$$

and

$$T''(t) + \sigma c^2 T(t) = 0. \tag{5.4}$$

The boundary value problem in Eq. (5.3) has been solved in the context of the heat equation. Equation (5.3) has nontrivial solutions when $\sigma = n^2\pi^2/L^2$ and the corresponding nontrivial solutions are

$$X(x) = A\sin\frac{n\pi x}{L} \tag{5.5}$$

for each $n \in \mathbb{N}$, where $A \neq 0$ is an arbitrary constant. As before, the constants $n^2\pi^2/L^2$ for $n \in \mathbb{N}$ are called the eigenvalues of the boundary value problem in Eq. (5.3) and the corresponding solutions of Eq. (5.5) are called the eigenfunctions associated with the eigenvalue. The eigenvalues and eigenfunctions can be indexed by the natural numbers and thus will be denoted as $\sigma_n = n^2\pi^2/L^2$ and $X_n(x) = \sin(n\pi x/L)$ for $n \in \mathbb{N}$.

When $\sigma_n = n^2\pi^2/L^2$, Eq. (5.4) becomes

$$T''(t) + \frac{n^2\pi^2 c^2}{L^2}T(t) = 0.$$

This is a linear ordinary differential equation of second order whose general solution is

$$T_n(t) = c_n \cos\frac{n\pi c\, t}{L} + d_n \sin\frac{n\pi c\, t}{L}$$

where c_n and d_n are arbitrary constants and $n \in \mathbb{N}$. These solutions are also indexed by the natural numbers. The product solutions of the boundary value problem of Eq. (5.2) can be expressed as

$$u_n(x,t) = \sin\frac{n\pi x}{L}\left(c_n \cos\frac{n\pi c\, t}{L} + d_n \sin\frac{n\pi c\, t}{L}\right). \tag{5.6}$$

These solutions are usually referred to as the **fundamental solutions** of Eq. (5.1). The reader should note that any such function $u_n(x,t)$ solves the partial differential equation and satisfies the boundary conditions in Eq. (5.1). The issue of satisfying the initial condition for the wave equation remains. The Principle of Superposition allows a formal solution to Eq. (5.1) to be constructed from these fundamental solutions.

5.1.2   Solution of the Initial Boundary Value Problem

Guided what has been done in the case of the one-dimensional heat equation, a formal solution to Eq. (5.1) of the following form is sought,

$$u(x,t) = \sum_{n=1}^{\infty} u_n(x,t) = \sum_{n=1}^{\infty}\left(c_n\cos\frac{n\pi c\, t}{L} + d_n\sin\frac{n\pi c\, t}{L}\right)\sin\frac{n\pi x}{L} \tag{5.7}$$

where c_n and d_n are constants to be determined. Clearly, the function defined in Eq. (5.7), if well-defined, satisfies the boundary conditions. However, can c_n and d_n be chosen such that $u(x,t)$ is well-defined (in other words, that the infinite series in Eq. (5.7) converges)? If so, is $u(x,t)$ still a solution of Eq. (5.1) and are both initial conditions satisfied? Specifically the following three questions need to be answered:

(1) Is $u(x,t)$ well-defined on $[0,L] \times [0,\infty)$?
(2) Can c_n and d_n be chosen such that the initial conditions are satisfied?

$$u(x,0) = f(x) \text{ and } u_t(x,0) = g(x) \text{ for } 0 < x < L. \qquad (5.8)$$

(3) Is the resulting $u(x,t)$ a solution of the wave equation?

Of course, the answers to all three questions are related. It is more convenient to answer the second question first under the assumption that $u(x,t)$ in Eq. (5.7) is well-defined. Setting $t = 0$ yields

$$u(x,0) = \sum_{n=1}^{\infty} c_n \sin \frac{n\pi x}{L}.$$

Based on the work done on Fourier series in Chap. 3, it is clear that c_n should be chosen to be the Fourier sine coefficients. To do so, treat $f(x)$ as its own odd $2L$-periodic function, and define

$$c_n = \frac{2}{L} \int_0^L f(x) \sin \frac{n\pi x}{L} \, dx \quad \text{for } n \in \mathbb{N}. \qquad (5.9)$$

Actually, by the convergence theorem of Fourier series, the first initial condition in Eq. (5.8) holds as long as $f(x)$ is piecewise smooth and $f(0) = f(L) = 0$, that is

$$u(x,0) = \sum_{n=1}^{\infty} c_n \sin \frac{n\pi x}{L} = f(x) \qquad (5.10)$$

for $x \in [0,L]$.

To accomplish the next step, assume that $u(x,t)$ in Eq. (5.7) is well-defined and that the series can be differentiated term by term with respect to t. Differentiating with respect to t and setting $t = 0$ in the result produce

$$u_t(x,0) = \sum_{n=1}^{\infty} \frac{n\pi c}{L} d_n \sin \frac{n\pi x}{L}.$$

Therefore, the expression $n\pi c \, d_n / L$ should be chosen as the Fourier sine coefficients of $g(x)$, that is

$$\frac{n\pi c}{L} d_n = \frac{2}{L} \int_0^L g(x) \sin \frac{n\pi x}{L} \, dx,$$

or equivalently,

$$d_n = \frac{2}{n\pi c} \int_0^L g(x) \sin \frac{n\pi x}{L} \, dx. \tag{5.11}$$

This produces, under appropriate assumptions on $g(x)$,

$$u_t(x,0) = \sum_{n=1}^{\infty} \frac{n\pi c}{L} d_n \sin \frac{n\pi x}{L} = g(x). \tag{5.12}$$

Substituting c_n and d_n defined in Eq. (5.9) and Eq. (5.11) into the infinite series in Eq. (5.7), produces a formal solution of the initial boundary value problem in Eq. (5.1). Of course it must be assumed that $f(x)$ and $g(x)$ satisfy the conditions necessary to guarantee the convergence of the series in Eq. (5.10) and Eq. (5.12) to the functions $f(x)$ and $g(x)$, respectively, where $f(x)$ and $g(x)$ have been considered as their own odd $2L$-periodic extensions.

The justification of the conclusions in the first and third questions posed above is quite different from the case of heat equation since there is no exponentially decaying term to guarantee the convergence of the series in Eq. (5.7). The issue of convergence will be addressed in Sec. 5.3.

Example 5.1. The ends of a stretched string of length $L = 1$ are fixed at $x = 0$ and $x = 1$. The string is set into motion by displacing the string in a piecewise linear fashion so that the initial displacement at position x is

$$f(x) = \begin{cases} 0 & \text{if } 0 \le x \le 1/4, \\ x - 1/4 & \text{if } 1/4 \le x \le 3/8, \\ 1/2 - x & \text{if } 3/8 \le x \le 1/2, \\ 0 & \text{if } 1/2 \le x \le 1. \end{cases}$$

The string is released from this displacement at $t = 0$. Assuming that $c = 1$, write down the initial boundary value problem for the motion of the string and solve the initial boundary value problem.

Solution. The assumptions given imply that the initial boundary value problem for the displacement of the string is:

$$u_{tt} = u_{xx} \text{ for } 0 < x < 1 \text{ and } t > 0,$$

$$u(0,t) = u(1,t) = 0 \text{ for } t > 0,$$

$$u(x,0) = f(x) \text{ and } u_t(x,0) = 0 \text{ for } 0 < x < 1.$$

A formal solution of the initial boundary value problem is

$$u(x,t) = \sum_{n=1}^{\infty} [c_n \cos(n\pi t) + d_n \sin(n\pi t)] \sin(n\pi x),$$

where

$$c_n = 2 \int_0^1 f(x) \sin(n\pi x)\, dx$$

$$= 2 \int_{1/4}^{3/8} (x - 1/4) \sin(n\pi x)\, dx + 2 \int_{3/8}^{1/2} (1/2 - x) \sin(n\pi x)\, dx$$

$$= \frac{2}{n\pi} \left[\left(\frac{1}{4} - x \right) \cos(n\pi x) + \frac{\sin(n\pi x)}{n\pi} \right]_{1/4}^{3/8}$$

$$+ \frac{2}{n\pi} \left[\left(x - \frac{1}{2} \right) \cos(n\pi x) - \frac{\sin(n\pi x)}{n\pi} \right]_{3/8}^{1/2}$$

$$= \frac{2}{n^2 \pi^2} \left(2 \sin \frac{3n\pi}{8} - \sin \frac{n\pi}{4} - \sin \frac{n\pi}{2} \right) = \frac{8}{n^2 \pi^2} \sin \frac{3n\pi}{8} \sin^2 \frac{n\pi}{16}.$$

The last equality is made possible by using the product-to-sum formulas for the sine function. The coefficients $d_n = 0$ since $g(x) = 0$. Therefore the formal solution can be expressed as

$$u(x, t) = \frac{8}{\pi^2} \sum_{n=1}^{\infty} \frac{1}{n^2} \sin \frac{3n\pi}{8} \sin^2 \frac{n\pi}{16} \cos(n\pi t) \sin(n\pi x).$$

Plots of the displacement of the string from equilibrium for several different values of t are shown in Fig. 5.1.

Note that the series in the last example certainly converges. However, taking the second derivative term by term results in the loss of the factor $1/n^2$. Thus it is not immediately obvious that the resulting series for the second partial derivative converges. This point will be investigated further in Sec. 5.3.

Example 5.2. The ends of a stretched string of length $L = 2$ are fixed at $x = 0$ and $x = 2$. The string is in its equilibrium position and is set into motion by striking the portion occupying the interval from $x = 1/2$ to $x = 1$. Assume the strike gives the string an initial velocity given by the following function:

$$g(x) = \begin{cases} 0 & \text{if } 0 \le x < 1/2, \\ 1 & \text{if } 1/2 \le x \le 1, \\ 0 & \text{if } 1 < x \le 2. \end{cases} \tag{5.13}$$

Write down the initial boundary value problems for the motion of the string and solve the initial boundary value problem.

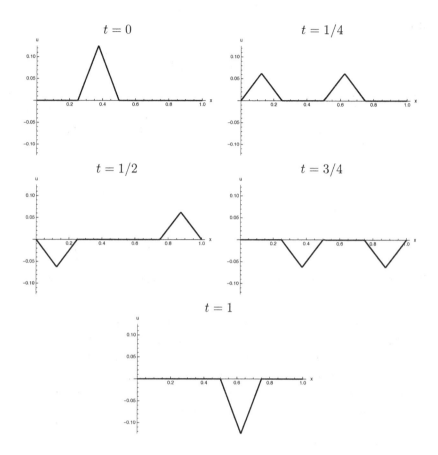

Fig. 5.1 Displacement of the string from equilibrium for the situation described in Example 5.1. The displacement is depicted at five different times $t = 0$, $1/4$, $1/2$, $3/4$, and 1. For each value of t the displacement of the string is piecewise linear.

Solution. With the information given, the initial boundary value problem for the string is the following:

$$u_{tt} = c^2 u_{xx} \text{ for } 0 < x < 2 \text{ and } t > 0,$$
$$u(0, t) = u(2, t) = 0 \text{ for } t > 0, \qquad (5.14)$$
$$u(x, 0) = 0 \text{ and } u_t(x, 0) = g(x) \text{ for } 0 < x < 2,$$

where $g(x)$ is given in Eq. (5.13) and $c > 0$ is a constant.

By the solution procedure described above, a formal solution of the initial boundary value problem is

$$u(x,t) = \sum_{n=1}^{\infty} \left[c_n \cos \frac{cn\pi t}{2} + d_n \sin \frac{cn\pi t}{2} \right] \sin \frac{n\pi x}{2},$$

where $c_n = 0$ since $u(x,0) = 0$ and

$$d_n = \frac{2}{n\pi c} \int_0^2 g(x) \sin \frac{n\pi x}{2} \, dx = \frac{2}{n\pi c} \int_{1/2}^1 \sin \frac{n\pi x}{2} \, dx = \frac{-4}{n^2\pi^2 c} \left[\cos \frac{n\pi x}{2} \right]_{1/2}^1$$

$$= \frac{4}{n^2\pi^2 c} \left[\cos \frac{n\pi}{4} - \cos \frac{n\pi}{2} \right].$$

Therefore, the formal solution of the initial boundary value problem in Eq. (5.14) can be expressed as

$$u(x,t) = \frac{4}{\pi^2 c} \sum_{n=1}^{\infty} \frac{1}{n^2} \left[\cos \frac{n\pi}{4} - \cos \frac{n\pi}{2} \right] \sin \frac{n\pi c t}{2} \sin \frac{n\pi x}{2}.$$

Plots of the displacement of the string from equilibrium for several different values of t are shown in Fig. 5.2.

Similar to the first example, the series in the second example certainly converges. However, upon taking the second derivative term by term, the factor $1/n^2$ is lost and thus, it is not immediately obvious that the resulting series for the second partial derivative converges.

In Example 5.1, the vibration of the string is caused by the initial displacement and such a string is called a **plucked string**. While in Example 5.2, the string is initially at rest and the motion was caused by the initial velocity of the string. Such a string is called a **struck string**. In general, the vibration of a string with fixed ends as described by the initial boundary value problem of Eq. (5.1) is a superposition of the motion of the following plucked string and struck string.

Plucked String

$$u_{tt} = c^2 u_{xx} \text{ for } 0 < x < L \text{ and } t > 0,$$

$$u(0,t) = u(L,t) = 0 \text{ for } t > 0,$$

$$u(x,0) = f(x) \text{ and } u_t(x,0) = 0 \text{ for } 0 < x < L.$$

Struck String

$$u_{tt} = c^2 u_{xx} \text{ for } 0 < x < L \text{ and } t > 0,$$

$$u(0,t) = u(L,t) = 0 \text{ for } t > 0,$$

$$u(x,0) = 0 \text{ and } u_t(x,0) = g(x) \text{ for } 0 < x < L.$$

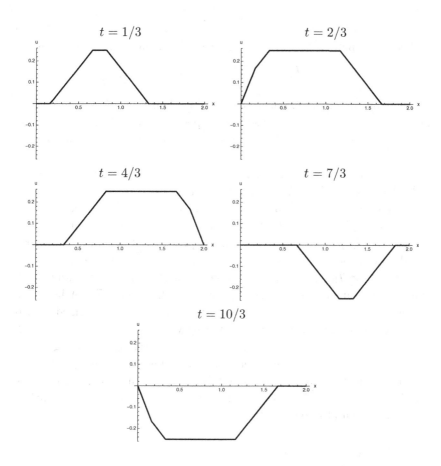

Fig. 5.2   Displacement of the string from equilibrium for the situation described in Example 5.2. Configurations of the string at several different values of t are shown. The value $c = 1$ was used to generate this solution.

5.1.3   *Another Interpretation of the Solution*

So far this section has described the solution to the initial boundary value problem for the wave equation model of a vibrating string stated in Eq. (5.1) as

$$u(x,t) = \sum_{n=1}^{\infty} \left(c_n \cos \frac{n\pi c\, t}{L} + d_n \sin \frac{n\pi c\, t}{L} \right) \sin \frac{n\pi x}{L},$$

with c_n and d_n given by Eq. (5.9) and Eq. (5.11). Many musical instruments (for examples, the guitar, violin, and piano) produce sound through the vibration of one or more strings whose ends are fixed just as in the mathematical model studied here. In musical language, the expression

$$u_n(x,t) = \left(c_n \cos \frac{n\pi c t}{L} + d_n \sin \frac{n\pi c t}{L} \right) \sin \frac{n\pi x}{L} \qquad (5.15)$$

is called the nth **normal mode** of the vibration. The formal solution to the initial boundary value problem is a superposition of the normal modes. The intensity (the "loudness") of the normal mode depends on its amplitude $\sqrt{c_n^2 + d_n^2}$.

The normal mode corresponding to $n = 1$ is called the **first harmonic** or the **fundamental mode**. This fundamental mode has circular frequency $\pi c/L$ which is called the **fundamental frequency**. The larger the frequency, the higher the pitch of the sound produced. Recall that the wave speed $c = \sqrt{T_0/\rho}$, where T_0 is the tension present in the string, and therefore the fundamental frequency $\pi c/L = (\pi/L)\sqrt{T_0/\rho}$ can be changed by adjusting L or T_0, since the mass density of the string is usually fixed. Thus the instrument can be tuned by varying the tension T_0. A larger T_0 produces a higher fundamental frequency. While playing a stringed instrument, the player can also vary the pitch by varying the effective length L of the string by pressing down on the string. Shortening the string makes the fundamental frequency greater and, as a result, produces higher notes.

The nth normal mode is also called the nth harmonic and has frequency $n\pi c/L = (n\pi/L)\sqrt{T_0/L}$. Each of these frequencies is an integer multiple of the fundamental frequency (Fig. 5.3).

5.2   d'Alembert's Approach

A clever, alternative method for solving the wave equation will be described in this section. The procedure begins by eliminating the boundary conditions and considering the string as infinitely long. Consider the following initial value problem for the one-dimensional wave equation:

$$\begin{aligned} u_{tt} = c^2 u_{xx} \text{ for } -\infty < x < \infty \text{ and } t > 0, \\ u(x,0) = f(x) \text{ and } u_t(x,0) = g(x) \text{ for } -\infty < x < \infty, \end{aligned} \qquad (5.16)$$

This initial value problem can be solved by an idea similar to the one used to solve an initial value problem of an ordinary differential equation (after all Eq. (5.16) is an initial value problem). First the general solution is found, then the initial conditions can be used to determine the specific solution. The method presented below is called d'Alembert's approach and the solution obtained is called **d'Alembert's solution**.

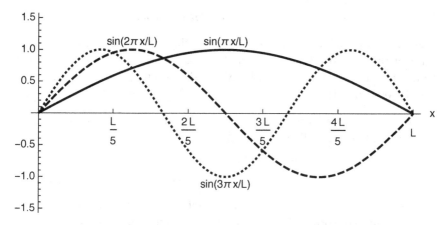

Fig. 5.3   Plots of the first three normal modes for the wave equation. These curves are generated by choosing $t = 0$ and assuming $c_1 = c_2 = c_3 = 1$. The first harmonic or fundamental mode is depicted as a solid curve.

5.2.1   *The General Solution of the Wave Equation*

To find the general solution of the wave equation in Eq. (5.16), make a change of variable by letting $\xi = x + ct$ and $\eta = x - ct$. The displacement of the string $u = u(x, t)$ then becomes the function $u = u(\xi, \eta)$. Calculating derivatives by means of the chain rule produces:

$$u_x = u_\xi \xi_x + u_\eta \eta_x = u_\xi + u_\eta,$$

$$u_t = u_\xi \xi_t + u_\eta \eta_t = c\, u_\xi - c\, u_\eta,$$

and

$$u_{xx} = u_{\xi\xi} + 2u_{\xi\eta} + u_{\eta\eta},$$

$$u_{tt} = c^2 u_{\xi\xi} - 2c^2 u_{\xi\eta} + c^2 u_{\eta\eta}.$$

Therefore in terms of the variables ξ and η, the wave equation in Eq. (5.16) can be written as

$$-4c^2 u_{\xi\eta} = 0 \iff u_{\xi\eta} = (u_\xi)_\eta = 0. \tag{5.17}$$

Integrating Eq. (5.17) with respect to η first, then with respect to ξ, produces

$$u_\xi(\xi, \eta) = \phi(\xi) \implies u = \int \phi(\xi)\, d\xi + \Psi(\eta) = \Phi(\xi) + \Psi(\eta),$$

where Φ and Ψ can be any smooth functions of ξ and η, respectively. Using the definitions of ξ and η the function $u(x, t)$ can be written as

$$u(x, t) = \Phi(x + ct) + \Psi(x - ct). \tag{5.18}$$

Since Φ and Ψ are arbitrary smooth functions, Eq. (5.18) is usually referred to as d'Alembert's solution or the general solution of the wave equation.

5.2.2   The Solution of the Initial Value Problem

Since $u(x,t)$ as defined in Eq. (5.18) solves the wave equation, $u_{tt} = c^2 u_{xx}$ for arbitrary, smooth functions Φ and Ψ, the initial conditions can be used to determine these functions. The two initial conditions in Eq. (5.16) will be treated separately.

Case $u(x,0) = f(x)$ and $u_t(x,0) = 0$: The function $u(x,t)$ of Eq. (5.18) must satisfy the following equations.

$$u(x,0) = \Phi(x) + \Psi(x) = f(x),$$
$$u_t(x,0) = c\,\Phi'(x) - c\,\Psi'(x) = 0.$$

Integrating the second equation yields $\Phi(x) = \Psi(x) + K$ where K is an arbitrary constant. Substituting this in the first equation produces

$$\Psi(x) = \frac{1}{2}(f(x) - K) \text{ and } \Phi(x) = \frac{1}{2}(f(x) + K)$$

which implies that

$$u(x,t) = \frac{1}{2}\left[f(x - ct) + f(x + ct)\right]. \tag{5.19}$$

If $f(x)$ is twice differentiable, $u(x,t)$ is a solution of the wave equation with the given initial conditions. This is the case of a plucked string.

Case $u(x,0) = 0$ and $u_t(x,0) = g(x)$: Function $u(x,t)$ must satisfy the equations,

$$\Phi(x) + \Psi(x) = 0,$$
$$c\,\Phi'(x) - c\,\Psi'(x) = g(x).$$

Differentiating the first equation and substituting into the second equation imply $2c\,\Phi'(x) = g(x)$ and thus

$$\Phi(x) = \frac{1}{2c}\int_0^x g(s)\,ds + K,$$

$$\Psi(x) = -\frac{1}{2c}\int_0^x g(s)\,ds - K,$$

where K is an arbitrary constant. Therefore

$$u(x,t) = \frac{1}{2c}\int_0^{x+ct} g(s)\,ds - \frac{1}{2c}\int_0^{x-ct} g(s)\,ds$$

$$= \frac{1}{2c}\int_{x-ct}^{x+ct} g(s)\,ds. \tag{5.20}$$

If $g(x)$ is continuously differentiable, then $u(x,t)$ is a solution of the wave equation. This is the case of a struck string.

For the general case described in Eq. (5.16) with both nonzero initial displacement and initial velocity, by the Principle of Superposition, the general solution is

$$u(x,t) = \frac{1}{2}\left[f(x - ct) + f(x + ct)\right] + \frac{1}{2c}\int_{x-ct}^{x+ct} g(s)\,ds. \qquad (5.21)$$

The solution formula in Eq. (5.21) reveals how the initial values influence the displacement of the string at location x and time t. Consider the displacement $u(x_0, t_0)$ at a fixed position x_0 and time t_0.

$$u(x_0,t_0) = \frac{1}{2}\left[f(x_0 - ct_0) + f(x_0 + ct_0)\right] + \frac{1}{2c}\int_{x_0-ct_0}^{x_0+ct_0} g(s)\,ds.$$

The value of u at (x_0, t_0) depends only on the values of f at the points $x_0 + ct_0$ and $x_0 - ct_0$ and the values of g on the interval $[x_0 - ct_0, x_0 + ct_0]$. The interval $[x_0 - ct_0, x_0 + ct_0]$ is called the **domain of dependence** for $u(x_0, t_0)$. Figure 5.4 illustrates the domain of dependence. Initial conditions outside of the domain of dependence have no effect on the displacement u at point (x_0, t_0).

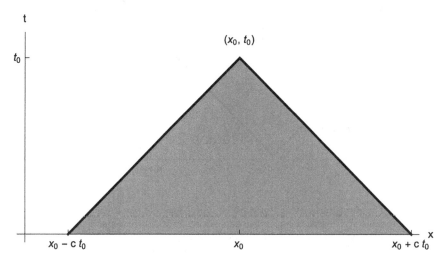

Fig. 5.4 The domain of dependence of the solution to the initial value problem of Eq. (5.16) at point (x_0, t_0).

Lines defined by $x - ct = K_1$ and $x + ct = K_2$ where K_1 and K_2 are constants are called **characteristics** of the wave equation. At the point with coordinates (x_0, t_0) two characteristics intersect, namely the lines $x - ct = x_0 - ct_0$ and $x + ct = x_0 + ct_0$. In turn these characteristics intersect the x-axis at the points $(x_0 - ct_0, 0)$ and $(x_0 + ct_0, 0)$ respectively. From each of these points emanate two characteristics, one with slope $1/c$ and one with slope $-1/c$. Together these four lines divide the xt-plane (with $t \geq 0$) into the six regions pictured in Fig. 5.5. The displacements at all points within Regions 1 and 3 are independent of the domain of dependence $[x_0 - ct_0, x_0 + ct_0]$ for point (x_0, t_0). The points in Region 2 have displacements which depend only on the values of the initial conditions within the domain of dependence of $u(x_0, t_0)$. Points within Regions 4 and 6 have displacements which depend on initial data in $[x_0 - ct_0, x_0 + ct_0]$ and other points as well. No characteristic intersecting the x-axis somewhere in the interval $[x_0 - ct_0, x_0 + ct_0]$ passes through Region 5, thus displacements at points in Region 5 are independent of the initial displacement of the string in the interval $[x_0 - ct_0, x_0 + ct_0]$, though they do depend on the initial velocity of the string in that interval.

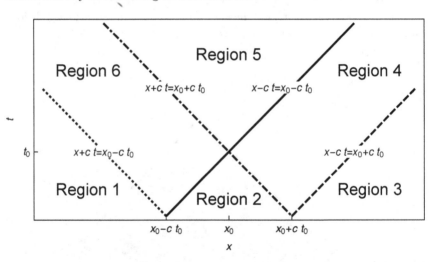

Fig. 5.5   The characteristics of the wave equation separate the xt-plane (with $t \geq 0$) into six regions. Within each region the dependence of the displacement u on the initial data in the interval $[x_0 - ct_0, x_0 + ct_0]$ can be determined.

On the other hand, given a point x_0, the values of the initial conditions at x_0 influence the value of the displacement u at point (x, t) only if the domain of dependence of $u(x, t)$ contains x_0, or equivalently only if

$$x - ct \leq x_0 \leq x + ct.$$

The values of the initial data at x_0 influence the displacement $u(x, t)$ at point (x, t) only if (x, t) falls in the **range of influence** defined as

$$\{(x, t) \,|\, t > 0 \text{ and } x_0 - ct \le x \le x_0 + ct\}.$$

See Fig. 5.6. The concept of the range of influence of a point can be generalized to an interval $[a, b]$. The values of the initial conditions in Eq. (5.16) on the interval $[a, b]$ influence the value of the displacement $u(x, t)$ at point (x, t) only if the domain of dependence of $u(x, t)$ intersects $[a, b]$. Thus the **range of influence of** $[a, b]$ is defined as the set

$$\{(x, t) \,|\, t > 0 \text{ and } x + ct \ge a \text{ and } x - ct \le b\}.$$

and is illustrated in Fig. 5.7.

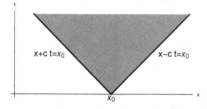

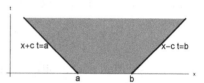

Fig. 5.6 The range of influence of the initial data at point x_0 on the solution to the initial value problem of Eq. (5.16).

Fig. 5.7 The range of influence of the initial data in interval $[a, b]$ on the solution to the initial value problem of Eq. (5.16).

5.2.3 *Propagation of Waves – The Plucked String*

In the case of a plucked string, there is an initial displacement of the string and no initial velocity. The initial value problem can be expressed as,

$$u_{tt} = c^2 u_{xx} \text{ for } -\infty < x < \infty \text{ and } t > 0,$$
$$u(x, 0) = f(x) \text{ and } u_t(x, 0) = 0 \text{ for } -\infty < x < \infty. \tag{5.22}$$

As derived above, the solution of Eq. (5.22) is given by the function in Eq. (5.19). The function $u(x, t)$ is just the average of $f(x + ct)$ and $f(x - ct)$. For each fixed $t > 0$, the graph of $f(x + ct)$ is just the graph of $f(x)$, the initial displacement, shifted to the left ct units. In other words $f(x + ct)$ represents a horizontal shift of the displacement $f(x)$, or equivalently, the initial wave is propagated to the left ct units at speed c. A similar explanation can be given for the $f(x - ct)$ term. Therefore, in a sense, the initial displacement $u(x, 0) = f(x)$ moves to the right at speed c while an identical copy moves to the left at speed c. The solution to the initial value problem in Eq. (5.22) is the average of these traveling initial displacements. Note that the $f(x + ct)$ portion of the solution of the plucked string problem is constant along the family of straight lines where $x + ct$ is a constant. Likewise the other portion $f(x - ct)$ is constant along the family of straight lines where $x - ct$ is a constant.

The initial displacement of the plucked string propagates along the characteristic lines. The displacement of the string at (x_0, t_0) with $t_0 > 0$ is related to the two characteristics which intersect at (x_0, t_0). These characteristics have equations $x_0 - ct_0 = x - ct$ and $x_0 + ct_0 = x + ct$, and intersect the x-axis at the points $(x_0 - ct_0, 0)$ and $(x_0 + ct_0, 0)$ respectively. By Eq. (5.19) the displacement of the string at x_0, at time t_0, is merely the average of the initial displacements of the string at $x_0 - ct_0$ and $x_0 + ct_0$.

Example 5.3. Let $f(x)$ be defined by

$$f(x) = \begin{cases} 2 & \text{if } -1 < x < 1, \\ 0 & \text{otherwise.} \end{cases} \tag{5.23}$$

Find the solution of the initial value problem in Eq. (5.16) using this initial displacement and zero initial velocity and discuss how the initial wave is propagated by graphing the solutions for different values of t.

Solution. The solution can be expressed as

$$u(x, t) = \frac{1}{2} [f(x - ct) + f(x + ct)]$$

where $f(x)$ is given in Eq. (5.23). Using the definition of $f(x)$ results in

$$f(x - ct) = \begin{cases} 2 & \text{if } -1 < x - ct < 1, \\ 0 & \text{otherwise} \end{cases}$$

and

$$f(x + ct) = \begin{cases} 2 & \text{if } -1 < x + ct < 1, \\ 0 & \text{otherwise.} \end{cases}$$

The definition of function $f(x)$ makes the characteristics of the form $x + ct = \pm 1$ and $x - ct = \pm 1$ of special interest. These four characteristics divide the xt-plane with $t \geq 0$ into six different regions as illustrated in Fig. 5.5. The regions can be precisely described as sets satisfying the following inequalities.

Region 1: $\{(x, t) \mid x + ct < -1\}$,
Region 2: $\{(x, t) \mid -1 < x - ct \text{ and } x + ct < 1\}$,
Region 3: $\{(x, t) \mid 1 < x - ct\}$,

Region 4: $\{(x,t) \mid 1 < x + ct \text{ and } -1 < x - ct < 1\}$,
Region 5: $\{(x,t) \mid 1 < x + ct \text{ and } x - ct < -1\}$,
Region 6: $\{(x,t) \mid -1 < x + ct < 1 \text{ and } x - ct < -1\}$.

The solution can be piecewise defined for these six regions.

$$u(x,t) = \begin{cases} 0 & \text{if } x + ct < -1, \\ 2 & \text{if } -1 < x - ct \text{ and } x + ct < 1, \\ 0 & \text{if } 1 < x - ct, \\ 1 & \text{if } 1 < x + ct \text{ and } -1 < x - ct < 1, \\ 0 & \text{if } 1 < x + ct \text{ and } x - ct < -1, \\ 1 & \text{if } -1 < x + ct < 1 \text{ and } x - ct < -1. \end{cases}$$

Figure 5.8 illustrates this solution as a density plot. It might be more revealing to consider the solution as a function of x for various fixed t values. Figure 5.9 illustrates the graph of $u(x,t)$ for $t = 0$, $t = 1/(2c)$, $t = 1/c$, and $t = 2/c$.

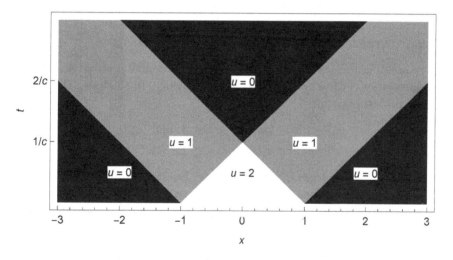

Fig. 5.8   A density plot of the solution to Example 5.3.

The piecewise constant initial displacement of the string given in Eq. (5.23) was used for convenience to demonstrate the phenomenon of wave propagation. A real, physical string could not assume this initial displacement.

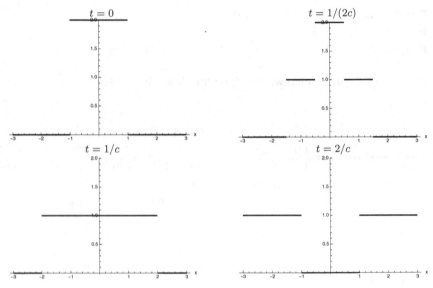

Fig. 5.9   A plot of the solution to Example 5.3 for $t = 0$, $t = 1/(2c)$, $t = 1/c$, and $t = 2/c$.

5.2.4   *Propagation of Waves – the Struck String*

Now consider the case in which the string starts oscillation from its equilibrium position with a nonzero initial velocity $g(x)$. In this case, the initial value problem becomes

$$u_{tt} = c^2 u_{xx} \text{ for } -\infty < x < \infty \text{ and } t > 0,$$
$$u(x,0) = 0 \text{ and } u_t(x,0) = g(x) \text{ for } -\infty < x < \infty. \tag{5.24}$$

Although, as in the case of plucked string, Eq. (5.20) gives an explicit expression for the solution, it is not as easy to see how the initial velocity influences the vibration of the string in general. In special cases, the vibration of the struck string can be illustrated easily.

Example 5.4. Consider the initial value problem in Eq. (5.24) with

$$g(x) = \begin{cases} 1 & \text{if } -1 < x < 1, \\ 0 & \text{elsewhere.} \end{cases}$$

Find an expression of the solution and discuss the motion of the string geometrically.

Solution. According to Eq. (5.20) the displacement of the struck string is given by

$$u(x,t) = \frac{1}{2c} \int_{x-ct}^{x+ct} g(s)\,ds = \frac{1}{2c}\left[G(x+ct) - G(x-ct)\right],$$

where

$$G(z) = \int_0^z g(w)\,dw = \begin{cases} -1 & \text{if } z < -1, \\ z & \text{if } -1 \le z \le 1, \\ 1 & \text{if } z > 1. \end{cases}$$

As before, the characteristics $x + ct = \pm 1$ and $x - ct = \pm 1$ divide the xt-plane for $t \ge 0$ into the same six regions illustrated in Fig. 5.8 and the solution to the initial value problem is given by the following piecewise-defined function:

$$\begin{aligned} u(x,t) &= \frac{1}{2c}\left[G(x+ct) - G(x-ct)\right] \\ &= \frac{1}{2c} \begin{cases} 0 & \text{if } x+ct < -1, \\ 2ct & \text{if } -1 < x-ct \text{ and } x+ct < 1, \\ 0 & \text{if } 1 < x-ct, \\ 1-x+ct & \text{if } -1 < x-ct < 1 \text{ and } 1 < x+ct, \\ 2 & \text{if } 1 < x+ct \text{ and } x-ct < -1, \\ 1+x+ct & \text{if } x-ct < -1 \text{ and } -1 < x+ct < 1. \end{cases} \end{aligned}$$

A plot of this solution is shown in Fig. 5.10. "Snapshots" of the function $u(x,t)$ for several values of t are graphed in Fig. 5.11.

5.3 Solving the Wave Equation – Revisited

D'Alembert's solution to the wave equation was developed for the unbounded, infinitely long string. In this section d'Alembert's solution will be seen also to solve the initial boundary value problem for the finite-length string and provides a way for the convergence of the Fourier series solutions developed earlier to be established. Consider again the vibrating string with finite length L as described in Sec. 5.1, where two cases for the initial conditions are explored.

Case $g(x) = 0$: this is the case of the plucked string. For this situation, the Fourier coefficients $d_n = 0$ and the formal solution from Eq. (5.7) becomes

$$u(x,t) = \sum_{n=1}^{\infty} c_n \sin\frac{n\pi x}{L} \cos\frac{n\pi c t}{L} \tag{5.25}$$

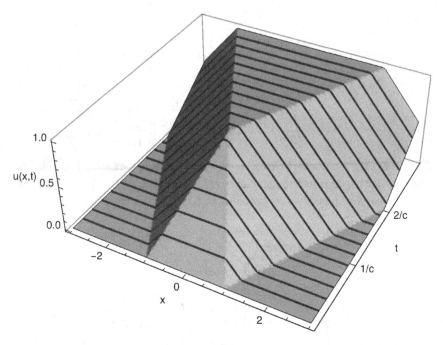

Fig. 5.10   A surface plot of the solution to Example 5.4. The upper half of the xt-plane is divided into six regions by the characteristics $x \pm ct = \pm 1$.

where c_n is given by the integral in Eq. (5.9). Using the product-to-sum trigonometric formula found in Eq. (3.8) the solution $u(x, t)$ given in Eq. (5.25) can be written as

$$u(x,t) = \frac{1}{2}\left[\sum_{n=1}^{\infty} c_n \sin\frac{n\pi(x+ct)}{L} + \sum_{n=1}^{\infty} c_n \sin\frac{n\pi(x-ct)}{L}\right]. \quad (5.26)$$

By Eq. (5.10) the two series in Eq. (5.26) converge and

$$u(x,t) = \frac{1}{2}\left[f(x+ct) + f(x-ct)\right]. \quad (5.27)$$

Therefore it has been established that the series in Eq. (5.7) converges and its sum has been found. In order to show that the function defined by Eq. (5.26) or Eq. (5.19) is actually a solution of the wave equation, it must be assumed that $f(x)$ is smooth enough. For example, if $f(0) = 0 = f(L)$ with $f'(x)$ and $f''(x)$ both continuous and if $f''(0) = 0 = f''(L)$, then the odd $2L$–periodic extension of $f(x)$ is twice differentiable and therefore by the results of the last section, $u(x,t)$ is a solution.

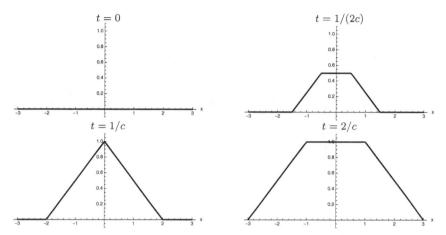

Fig. 5.11   A plot of the solution to Example 5.4 for $t = 0$, $t = 1/(2c)$, $t = 1/c$, and $t = 2/c$.

Case $f(x) = 0$: this is the case of the struck string. The formal solution in this case is given by

$$u(x,t) = \sum_{n=1}^{\infty} d_n \sin \frac{n\pi x}{L} \sin \frac{n\pi c t}{L} \tag{5.28}$$

where the d_n is given by the formula in Eq. (5.11). Once again by employing the product-to-sum identity found in Eq. (3.9), the series in Eq. (5.28) can be written as

$$u(x,t) = \frac{1}{2} \sum_{n=1}^{\infty} d_n \left[\cos \frac{n\pi(x - ct)}{L} - \cos \frac{n\pi(x + ct)}{L} \right].$$

Assuming the order of integration and summation can be interchanged, the previous equation implies that

$$u(x,t) = \frac{1}{2} \sum_{n=1}^{\infty} d_n \frac{n\pi}{L} \int_{x-ct}^{x+ct} \sin \frac{n\pi s}{L} \, ds$$

$$= \frac{1}{2} \int_{x-ct}^{x+ct} \left(\sum_{n=1}^{\infty} \left[d_n \frac{n\pi}{L} \right] \sin \frac{n\pi s}{L} \right) ds.$$

By Eq. (5.12),

$$
\begin{aligned}
u(x,t) &= \frac{1}{2} \int_{x-ct}^{x+ct} \left(\sum_{n=1}^{\infty} \left[\frac{2}{n\pi c} \int_0^L g(x) \sin \frac{n\pi x}{L} \, dx \frac{n\pi}{L} \right] \sin \frac{n\pi s}{L} \right) ds \\
&= \frac{1}{2c} \int_{x-ct}^{x+ct} \left(\sum_{n=1}^{\infty} \left[\frac{2}{L} \int_0^L g(x) \sin \frac{n\pi x}{L} \, dx \right] \sin \frac{n\pi s}{L} \right) ds \\
&= \frac{1}{2c} \int_{x-ct}^{x+ct} g(s) \, ds.
\end{aligned}
\tag{5.29}
$$

Again, if $g(x)$ is smooth enough, the function defined in Eq. (5.29) is a solution of the wave equation by the results of the last section.

For the initial boundary value problem of Example 5.1 the solution was found in the form of a Fourier series (repeated below):

$$
u(x,t) = \frac{8}{\pi^2} \sum_{n=1}^{\infty} \frac{1}{n^2} \sin \frac{3n\pi}{8} \sin^2 \frac{n\pi}{16} \cos(n\pi t) \sin(n\pi x).
$$

The nature of the motion of the string is all but impossible to determine from a glance at this infinite series. However, if d'Alembert's method is used to find the solution, the vibration phenomenon can be better understood.

First, $f(x)$ is extended as an odd 2-periodic function whose graph is shown in Fig. 5.12. The solution to the initial boundary value problem can then be expressed as

$$
u(x,t) = \frac{1}{2} \left[f(x+t) + f(x-t) \right].
$$

For fixed values of t, the solution can be plotted to illustrate the propagation of the initial displacement. At $t = 1/6$ half height copies of the initial displacement have moved $1/6$ units left and right as shown in Fig. 5.13. Similar graphs for $t = 1/3$ and $t = 2/3$ are shown in Fig. 5.14 and Fig. 5.15 respectively. Since f has been made 2-periodic the string returns to its initial configuration when

$$
\frac{1}{2} \left[f(x-t) + f(x+t) \right] = f(x).
$$

This occurs when $t = 2$ and again for any integer multiple of 2.

In general the d'Alembertian solution to the wave equation on a bounded string of length L can be expressed in the form of Eq. (5.21) where f and g are extended as odd, $2L$-periodic versions of the initial conditions. The notion of the propagation of the displacement along characteristics is helpful again for describing the solution. The characteristics are straight lines of

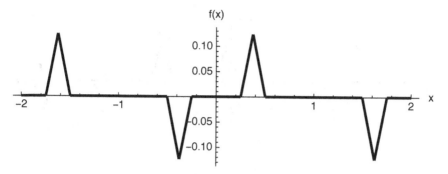

Fig. 5.12   A plot of the odd, period 2 extension of function $f(x)$ found in Example 5.1.

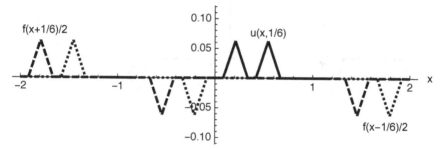

Fig. 5.13   A plot of $u(x, 1/6)$ found in Example 5.1. The 2–periodic odd extensions of the initial displacement have been shifted 1/6 units horizontally both left and right and scaled by a factor of 1/2. The graph of $u(x, 1/6)$ is the sum of these shifted periodic curves.

the form $x - ct = K_1$ and $x + ct = K_2$ where K_1 and K_2 are constants. To determine the displacement u at the point with coordinates (x_0, t_0) where $t_0 > 0$ consider the parallelogram formed by the characteristics. Let the point with coordinates (x_0, t_0) be the highest vertex (the vertex with the greatest t-coordinate) of the parallelogram, the two adjacent vertices are denoted as (x_a, t_a) and (x_c, t_c) and lie on the characteristics $x - ct = x_0 - ct_0$ and $x + ct = x_0 + ct_0$ respectively, and let the remaining vertex be denoted as (x_b, t_b). See Fig. 5.16 for a typical parallelogram. The vertices (x_a, t_a) and (x_c, t_c) are chosen so that the parallelogram lies in the region of the xt-plane where $0 \leq x \leq L$ and $t \geq 0$. As labeled in Fig. 5.16 the value of the displacement at $u(x_0, t_0)$ satisfies the equation

$$u(x_0, t_0) = u(x_a, t_a) + u(x_c, t_c) - u(x_b, t_b).$$

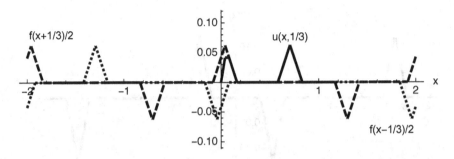

Fig. 5.14   A plot of $u(x, 1/3)$ found in Example 5.1.

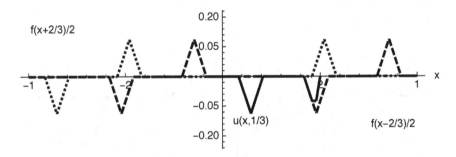

Fig. 5.15   A plot of $u(x, 2/3)$ found in Example 5.1.

In fact, recall from Eq. (5.18) that

$$
\begin{aligned}
u(x_a, t_a) &+ u(x_c, t_c) - u(x_b, t_b) \\
&= \Phi(x_a + c\,t_a) + \Psi(x_a - c\,t_a) + \Phi(x_c + c\,t_c) + \Psi(x_c - c\,t_c) \\
&\quad - \Phi(x_b + c\,t_b) - \Psi(x_b - c\,t_b) \\
&= \Phi(x_b + c\,t_b) + \Psi(x_0 - c\,t_0) + \Phi(x_0 + c\,t_0) + \Psi(x_b - c\,t_b) \\
&\quad - \Phi(x_b + c\,t_b) - \Psi(x_b - c\,t_b) \\
&= \Phi(x_0 + c\,t_0) + \Psi(x_0 - c\,t_0) = u(x_0, t_0).
\end{aligned}
$$

Since the points with coordinates (x_a, t_a), (x_b, t_b), and (x_c, t_c) all have t-coordinates smaller than t_0 then the displacement of the string at (x_0, t_0) can be determined from the displacement of the string "earlier" times.

This section ends with two examples of struck strings.

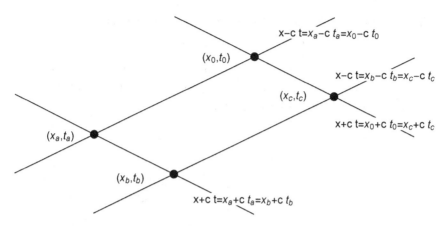

Fig. 5.16 Four points in the xt-plane lying on characteristics of the wave equation. If the value of the displacement u is known at any three of the points, the displacement can be determined at the fourth point.

Example 5.5. Find the solution to the initial boundary value problem

$$u_{tt} = c^2 u_{xx} \text{ for } 0 < x < L \text{ and } t > 0,$$

$$u(0,t) = u(L,t) = 0 \text{ for } t > 0,$$

$$u(x,0) = 0 \text{ and } u_t(x,0) = g(x) \text{ for } 0 < x < L, \text{ where}$$

$$g(x) = \begin{cases} 0 & \text{if } 0 < x < L/4, \\ 1 & \text{if } L/4 \le x \le 3L/4, \\ 0 & \text{if } 3L/4 < x < L \end{cases}$$

and graph the solution for several different values of t to show the behavior of the solution.

Solution. Since the initial displacement is zero and the initial velocity is nonzero, this initial boundary value problem may be interpreted as a model of a string struck by a broad hammer across the middle of its length. The solution is

$$u(x,t) = \frac{1}{2c} \int_{x-ct}^{x+ct} g_o(s)\, ds = \frac{1}{2c}\left[G(x+ct) - G(x-ct) \right]$$

where g_o is the odd, $2L$-periodic extension of g to the real number line and the function $G(x)$ is $2L$-periodic and defined for $0 \le x \le 2L$ as

$$G(x) = \int_0^x g_o(s)\, ds = \begin{cases} 0 & \text{if } 0 \le x < L/4, \\ x - L/4 & \text{if } L/4 \le x \le 3L/4, \\ L/2 & \text{if } 3L/4 < x < 5L/4, \\ -x + 7L/4 & \text{if } 5L/4 \le x \le 7L/4, \\ 0 & \text{if } 7L/4 < x \le 2L. \end{cases}$$

Therefore the solution may be expressed using modular arithmetic as

$$u(x,t) = \frac{1}{2c}\left[G((x+ct) \bmod 2L) - G((x-ct) \bmod 2L)\right],$$

for $0 \le x \le L$ and $t \ge 0$. Figure 5.17 graphs the solution for $0 \le t \le L/(2c)$. Readers trying to graph the solution for additional values of t may benefit from choosing $c = 1$ and $L = 1$ for simplicity.

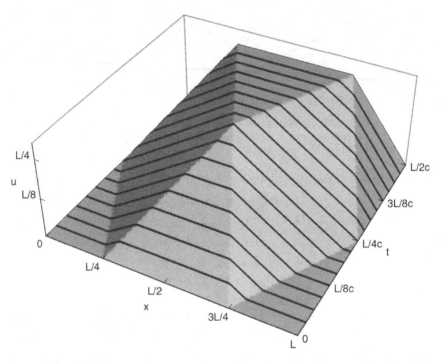

Fig. 5.17   A plot of $u(x,t)$, the solution to the initial boundary value problem found in Example 5.5. For convenience c and L have both been chosen to be 1.

Example 5.6. Assume that a string of length π with fixed ends is set into motion from rest with an initial velocity of $g(x) = x(1 - x/\pi)$. Find the solution to the initial boundary value problem assuming the wave speed $c = 1$.

Solution. The mathematical model of this situation can be stated as,

$$u_{tt} = u_{xx} \text{ for } 0 < x < \pi \text{ and } t > 0,$$
$$u(0,t) = u(L,t) = 0 \text{ for } t > 0,$$
$$u(x,0) = 0 \text{ and } u_t(x,0) = g(x) \text{ for } 0 < x < \pi.$$

The formal solution can be expressed as a Fourier series,

$$u(x,t) = \sum_{n=1}^{\infty} d_n \sin(nx) \sin(nt)$$

with d_n given by

$$d_n = \frac{2}{n\pi} \int_0^\pi x(1 - x/\pi) \sin(2nx)\, dx = \begin{cases} 0 & \text{if } n \text{ is even,} \\ 8/(n^4\pi^2) & \text{if } n \text{ is odd.} \end{cases}$$

Therefore the formal solution can be stated as

$$u(x,t) = \frac{8}{\pi^2} \sum_{n=1}^{\infty} \frac{\sin(nx)\sin(nt)}{(2n-1)^4}.$$

If $g_o(x)$ is thought of as the odd, 2π-periodic extension of $g(x)$ to the whole real line then the solution to the initial value problem can also be expressed using d'Alembert's approach as

$$u(x,t) = \frac{1}{2} \int_{x-t}^{x+t} g_o(s)\, ds.$$

Define function $G(x)$ as

$$G(x) = \int_0^x g_o(s)\, ds = \begin{cases} x^2 \left(\dfrac{1}{2} - \dfrac{x}{3\pi} \right) & \text{if } 0 \le x \le \pi, \\ \dfrac{(2x - \pi)(x - 2\pi)^2}{6\pi} & \text{if } \pi < x \le 2\pi \end{cases}$$

and extended 2π-periodically to the entire real line. Using this definition of G, the solution to the initial boundary value problem can be written as

$$u(x,t) = \frac{1}{2} \left[G((x+t) \bmod 2\pi) - G((x-t) \bmod 2\pi) \right],$$

for $0 \le x \le L$ and $t \ge 0$. A surface plot of the solution is shown in Fig. 5.18 for various values of $t > 0$.

5.4 Nonhomogeneous Cases

All of the theory and examples of the wave equation for bounded strings presented in this chapter have concerned the homogeneous wave equation with homogeneous Dirichlet boundary conditions. In this section more general initial boundary value problems will be considered, namely versions of the wave equation containing "forcing" terms and/or boundary conditions

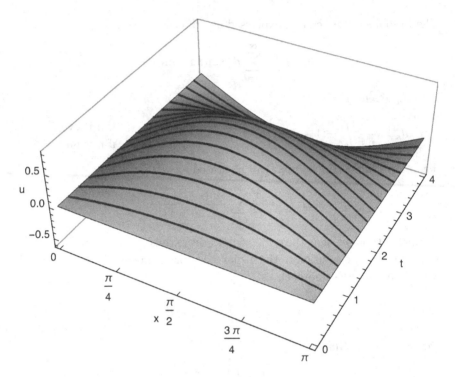

Fig. 5.18   A plot of $u(x,t)$, the solution to the initial boundary value problem found in Example 5.6.

which are time dependent. The general case can be stated in the form of the following initial boundary value problem:

$$u_{tt} = c^2 u_{xx} + F(x,t) \text{ for } 0 < x < L \text{ and } t > 0,$$
$$u(0,t) = \phi(t) \text{ and } u(L,t) = \psi(t) \text{ for } t > 0, \qquad (5.30)$$
$$u(x,0) = f(x) \text{ and } u_t(x,0) = g(x) \text{ for } 0 < x < L.$$

The solution to the initial boundary value problem in Eq. (5.30) can be found by solving two simpler initial boundary value problems and using the Principle of Superposition to reconstruct the solution. The two problems to be solved are:

$$u_{tt} = c^2 u_{xx} + F(x,t) \text{ for } 0 < x < L \text{ and } t > 0,$$
$$u(0,t) = u(L,t) = 0 \text{ for } t > 0, \qquad (5.31)$$
$$u(x,0) = u_t(x,0) = 0 \text{ for } 0 < x < L$$

and

$$u_{tt} = c^2 u_{xx} \text{ for } 0 < x < L \text{ and } t > 0,$$

$$u(0,t) = \phi(t) \text{ and } u(L,t) = \psi(t) \text{ for } t > 0, \tag{5.32}$$

$$u(x,0) = f(x) \text{ and } u_t(x,0) = g(x) \text{ for } 0 < x < L.$$

The reader should note how the nonhomogeneous elements of Eq. (5.30) have been split between the two new initial boundary value problems. The problem in Eq. (5.31) contains the nonhomogeneous partial differential equation of Eq. (5.30) but has homogeneous Dirichlet boundary conditions and zero initial displacement and initial velocity conditions. The other problem in Eq. (5.32) contains a homogeneous version of the partial differential equation from Eq. (5.30) and the nonhomogeneous boundary conditions and nontrivial initial conditions. Due to the presence of the nonhomogeneous terms in the two problems, the method of separation of variables cannot be used directly.

The initial boundary value problem in Eq. (5.31) is solved first. Guided by the method of variation of parameters used to solve nonhomogeneous ordinary differential equations, a solution to the partial differential equation in Eq. (5.31) is hypothesized to have the form

$$u(x,t) = \sum_{n=1}^{\infty} T_n(t) \sin \frac{n\pi x}{L}, \tag{5.33}$$

where the coefficient functions $T_n(t)$ are to be determined. Substituting $u(x,t)$ from Eq. (5.33) into the wave equation produces

$$\sum_{n=1}^{\infty} T_n''(t) \sin \frac{n\pi x}{L} = -c^2 \sum_{n=1}^{\infty} \left(\frac{n\pi}{L}\right)^2 T_n(t) \sin \frac{n\pi x}{L} + F(x,t),$$

or equivalently

$$\sum_{n=1}^{\infty} \left[T_n''(t) + \left(\frac{n\pi c}{L}\right)^2 T_n(t) \right] \sin \frac{n\pi x}{L} = F(x,t).$$

Multiplying the above equation by $\sin(m\pi x/L)$ and integrating both sides with respect to x over $[0, L]$ (assuming the order of the operations of integration and summation can be interchanged) yield

$$T_m''(t) + \left(\frac{m\pi c}{L}\right)^2 T_m(t) = F_m(t) \tag{5.34}$$

for $m \in \mathbb{N}$, where

$$F_m(t) = \frac{2}{L} \int_0^L F(x,t) \sin \frac{m\pi x}{L} \, dx.$$

Further, from the initial conditions in Eq. (5.31) the following initial conditions for the ordinary differential equation in Eq. (5.34) are imposed:

$$T_m(0) = T_m'(0) = 0. \qquad (5.35)$$

The initial value problem consisting of Eqs. (5.34) and (5.35) can be solved by the method of variation of parameters since the solutions of the corresponding homogeneous version of the ordinary differential equation in Eq. (5.34) are known. The reader is asked to provide the details in Exercise 18. The solution approach leads to the concept of a **Green's function**. The functions $T_m(t)$ for $m \in \mathbb{N}$ thus obtained are used to form the series solution to Eq. (5.31) as long as $F(x, t)$ is smooth enough.

Attention can now be turned to solving Eq. (5.32) with its nonhomogeneous boundary conditions. Again, since the boundary conditions are not homogeneous, the method of separation of variables is not applicable. This initial boundary value problem can be transformed into one with homogeneous boundary conditions. Let function $v(x, t)$ be such that

$$v(0, t) = \phi(t) \text{ and } v(L, t) = \psi(t) \qquad (5.36)$$

and set $w(x, t) = u(x, t) - v(x, t)$, where $u(x, t)$ is the as yet unknown solution to the wave equation in the initial boundary value problem in Eq. (5.32). Function $w(x, t)$ will satisfy the homogeneous boundary conditions,

$$w(0, t) = 0 = w(L, t).$$

There are many functions $v(x, t)$ that satisfy Eq. (5.36), for example,

$$v(x, t) = \frac{x}{L}[\psi(t) - \phi(t)] + \phi(t).$$

For this choice, $w(x, t) = u(x, t) - v(x, t)$ is a solution of the following initial boundary value problem:

$$w_{tt} = c^2 w_{xx} - \frac{x}{L}[\psi''(t) - \phi''(t)] - \phi''(t) \text{ for } 0 < x < L, \ t > 0,$$

$$w(0, t) = w(L, t) = 0 \text{ for } t > 0,$$

$$w(x, 0) = f(x) - \frac{x}{L}[\psi(0) - \phi(0)] - \phi(0) \text{ for } 0 < x < L,$$

$$w_t(x, 0) = g(x) - \frac{x}{L}[\psi'(0) - \phi'(0)] - \phi'(0) \text{ for } 0 < x < L.$$

$$(5.37)$$

The initial boundary value problem in Eq. (5.37) can be decomposed into two initial boundary value problems, one with a nonhomogeneous wave equation and zero initial conditions, and the other with a homogeneous wave equation and initial values as in Eq. (5.37). The solution technique will be illustrated in the following example.

Example 5.7. Find the solution of the following initial boundary value problem:

$$u_{tt} = c^2 u_{xx} \text{ for } 0 < x < L \text{ and } t > 0,$$
$$u(0,t) = 0 \text{ and } u(L,t) = \sin(\omega t) \text{ for } t > 0, \qquad (5.38)$$
$$u(x,0) = u_t(x,0) = 0 \text{ for } 0 < x < L.$$

Solution. Introduce the new unknown function

$$w(x,t) = u(x,t) - \frac{x}{L}\sin(\omega t)$$

then $w(x,t)$ is a solution to the following initial boundary value problem with homogeneous Dirichlet boundary conditions:

$$w_{tt} = c^2 w_{xx} + \frac{\omega^2 x}{L}\sin(\omega t) \text{ for } 0 < x < L \text{ and } t > 0,$$
$$w(0,t) = w(L,t) = 0 \text{ for } t > 0, \qquad (5.39)$$
$$w(x,0) = 0 \text{ and } w_t(x,0) = -\frac{\omega x}{L} \text{ for } 0 < x < L.$$

The solution to the initial boundary value problem in Eq. (5.39) can be thought of as the sum of two functions, *i.e.*, $w = w_1 + w_2$, where w_1 is the solution to the initial boundary value problem,

$$w_{tt} = c^2 w_{xx} + \frac{\omega^2 x}{L}\sin(\omega t) \text{ for } 0 < x < L \text{ and } t > 0,$$
$$w(0,t) = w(L,t) = 0 \text{ for } t > 0, \qquad (5.40)$$
$$w(x,0) = w_t(x,0) = 0 \text{ for } 0 < x < L,$$

and w_2 is the solution to the following initial boundary value problem:

$$w_{tt} = c^2 w_{xx} \text{ for } 0 < x < L \text{ and } t > 0,$$
$$w(0,t) = w(L,t) = 0 \text{ for } t > 0, \qquad (5.41)$$
$$w(x,0) = 0 \text{ and } w_t(x,0) = -\frac{\omega x}{L} \text{ for } 0 < x < L.$$

The solution w_1 to Eq. (5.40) is found using the method of variation of parameters. Let

$$w_1(x,t) = \sum_{n=1}^{\infty} T_n(t)\sin\frac{n\pi x}{L},$$

then

$$\sum_{n=1}^{\infty} T_n''(t)\sin\frac{n\pi x}{L} = -\sum_{n=1}^{\infty}\left(\frac{n\pi c}{L}\right)^2 T_n(t)\sin\frac{n\pi x}{L} + \frac{\omega^2 x}{L}\sin(\omega t). \quad (5.42)$$

The linear expression x on the right-hand side of the last equation can be expanded as a Fourier series of the form

$$x \sim \frac{2L}{\pi} \sum_{n=1}^{\infty} \frac{(-1)^{n+1}}{n} \sin \frac{n\pi x}{L}.$$

Substituting this Fourier series expansion in Eq. (5.42) and rearranging terms produce the following equation,

$$\sum_{n=1}^{\infty} \left[T_n''(t) + \left(\frac{n\pi c}{L}\right)^2 T_n(t) + \frac{2(-1)^n \omega^2}{n\pi} \sin(\omega t) \right] \sin \frac{n\pi x}{L} = 0.$$

Thus for $n \in \mathbb{N}$,

$$T_n''(t) + \left(\frac{n\pi c}{L}\right)^2 T_n(t) + \frac{2(-1)^n \omega^2}{n\pi} \sin(\omega t) = 0. \tag{5.43}$$

From the initial conditions imposed on $w_1(x,t)$ in Eq. (5.39), initial conditions for $T_n(t)$ are

$$T_n(0) = T_n'(0) = 0. \tag{5.44}$$

The general solution of Eq. (5.43) is

$$T_n(t) = C_n \cos \frac{n\pi c t}{L} + D_n \sin \frac{n\pi c t}{L} + Y_n(t),$$

where $Y_n(t)$ is a particular solution to Eq. (5.43) which can be found by the method of undetermined coefficients. More specifically, setting $\omega_n = n\pi c/L$ and $Y_n(t) = A_n \sin(\omega t)$, differentiating this function and substituting into Eq. (5.43) result in the equation,

$$-A_n \omega^2 \sin(\omega t) + A_n \omega_n^2 \sin(\omega t) + \frac{2(-1)^n \omega^2}{n\pi} \sin(\omega t) = 0.$$

If ω is such that $\omega \neq \omega_n$ for every $n \in \mathbb{N}$, then

$$A_n = \frac{2(-1)^n \omega^2}{n\pi(\omega^2 - \omega_n^2)} \quad \text{and} \quad Y_n(t) = \frac{2(-1)^n \omega^2}{n\pi(\omega^2 - \omega_n^2)} \sin(\omega t). \tag{5.45}$$

If there exists an $N \in \mathbb{N}$ for which $\omega = \omega_N$, then the particular solution to Eq. (5.43) takes on the form $Y_N(t) = A_N t \cos(\omega t)$. Differentiating this function and substituting it in Eq. (5.43) yield the equation:

$$0 = -2A_N \omega \sin(\omega t) - A_N \omega^2 t \cos(\omega t) + A_N \omega^2 t \cos(\omega t) + \frac{2(-1)^N \omega^2}{N\pi} \sin(\omega t)$$

$$= -2A_N \omega \sin(\omega t) + \frac{2(-1)^N \omega^2}{N\pi} \sin(\omega t),$$

which implies

$$A_N = \frac{(-1)^N \omega}{N\pi} \text{ and thus } Y_N(t) = \frac{(-1)^N \omega t}{N\pi} \cos(\omega t). \tag{5.46}$$

For all $n \neq N$ the particular solution to Eq. (5.43) retains the form shown in Eq. (5.45). Therefore, the general solution of Eq. (5.43) is

$$T_n(t) = C_n \cos\frac{n\pi c t}{L} + D_n \sin\frac{n\pi c t}{L}$$

$$+ \begin{cases} \dfrac{2(-1)^n \omega^2}{n\pi(\omega^2 - \omega_n^2)} \sin(\omega t) & \text{if } n \neq \dfrac{\omega L}{c\pi}, \\[3mm] \dfrac{(-1)^n \omega t}{n\pi} \cos(\omega t) & \text{if } n = \dfrac{\omega L}{c\pi}. \end{cases}$$

Using the initial conditions in Eq. (5.44) it is seen that $C_n = 0$ for all $n \in \mathbb{N}$. Taking the derivative of $T_n(t)$ and setting $t = 0$ produce

$$T_n'(0) = 0 = D_n \frac{n\pi c}{L} + \begin{cases} \dfrac{2(-1)^n \omega^3}{n\pi(\omega^2 - \omega_n^2)} & \text{if } n \neq \dfrac{\omega L}{c\pi}, \\[3mm] \dfrac{(-1)^n \omega}{n\pi} & \text{if } n = \dfrac{\omega L}{c\pi}. \end{cases}$$

Consequently the coefficient D_n takes the form,

$$D_n = \begin{cases} \dfrac{2(-1)^n L \omega^3}{c(n\pi)^2(\omega_n^2 - \omega^2)} & \text{if } n \neq \dfrac{\omega L}{c\pi}, \\[3mm] \dfrac{(-1)^{n+1} L \omega}{c(n\pi)^2} & \text{if } n = \dfrac{\omega L}{c\pi}. \end{cases}$$

After simplifying the results derived above,

$$T_n(t) = \begin{cases} \dfrac{2(-1)^n \omega^2}{n\pi\omega_n(\omega_n^2 - \omega^2)}[\omega \sin(\omega_n t) - \omega_n \sin(\omega t)] & \text{if } n \neq \dfrac{\omega L}{c\pi}, \\[3mm] \dfrac{(-1)^n c t}{L} \cos(\omega t) - \dfrac{(-1)^n}{n\pi} \sin(\omega t) & \text{if } n = \dfrac{\omega L}{c\pi}. \end{cases}$$

Finally the formal solution to Eq. (5.39) can be expressed as

$$w_1(x,t) = \sum_{n=1}^{\infty} \frac{2(-1)^n \omega^2}{n\pi\omega_n(\omega_n^2 - \omega^2)}[\omega \sin(\omega_n t) - \omega_n \sin(\omega t)] \sin\frac{n\pi x}{L},$$

if $\omega L/(c\pi) \notin \mathbb{N}$. If there exists $N \in \mathbb{N}$ for which $\omega = N\pi c/L$, then

$$w_1(x,t) = \left(\frac{(-1)^N c t}{L} \cos(\omega t) - \frac{(-1)^N}{N\pi} \sin(\omega t)\right) \sin\frac{N\pi x}{L}$$

$$+ \sum_{n=1, \, n \neq N}^{\infty} \frac{2(-1)^n \omega^2}{n\pi\omega_n(\omega_n^2 - \omega^2)}[\omega \sin(\omega_n t) - \omega_n \sin(\omega t)] \sin\frac{n\pi x}{L}.$$

After the hard work done to find the solution to the nonhomogeneous initial boundary value problem in Eq. (5.39), the solution $w_2(x, t)$ to Eq. (5.41) is easier to find. Since the initial boundary value problem is homogeneous and linear and one of the initial conditions (the initial displacement condition) is zero, $w_2(x, t)$ must be of the form

$$w_2(x, t) = \sum_{n=1}^{\infty} d_n \sin \frac{n\pi ct}{L} \sin \frac{n\pi x}{L}.$$

Use of the remaining initial condition in Eq. (5.41) and the formula for d_n in Eq. (5.11) produce

$$w_2(x, t) = \sum_{n=1}^{\infty} \frac{2(-1)^n \omega L}{c(n\pi)^2} \sin \frac{n\pi ct}{L} \sin \frac{n\pi x}{L}.$$

Combining these two solutions as $w(x, t) = w_1(x, t) + w_2(x, t)$ the solution to Eq. (5.38) can be written as

$$u(x, t) = w(x, t) + \frac{x}{L} \sin(\omega t).$$

5.5   The Energy Integral and Uniqueness of Solutions

The earlier sections of this chapter have explained and explored methods for finding solutions to the wave equation under various conditions: homogeneous partial differential equation and homogeneous Dirichlet boundary conditions, nonhomogeneous partial differential equation, and nonhomogeneous boundary conditions. The solutions have been expressed as Fourier series and as d'Alembertian functions. In this final section the issue of the uniqueness of solutions is considered. If the solution to an initial boundary value problem is unique then it is merely a matter of convenience as to whether the solution is expressed as a Fourier series or function in d'Alembertian form. Consider the initial boundary value problem:

$$u_{tt} = c^2 u_{xx} \text{ for } 0 < x < L \text{ and } t > 0,$$

$$u(0, t) = u(L, t) = 0 \text{ for } t > 0, \qquad (5.47)$$

$$u(x, 0) = f(x) \text{ and } u_t(x, 0) = g(x) \text{ for } 0 < x < L,$$

describing the vibrating string with fixed ends and no external forces. Recall that $c^2 = T_0/\rho$, where T_0 and ρ are the tension force in the string and mass density of the string respectively. The energy of the mechanical system consists of kinetic and potential energies. For a point mass m moving at velocity $\mathbf{v}$ the kinetic energy is $m\|\mathbf{v}\|^2/2$. Since the vibrating string has linear density ρ the kinetic energy can be expressed as the integral

$$K = \frac{1}{2} \int_0^L \rho(u_t)^2 \, dx.$$

The potential energy is related to the work done to stretch the string as it vibrates. Consider the segment of the string in a short interval $[x, x + \Delta x]$. In its equilibrium configuration this segment has length Δx. When the string is displaced from equilibrium this segment has length approximated by $ds = \sqrt{1 + (u_x(x,t))^2} \Delta x$. Under the usual assumption that the displacement of the string is small, then

$$\sqrt{1 + (u_x(x,t))^2} \Delta x \approx \left(1 + \frac{1}{2}(u_x(x,t))^2 \right) \Delta x.$$

Thus the amount by which this segment of the string has been elongated is

$$\left(1 + \frac{1}{2}(u_x(x,t))^2 \right) \Delta x - \Delta x = \frac{1}{2}(u_x(x,t))^2 \Delta x.$$

The string segment has been elongated by working against the tension T_0 and thus the potential energy can be expressed as

$$P = \frac{T_0}{2} \int_0^L (u_x)^2 \, dx.$$

The total energy in the system is

$$E(t) = \frac{1}{2} \int_0^L [\rho(u_t)^2 + T_0(u_x)^2] \, dx. \tag{5.48}$$

For the case of the infinitely long string, the integral can be taken over the whole real number line with the understanding that the resulting improper integral converges since for every t the displacement and velocity of the string must vanish outside the finite range of influence. Differentiating $E(t)$, assuming that u_t and u_x are smooth functions so that the derivative can be taken inside the integral, produces

$$E'(t) = \frac{1}{2} \int_0^L (2\rho u_t u_{tt} + 2T_0 u_x u_{xt}) \, dx = \int_0^L (\rho c^2 u_t u_{xx} + T_0 u_x u_{xt}) \, dx$$

$$= T_0 \int_0^L (u_t u_{xx} + u_x u_{xt}) \, dx = T_0 \int_0^L \frac{d}{dx}(u_t u_x) dx = [T_0 u_t u_x]_0^L$$

$$= T_0[u_t(L,t)u_x(L,t) - u_t(0,t)u_x(0,t)] = 0.$$

Therefore $E(t)$ is a constant for all t. Consequently,

$$E(t) = E(0) = \frac{1}{2} \int_0^L (\rho[u_t(x,0)]^2 + T_0[u_x(x,0)]^2) \, dx$$

$$= \frac{1}{2} \int_0^L (\rho[g(x)]^2 + T_0[f'(x)]^2) \, dx.$$

The constant energy in the system is the energy the system possesses at $t = 0$. This enables the uniqueness of solutions for initial boundary value problems to be established for the wave equation.

Theorem 5.1 (Uniqueness). *The solution to an initial boundary value problem for the wave equation of the form shown in Eq. (5.47) is unique.*

Proof. If there exist two solutions $u_1(x,t)$ and $u_2(x,t)$ to Eq. (5.47), then $v(x,t) = u_1(x,t) - u_2(x,t)$ is a solution to the initial boundary value problem

$$v_{tt} = c^2 v_{xx} \text{ for } 0 < x < L \text{ and } t > 0,$$

$$v(0,t) = v(L,t) = 0 \text{ for } t > 0,$$

$$v(x,0) = v_t(x,0) = 0 \text{ for } 0 < x < L.$$

Let $E(t)$ be defined as in Eq. (5.48), then $E(t) = E(0) = 0$ which implies $v_t = 0$ and $v_x = 0$. Thus $v(x,t)$ must be a constant and therefore must be zero everywhere. Hence $u_1(x,t) = u_2(x,t)$ everywhere. $\qquad\qquad\square$

The method described above can also be used to show that the solution of Eq. (5.47) is continuous with respect to the initial and boundary conditions.

5.6   Exercises

(1) Find the solution to the initial boundary value problem of Eq. (5.1) where $L = \pi$, $c = 1$ and

$$f(x) = \sin x \cos x,$$

$$g(x) = 0$$

for $0 < x < L$.

(2) Find the solution to the initial boundary value problem of Eq. (5.1) where $L = \pi$, $c = 1$ and

$$f(x) = \sin x + 3\sin(2x),$$

$$g(x) = 0$$

for $0 < x < L$.

(3) Find the solution to the initial boundary value problem of Eq. (5.1) where $L = \pi$, $c = 1$ and

$$f(x) = 0,$$

$$g(x) = \sin(3x)$$

for $0 < x < L$.

(4) Find the solution to the initial boundary value problem of Eq. (5.1) where $L = 1$, $c = 1$ and

$$f(x) = \sin(\pi x) + \frac{1}{2}\sin(3\pi x) + 3\sin(7\pi x),$$
$$g(x) = \sin(2\pi x)$$

for $0 < x < L$.

(5) Find the solution to the initial boundary value problem of Eq. (5.1) where $L = \pi$, $c = 1$ and

$$f(x) = 0,$$
$$g(x) = x(x - \pi)$$

for $0 < x < L$.

(6) Find the solution to the initial boundary value problem of Eq. (5.1) where $L = \pi$, $c = 1$ and

$$f(x) = \sin(2x),$$
$$g(x) = x\cos(x/2)$$

for $0 < x < L$.

(7) Find the solution to the initial boundary value problem of Eq. (5.1) where $L = \pi$, $c = 1$ and

$$f(x) = x(x - \pi),$$
$$g(x) = 1$$

for $0 < x < L$.

(8) Find the solution to the initial boundary value problem of Eq. (5.1) where $L = 1$, $c = 1/\pi$ and

$$f(x) = \begin{cases} x & \text{if } 0 \le x \le 1/2, \\ 1 - x & \text{if } 1/2 < x \le 1, \end{cases}$$
$$g(x) = 0$$

for $0 < x < L$.

(9) Suppose an elastic string is fixed at $x = 0$ and is free at $x = L$, so that the boundary conditions are $u(0, t) = 0$ and $u_x(L, t) = 0$.

(a) Show that the product solutions to the wave equation with the given boundary conditions are

$$u_n(x, t) = (c_n \cos(\lambda_n ct) + d_n \sin(\lambda_n ct)) \sin(\lambda_n x)$$

where $\lambda_n = (2n - 1)\pi/(2L)$ for $n \in \mathbb{N}$.

218   A First Course in Partial Differential Equations

(b) Derive the formal solution of the initial boundary value problem
assuming initial conditions $u(x,0) = f(x)$ and $u_t(x,0) = g(x)$.

(c) Find the formal solution to the initial value problem if $L = 1$,
$g(x) = 0$, and

$$f(x) = \begin{cases} 0 & \text{if } 0 \leq x \leq 1/4, \\ x - 1/4 & \text{if } 1/4 \leq x \leq 1/2, \\ 3/4 - x & \text{if } 1/2 \leq x \leq 3/4, \\ 0 & \text{if } 3/4 \leq x \leq 1. \end{cases}$$

(10) Consider the nth normal mode of the solution to the wave equation
as expressed in Eq. (5.15).

(a) Show that the nth normal mode is periodic in t with period
$2L/(nc)$.

(b) Show that the solution to the wave equation given in Eq. (5.7)
is periodic in t with period $2L/c$ (assuming that the series converges).

(c) Show that for $u(x,t)$ as defined in Eq. (5.7)

$$u(x,t + L/c) = -u(L - x,t).$$

What does this imply about the symmetry of the vibrating string
at half of its temporal period?

(11) Solve the initial value problem

$$u_{tt} = c^2 u_{xx} \text{ for } -\infty < x < \infty \text{ and } t > 0,$$
$$u(x,0) = e^{-x^2} \text{ and } u_t(x,0) = 0 \text{ for } -\infty < x < \infty.$$

(12) Solve the initial value problem

$$u_{tt} = c^2 u_{xx} \text{ for } -\infty < x < \infty \text{ and } t > 0,$$
$$u(x,0) = 0 \text{ and } u_t(x,0) = \sinh x \text{ for } -\infty < x < \infty.$$

(13) Use d'Alembert's solution from Eq. (5.21) to plot the solution to
Eq. (5.1) with $L = 1$, $c = 1$, and

$$f(x) = \sin(2\pi x),$$
$$g(x) = 0$$

for $0 < x < L$ and $t = 0$, $t = 1/4$, $t = 1/2$, and $t = 3/4$.

(14) Use d'Alembert's solution from Eq. (5.21) to plot the solution to
Eq. (5.1) with $L = 1$, $c = 1$, and

$$f(x) = 0,$$
$$g(x) = \sin(\pi x)$$

for $0 < x < L$ and $t = 0$, $t = 1/4$, $t = 1/2$, and $t = 3/4$.

(15) Use d'Alembert's solution from Eq. (5.21) to plot the solution to Eq. (5.1) with $L = 1$, $c = 1$, and

$$f(x) = \sin(3\pi x),$$
$$g(x) = \sin(\pi x)$$

for $0 < x < L$ and $t = 0$, $t = 1/4$, $t = 1/2$, and $t = 3/4$.

(16) Let $u(x,t) = \sum_{n=1}^{\infty} c_n \cos\left(\frac{n\pi ct}{L}\right) \sin\left(\frac{n\pi x}{L}\right)$ be a solution of the vibrating string problem *i.e.*, $u(x,t)$ is a solution of the equation:

$$u_{tt} = c^2 u_{xx}$$

and satisfies $u(0,t) = u(L,t) = 0$. If the string is further constrained at the midpoint so that $u(L/2,t) = 0$ for all t, what condition does this impose on the coefficients c_n?

(17) Consider the initial boundary value problem:

$$u_{tt} = u_{xx} \text{ for } 0 < x < 1 \text{ and } t > 0,$$
$$u(0,t) = u(1,t) = 0 \text{ for } t > 0,$$
$$u(x,0) = \begin{cases} 4x & \text{if } 0 < x \le 1/4, \\ 4(1/2 - x) & \text{if } 1/4 < x < 1/2, \\ 0 & \text{if } 1/2 \le x < 1, \end{cases}$$
$$u_t(x,0) = 0 \text{ for } 0 < x < 1.$$

(a) Use d'Alembert's approach to plot the string at times $t = 0$, $t = 1/4$, and $t = 1/2$.

(b) For $t = 1/4$, identify the points on the string that are still in rest position.

(c) Take any point x on the string with zero initial displacement ($1/2 < x < 1$), how long does it take before the point x starts to vibrate?

(18) Use variation of parameters to solve the following initial value problem:

$$T_m''(t) + \left(\frac{m\pi c}{L}\right)^2 T_m(t) = F_m(t),$$
$$T_m(0) = T_m'(0) = 0.$$

(19) Solve the following initial boundary value problem for the wave equation:

$$u_{tt} = u_{xx} + t \text{ for } 0 < x < 1 \text{ and } t > 0,$$
$$u(0,t) = u(1,t) = 0 \text{ for } t > 0,$$
$$u(x,0) = u_t(x,0) = 0 \text{ for } 0 < x < 1.$$

(20) The partial differential equation $u_{tt} + 2au_t + bu = c^2 u_{xx}$, where a, b, and c are positive constants is called the telegraph equation, arising from the modeling the flow of electricity in a cable.

   (a) Show that if $b = a^2$ the function $v(x,t) = u(x,t)e^{at}$ satisfies the wave equation $v_{tt} = c^2 v_{xx}$.

   (b) Use d'Alembert's approach to find a solution of the initial value problem,

$$u_{tt} + 2au_t + a^2 u = c^2 u_{xx} \text{ for } -\infty < x < \infty \text{ and } t > 0,$$

$$u(x,0) = f(x) \text{ for } -\infty < x < \infty,$$

$$u_t(x,0) = g(x) \text{ for } -\infty < x < \infty.$$

(21) Solve the following initial boundary value problem for the wave equation:

$$u_{tt} = u_{xx} \text{ for } 0 < x < 1 \text{ and } t > 0,$$

$$u(0,t) = u(1,t) = t \text{ for } t > 0,$$

$$u(x,0) = u_t(x,0) = 0 \text{ for } 0 < x < 1.$$

(22) Solve the following initial boundary value problem for the wave equation:

$$u_{tt} = u_{xx} - g \text{ for } 0 < x < 1 \text{ and } t > 0,$$

$$u(0,t) = u(1,t) = 0 \text{ for } t > 0,$$

$$u(x,0) = u_t(x,0) = 0 \text{ for } 0 < x < 1.$$

The constant g can be thought of as gravitational acceleration.

(23) Solve the following initial boundary value problem for the wave equation:

$$u_{tt} = u_{xx} \text{ for } 0 < x < 1 \text{ and } t > 0,$$

$$u(0,t) = \sin(\pi t) \text{ and } u(1,t) = 0 \text{ for } t > 0,$$

$$u(x,0) = u_t(x,0) = 0 \text{ for } 0 < x < 1.$$

(24) Consider the total energy $E(t)$ given in Eq. (5.48) for a vibrating string. How does the energy change over time if $u_x(0,t) = u(L,t) = 0$ for all $t > 0$?

(25) Consider the nth normal mode of the vibration of a finite string with fixed ends:

$$u_n(x,t) = \left(c_n \cos \frac{n\pi ct}{L} + d_n \sin \frac{n\pi ct}{L} \right) \sin \frac{n\pi x}{L}.$$

Show the energy of $u_n(x,t)$ as defined by Eq. (5.48) is given by

$$E_n(t) = \frac{T_0}{4}\lambda_n^2(c_n^2 + d_n^2) = \frac{T_0}{4}\left(\frac{n\pi}{L}\right)^2(c_n^2 + d_n^2).$$

(26) Show that the solution to the following initial boundary value problem is unique:

$$u_{tt} = c^2 u_{xx} + F(x,t) \text{ for } 0 < x < L \text{ and } t > 0,$$
$$u(0,t) = u(L,t) = 0 \text{ for } t > 0,$$
$$u(x,0) = f(x) \text{ for } 0 < x < L,$$
$$u_t(x,0) = g(x) \text{ for } 0 < x < L.$$

Chapter 6

Laplace's Equation

This chapter is devoted to the discussion of Laplace's equation, which is an example of a second-order linear elliptic partial differential equation. Preliminary discussion of this type of equation was covered in Sec. 1.5. Laplace's equation and the more general Poisson's equation are of great importance in mathematics, mathematical physics, and various other areas. The steady-state solutions of the heat and wave equations satisfy Laplace's equation (see Chap. 4 and Chap. 5).

This chapter will concentrate on the basic theory and solution techniques related to the boundary value problems of Laplace's equation. Section 6.1 will introduce some notation for the Laplacian operator and define some topological concepts useful for describing the domain of the solution to Laplace's equation. Section 6.2 will explore the solution to boundary value problems on rectangular domains with Dirichlet boundary conditions imposed while Sec. 6.3 introduces polar coordinates useful for solving Dirichlet boundary value problems for Laplace's equation on disks. Section 6.4 concludes the discussion of Dirichlet boundary value problems by extending the methods for solving Laplace's equation on disks to related geometries such as annuli, circular sectors, and the pierced plane. Sections 6.5 and 6.6 explore solutions to Laplace's equation with imposed Neumann boundary conditions on rectangular and circular disk domains respectively, though as the reader will see, the solutions in these cases are not unique. Section 6.7 discusses solutions to Laplace's equation with a mixture of Dirichlet and Neumann boundary conditions imposed on the boundary of the domain. Section 6.8 introduces Poisson's formula which enables a closed form solution to Laplace's equation on the disk to be given. This section also develops results regarding various properties of solutions to Laplace's equation. The chapter concludes in Sec. 6.9 with the statement and proof of the Maximum Principle and uniqueness theorems pertaining to solutions to Laplace's equation.

6.1   Boundary Value Problems of Laplace's Equation

The stage is set for the study of Laplace's equation with the introduction of some common mathematical notation and terminology. Let $\mathbf{x} = (x_1, x_2, \ldots, x_n) \in \mathbb{R}^n$. Note that $\mathbf{x}$ is sometimes called an n-tuple and $\mathbb{R}^n$ is often referred to as n-dimensional Euclidean space. The bold notation $\mathbf{x}$ is employed to distinguish this symbol from the scalar x. In some texts $\mathbf{x}$ is used to refer to a vector of n components in an n-dimensional vector space. Since n-dimensional Euclidean space and an n-dimensional vector space equipped with an inner product are isomorphic, the distinction is merely one of interpretation and thus the re-appropriation of notation will not cause any confusion here. If u is a function of $\mathbf{x}$, then the **Laplacian** of u, denoted as $\triangle u$ or $\nabla^2 u$, is defined as

$$\triangle u = \nabla^2 u = \frac{\partial^2 u}{\partial x_1^2} + \frac{\partial^2 u}{\partial x_2^2} + \cdots + \frac{\partial^2 u}{\partial x_n^2}.$$

When $n = 2$ or 3, the components of $\mathbf{x}$ are often denoted as x and y in place of x_1 and x_2 (in the case of $n = 2$) and as x, y, and z in place of x_1, x_2, and x_3 (in the case of $n = 3$). This facilitates the connection of the independent variables with 2- or 3-dimensional Euclidean space. With the notation above, Laplace's equation can be written as

$$\triangle u = 0.$$

Laplace's equation is usually defined on some region in $\mathbb{R}^n$ referred to as a **domain**. Domains in $\mathbb{R}^n$ are open, connected subsets. The concepts of openness and connectedness will not be defined in full rigor, but merely intuitively for $\mathbb{R}^2$. A subset Ω of $\mathbb{R}^2$ is **open** if for every ordered pair $(x, y) \in \Omega$ there is a disk with positive radius centered at (x, y) which is completely contained within Ω. A subset Ω of $\mathbb{R}^2$ is **connected** if for every pair of ordered pairs (x_1, y_1), $(x_2, y_2) \in \Omega$ there exists a continuous path from (x_1, y_1) to (x_2, y_2) which lies completely in Ω. In order to prove many of the theorems in this chapter, the domain Ω will also be assumed to be **bounded**. A set $\Omega \subset \mathbb{R}^2$ is bounded if there exists a finite real number R such that $\Omega \subset \{(x, y) \mid x^2 + y^2 < R\}$. See Fig. 6.1 and Fig. 6.2 for typical examples of open, connected and bounded sets. The reader should be able to generalize these intuitive descriptions to open, connected subsets of $\mathbb{R}^n$.

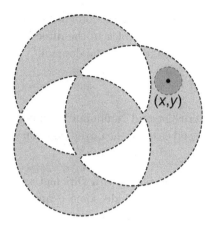

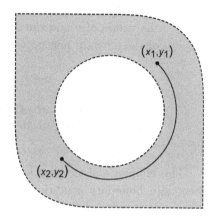

Fig. 6.1 An open, bounded subset of $\mathbb{R}^2$.

Fig. 6.2 An open, connected and bounded subset of $\mathbb{R}^2$.

Laplace's equation is assumed to hold on a domain $\Omega \subset \mathbb{R}^n$. Any smooth function satisfying Laplace's equation on Ω is called a **harmonic function** on Ω. If f is a given function of $\mathbf{x}$ the partial differential equation,

$$\triangle u = f$$

is known as **Poisson's equation**. Just as in the case of Laplace's equation, Poisson's equation is a linear elliptic partial differential equation as defined in Sec. 1.2.

If Ω is any domain in $\mathbb{R}^n$ then $\partial\Omega$ will denote the boundary of Ω. Note that the points in $\partial\Omega$ are not members of Ω itself. Poisson's or Laplace's equation defined on Ω is usually supplemented with boundary conditions on the boundary $\partial\Omega$. There are three typical types of boundary conditions.

Dirichlet: boundary conditions specify the values of a solution on the boundary of the domain,

$$u = \phi(\mathbf{x}) \text{ for } \mathbf{x} \in \partial\Omega.$$

Neumann: boundary conditions specify the values of the directional derivative of solutions in the direction of the outward normal vector on the boundary of the domain,

$$\frac{\partial u}{\partial \mathbf{n}} = \nabla u \cdot \mathbf{n} = \phi(\mathbf{x}) \text{ for } \mathbf{x} \in \partial\Omega,$$

where $\mathbf{n}$ represents the unit outward normal vector to $\partial\Omega$.

Mixed: boundary conditions specify the values of a linear combination of the values of u and the directional derivative of u in the direction of the outward normal vector on the boundary of the domain,

$$u + \alpha \frac{\partial u}{\partial \mathbf{n}} = \phi(\mathbf{x}) \text{ for } \mathbf{x} \in \partial\Omega,$$

where α is a constant. Clearly, Dirichlet and Neumann boundary conditions can be considered as special cases of the mixed boundary condition.

This chapter will focus on boundary value problems with Dirichlet and Neumann boundary conditions. Section 6.7 will include some examples of boundary value problems on rectangular domains with mixed boundary conditions.

Example 6.1. Verify each of the following functions is harmonic on the domain indicated.

- $u(x, y) = \ln((x - x_0)^2 + (y - y_0)^2)$, where (x_0, y_0) is an arbitrary fixed point in $\mathbb{R}^2$ and $\Omega = \mathbb{R}^2 - \{(x_0, y_0)\}$.
- $u(x, y) = e^x \sin y$ on $\mathbb{R}^2$.
- $u(x, y, z) = \frac{1}{x^2 + y^2 + z^2}$ on $\Omega = \mathbb{R}^3 - \{(0, 0, 0)\}$.

Solution. Verification can be made directly by differentiation and simplification of the results.

6.2 Dirichlet Problems on Rectangles

The first case of Laplace's equation to be examined in detail will involve a rectangular domain Ω in $\mathbb{R}^2$ and Dirichlet boundary conditions. Without loss of generality assume that Ω is given as

$$\Omega = (0, a) \times (0, b) = \{(x, y) \mid 0 < x < a, \, 0 < y < b\},$$

where $a > 0$ and $b > 0$ are constants. The boundary of this open set is

$$\partial\Omega = \{(x, y) \mid x = 0 \text{ or } a \text{ and } 0 \leq y \leq b, \text{ or } y = 0 \text{ or } b \text{ and } 0 \leq x \leq a\}.$$

Consider the following Dirichlet boundary value problem,

$$\triangle u = 0 \text{ for } (x, y) \in \Omega,$$
$$u = \phi(x, y) \text{ for } (x, y) \in \partial\Omega. \tag{6.1}$$

The boundary value problem of Eq. (6.1) is further simplified by restricting the function $\phi(x, y)$ to have the form specified below,

$$\triangle u = 0 \text{ for } (x, y) \in \Omega,$$

$$u(0, y) = 0 \text{ and } u(a, y) = f(y) \text{ for } 0 \leq y \leq b, \quad (6.2)$$

$$u(x, 0) = 0 \text{ and } u(x, b) = 0 \text{ for } 0 \leq x \leq a.$$

Boundary value problems of the form shown in Eq. (6.1) may be thought of as superpositions of four boundary value problems, each in a form similar to the boundary value problem shown in Eq. (6.2). The geometry of a rectangular domain makes it possible to solve Eq. (6.2) using the method of separation of variables. The solution to Laplace's equation will be assumed to be a product solution of the form $u(x, y) = X(x)Y(y)$ that satisfies the three homogeneous boundary conditions in Eq. (6.2). After the experience gained finding product solutions for the heat and wave equations, it is routine to see that $X(x)$ and $Y(y)$ must be solutions of the following boundary value problems,

$$Y'' + \sigma Y = 0 \text{ for } 0 < y < b,$$
$$Y(0) = Y(b) = 0, \quad (6.3)$$

and

$$X'' - \sigma X = 0 \text{ for } 0 < x < a,$$
$$X(0) = 0, \quad (6.4)$$

where σ is a constant. Earlier in Sec. 1.6 the boundary value problem of Eq. (6.3) was shown to have nontrivial solutions if and only if $\sigma = (n\pi/b)^2$ for $n \in \mathbb{N}$ and the corresponding nontrivial solutions can be expressed as

$$Y(y) = c_n \sin \frac{n\pi y}{b},$$

where $c_n \neq 0$ is an arbitrary constant. For the sake of notation let $\lambda_n = n\pi/b$ and $\sigma_n = \lambda_n^2$ and $Y_n(y) = \sin(\lambda_n y)$ for $n \in \mathbb{N}$. Recall that σ_n and $Y_n(y)$ are called the eigenvalues and eigenfunctions (respectively) of the boundary value problem of Eq. (6.3). Next replace σ with σ_n in Eq. (6.4) and solve the resulting ordinary differential equation to obtain

$$X_n(x) = d_n \sinh \frac{n\pi x}{b},$$

where again, d_n is an arbitrary constant. Therefore the product solutions of Laplace's equation which satisfy the three homogeneous boundary conditions take the form

$$u_n(x, y) = c_n \sinh \frac{n\pi x}{b} \sin \frac{n\pi y}{b}, \quad (6.5)$$

where c_n is an arbitrary constant for $n \in \mathbb{N}$. As was done earlier, the functions $u_n(x, y)$ are called the **fundamental solutions** of Eq. (6.2). A formal solution to the boundary value problem of Eq. (6.2) is sought in the form

$$u(x, y) = \sum_{n=1}^{\infty} u_n(x, y) = \sum_{n=1}^{\infty} c_n \sinh \frac{n\pi x}{b} \sin \frac{n\pi y}{b}. \qquad (6.6)$$

In order to establish such a solution, the following questions require answers.

(1) Is $u(x, y)$ as described in Eq. (6.6) well-defined on Ω?
(2) Can the constants c_n be chosen so that $u(x, y)$ satisfies the necessary boundary conditions?
(3) Is the function $u(x, y)$ a solution to Laplace's equation?

This section will mainly address the second question and comment on the other two. Assuming that $u(x, y)$ is well-defined it is readily seen that for any choice of c_n for $n \in \mathbb{N}$,

$$0 = u(0, y) = u(x, 0) = u(x, b).$$

Thus the choice of values for c_n is dictated by the necessity that $u(x, y)$ defined by Eq. (6.6) satisfy the boundary condition $u(a, y) = f(y)$, that is, c_n must be chosen to satisfy

$$u(a, y) = \sum_{n=1}^{\infty} c_n \sinh \frac{n\pi a}{b} \sin \frac{n\pi y}{b} = f(y).$$

This can be achieved by extending $f(y)$ as a $2b$-periodic, odd function and choosing c_n such that $c_n \sinh(n\pi a/b)$ is the Fourier sine coefficient of $f(y)$, that is,

$$c_n \sinh \frac{n\pi a}{b} = \frac{2}{b} \int_0^b f(y) \sin \frac{n\pi y}{b} \, dy.$$

This is equivalent to defining c_n for $n \in \mathbb{N}$ as

$$c_n = \frac{2}{b \sinh \frac{n\pi a}{b}} \int_0^b f(y) \sin \frac{n\pi y}{b} \, dy. \qquad (6.7)$$

To address the issue of convergence of the series in Eq. (6.6), the reader should note that

$$\sinh \frac{n\pi x}{b} = \frac{1}{2} \left(e^{n\pi x/b} - e^{-n\pi x/b} \right) \approx \frac{1}{2} e^{n\pi x/b}$$

for large n and all $x > 0$. Using this approximation in Eq. (6.6) and Eq. (6.7) results in the approximation

$$c_n \sinh \frac{n\pi x}{b} \approx \frac{2 e^{-n\pi(a-x)/b}}{b} \int_0^b f(y) \sin \frac{n\pi y}{b} \, dy$$

for all $x \in (0, a)$ and large $n \in \mathbb{N}$. This asymptotic estimate is sufficient to establish that the infinite series in Eq. (6.6) converges. If $f(y)$ is piecewise smooth then it can be shown that $u(x, y)$ along with its second partial derivatives are well-defined and satisfy Laplace's equation on Ω.

The general case of Eq. (6.1) can be decomposed into four boundary value problems for Laplace's equation with boundary conditions on $\partial\Omega$ given as follows and illustrated in Fig. 6.3:

$$u_1(x, y) = \begin{cases} \phi(0, y) & \text{if } 0 \le y \le b, \\ 0 & \text{otherwise}, \end{cases}$$

$$u_2(x, y) = \begin{cases} \phi(a, y) & \text{if } 0 \le y \le b, \\ 0 & \text{otherwise}, \end{cases}$$

$$u_3(x, y) = \begin{cases} \phi(x, 0) & \text{if } 0 \le x \le a, \\ 0 & \text{otherwise}, \end{cases}$$

$$u_4(x, y) = \begin{cases} \phi(x, b) & \text{if } 0 \le x \le a, \\ 0 & \text{otherwise}. \end{cases}$$

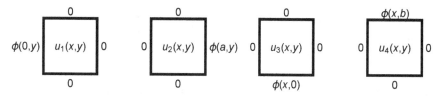

Fig. 6.3 Laplace's equation on a rectangle with general Dirichlet boundary conditions can be decomposed into four boundary value problems each with homogeneous Dirichlet boundary conditions on three sides of the rectangle.

Each of these four boundary value problems can be solved by similar method used to solve the boundary value problem of Eq. (6.2). Assuming the solutions to the four problems are $u_1(x, y)$, $u_2(x, y)$, $u_3(x, y)$, and $u_4(x, y)$ respectively, then by the Principle of Superposition, a solution to Eq. (6.1) is

$$u(x, y) = u_1(x, y) + u_2(x, y) + u_3(x, y) + u_4(x, y).$$

It has already been shown that

$$u_2(x, y) = \sum_{n=1}^{\infty} \left(\frac{2}{b \sinh \frac{n\pi a}{b}} \int_0^b \phi(a, y) \sin \frac{n\pi y}{b} \, dy \right) \sinh \frac{n\pi x}{b} \sin \frac{n\pi y}{b}.$$

Without formally repeating the steps used to find $u_2(x, y)$, the remaining three solutions can be found to be

$$u_1(x, y) = \sum_{n=1}^{\infty} \left(\frac{2}{b \sinh \frac{n\pi a}{b}} \int_0^b \phi(0, y) \sin \frac{n\pi y}{b} \, dy \right) \sinh \frac{n\pi(a - x)}{b} \sin \frac{n\pi y}{b},$$

$$u_3(x, y) = \sum_{n=1}^{\infty} \left(\frac{2}{a \sinh \frac{n\pi b}{a}} \int_0^a \phi(x, 0) \sin \frac{n\pi x}{a} \, dx \right) \sinh \frac{n\pi(b - y)}{a} \sin \frac{n\pi x}{a},$$

$$u_4(x, y) = \sum_{n=1}^{\infty} \left(\frac{2}{a \sinh \frac{n\pi b}{a}} \int_0^a \phi(x, b) \sin \frac{n\pi x}{a} \, dx \right) \sinh \frac{n\pi y}{a} \sin \frac{n\pi x}{a}.$$

This section ends with two examples demonstrating the procedure described above.

Example 6.2. Determine the steady-state temperature in a one-by-one square plate where one edge is kept at $100°C$ while the other edges are kept at $0°C$.

Solution. The coordinate system may be oriented so that the temperature distribution on the square plate is described by the two-dimensional heat equation:

$$u_t = k(u_{xx} + u_{yy}) \text{ for } 0 < x < 1 \text{ and } 0 < y < 1$$

with the boundary condition,

$$u(x, y, t) = \begin{cases} 100 & \text{if } 0 \leq x \leq 1 \text{ and } y = 1, \\ 0 & \text{otherwise on the boundary of the square.} \end{cases}$$

Therefore the steady state temperature distribution satisfies

$$u_{xx} + u_{yy} = 0 \text{ for } 0 < x < 1 \text{ and } 0 < y < 1,$$

with the following boundary condition:

$$u(x, y) = \begin{cases} 100 & \text{if } 0 \leq x \leq 1 \text{ and } y = 1, \\ 0 & \text{otherwise.} \end{cases}$$

The steady-state solution is

$$u(x, y) = \sum_{n=1}^{\infty} \frac{c_n}{\sinh n\pi} \sinh(n\pi y) \sin(n\pi x),$$

with

$$c_n = 2 \int_0^1 100 \sin(n\pi x) \, dx = \frac{200}{n\pi}(1 - \cos n\pi) = \begin{cases} 400/(n\pi) & \text{if } n \text{ is odd,} \\ 0 & \text{if } n \text{ is even.} \end{cases}$$

Therefore the solution to the boundary value problem can be written as

$$u(x, y) = \frac{400}{\pi} \sum_{n=1}^{\infty} \frac{\sin((2n - 1)\pi x) \sinh((2n - 1)\pi y)}{(2n - 1) \sinh((2n - 1)\pi)}.$$

A contour plot of the solution is illustrated in Fig. 6.4.

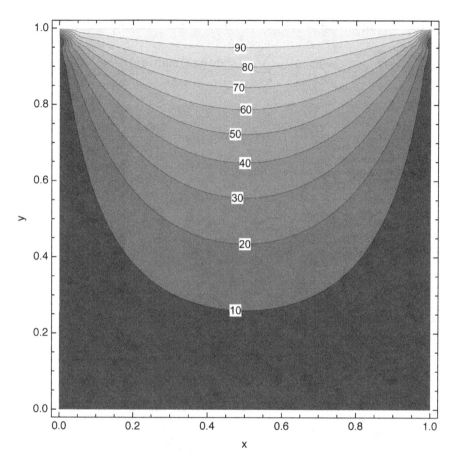

Fig. 6.4   A contour plot of the solution to the steady-state heat distribution on a rectangular plate described in Example 6.2.

If the orientation of the coordinate systems is set differently, a different expression for the solution would result.

Example 6.3. Find a solution of the boundary value problem:

$$u_{xx} + u_{yy} = 0 \text{ for } 0 < x < 2 \text{ and } 0 < y < 1,$$
$$u(x,0) = 50x \text{ and } u(x,1) = 0 \text{ for } 0 < x < 2,$$
$$u(0,y) = 0 \text{ and } u(2,y) = 100 \text{ for } 0 < y < 1.$$

Solution. This is a boundary value problem of the type associated with the second and third solution components $u_2(x,y)$ and $u_3(x,y)$ described in the decomposition process above. The formal solution is

$$u_2(x, y) = \sum_{n=1}^{\infty} c_n \sinh(n\pi x) \sin(n\pi y),$$

where

$$c_n = \frac{2}{\sinh(2n\pi)} \int_0^1 100 \sin(n\pi y)\, dy = \frac{200(1 - (-1)^n)}{n\pi \sinh(2n\pi)}.$$

The other component of the solution is

$$u_3(x, y) = \sum_{n=1}^{\infty} d_n \sinh \frac{n\pi(1 - y)}{2} \sin \frac{n\pi x}{2},$$

where

$$d_n = \frac{1}{\sinh \frac{n\pi}{2}} \int_0^2 50x \sin \frac{n\pi x}{2}\, dx = -\frac{200(-1)^n}{n\pi \sinh \frac{n\pi}{2}}.$$

Thus the series solution takes the form

$$u(x, y) = u_2(x, y) + u_3(x, y).$$

A contour plot of the solution is shown in Fig. 6.5.

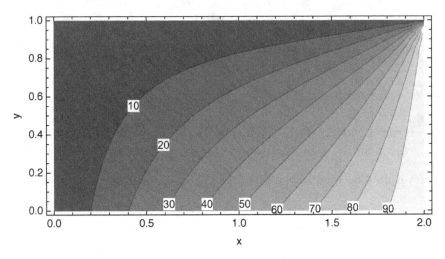

Fig. 6.5 A contour plot of the solution to the boundary value problem on a rectangle described in Example 6.3.

6.3 Dirichlet Problems on Disks

In this section the Dirichlet boundary value problems of Laplace's equation on a circular disk will be discussed. This change in the domain of the partial differential equation brings about several fundamental changes in the treatment of the problem, namely the eigenfunctions, fundamental solutions, and even the choice of coordinate system to use. Let $a > 0$ be a constant and consider the boundary value problem

$$
\begin{aligned}
&\triangle u = 0 \text{ for } x^2 + y^2 < a^2, \\
&u(x,y) = \phi(x,y) \text{ for } x^2 + y^2 = a^2.
\end{aligned}
\tag{6.8}
$$

Unlike the case when Ω is a rectangle, the method of separation of variables cannot be applied directly. However, due to the geometry of a disk, it is natural to consider the disk in polar coordinates. This requires that rectangular coordinate variables (x, y) be mapped to polar coordinate variables (r, θ). Recall the relation between these two coordinate systems in the plane:

$$
\begin{cases} x = r\cos\theta \\ y = r\sin\theta \end{cases} \text{ or equivalently } \begin{cases} r = \sqrt{x^2 + y^2}, \\ \theta = \arctan(y/x). \end{cases}
\tag{6.9}
$$

Under this transformation, the solution $u(x, y)$ of Laplace's equation is converted to a function $v(r, \theta)$ where

$$
v(r, \theta) = u(x, y) = u(r\cos\theta, r\sin\theta).
$$

The function v must solve the Laplacian partial differential equation in polar coordinates. The chain rule for derivatives can be used to express the Laplacian operator in terms of r and θ. By Eq. (6.9), $r_x = x/r$, $r_y = y/r$, $\theta_x = -y/r^2$, and $\theta_y = x/r^2$. This implies the first partial derivatives of u and v are related through the following equations:

$$
\begin{aligned}
u_x &= v_r r_x + v_\theta \theta_x = \frac{x}{r} v_r - \frac{y}{r^2} v_\theta, \\
u_y &= v_r r_y + v_\theta \theta_y = \frac{y}{r} v_r + \frac{x}{r^2} v_\theta.
\end{aligned}
$$

Differentiating again and making use of the product, quotient, and chain rules for derivatives yield

$$u_{xx} = \frac{\partial}{\partial x}\left[\frac{x}{r}\right]v_r + \frac{x}{r}\frac{\partial v_r}{\partial x} - \frac{\partial}{\partial x}\left[\frac{y}{r^2}\right]v_\theta - \frac{y}{r^2}\frac{\partial v_\theta}{\partial x}$$

$$= \frac{r - xr_x}{r^2}v_r + \frac{x}{r}(v_{rr}r_x + v_{r\theta}\theta_x) + \frac{2y}{r^3}r_x v_\theta - \frac{y}{r^2}(v_{\theta r}r_x + v_{\theta\theta}\theta_x)$$

$$= \left(\frac{1}{r} - \frac{x^2}{r^3}\right)v_r + v_{rr}\frac{x^2}{r^2} - \frac{xy}{r^3}v_{r\theta} + \frac{2xy}{r^4}v_\theta - \frac{xy}{r^3}v_{\theta r} + \frac{y^2}{r^4}v_{\theta\theta}$$

$$= \frac{y^2}{r^3}v_r + \frac{x^2}{r^2}v_{rr} - \frac{2xy}{r^3}v_{r\theta} + \frac{2xy}{r^4}v_\theta + \frac{y^2}{r^4}v_{\theta\theta},$$

$$u_{yy} = \frac{\partial}{\partial y}\left[\frac{y}{r}\right]v_r + \frac{y}{r}\frac{\partial v_r}{\partial y} + \frac{\partial}{\partial y}\left[\frac{x}{r^2}\right]v_\theta + \frac{x}{r^2}\frac{\partial v_\theta}{\partial y}$$

$$= \frac{r - yr_y}{r^2}v_r + \frac{y}{r}(v_{rr}r_y + v_{r\theta}\theta_y) - \frac{2x}{r^3}r_y v_\theta + \frac{x}{r^2}(v_{\theta r}r_y + v_{\theta\theta}\theta_y)$$

$$= \frac{x^2}{r^3}v_r + \frac{y^2}{r^2}v_{rr} + \frac{2xy}{r^3}v_{r\theta} - \frac{2xy}{r^4}v_\theta + \frac{x^2}{r^4}v_{\theta\theta}.$$

Adding these two second partial derivatives to form the Laplacian operator and using the relationships in Eq. (6.9) result in the conclusion that $u(x, y)$ is a solution to Laplace's equation if and only if $v(r, \theta)$ is a solution to the equation

$$v_{rr} + \frac{1}{r}v_r + \frac{1}{r^2}v_{\theta\theta} = 0. \tag{6.10}$$

In polar coordinates (r, θ) and $(r, \theta + 2k\pi)$ represent the same point for any integer k, therefore any solution to Laplace's equation on a disk must satisfy the following condition:

$$v(r, \theta) = v(r, \theta + 2\pi). \tag{6.11}$$

Furthermore, the boundary condition expressed in rectangular coordinates in Eq. (6.8) becomes

$$v(a, \theta) = \phi(a\cos\theta, a\sin\theta) = f(\theta) \tag{6.12}$$

in polar coordinates.

To summarize, the change of variables from rectangular coordinates to polar coordinates has converted the boundary value problem of Eq. (6.8) to the following boundary value problem:

$$v_{rr} + \frac{1}{r}v_r + \frac{1}{r^2}v_{\theta\theta} = 0 \text{ for } 0 < r < a \text{ and } -\infty < \theta < \infty,$$

$$v(a, \theta) = f(\theta) \text{ for } -\infty < \theta < \infty. \tag{6.13}$$

It is assumed that $f(\theta) = \phi(a\cos\theta, a\sin\theta)$ and that $f(\theta + 2\pi) = f(\theta)$ for all θ.

While separation of variables was not feasible for Laplace's equation in rectangular coordinates on a circular disk, product solutions of Eqs. (6.10) and (6.11) can be found. Suppose the solution takes the form $v(r,\theta) = R(r)T(\theta)$. In Example 1.11 the bounded product solutions to the boundary value problem on the disk were found to have the form

$$v_n(r,\theta) = r^n(c_n \cos(n\,\theta) + d_n \sin(n\,\theta)),$$

for $n = 0, 1, 2, \ldots$. By the Principle of Superposition the solution to Laplace's equation on the disk can be written formally as the infinite series,

$$v(r,\theta) = c_0 + \sum_{n=1}^{\infty} r^n[c_n \cos(n\theta) + d_n \sin(n\theta)]. \tag{6.14}$$

Just as in the case of Laplace's equation on a rectangular domain, the following three familiar questions must be answered.

(1) Is $v(r,\theta)$ well-defined for $0 < r < a$ and $-\infty < \theta < \infty$?
(2) Can the constants c_0, c_n, and d_n be chosen so that the boundary condition is satisfied?
(3) Is $v(r,\theta)$ a solution to Laplace's equation on the disk?

As was done earlier, attention is focused on the second question. The constants c_0, c_n, and d_n for $n \in \mathbb{N}$ must be determined so that

$$v(a,\theta) = f(\theta). \tag{6.15}$$

That is,

$$c_0 + \sum_{n=1}^{\infty} a^n[c_n \cos(n\theta) + d_n \sin(n\theta)] = f(\theta).$$

It is evident that

$$c_0 = \frac{1}{2\pi} \int_{-\pi}^{\pi} f(\theta)\, d\theta,$$

$$c_n = \frac{1}{a^n \pi} \int_{-\pi}^{\pi} f(\theta) \cos(n\theta)\, d\theta,$$

$$d_n = \frac{1}{a^n \pi} \int_{-\pi}^{\pi} f(\theta) \sin(n\theta)\, d\theta.$$

With the above choices for the constants c_0, c_n, and d_n the solution to the boundary value problem of Laplace's equation on the disk can be formally written as

$$v(r,\theta) = \frac{\alpha_0}{2} + \sum_{n=1}^{\infty} \left(\frac{r}{a}\right)^n [\alpha_n \cos(n\theta) + \beta_n \sin(n\theta)] \tag{6.16}$$

where α_n and β_n are the Fourier coefficients of $f(\theta)$:

$$\alpha_n = \frac{1}{\pi} \int_{-\pi}^{\pi} f(\theta) \cos(n\theta)\, d\theta \text{ for } n = 0, 1, 2, \ldots,$$

$$\beta_n = \frac{1}{\pi} \int_{-\pi}^{\pi} f(\theta) \sin(n\theta)\, d\theta \text{ for } n \in \mathbb{N}.$$

Note that the function defined in Eq. (6.16) converges since $0 \le r/a < 1$ for any $0 \le r < a$ and α_n and β_n are bounded as long as $f(\theta)$ is piecewise smooth. Using the same reasoning it can be shown that v_r, v_{rr}, and $v_{\theta\theta}$ all exist and that Laplace's equation in polar coordinates is satisfied. The justifying details are omitted.

If a solution is independent of θ, the solution is commonly referred to as a **radial solution**.

Finally, the solution to Laplace's equation on a disk in rectangular coordinates as given in Eq. (6.8) is

$$u(x, y) = v(\sqrt{x^2 + y^2}, \tan^{-1}(y/x)).$$

Example 6.4. Solve the Dirichlet problem on the unit disk with boundary values,

$$f(\theta) = \frac{1}{2}(\pi - \theta) \text{ for } -\pi < \theta < \pi.$$

Solution. The solution is

$$v(r, \theta) = \frac{\alpha_0}{2} + \sum_{n=1}^{\infty} r^n [\alpha_n \cos(n\theta) + \beta_n \sin(n\theta)]$$

where

$$\alpha_0 = \frac{1}{\pi} \int_{-\pi}^{\pi} \frac{1}{2}(\pi - \theta)\, d\theta = \pi,$$

$$\alpha_n = \frac{1}{\pi} \int_{-\pi}^{\pi} \frac{1}{2}(\pi - \theta) \cos(n\theta)\, d\theta = 0,$$

$$\beta_n = \frac{1}{\pi} \int_{-\pi}^{\pi} \frac{1}{2}(\pi - \theta) \sin(n\theta)\, d\theta = \frac{(-1)^n}{n}.$$

Thus in polar coordinates the solution can be expressed as

$$v(r, \theta) = \frac{\pi}{2} + \sum_{n=1}^{\infty} \frac{(-1)^n r^n}{n} \sin(n\theta).$$

Figure 6.6 shows a contour plot of the solution to this example in Cartesian coordinates.

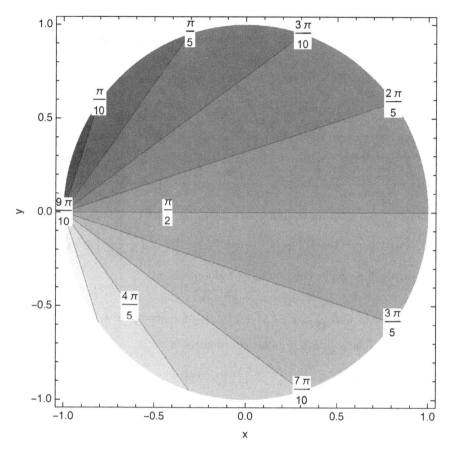

Fig. 6.6   A contour plot of the solution to the boundary value problem on the unit disk described in Example 6.4.

The solution to the boundary value problem of the last example can also be determined in closed form (as opposed to an infinite series form). Suppose $z = re^{i\theta}$ is a complex number with $r < 1$ and consider the Taylor series expansion of $f(z) = \ln(1 + z)$ about $z_0 = 0$. Since $1/(1 + z)$ can be expressed as a geometric series, then

$$\frac{1}{1 + z} = 1 - z + z^2 - z^3 + \cdots$$

and consequently term-by-term integration produces

$$\ln(1 + z) = z - \frac{z^2}{2} + \frac{z^3}{3} - \frac{z^4}{4} + \cdots,$$

so long as $|z| < 1$.

The real and imaginary parts of $\ln(1 + z)$ can be computed using the Euler identity of Eq. (3.40). The reader is asked to provide the details in Exercises 8 and 9 which show that

$$\text{Im}\left(\ln(1 + z)\right) = \text{Im}\left(\ln(1 + re^{i\theta})\right) = \text{Im}\left(\ln(1 + r\cos\theta + ir\sin\theta)\right)$$

$$= \tan^{-1}\left(\frac{r\sin\theta}{1 + r\cos\theta}\right). \tag{6.17}$$

On the other hand, the imaginary part of $z - z^2/2 + z^3/3 - z^4/4 + \cdots$ can be calculated as

$$\text{Im}\left(-\sum_{n=1}^{\infty} \frac{(-1)^n z^n}{n}\right) = -\sum_{n=1}^{\infty} \frac{(-1)^n r^n}{n} \text{Im}\left(e^{in\theta}\right) = -\sum_{n=1}^{\infty} \frac{(-1)^n r^n}{n} \sin n\theta.$$

Therefore the solution to Example 6.4 can be written as

$$v(r,\theta) = \frac{\pi}{2} - \tan^{-1}\left(\frac{r\sin\theta}{1 + r\cos\theta}\right) = \frac{\pi}{2} - \tan^{-1}\left(\frac{y}{1 + x}\right) = v(x,y).$$

Example 6.5. Solve the boundary value problem:

$$\triangle u = 0 \text{ for } x^2 + y^2 < 4,$$

$$u(x,y) = x^4 - y^4 + x^2y^2 + y \text{ for } x^2 + y^2 = 4.$$

Solution. The solution in polar coordinates is

$$v(r,\theta) = \frac{\alpha_0}{2} + \sum_{n=1}^{\infty} \left(\frac{r}{2}\right)^n [\alpha_n \cos(n\theta) + \beta_n \sin(n\theta)],$$

where α_n and β_n are the Fourier coefficients of

$$v(2,\theta) = [(x^2 + y^2)(x^2 - y^2) + x^2y^2 + y]_{x=2\cos\theta,\, y=2\sin\theta}$$

$$= 16(\cos^2\theta - \sin^2\theta) + 16\cos^2\theta\sin^2\theta + 2\sin\theta$$

$$= 16\cos(2\theta) + 4\sin^2(2\theta) + 2\sin\theta$$

$$= 2 + 2\sin\theta + 16\cos(2\theta) - 2\cos(4\theta).$$

Since the boundary condition is a linear combination of eigenfunctions, there is no need to resort to the Euler-Fourier coefficient formulas to determine the coefficients of the solution. In this case by equating coefficients $\alpha_0 = 4$, $\alpha_2 = 16$, $\alpha_4 = -2$, $\beta_1 = 2$, and all other coefficients are 0. This produces the solution

$$v(r,\theta) = 2 + \frac{r}{2}(\sin\theta) + \left(\frac{r}{2}\right)^2 [16\cos(2\theta)] - \left(\frac{r}{2}\right)^4 [2\cos(4\theta)]$$

$$= 2 + r\sin\theta + 4r^2(\cos^2\theta - \sin^2\theta) - \frac{1}{8}r^4[1 - 2\sin^2(2\theta)]$$

$$= 2 + r\sin\theta + 4r^2(\cos^2\theta - \sin^2\theta) - \frac{1}{8}(r^2)^2 + (r\sin\theta)^2(r\cos\theta)^2.$$

Thus it is possible to express the solution in terms of the original rectangular coordinates,

$$u(x,y) = 2 + y + 4(x^2 - y^2) - \frac{1}{8}(x^2 + y^2)^2 + x^2 y^2.$$

The reader can check by differentiating directly that this function is harmonic.

For the examples above, the solution in terms of rectangular coordinates can be recovered. This cannot always be done easily. Fortunately it is not necessary to express the solution in rectangular coordinates in most cases.

Example 6.6. Find the steady-state temperature distribution in a disk of radius 1 with the upper half of its circumference kept at 100°C and the lower half at 0°C.

Solution. The steady-state temperature distribution is a solution of the Laplace boundary value problem:

$$\triangle u = 0 \text{ for } x^2 + y^2 < 1,$$

$$u(x,y) = \begin{cases} 100 & \text{if } x^2 + y^2 = 1 \text{ and } y \geq 0, \\ 0 & \text{if } x^2 + y^2 = 1 \text{ and } y < 0. \end{cases}$$

The solution in polar coordinates is

$$v(r, \theta) = \frac{\alpha_0}{2} + \sum_{n=1}^{\infty} r^n [\alpha_n \cos(n\theta) + \beta_n \sin(n\theta)],$$

where the Fourier coefficients are calculated as

$$\alpha_0 = \frac{1}{\pi} \int_0^\pi 100 \, d\theta = 100,$$

$$\alpha_n = \frac{1}{\pi} \int_0^\pi 100 \cos(n\theta) \, d\theta = 0,$$

$$\beta_n = \frac{1}{\pi} \int_0^\pi 100 \sin(n\theta) \, d\theta = \frac{100(1 - \cos(n\pi))}{n\pi}.$$

Hence the steady-state temperature distribution can be represented as

$$v(r, \theta) = 50 + \frac{100}{\pi} \sum_{n=1}^{\infty} \frac{r^n}{n}(1 - \cos(n\pi)) \sin(n\theta).$$

Rearranging terms in this infinite series enables it to be re-written as

$$v(r, \theta) = 50 + \frac{100}{\pi} \sum_{n=1}^{\infty} \frac{r^n}{n} \sin(n\theta) - \frac{100}{\pi} \sum_{n=1}^{\infty} \frac{r^n}{n} \cos(n\pi) \sin(n\theta).$$

The first summation may be expressed in closed form using Eq. (6.17). The second summation may also be expressed in closed form using the same formula after first applying the trigonometric identity,

$$\cos(n\pi)\sin(n\theta) = \sin(n(\theta - \pi)).$$

Consequently the sum of the infinite series solution can be written relatively compactly as

$$v(r,\theta) = 50 + \frac{100}{\pi}\left[\tan^{-1}\left(\frac{r\sin\theta}{1 - r\cos\theta}\right) + \tan^{-1}\left(\frac{r\sin\theta}{1 + r\cos\theta}\right)\right] \quad \text{or}$$

$$u(x,y) = 50 + \frac{100}{\pi}\left[\tan^{-1}\left(\frac{y}{1 - x}\right) + \tan^{-1}\left(\frac{y}{1 + x}\right)\right].$$

Figure 6.7 shows a contour plot of the solution to the boundary value problem in Example 6.6.

6.4   Dirichlet Problems on Domains Related to Disks

This section will explore Dirichlet boundary value problems on domains whose geometries are related to disks, namely the plane with a disk-shaped "hole", an annular or "ring"-shaped domain, or others, see Fig. 6.8. The solutions will be found by methods similar to those of the previous section.

First consider the case of the infinite two-dimensional domain missing a disk-shaped region of radius $a > 0$ centered at the origin. On this region the boundary value problem is expressed as follows:

$$\begin{aligned} \triangle u &= 0 \text{ for } x^2 + y^2 > a^2, \\ u(x,y) &= \phi(x,y) \text{ for } x^2 + y^2 = a^2. \end{aligned} \tag{6.18}$$

A desirable property of the solution to Eq. (6.18) will be that it is bounded on the domain $\{(x,y)\,|\,x^2 + y^2 > a^2\}$. As was the case for the circular disk, the boundary value problem of Eq. (6.18) is treated in polar coordinates. After the change of variables is performed, the boundary value problem becomes,

$$\begin{aligned} v_{rr} + \frac{1}{r}v_r + \frac{1}{r^2}v_{\theta\theta} &= 0 \text{ for } r > a \text{ and } -\infty < \theta < \infty, \\ v(a,\theta) &= \phi(a\cos\theta, a\sin\theta) = f(\theta) \text{ for } -\infty < \theta < \infty. \end{aligned} \tag{6.19}$$

As seen in Example 1.11, the product solutions to this boundary value problem take the form,

$$v_n(r,\theta) = \begin{cases} c_n \ln r + d_n & \text{if } n = 0, \\ (c_n r^{-n} + d_n r^n)(a_n \cos n\theta + b_n \sin n\theta) & \text{if } n \in \mathbb{N}. \end{cases} \tag{6.20}$$

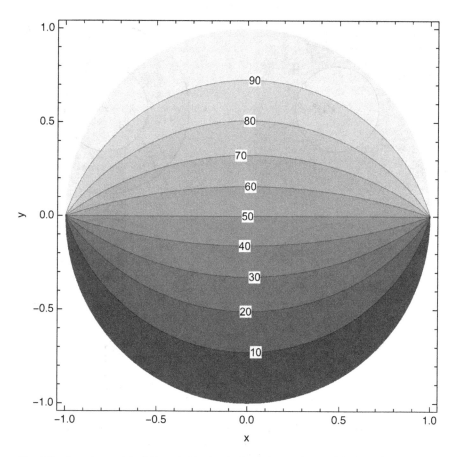

Fig. 6.7   A contour plot of the solution to the boundary value problem on the unit disk described in Example 6.6.

A superposition of these types of product solutions will be used to find the solutions to the boundary value problem expressed in Eq. (6.19). Again, it is desirable to find a bounded solution. To avoid the solution becoming unbounded as $r \to \infty$, the coefficients of r^n and $\ln r$ are chosen to be zero This leads to the following formal solution to the problem:

$$v(r, \theta) = d_0 + \sum_{n=1}^{\infty} r^{-n}[c_n \cos(n\theta) + d_n \sin(n\theta)],$$

with d_0, c_n, and d_n chosen so that

$$v(a, \theta) = d_0 + \sum_{n=1}^{\infty} a^{-n}[c_n \cos(n\theta) + d_n \sin(n\theta)] = f(\theta).$$

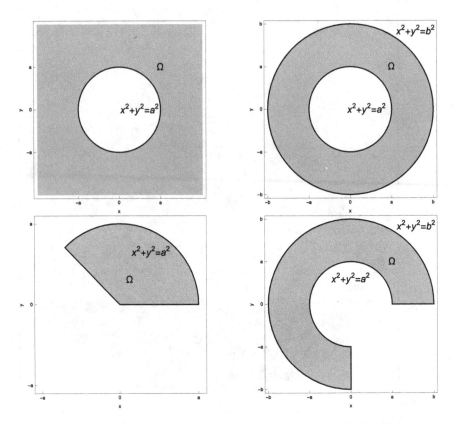

Fig. 6.8   The geometries of domains for boundary value problems with Dirichlet boundary conditions in the xy-plane. These domains are commonly referred to as the infinite plane with a "hole", an annulus or "ring", and a sector.

The constants d_0, c_n, and d_n should be chosen as $d_0 = \alpha_0/2$, $c_n = a^n \alpha_n$, and $d_n = a^n \beta_n$ where α_0, α_n, and β_n are the Fourier coefficients of $f(\theta)$.

Example 6.7. Find a bounded solution to Laplace's equation on the domain $\{(r, \theta) \, | \, r > a\}$ with the boundary condition $u(a, \theta) = 1$ for $0 < \theta < \pi$ and $u(a, \theta) = 0$ for $-\pi < \theta < 0$.

Solution. As derived above, the infinite series solution has the form

$$u(r, \theta) = d_0 + \sum_{n=1}^{\infty} r^{-n} [c_n \cos(n\theta) + d_n \sin(n\theta)],$$

where d_0, c_n, and d_n are constants such that

$$u(a, \theta) = d_0 + \sum_{n=1}^{\infty} a^{-n} [c_n \cos(n\theta) + d_n \sin(n\theta)] = \begin{cases} 1 & \text{if } 0 < \theta < \pi, \\ 0 & \text{if } -\pi < \theta < 0. \end{cases}$$

Applying the Euler-Fourier formulas for the coefficients results in

$$d_0 = \frac{1}{2\pi} \int_0^{2\pi} u(a,\theta)\, d\theta = \frac{1}{2\pi} \int_0^{\pi} (1)\, d\theta = \frac{1}{2},$$

$$c_n = \frac{a^n}{\pi} \int_0^{\pi} (1) \cos(n\theta)\, d\theta = 0,$$

$$d_n = \frac{a^n}{\pi} \int_0^{\pi} (1) \sin(n\theta)\, d\theta = \frac{a^n}{n\pi}(1 - \cos(n\pi)).$$

A similar procedure as used in Example 6.6 yields the following,

$$u(r,\theta) = \frac{1}{2} + \frac{1}{\pi} \sum_{n=1}^{\infty} (1 - \cos(n\pi)) \left(\frac{a}{r}\right)^n \frac{\sin(n\theta)}{n}$$

$$= \frac{1}{2} + \frac{1}{\pi} \left[\tan^{-1}\left(\frac{a \sin\theta}{r - a \cos\theta}\right) + \tan^{-1}\left(\frac{a \sin\theta}{r + a \cos\theta}\right) \right].$$

Therefore,

$$u(x,y) = \frac{1}{2} + \frac{1}{\pi} \left[\tan^{-1}\left(\frac{ay}{x^2 + y^2 - ax}\right) + \tan^{-1}\left(\frac{ay}{x^2 + y^2 + ax}\right) \right].$$

Figure 6.9 shows a contour plot of the solution in Cartesian coordinates.

An **annulus** is a common geometrical figure in applied mathematics, physics, and engineering. The annulus or "ring" is the next geometry explored as a domain for Laplace's equation in the plane. Consider the boundary value problem of Laplace's equation on a ring-shaped domain:

$$\triangle u = 0 \text{ for } a^2 < x^2 + y^2 < b^2,$$
$$u(x,y) = \phi_1(x,y) \text{ for } x^2 + y^2 = a^2, \tag{6.21}$$
$$u(x,y) = \phi_2(x,y) \text{ for } x^2 + y^2 = b^2.$$

After converting to polar coordinates this boundary value problem becomes,

$$v_{rr} + \frac{1}{r}v_r + \frac{1}{r^2}v_{\theta\theta} = 0 \text{ for } a < r < b \text{ and } -\infty < \theta < \infty,$$
$$v(a,\theta) = \phi_1(a \cos\theta, a \sin\theta) = f(\theta) \text{ for } -\infty < \theta < \infty, \quad (6.22)$$
$$v(b,\theta) = \phi_2(b \cos\theta, b \sin\theta) = g(\theta) \text{ for } -\infty < \theta < \infty.$$

Once again, the product solutions of Eqs. (6.10) and (6.11) given by Eq. (6.20) are used to build solutions to the boundary value problem in Eq. (6.22). Therefore a formal solution to Eq. (6.22) is sought in the form of

$$v(r,\theta) = c_0 \ln r + d_0 + \sum_{n=1}^{\infty} [(c_n^+ r^n + c_n^- r^{-n}) \cos(n\theta) + (d_n^+ r^n + d_n^- r^{-n}) \sin(n\theta)].$$

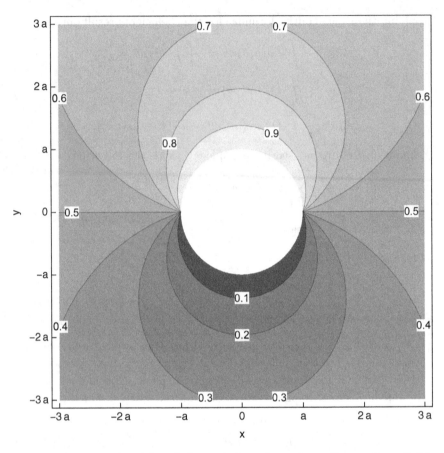

Fig. 6.9   A contour plot of the solution to the boundary value problem on the unit disk described in Example 6.7.

The two boundary conditions imply that the constant coefficients must satisfy simultaneously the following equations:

$$f(\theta) = c_0 \ln a + d_0 + \sum_{n=1}^{\infty} \left[\left(c_n^+ a^n + \frac{c_n^-}{a^n} \right) \cos(n\theta) + \left(d_n^+ a^n + \frac{d_n^-}{a^n} \right) \sin(n\theta) \right],$$

$$g(\theta) = c_0 \ln b + d_0 + \sum_{n=1}^{\infty} \left[\left(c_n^+ b^n + \frac{c_n^-}{b^n} \right) \cos(n\theta) + \left(d_n^+ b^n + \frac{d_n^-}{b^n} \right) \sin(n\theta) \right].$$

Thus coefficients c_0, d_0, c_n^+, c_n^-, d_n^+, and d_n^- must be chosen so that

$$c_0 \ln a + d_0 = \frac{\alpha_0}{2} \text{ and } c_0 \ln b + d_0 = \frac{\overline{\alpha}_0}{2}, \tag{6.23}$$

$$c_n^+ a^n + c_n^- a^{-n} = \alpha_n \text{ and } c_n^+ b^n + c_n^- b^{-n} = \overline{\alpha}_n, \tag{6.24}$$

$$d_n^+ a^n + d_n^- a^{-n} = \beta_n \text{ and } d_n^+ b^n + d_n^- b^{-n} = \overline{\beta}_n, \tag{6.25}$$

where α_n, β_n, $\overline{\alpha}_n$, and $\overline{\beta}_n$ are the Fourier coefficients of $f(\theta)$ and $g(\theta)$ respectively. It is not difficult to show that Eqs. (6.23)–(6.25) have a unique solution. Therefore the boundary value problem of Eq. (6.22) is solved formally. As noted earlier, if the solution is independent of θ, it is called a radial solution.

Example 6.8. Solve the boundary value problem:

$$\triangle u = 0 \text{ for } R_1^2 < x^2 + y^2 < R_2^2,$$

$$u(x, y) = T_1 \text{ for } x^2 + y^2 = R_1^2,$$

$$u(x, y) = T_2 \text{ for } x^2 + y^2 = R_2^2,$$

where T_1 and T_2 are constants.

Solution. The solution of this boundary value problem is a radial solution of the form $v(r, \theta) = v(r) = A \ln r + B$ with A and B determined by the pair of equations,

$$A \ln R_1 + B = T_1,$$

$$A \ln R_2 + B = T_2.$$

Solving the system of equations yields

$$A = \frac{T_2 - T_1}{\ln R_2 - \ln R_1} \text{ and } B = T_1 - \frac{(T_2 - T_1) \ln R_1}{\ln R_2 - \ln R_1}.$$

Example 6.9. Solve Laplace's equation on the domain $\{(r, \theta) \mid 1 < r < 2\}$ with boundary conditions $u(1, \theta) = 0$ for $\theta \in [-\pi, \pi]$ and $u(2, \theta) = 1$ if $0 < \theta < \pi$ and $u(2, \theta) = -1$ if $-\pi < \theta < 0$.

Solution. The solution is of the form

$$u(r, \theta) = c_0 \ln r + d_0 + \sum_{n=1}^{\infty} [(c_n^+ r^n + c_n^- r^{-n}) \cos(n\theta) + (d_n^+ r^n + d_n^- r^{-n}) \sin(n\theta)],$$

where c_0, d_0, c_n^+, c_n^-, d_n^+, and d_n^- are constants determined below. The constants c_0 and d_0 must satisfy:

$$c_0 \ln 1 + d_0 = \frac{1}{2\pi} \int_{-\pi}^{\pi} 0 \, d\theta = 0,$$

$$c_0 \ln 2 + d_0 = \frac{1}{2\pi} \int_{-\pi}^{0} (-1) \, d\theta + \frac{1}{2\pi} \int_{0}^{\pi} 1 \, d\theta = 0.$$

Consequently $c_0 = d_0 = 0$. The coefficients c_n^+ and c_n^- must solve the following for each $n \in \mathbb{N}$,

$$c_n^+(1^n) + c_n^-(1^{-n}) = \frac{1}{\pi} \int_{-\pi}^{\pi} 0 \, d\theta = 0,$$

$$c_n^+(2^n) + c_n^-(2^{-n}) = \frac{1}{\pi} \int_{-\pi}^{0} (-\cos(n\theta)) \, d\theta + \frac{1}{\pi} \int_{0}^{\pi} \cos(n\theta) \, d\theta = 0.$$

Fortunately this system of equations can also be solved by inspection and results in $c_n^+ = c_n^- = 0$ for all $n \in \mathbb{N}$. Finally constants d_n^+ and d_n^- must be solutions to the following system of equations:

$$d_n^+ + d_n^- = \frac{1}{\pi} \int_{-\pi}^{\pi} 0 \, d\theta = 0,$$

$$2^n d_n^+ + \frac{d_n^-}{2^n} = \frac{1}{\pi} \int_{-\pi}^{0} (-\sin(n\theta)) \, d\theta + \frac{1}{\pi} \int_{0}^{\pi} \sin(n\theta) \, d\theta = \frac{2}{n\pi}(1 - (-1)^n).$$

Solving this linear system yields:

$$d_n^+ = \frac{2(1 - (-1)^n)}{n\pi(2^n - 2^{-n})} = -d_n^-.$$

Consequently the solution to the boundary value problem is

$$u(r, \theta) = \frac{2}{\pi} \sum_{n=1}^{\infty} \left[\frac{(1 - (-1)^n)r^n}{n(2^n - 2^{-n})} - \frac{(1 - (-1)^n)r^{-n}}{n(2^n - 2^{-n})} \right] \sin n\theta$$

$$= \frac{2}{\pi} \sum_{n=1}^{\infty} (1 - (-1)^n) \frac{r^n - r^{-n}}{(2^n - 2^{-n})n} \sin n\theta.$$

Figure 6.10 shows a contour plot of the solution in Cartesian coordinates.

Yet another variation on the boundary value problem of Laplace's equation involves the spatial domain called a **sector** which is the portion of a circular disk or annulus enclosed by a central angle of less than 2π radians. In the following discussion the case of the sector of a disk will be explored. The case of a sector of an annulus is similar and left as an exercise. As has been demonstrated throughout this chapter, the solution to this type of problem is often most conveniently expressed as a Fourier series of product solutions depending on the polar coordinates r and θ. The typical approach to solving Laplace's equation on a circular sector will be illustrated in the following example.

Let $0 < \Theta < 2\pi$ and consider the boundary value problem:

$$u_{rr} + \frac{1}{r} u_r + \frac{1}{r^2} u_{\theta\theta} = 0 \text{ for } 0 < r < a \text{ and } 0 < \theta < \Theta,$$

$$u(a, \theta) = f(\theta) \text{ for } 0 < \theta < \Theta,$$

$$u(r, 0) = u(r, \Theta) = 0 \text{ for } 0 < r < a.$$

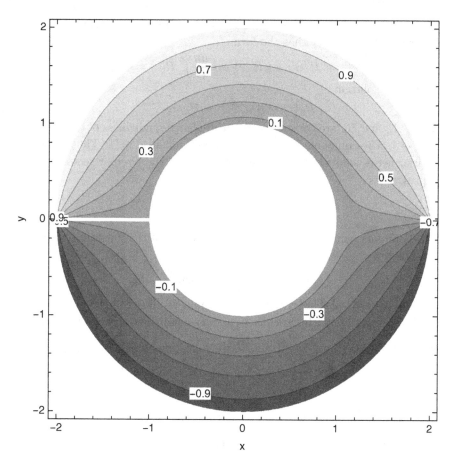

Fig. 6.10   A contour plot of the solution to the boundary value problem on the annulus described in Example 6.9.

The nontrivial product solutions to Laplace's equation take the form $u(r,\theta) = R(r)T(\theta)$ where $R(r)$ solves an Euler equation and the homogeneous Dirichlet boundary conditions at $\theta = 0$ and $\theta = \Theta$ imply that $T(\theta)$ must solve the boundary problem:

$$T''(\theta) + \sigma T(\theta) = 0,$$
$$T(0) = 0 = T(\Theta).$$

This boundary value problem is the same as the one given in Eqs. (1.52)–(1.54) and has eigenvalues $\sigma_n = \lambda_n^2 = (n\pi/\Theta)^2$ and eigenfunctions

$$T_n(\theta) = c_n \sin\frac{n\pi\theta}{\Theta},$$

where c_n is a constant for all $n \in \mathbb{N}$. Having found the eigenvalues, the Euler equation for $R(r)$ has the form

$$r^2 R''(r) + r R'(r) - \left(\frac{n\pi}{\Theta}\right)^2 R(r) = 0$$

which has solution

$$R_n(r) = A_n r^{n\pi/\Theta} + B_n r^{-n\pi/\Theta}$$

where A_n and B_n are constants. To ensure the boundedness of the solution as $r \to 0$, the constant B_n is chosen to be 0 for all $n \in \mathbb{N}$. Thus a formal Fourier series solution to Laplace's equation on the circular sector can be expressed as

$$u(r, \theta) = \sum_{n=1}^{\infty} c_n r^{n\pi/\Theta} \sin \frac{n\pi\theta}{\Theta}.$$

The boundary condition $u(a, \theta) = f(\theta)$ dictates that the constants c_n should be chosen as

$$c_n = \frac{2}{\Theta \, a^{n\pi/\Theta}} \int_0^{\Theta} f(\theta) \sin \frac{n\pi\theta}{\Theta} \, d\theta.$$

This section concludes with an example to illustrate the solution of Laplace's equation with Dirichlet boundary conditions on a circular sector.

Example 6.10. Find the solution to the following boundary value problem:

$$v_{rr} + \frac{1}{r} v_r + \frac{1}{r^2} v_{\theta\theta} = 0 \text{ for } 0 < r < 1 \text{ and } 0 < \theta < \pi/2,$$

$$v(1, \theta) = \theta(\pi/2 - \theta) \text{ for } 0 < \theta < \pi/2,$$

$$v(r, 0) = v(r, \pi/2) = 0 \text{ for } 0 < r < 1.$$

Solution. According to the theory laid out above, the eigenvalues of this boundary value problem are $\lambda_n = (n\pi/(\pi/2))^2 = 4n^2$ for $n \in \mathbb{N}$. The solution should be of the form

$$v(r, \theta) = \sum_{n=1}^{\infty} a_n r^{2n} \sin(2n\theta).$$

The Fourier sine coefficients are

$$a_n = \frac{2}{\pi/2} \int_0^{\pi/2} \theta(\pi/2 - \theta) \sin(2n\theta) \, d\theta = \frac{1 - \cos(n\pi)}{n^3 \pi}$$

$$= \begin{cases} 0 & \text{if } n \text{ is even,} \\ 2/(n^3 \pi) & \text{if } n \text{ is odd.} \end{cases}$$

Thus the solution to Laplace's equation can be written as

$$v(r, \theta) = \frac{2}{\pi} \sum_{n=1}^{\infty} \frac{r^{4n-2}}{(2n-1)^3} \sin(4n - 2)\theta.$$

Figure 6.11 shows a contour plot of the solution in Cartesian coordinates.

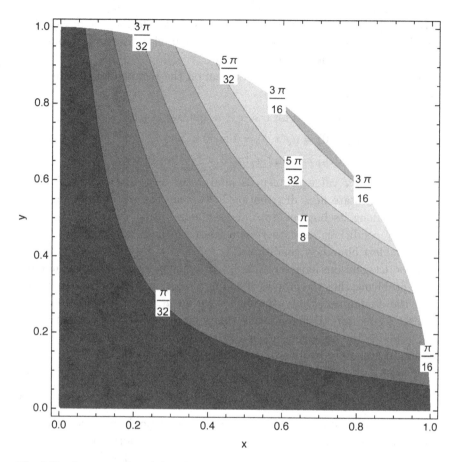

Fig. 6.11   A contour plot of the solution to the boundary value problem on the circular sector described in Example 6.10.

While the two most recent sections have explored solutions to Laplace's equation in a variety of domain geometries and with a variety of Dirichlet boundary conditions, much remains to be considered. Some of the other cases are left as exercises. In the next section, a different type of boundary condition for Laplace's equation will be explored.

6.5   Neumann Problems on Rectangles

This section will treat the Neumann boundary value problem of Laplace's equation on rectangular domains. Much of the analysis is similar to the earlier sections of this chapter in that product solutions in rectangular coordinates coupled with Fourier series will be used to find formal solutions to the boundary value problems.

Consider the boundary value problem on the rectangular domain $\Omega = \{(x, y) \mid 0 < x < a, \, 0 < y < b\}$:

$$\triangle u = 0 \text{ for } 0 < x < a \text{ and } 0 < y < b,$$

$$u_y(x, 0) = u_y(x, b) = 0 \text{ for } 0 < x < a, \qquad (6.26)$$

$$u_x(0, y) = 0 \text{ and } u_x(a, y) = f(y) \text{ for } 0 < y < b.$$

If this boundary value problem is interpreted as a mathematical model of the steady-state heat distribution in Ω, then the rectangular region is insulated along its bottom, top, and left edges (first three homogeneous boundary conditions) and there is a source of heat applied to the right edge (the last boundary condition).

Product solutions have served well as starting points in solving boundary value problems, thus the process of finding a solution in this situation begins with the assumption that a product solution $u(x, y) = X(x)Y(y)$ solves Laplace's equation. Differentiating this solution, separating the variables, and applying the homogeneous boundary conditions produce the following system of boundary value ordinary differential equations:

$$X''(x) - \sigma X(x) = 0 \text{ with } X'(0) = 0, \qquad (6.27)$$

$$Y''(y) + \sigma Y(y) = 0 \text{ with } Y'(0) = 0 = Y'(b), \qquad (6.28)$$

where σ is a constant. Solving these types of boundary value problems has become familiar and the reader can check that nontrivial solutions to Eq. (6.28) exist only if $\sigma = \sigma_n = (n\pi/b)^2$ and $Y(y) = Y_n(y) = \cos(n\pi y/b)$ for $n = 0, 1, 2, \ldots$. See Exercise 18 in Chap. 1. Replacing σ in Eq. (6.27) with σ_n and solving the resulting boundary value problem produce solutions $X(x) = X_n(x) = \cosh(n\pi x/b)$ for $n = 0, 1, 2, \ldots$. Therefore the functions

$$u_n(x, y) = \cosh\frac{n\pi x}{b} \cos\frac{n\pi y}{b}$$

solve Laplace's equation and satisfy the three homogeneous boundary conditions of Eq. (6.26). This leads to the following formal infinite series solution to the boundary value problem motivated by the Principle of Superposition,

$$u(x, y) = c_0 + \sum_{n=1}^{\infty} c_n \cosh\frac{n\pi x}{b} \cos\frac{n\pi y}{b}. \qquad (6.29)$$

This formal solution satisfies the homogeneous boundary conditions for any choice of coefficients c_n. Assuming for the moment that the series in

Eq. (6.29) is well-defined and can be differentiated term by term, the coefficients can be chosen to satisfy the remaining nonhomogeneous boundary condition, that is, under appropriate conditions,

$$u_x(a, y) = \sum_{n=1}^{\infty} c_n \left(\frac{n\pi}{b} \right) \sinh \frac{n\pi a}{b} \cos \frac{n\pi y}{b} = f(y). \qquad (6.30)$$

The infinite series can be regarded as a cosine series representation of $f(y)$ if the integral of f over $[0, b]$ vanishes. In fact, if $\int_0^b f(y)\, dy \neq 0$ a solution cannot be determined since Eq. (6.30) will not be valid. Physical intuition can account for this restriction as well. If the definite integral vanishes then there is no net flux of heat across the boundary at $x = a$ and hence a steady-state (time independent) heat distribution can evolve. If the definite integral does not vanish, then there is a net flux of heat in or out of Ω and no time independent temperature distribution can exist. Another subtle issue related to solving Laplace's equation on a rectangle with Neumann boundary conditions on all four edges is connected with the constant c_0 of Eq. (6.29). Since this constant vanishes upon differentiation, then its value cannot be determined using an Euler-Fourier integral formula. Thus Laplace's equation on a rectangle with Neumann boundary conditions on all four edges has no unique solution since any constant is a solution to Laplace's equation satisfying the zero Neumann boundary condition. This type of boundary value problem is ill-posed.

Making the assumption that the integral of f vanishes over $[0, b]$, then the constants c_n are

$$c_n = \frac{2}{n\pi \sinh \frac{n\pi a}{b}} \int_0^b f(y) \cos \frac{n\pi y}{b}\, dy, \qquad (6.31)$$

for $n \in \mathbb{N}$.

Example 6.11. Find a solution to the Neumann boundary value problem on the unit square:

$$\triangle u = 0 \text{ for } 0 < x < 1 \text{ and } 0 < y < 1,$$
$$u_y(x, 0) = u_y(x, 1) = 0 \text{ for } 0 < x < 1,$$
$$u_x(0, y) = 0 \text{ and } u_x(1, y) = y - 1/2 \text{ for } 0 < y < 1.$$

Solution. Since the definite integral of $f(y) = y - 1/2$ vanishes on $[0, 1]$ the solution should be an infinite series of the form given in Eq. (6.29) with coefficients as defined in Eq. (6.31). Computing the coefficients yields

$$c_n = \frac{2}{n\pi \sinh(n\pi)} \int_0^1 (y - 1/2) \cos(n\pi y)\, dy = \frac{2((-1)^n - 1)}{n^3 \pi^3 \sinh(n\pi)}$$

for $n \in \mathbb{N}$ and thus

$$u(x,y) = c_0 - \frac{4}{\pi^3} \sum_{n=1}^{\infty} \frac{\cosh((2n-1)\pi x)\cos((2n-1)\pi y)}{(2n-1)^3 \sinh((2n-1)\pi)}$$

where c_0 is an arbitrary constant. Figure 6.12 shows a contour plot of the solution with c_0 chosen to be 0.

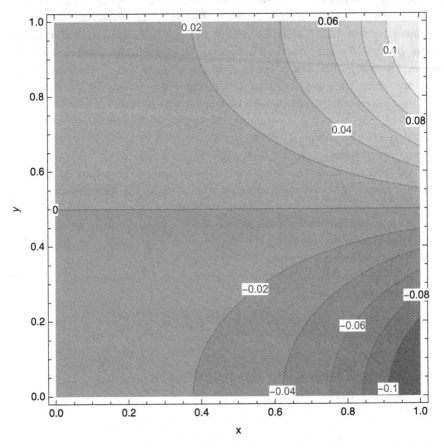

Fig. 6.12   A contour plot of the solution to the boundary value problem on the unit square described in Example 6.11. The arbitrary constant c_0 present in the solution was chosen to be 0.

The nonhomogeneous boundary condition could have been placed at any one of the four edges of the rectangle with suitable modifications being made to the solution. Likewise as in the case of the Dirichlet boundary conditions, if more than one edge is subject to a nonhomogeneous Neumann boundary condition, the boundary value problem can be decomposed into multiple boundary value problems each having homogeneous Neumann boundary conditions along three sides of the rectangle and a nonhomogeneous Neumann boundary condition on the remaining side. Once the individual boundary value problems are solved, the sum of their solutions solves the original Neumann boundary value problem.

6.6 Neumann Problems on Disks

To accompany the discussion of Laplace's equation on a rectangle with Neumann boundary conditions, this section treats the Neumann boundary problem on a disk:

$$\triangle u = 0 \text{ for } x^2 + y^2 < a^2,$$
$$\frac{\partial u}{\partial \mathbf{n}}(x, y) = \phi(x, y) \text{ for } x^2 + y^2 = a^2. \tag{6.32}$$

The operator $\partial u / \partial \mathbf{n}$ denotes the derivative in the direction of unit vector $\mathbf{n}$. In this situation the vector $\mathbf{n}$ will be assumed to be the unit outward normal vector to the circle $x^2 + y^2 = a^2$. Naturally, the problem is more easily studied in the polar coordinate system,

$$v_{rr} + \frac{1}{r} v_r + \frac{1}{r^2} v_{\theta\theta} = 0 \text{ for } 0 < r < a \text{ and } -\infty < \theta < \infty,$$
$$\frac{\partial v}{\partial r}(a, \theta) = \phi(a\cos\theta, a\sin\theta) = f(\theta) \text{ for } -\infty < \theta < \infty. \tag{6.33}$$

Note that the radial direction of the polar coordinate r is the outward normal direction of the circle $x^2 + y^2 = a^2$. The product solutions of Laplace's equation were found in Eq. (6.20). Thus a formal solution to Laplace's equation with Neumann boundary condition on a disk of radius $a > 0$ can be formulated as

$$v(r, \theta) = d_0 + \sum_{n=1}^{\infty} r^n [c_n^+ \cos(n\theta) + d_n^+ \sin(n\theta)],$$

with coefficients d_0, c_n^+, and d_n^+ chosen such that

$$v_r(a, \theta) = \sum_{n=1}^{\infty} n a^{n-1} [c_n^+ \cos(n\theta) + d_n^+ \sin(n\theta)] = f(\theta).$$

Thus a necessary condition for the solution to Eq. (6.32) to exist is that

$$\int_{-\pi}^{\pi} f(\theta) \, d\theta = \int_{-\pi}^{\pi} \phi(a\cos\theta, a\sin\theta) \, d\theta = 0. \tag{6.34}$$

Just as in the case of the rectangular domain with Neumann boundary conditions, this integral condition can be interpreted physically. The net flux of heat energy across the boundary of the disk must be zero in order for a steady-state solution to the heat equation to exist, otherwise no steady-state temperature distribution exists.

The constants c_n^+ and d_n^+ can be chosen as

$$c_n^+ = \frac{a^{1-n}}{n\pi} \int_{-\pi}^{\pi} f(\theta) \cos(n\theta) \, d\theta,$$

$$d_n^+ = \frac{a^{1-n}}{n\pi} \int_{-\pi}^{\pi} f(\theta) \sin(n\theta) \, d\theta.$$

However, the constant term d_0 was lost during the differentiation operation. Thus d_0 can be chosen arbitrarily which implies the solution to Laplace's equation on a disk with Neumann boundary conditions is not unique.

Example 6.12. Find a bounded solution to Laplace's equation on $\Omega = \{(r, \theta) \,|\, 0 \le r < a\}$ that satisfies the Neumann boundary condition,

$$u_r(a, \theta) = f(\theta) = \begin{cases} 1 & \text{if } 0 < \theta < \pi, \\ -1 & \text{if } -\pi < \theta < 0. \end{cases}$$

Solution. The solution should have the form

$$u(r, \theta) = d_0 + \sum_{n=1}^{\infty} r^n [c_n^+ \cos(n\theta) + d_n^+ \sin(n\theta)]$$

where d_0 is arbitrary. The boundary condition implies

$$u_r(a, \theta) = \sum_{n=1}^{\infty} n a^{n-1} [c_n^+ \cos(n\theta) + d_n^+ \sin(n\theta)] = f(\theta).$$

Applying the Euler-Fourier formulas produces

$$c_n^+ = \frac{a^{1-n}}{n\pi} \int_{-\pi}^{\pi} u(a, \theta) \cos(n\theta) \, d\theta = 0,$$

$$d_n^+ = \frac{a^{1-n}}{n\pi} \int_{-\pi}^{\pi} u(a, \theta) \sin(n\theta) \, d\theta = \frac{2a^{1-n}}{n^2\pi} (1 - (-1)^n)$$

$$= \begin{cases} 4a^{1-n}/(n^2\pi) & \text{if } n \text{ is odd}, \\ 0 & \text{if } n \text{ is even}. \end{cases}$$

This yields the solution

$$u(r, \theta) = d_0 + \frac{4a}{\pi} \sum_{n=1}^{\infty} \left(\frac{r}{a}\right)^{2n-1} \frac{\sin((2n-1)\theta)}{(2n-1)^2}.$$

where d_0 is arbitrary. Figure 6.13 shows a contour plot of the formal solution with d_0 chosen to be 0.

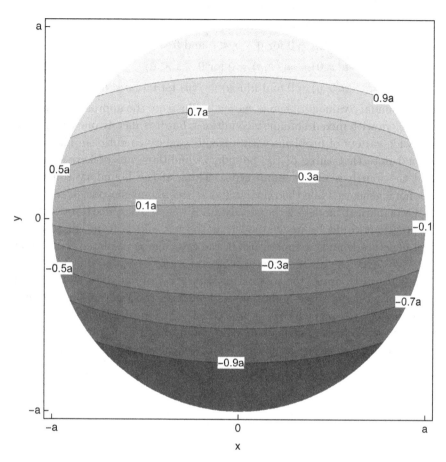

Fig. 6.13 A contour plot of the solution to the boundary value problem on the disk of radius $a > 0$ described in Example 6.12. For the sake of simplicity, the arbitrary constant d_0 was set to 0.

6.7 Mixed Boundary Conditions on Rectangles

Many applications of partial differential equations involve situations with both Dirichlet and Neumann boundary conditions. The reader will see that in most cases the solution to such problems can be found using product solutions, the appropriate eigenvalues and eigenfunctions, and Fourier series. To start, consider the following mixed boundary value problem on a rectangle in the xy-plane:

$$\Delta u = 0 \text{ for } 0 < x < a \text{ and } 0 < y < b,$$
$$u(x,0) = u_y(x,b) = 0 \text{ for } 0 < x < a, \tag{6.35}$$
$$u(0,y) = 0 \text{ and } u(a,y) = f(y) \text{ for } 0 < y < b.$$

This boundary value problem is chosen to illustrate the approach to take to a problem with mixed boundary conditions, but it is merely one example of the wide variety of problems with mixed boundary conditions. The reader should note that three of the boundary conditions are of Dirichlet type and the fourth is of Neumann type. Three of the boundary conditions are homogeneous while the remaining one is nonhomogeneous. If this were a mathematical model of the steady-state temperature distribution on a rectangle, it could be understood that the left and bottom edges of the rectangle are kept at temperature 0, the top edge is insulated, and along the right edge the distribution of temperature has been specified as $f(y)$. It is natural to apply the method of separation of variables to Laplace's equation with the three homogeneous boundary conditions. The product function $u(x,y) = X(x)Y(y)$ is a product solution of Laplace's equation with the three homogeneous boundary conditions if and only if $X(x)$ and $Y(y)$ are solutions of the following boundary value problems:

$$X''(x) - \sigma X(x) = 0 \text{ for } 0 < x < a,$$
$$X(0) = 0, \tag{6.36}$$

and

$$Y''(y) + \sigma Y(y) = 0 \text{ for } 0 < y < b,$$
$$Y(0) = Y'(b) = 0, \tag{6.37}$$

where σ is a constant. The boundary value problem of Eq. (6.37) was encountered earlier, see Exercise 20 in Chap. 1 or Exercise 6 in Chap. 4, and has nontrivial solutions only when $\sigma = \sigma_n = ((2n-1)\pi/(2b))^2$ and the corresponding nontrivial eigenfunctions are

$$Y_n(y) = c_n \sin \frac{(2n-1)\pi y}{2b},$$

for $n \in \mathbb{N}$ and any constants $c_n \neq 0$. Moreover, for each σ_n, a solution of Eq. (6.36) is

$$X_n(x) = \sinh \frac{(2n-1)\pi x}{2b}.$$

This implies that for $n \in \mathbb{N}$ the functions

$$u_n(x,y) = c_n \sinh \frac{(2n-1)\pi x}{2b} \sin \frac{(2n-1)\pi y}{2b} \tag{6.38}$$

are product solutions of Laplace's equation which satisfy the three homogeneous boundary conditions. To find the solution of the boundary value problem in Eq. (6.35), formally set

$$u(x,y) = \sum_{n=1}^{\infty} c_n \sinh \frac{(2n-1)\pi x}{2b} \sin \frac{(2n-1)\pi y}{2b}, \tag{6.39}$$

and determine the constants c_n such that

$$u(a,y) = \sum_{n=1}^{\infty} c_n \sinh \frac{(2n-1)\pi a}{2b} \sin \frac{(2n-1)\pi y}{2b} = f(y). \tag{6.40}$$

In order to determine the coefficients c_n such that Eq. (6.40) holds, $f(y)$ must be extended in such a way that the even-indexed Fourier sine coefficients vanish. Section 3.5 explored this type of extension. The odd extension to the interval $[-2b, 2b]$ based on $f(y)$ can be defined as $f_o(y)$ where

$$f_o(y) = \begin{cases} -f(2b+y) & \text{if } -2b \le y \le -b, \\ -f(-y) & \text{if } -b \le y \le 0, \\ f(y) & \text{if } 0 \le y \le b, \\ f(2b-y) & \text{if } b \le y \le 2b. \end{cases} \tag{6.41}$$

The function $f_o(y)$ is then extended to a $4b$-periodic function on $(-\infty, \infty)$. The Fourier sine series representation of $f_o(y)$ is calculated as

$$f_o(y) \sim \sum_{n=1}^{\infty} \beta_n \sin \frac{(2n-1)\pi y}{2b},$$

where

$$\beta_n = \frac{2}{b} \int_0^b f(y) \sin \frac{(2n-1)\pi y}{2b} \, dy.$$

Thus the coefficients c_n of the series representation of $f(y)$ in Eq. (6.40) should be chosen as

$$c_n = \frac{2}{b \sinh \frac{(2n-1)\pi a}{2b}} \int_0^b f(y) \sin \frac{(2n-1)\pi y}{2b} \, dy. \tag{6.42}$$

With this selection of coefficients $u(x,y)$ defined by Eq. (6.39) is well-defined and is a solution to Laplace's equation. This procedure can be made more transparent with the following example.

Example 6.13. Find the solution to Laplace's equation with mixed boundary conditions below. Interpret this boundary value problem in the context of a steady-state solution to the heat equation on a rectangular plate.

$$\triangle u = 0 \text{ for } 0 < x < 2 \text{ and } 0 < y < 1,$$

$$u(0,y) = 0 \text{ and } u(2,y) = y(1-y) \text{ for } 0 < y < 1,$$

$$u(x,0) = u_y(x,1) = 0 \text{ for } 0 < x < 2.$$

Solution. The domain $\Omega = \{(x,y)\,|\,0 < x < 2,\, 0 < y < 1\}$ is a rectangle whose left and bottom edges are kept at constant temperature 0 (perhaps by being in perfect thermal contact with ice). The top edge is insulated so that no heat flows out the top of the rectangle. The right edge of the rectangle is in perfect thermal contact with a heat source with temperature distribution given by $f(y) = y(1-y)$.

The solution to this boundary value problem is given by the expression in Eq. (6.39) with $a = 2$ and $b = 1$. The coefficients of the infinite series must satisfy the equation,

$$u(2,y) = \sum_{n=1}^{\infty} c_n \sinh((2n-1)\pi) \sin \frac{(2n-1)\pi y}{2} = y(1-y).$$

The coefficients c_n are calculated as

$$c_n \sinh((2n-1)\pi) = 2 \int_0^1 f(y) \sin \frac{(2n-1)\pi y}{2}\, dy$$

$$= 2 \int_0^1 y(1-y) \sin \frac{(2n-1)\pi y}{2}\, dy$$

$$= \frac{8(4 + (-1)^n (2n-1)\pi)}{(2n-1)^3 \pi^3}.$$

Consequently the solution to the boundary value problem can be expressed as

$$u(x,y) = \frac{8}{\pi^3} \sum_{n=1}^{\infty} \frac{4 + (-1)^n (2n-1)\pi}{(2n-1)^3 \sinh((2n-1)\pi)} \sinh \frac{(2n-1)\pi x}{2} \sin \frac{(2n-1)\pi y}{2}.$$

Figure 6.14 shows a contour plot of the solution.

As mentioned earlier, there are many different ways mixed boundary conditions can be imposed. The reader will be asked to explore some of them in the exercises, along with some boundary value problems for Poisson's equation. Before returning to the theory surrounding Laplace's equation, this section concludes with one more example of a boundary value problem with mixed boundary conditions.

Example 6.14. Consider the boundary value problem:

$$\triangle u = 0 \text{ for } 0 < x < 1 \text{ and } 0 < y < 1,$$

$$u(x,0) = u_y(x,1) = 0 \text{ for } 0 < x < 1, \tag{6.43}$$

$$u(0,y) = 0 \text{ and } u_x(1,y) = y \text{ for } 0 < y < 1.$$

Find a solution to Laplace's equation with these mixed boundary conditions.

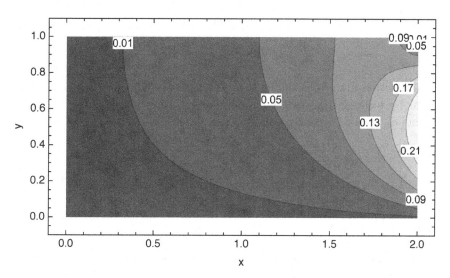

Fig. 6.14   A contour plot of the solution to the boundary value problem on the rectangle described in Example 6.13.

Solution. In this example three of the boundary conditions are homogeneous and one is nonhomogeneous. Two of the boundary conditions are of Dirichlet type while the remaining two are Neumann boundary conditions. As has been the practice, a product solution satisfying the homogeneous boundary conditions is sought. If $u(x,y) = X(x)Y(y)$ then differentiating this solution and separating the variables produce the equation

$$\frac{X''(x)}{X(x)} = -\frac{Y''(y)}{Y(y)} = \sigma,$$

where σ is constant. This implies the following two ordinary differential equations and sets of boundary conditions:

$$X''(x) - \sigma X(x) = 0 \text{ with } X(0) = 0,$$
$$Y''(y) + \sigma Y(y) = 0 \text{ with } Y(0) = 0 \text{ and } Y'(1) = 0.$$

The eigenvalues and eigenfunctions of the latter equation are $\sigma_n = ((2n-1)\pi/2)^2$ and $Y_n(y) = \sin((2n-1)\pi y/2)$ respectively for $n \in \mathbb{N}$. Replacing σ with σ_n in the first equation and solving the resulting equation along with the boundary conditions yield $X_n(x) = \sinh((2n-1)\pi x/2)$. Hence the product solutions satisfying the three homogeneous boundary conditions imposed in this example are

$$u_n(x,y) = \sinh\frac{(2n-1)\pi x}{2}\sin\frac{(2n-1)\pi y}{2} \text{ for } n \in \mathbb{N}.$$

A formal solution of Laplace's equation can be written as

$$u(x,y) = \sum_{n=1}^{\infty} c_n \sinh \frac{(2n-1)\pi x}{2} \sin \frac{(2n-1)\pi y}{2}.$$

The function $u(x,y)$, assuming it is well-defined and can be differentiated term by term, satisfies the homogeneous boundary conditions. The constants c_n should be chosen so that

$$u_x(1,y) = \sum_{n=1}^{\infty} \frac{(2n-1)\pi}{2} c_n \cosh \frac{(2n-1)\pi}{2} \sin \frac{(2n-1)\pi y}{2} = y,$$

to satisfy the remaining, nonhomogeneous boundary condition. Therefore, c_n must be chosen so that

$$\frac{(2n-1)\pi}{2} c_n \cosh \frac{(2n-1)\pi}{2} = 2 \int_0^1 y \sin \frac{(2n-1)\pi y}{2} \, dy = \frac{8(-1)^{n+1}}{(2n-1)^2 \pi^2}$$

for $n \in \mathbb{N}$. Thus the formal solution to the boundary value problem can be expressed as

$$u(x,y) = \frac{16}{\pi^3} \sum_{n=1}^{\infty} \frac{(-1)^{n+1}}{(2n-1)^3 \cosh \frac{(2n-1)\pi}{2}} \sinh \frac{(2n-1)\pi x}{2} \sin \frac{(2n-1)\pi y}{2}.$$

Figure 6.15 shows a contour plot of the solution.

6.8   Poisson's Formula and Mean Value Property

As mentioned earlier, solutions of Laplace's equation are called harmonic functions, a very important class of functions in mathematics and its applications. In this section, two important properties of harmonic functions are presented: **Poisson's formula** and the **mean value property**.

The solution of Laplace's equation on a disk with Dirichlet boundary conditions is

$$v(r,\theta) = \frac{a_0}{2} + \sum_{n=1}^{\infty} \left(\frac{r}{a}\right)^n [a_n \cos(n\theta) + b_n \sin(n\theta)], \qquad (6.44)$$

where a_0, a_n and b_n are the Fourier coefficients of the boundary condition function $f(\theta)$:

$$a_0 = \frac{1}{\pi} \int_{-\pi}^{\pi} f(\theta) \, d\theta,$$

$$a_n = \frac{1}{\pi} \int_{-\pi}^{\pi} f(\theta) \cos(n\theta) \, d\theta,$$

$$b_n = \frac{1}{\pi} \int_{-\pi}^{\pi} f(\theta) \sin(n\theta) \, d\theta.$$

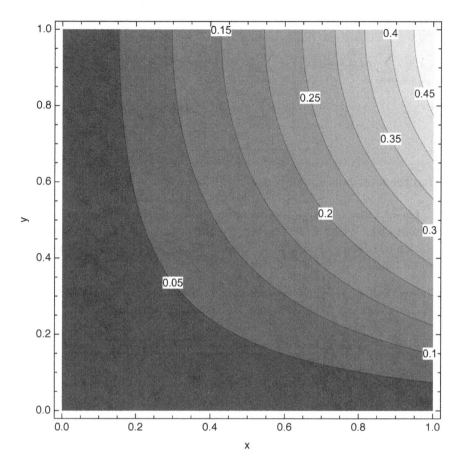

Fig. 6.15   A contour plot of the solution to the boundary value problem on the rectangle described in Example 6.14.

The formula in Eq. (6.44) is an infinite series and it would be nice to derive a "closed form" version of the solution. This was done with some examples earlier. The following lemma is a step toward finding a closed form solution in general.

Lemma 6.1. *If* $|\rho| < 1$, *then for any* τ,

$$\frac{1}{2} + \sum_{n=1}^{\infty} \rho^n \cos(n\tau) = \frac{1 - \rho^2}{2(1 - 2\rho\cos\tau + \rho^2)}.$$

Proof. By Euler's formula in Eq. (3.40), $\cos(n\tau) = (e^{in\tau} + e^{-in\tau})/2$ and therefore,

$$1 + 2\sum_{n=1}^{\infty} \rho^n \cos(n\tau) = 1 + \sum_{n=1}^{\infty}(\rho^n e^{in\tau} + \rho^n e^{-in\tau})$$

$$= \left[1 + \sum_{n=1}^{\infty}(\rho e^{i\tau})^n\right] + \left[1 + \sum_{n=1}^{\infty}(\rho e^{-i\tau})^n\right] - 1$$

$$= \frac{1}{1 - \rho e^{i\tau}} + \frac{1}{1 - \rho e^{-i\tau}} - 1 = \frac{1 - \rho^2}{1 - 2\rho\cos\tau + \rho^2}.$$

Thus the formula is established. $\square$

The quantity $(1-\rho^2)/(1-2\rho\cos\tau+\rho^2)$ is called the **Poisson kernel** on the unit disk. With this lemma in a supporting role, a closed form integral solution to Laplace's equation on a disk with Dirichlet boundary conditions can be found.

Theorem 6.1. *If $v(r,\theta)$ is a solution to the boundary value problem for Laplace's equation in polar coordinates as in Eq. (6.13), then*

$$v(r,\theta) = \frac{1}{2\pi}\int_{-\pi}^{\pi} \frac{(a^2 - r^2)f(s)}{a^2 - 2ar\cos(s-\theta) + r^2}\, ds. \qquad (6.45)$$

Proof. For any $0 \le r < a$ and θ, the solution $v(r,\theta)$ to Eq. (6.13) is given by the formula in Eq. (6.44) and can be expressed as

$$v(r,\theta) = \frac{1}{2\pi}\int_{-\pi}^{\pi} f(s)\, ds + \frac{1}{\pi}\sum_{n=1}^{\infty}\left(\frac{r}{a}\right)^n \left[\int_{-\pi}^{\pi} f(s)\cos(ns)\, ds\right]\cos(n\theta)$$

$$+ \frac{1}{\pi}\sum_{n=1}^{\infty}\left(\frac{r}{a}\right)^n \left[\int_{-\pi}^{\pi} f(s)\sin(ns)\, ds\right]\sin(n\theta).$$

The variable of integration in the definite integrals above is s and therefore expressions involving $\cos(n\theta)$ and $\sin(n\theta)$ can be brought inside the integrals and the summations can be combined in order to create a single integral in the summand. Applying the difference of angles formula for the cosine function produces

$$v(r,\theta) = \frac{1}{2\pi}\int_{-\pi}^{\pi} f(s)\, ds + \frac{1}{\pi}\sum_{n=1}^{\infty}\left(\frac{r}{a}\right)^n \int_{-\pi}^{\pi} f(s)\cos(n(s-\theta))\, ds$$

$$= \frac{1}{\pi}\int_{-\pi}^{\pi} f(s)\left[\frac{1}{2} + \sum_{n=1}^{\infty}\left(\frac{r}{a}\right)^n \cos(n(s-\theta))\right]\, ds.$$

Applying Lemma 6.1 with $\rho = r/a$ and $\tau = s - \theta$ results in

$$v(r, \theta) = \frac{1}{2\pi} \int_{-\pi}^{\pi} \frac{(a^2 - r^2)f(s)}{a^2 - 2ar\cos(s - \theta) + r^2} \, ds.$$

Hence $v(r, \theta)$ defined in Eq. (6.45) solves the boundary value problem described in Eq. (6.13). $\qquad\square$

The expression for $v(r, \theta)$ in Eq. (6.45) is called the **Poisson integral formula.** Even though Eq. (6.45) is rarely used to find the solutions of boundary value problems of Laplace's equation on disks, it has profound theoretical value and implications. For example, Eq. (6.45) reveals an interesting relationship between the Dirichlet boundary condition on the disk and the value of the solution at the center of the disk (typically the origin of the coordinate system).

Corollary 6.1. *If $v(r, \theta)$ is a solution to Eq. (6.13), then*

$$v(0, \theta) = \frac{1}{2\pi} \int_{-\pi}^{\pi} f(s) \, d\theta = \frac{1}{2\pi} \int_{-\pi}^{\pi} \phi(a\cos s, a\sin s) \, ds.$$

That is, $v(0, \theta)$ (which in Cartesian coordinates is $u(0, 0)$) is the mean value of the Dirichlet boundary condition function $f(\theta) = v(a, \theta)$ on the circle $r = a$.

The proof of this corollary is immediate by setting $r = 0$ in Eq. (6.45). In fact a stronger result than the statement in Corollary 6.1 is valid. According to Theorem 6.1 for any $\rho \in (0, a)$,

$$v(0, \theta) = \frac{1}{2\pi} \int_{-\pi}^{\pi} v(\rho\cos s, \rho\sin s) \, ds.$$

Hence the average value of the solution to the boundary value problem of Laplace's equation on any disk of radius ρ centered at $(0, 0)$ is the value of the solution at $(0, 0)$. This corollary can be further generalized to hold for any harmonic function on any domain Ω.

Let $u(x, y)$ be a harmonic function on Ω, an open, connected subset of $\mathbb{R}^2$. For any $(x_0, y_0) \in \Omega$, let $a > 0$ be such that the disk centered at (x_0, y_0) with radius a is contained entirely in the domain Ω. Let $D_a(x_0, y_0)$ denote the open disk of radius a centered at (x_0, y_0) while $C_a(x_0, y_0)$ denotes the circle of radius a centered at (x_0, y_0), that is

$$D_a(x_0, y_0) = \left\{ (x, y) \mid (x - x_0)^2 + (y - y_0)^2 < a^2 \right\},$$
$$C_a(x_0, y_0) = \left\{ (x, y) \mid (x - x_0)^2 + (y - y_0)^2 = a^2 \right\}.$$

The generalization of Corollary 6.1 is stated in the following theorem.

Theorem 6.2. *Let $u(x, y)$ be a harmonic function on Ω, an open, connected subset of $\mathbb{R}^2$. For any $(x_0, y_0) \in \Omega$ if $a > 0$ is chosen so that $D_a(x_0, y_0) \subset \Omega$ and $C_a(x_0, y_0) \subset \Omega$, then*

$$u(x_0, y_0) = \frac{1}{2\pi} \int_{-\pi}^{\pi} u(x_0 + a\cos\theta, y_0 + a\sin\theta)\, d\theta.$$

That is $u(x_0, y_0)$ is the mean value of u on $C_a(x_0, y_0)$, the boundary of $D_a(x_0, y_0)$.

Note that the radius of the disk and circle $a > 0$ can be any number as long as $C_a(x_0, y_0)$ and $D_a(x_0, y_0)$ are contained entirely in Ω.

Proof. A change of variables is used to translate the point (x_0, y_0) to the origin of the coordinate system. Let $\xi = x - x_0$ and $\eta = y - y_0$, then $D_a(x = x_0, y = y_0) = D_a(\xi = 0, \eta = 0)$. Define the function $U(\xi, \eta) = u(x_0 + \xi, y_0 + \eta)$, then

$$U_{\xi\xi} + U_{\eta\eta} = 0.$$

Assume in polar coordinates $U(\xi, \eta) = V(r, \theta)$ then by Poisson's formula in Eq. (6.45),

$$
\begin{aligned}
V(r, \theta) &= \frac{1}{2\pi} \int_{-\pi}^{\pi} \frac{(a^2 - r^2)V(a\cos s, a\sin s)}{a^2 - 2ar\cos(s - \theta) + r^2}\, ds \\
&= \frac{1}{2\pi} \int_{-\pi}^{\pi} \frac{(a^2 - r^2)u(x_0 + a\cos s, y_0 + a\sin s)}{a^2 - 2ar\cos(s - \theta) + r^2}\, ds
\end{aligned}
$$

where $r < a$. In particular,

$$U(0, 0) = u(x_0, y_0) = \frac{1}{2\pi} \int_{-\pi}^{\pi} u(x_0 + a\cos s, y_0 + a\sin s)\, ds.$$

Hence the theorem is proved. □

As a consequence of Theorem 6.2, if $(x_0, y_0) \in \Omega$ then $u(x_0, y_0)$ cannot be the strict maximum or minimum of u on $D_a(x_0, y_0) \subset \Omega$. More generally it can be shown that if $(x_0, y_0) \in \Omega$ and Ω is a connected domain, then $u(x_0, y_0)$ cannot be a strict maximum or minimum on Ω. It can also be shown that if $u(x, y) \in \mathcal{C}^2(\Omega)$ satisfies the mean value property on Ω, then u must be harmonic on Ω.

Having derived the solutions to Laplace's equation on various domains in the plane, attention now turns to some properties of these solutions. In particular, the extreme values of the solutions and their locations, the uniqueness of solutions to Laplace's and Poisson's equations, and the stability of solutions in response to changes in boundary conditions. These issues are explored in the next section.

6.9 Maximum Principle and Uniqueness

Recall that any steady-state solution of the heat equation is a harmonic function. Heat must flow from a region with high temperature to a region with lower temperature. If u is a steady-state solution of a heat equation, it cannot have a strict maximum or minimum inside the physical domain. Moreover, the mean value property for harmonic functions suggests the same is true for any harmonic function. This is stated in the following Maximum/Minimum Principle. For simplicity, only the two-dimensional case is presented.

Theorem 6.3 (Maximum/Minimum Principle). *Let Ω be an open, connected, and bounded domain in $\mathbb{R}^2$ with boundary $\partial\Omega$. Assume that $u(x,y) \in \mathcal{C}^2(\Omega) \cap \mathcal{C}(\overline{\Omega})$ where $\overline{\Omega} = \Omega \cup \partial\Omega$, then*

- *if $u_{xx} + u_{yy} \geq 0$ for all $(x,y) \in \Omega$, then*

$$\max_{(x,y)\in\overline{\Omega}} u(x,y) = \max_{(x,y)\in\partial\Omega} u(x,y),$$

- *if $u_{xx} + u_{yy} \leq 0$ for all $(x,y) \in \Omega$, then*

$$\min_{(x,y)\in\overline{\Omega}} u(x,y) = \min_{(x,y)\in\partial\Omega} u(x,y).$$

The reader should note that if $(x_0, y_0) \in \Omega$ is the location of a local maximum for u, then

$$u_{xx}(x_0, y_0) \leq 0 \quad \text{and} \quad u_{yy}(x_0, y_0) \leq 0.$$

These conditions are still compatible with the assumption of $u_{xx} + u_{yy} \geq 0$ for the case of the maximum stated in Theorem 6.3. A rigorous proof can be provided using the mean value property discussed in the previous section; however, the proof below is more straight forward and follows the same idea as the proof of the Maximum Principle in the case of the heat equation. A proof will be given only for the case of the maximum, the proof for the case of the minimum is similar.

Proof. Assume to the contrary that

$$\max_{(x,y)\in\overline{\Omega}} u(x,y) > \max_{(x,y)\in\partial\Omega} u(x,y).$$

Then there must be a point $(x_0, y_0) \in \Omega$ such that

$$u(x_0, y_0) > \max_{(x,y)\in\partial\Omega} u(x,y).$$

For $\epsilon > 0$, define a function $v^\epsilon : \Omega \subset \mathbb{R}^2 \to \mathbb{R}$ as

$$v^\epsilon(x, y) = u(x, y) + \epsilon[(x - x_0)^2 + (y - y_0)^2].$$

Since Ω is bounded, $\epsilon > 0$ can be chosen small enough that

$$\max_{(x,y)\in\partial\Omega} v^\epsilon(x, y) \le \max_{(x,y)\in\partial\Omega} u(x, y) + \max_{(x,y)\in\partial\Omega} \epsilon[(x - x_0)^2 + (y - y_0)^2]$$

$$< u(x_0, y_0) = v^\epsilon(x_0, y_0).$$

This implies that the maximum of v^ϵ on $\overline{\Omega}$ must occur in Ω. There must be a point $(\overline{x}, \overline{y}) \in \Omega$ such that

$$v^\epsilon(\overline{x}, \overline{y}) = \max_{(x,y)\in\overline{\Omega}} v^\epsilon(x, y).$$

Therefore,

$$v^\epsilon_{xx}(\overline{x}, \overline{y}) + v^\epsilon_{yy}(\overline{x}, \overline{y}) \le 0. \tag{6.46}$$

However, the definition of function v^ϵ implies

$$v^\epsilon_{xx}(x, y) + v^\epsilon_{yy}(x, y) = u_{xx}(x, y) + u_{yy}(x, y) + 4\epsilon > 0$$

which contradicts Eq. (6.46) and the statement regarding the maximum of u is proved. $\square$

The following immediate corollary of Theorem 6.3 holds for solutions of Laplace's equation.

Corollary 6.2. *Let Ω be an open, connected, and bounded domain in $\mathbb{R}^2$. Assume that $u(x, y) \in \mathcal{C}^2(\Omega) \cap \mathcal{C}(\overline{\Omega})$ where $\overline{\Omega} = \Omega \cup \partial\Omega$ and $u(x, y)$ satisfies Laplace's equation on Ω, then*

$$\max_{(x,y)\in\overline{\Omega}} u(x, y) = \max_{(x,y)\in\partial\Omega} u(x, y)$$

and

$$\min_{(x,y)\in\overline{\Omega}} u(x, y) = \min_{(x,y)\in\partial\Omega} u(x, y).$$

The final applications of Theorem 6.3 will be in establishing the uniqueness and continuous dependence of solutions to Poisson's equation. The proofs are left as exercises.

Theorem 6.4 (Uniqueness). *Suppose that Ω is an open, connected, and bounded domain in $\mathbb{R}^2$ with $f(x, y) \in \mathcal{C}(\Omega)$ and $\phi(x, y) \in \mathcal{C}(\partial\Omega)$. If $u(x, y) \in \mathcal{C}^2(\Omega) \cap \mathcal{C}(\overline{\Omega})$ where $\overline{\Omega} = \Omega \cup \partial\Omega$ is a solution to the boundary value problem,*

$$\triangle u = f(x, y) \text{ for } (x, y) \in \Omega,$$

$$u(x, y) = \phi(x, y) \text{ for } (x, y) \in \partial\Omega,$$

then this solution is unique.

Proof. See Exercise 30. □

The solution to Poisson's equation also depends continuously on its boundary conditions.

Theorem 6.5 (Continuous Dependence). *Suppose that Ω is an open, connected, and bounded domain in $\mathbb{R}^2$ with $f(x,y) \in \mathcal{C}(\Omega)$ and $\phi_i(x,y) \in \mathcal{C}(\partial\Omega)$ for $i = 1, 2$. If $u_i(x,y) \in \mathcal{C}^2(\Omega) \cap \mathcal{C}(\overline{\Omega})$ where $\overline{\Omega} = \Omega \cup \partial\Omega$ is a solution to the boundary value problem,*

$$\triangle u = f(x,y) \text{ for } (x,y) \in \Omega,$$

$$u(x,y) = \phi_i(x,y) \text{ for } (x,y) \in \partial\Omega,$$

for $i = 1, 2$ respectively, then

$$|u_1(x,y) - u_2(x,y)| \leq \max_{(x,y)\in\partial\Omega} |\phi_1(x,y) - \phi_2(x,y)|.$$

Proof. See Exercise 32. □

This chapter has explored Laplace's equation and its solution on a variety of domains in two-dimensional space. Boundary value problems were treated in Cartesian and polar coordinates. Dirichlet, Neumann, and mixed boundary conditions were treated in the different examples included in this chapter. A boundary value problem with only Neumann-type boundary conditions is ill-posed and thus may have no solution or infinitely many solutions. As Laplace's equation is linear and homogeneous, the reader should note that the method of separation of variables provided a means of solving the various examples provided the appropriate eigenfunctions and eigenvalues could be found. This chapter also outlined results which provide bounds for solutions to Laplace's equation including versions of the Maximum and Minimum Principles. The solutions to Poisson's equations (which includes Laplace's equation as a special case) were also shown to be unique and to depend continuously on their boundary conditions. The steps taken to solve Laplace's equation in this chapter can be generalized to solving boundary value problems on domains in three-dimensional space in Cartesian, cylindrical, and spherical coordinates. Some examples of these boundary value problems are found in Chap. 9.

6.10   Exercises

(1) Sketch examples of subsets of $\mathbb{R}^2$ with the following properties.

   (a) An open subset that is not connected.
   (b) A subset which is not open.
   (c) A subset that is not open and also not connected.

For Exercises 2–7, refer to the following diagram for the assignment of boundary conditions and dimensions of the rectangular domain.

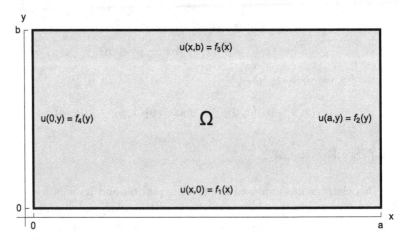

(2) Find the solution to Laplace's equation in rectangular coordinates on domain Ω where the Dirichlet boundary conditions are specified as $f_2(y) = 1$ and $f_1 = f_3 = f_4 = 0$ with $a = 1 = b$.

(3) Find the solution to Laplace's equation in rectangular coordinates on domain Ω where the Dirichlet boundary conditions are specified as $f_2(y) = y(1 - y)$ and $f_1 = f_3 = f_4 = 0$ with $a = 1 = b$.

(4) Find the solution to Laplace's equation in rectangular coordinates on domain Ω where the Dirichlet boundary conditions are specified as $f_1(x) = 100 = f_3(x)$ and $f_2 = f_4 = 0$ with $a = 1 = b$.

(5) Find the solution to Laplace's equation in rectangular coordinates on domain Ω where the Dirichlet boundary conditions are specified as $f_1(x) = \sin(\pi x)$, $f_3(x) = 1$ and $f_2 = f_4 = 0$ with $a = 1 = b$.

(6) Find the solution to Laplace's equation in rectangular coordinates on domain Ω where the Dirichlet boundary conditions are specified as $f_1(x) = x$, $f_2(y) = 1 - y$ and $f_3 = f_4 = 0$ with $a = 1 = b$.

(7) Find the solution to Laplace's equation in rectangular coordinates on

domain Ω where the Dirichlet boundary conditions are specified as $f_1(x) = 1 = f_3(x)$, $f_2(y) = \cos(2\pi y)$ and $f_4 = 0$ with $a = 1 = b$.

(8) Suppose that $z = re^{i\theta}$ and use the Euler identity to show that $\text{Re}\ln z = \ln r$.

(9) Suppose that $z = re^{i\theta} = x + iy$ and show that $\text{Im}(\ln z) = \tan^{-1}(y/x)$.
For Exercises 10–15, refer to the following diagram for the assignment of boundary conditions and dimensions of the circular domain.

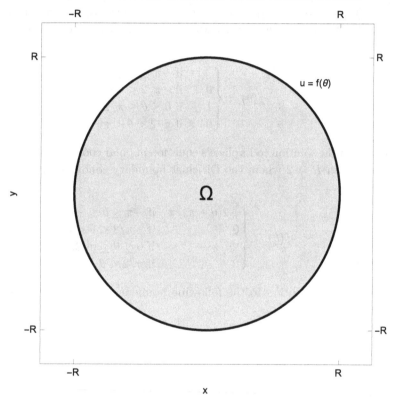

(10) Find the solution to Laplace's equation in polar coordinates on domain Ω where the Dirichlet boundary conditions are specified as $f(\theta) = \cos(2\theta)$ with $R = 1$.

(11) Find the solution to Laplace's equation in polar coordinates on domain Ω where the Dirichlet boundary conditions are specified as $f(\theta) = \sin(3\theta)$ with $R = 2$.

(12) Find the solution to Laplace's equation in polar coordinates on domain Ω where the Dirichlet boundary conditions are specified as $f(\theta) = \pi^2 - \theta^2$ for $-\pi < \theta < \pi$ with $R = 1$.

(13) Find the solution to Laplace's equation in polar coordinates on domain Ω with $R = 1$ where the Dirichlet boundary conditions are specified as

$$f(\theta) = \begin{cases} 0 & \text{if } -\pi/2 < \theta < 0, \\ 1 & \text{if } 0 \le \theta \le \pi/2, \\ 0 & \text{if } \pi/2 < \theta < \pi. \end{cases}$$

(14) Find the solution to Laplace's equation in polar coordinates on domain Ω with $R = 2$ where the Dirichlet boundary conditions are specified as

$$f(\theta) = \begin{cases} -1 & \text{if } -\pi \le \theta \le -\pi/2, \\ 0 & \text{if } -\pi/2 < \theta < 0, \\ 1 & \text{if } 0 \le \theta \le \pi/2, \\ 0 & \text{if } \pi/2 < \theta < \pi. \end{cases}$$

(15) Find the solution to Laplace's equation in polar coordinates on domain Ω with $R = 2$ where the Dirichlet boundary conditions are specified as

$$f(\theta) = \begin{cases} -2(\theta + \pi)/\pi & \text{if } -\pi \le \theta \le -\pi/2, \\ 0 & \text{if } -\pi/2 < \theta < 0, \\ 2\theta/\pi & \text{if } 0 \le \theta \le \pi/2, \\ 0 & \text{if } \pi/2 < \theta < \pi. \end{cases}$$

(16) Find the solution to the following boundary value problem on a circular sector:

$$v_{rr} + \frac{1}{r}v_r + \frac{1}{r^2}v_{\theta\theta} = 0 \text{ for } 0 < r < 1 \text{ and } 0 < \theta < \pi,$$

$$v(1, \theta) = \begin{cases} 1 & \text{if } 0 < \theta < \pi/2, \\ 0 & \text{if } \pi/2 < \theta < \pi, \end{cases}$$

$$v(r, 0) = v(r, \pi) = 0 \text{ for } 0 < r < 1.$$

(17) Find the solution to the following boundary value problem on a circular sector:

$$v_{rr} + \frac{1}{r}v_r + \frac{1}{r^2}v_{\theta\theta} = 0 \text{ for } 0 < r < 1 \text{ and } 0 < \theta < 3\pi/2$$

$$v(1, \theta) = \begin{cases} \pi\theta & \text{if } 0 < \theta < \pi, \\ 0 & \text{if } \pi < \theta < 3\pi/2, \end{cases}$$

$$v(r, 0) = v(r, 3\pi/2) = 0 \text{ for } 0 < r < 1.$$

(18) Find the solution to the following boundary value problem on a sector of an annulus:

$$v_{rr} + \frac{1}{r}v_r + \frac{1}{r^2}v_{\theta\theta} = 0 \text{ for } 1 < r < 2 \text{ and } 0 < \theta < \pi,$$

$$v(1,\theta) = \begin{cases} 1 & \text{if } 0 < \theta < \pi/2, \\ 0 & \text{if } \pi/2 < \theta < \pi, \end{cases}$$

$$v(2,\theta) = 0 \text{ for } 0 < \theta < \pi,$$

$$v(r,0) = v(r,\pi) = 0 \text{ for } 1 < r < 2.$$

(19) Find the solution to the following boundary value problem on a sector of an annulus:

$$v_{rr} + \frac{1}{r}v_r + \frac{1}{r^2}v_{\theta\theta} = 0 \text{ for } 1 < r < 2 \text{ and } 0 < \theta < 3\pi/2,$$

$$v(1,\theta) = 1 \text{ for } 0 < \theta < 3\pi/2,$$

$$v(2,\theta) = \begin{cases} \pi\theta & \text{if } 0 < \theta < \pi, \\ 0 & \text{if } \pi < \theta < 3\pi/2, \end{cases}$$

$$v(r,0) = v(r,3\pi/2) = 0 \text{ for } 1 < r < 2.$$

(20) Find the solution (up to an additive constant) to Laplace's equation in rectangular coordinates on the domain $\Omega = \{(x,y) \,|\, 0 < x < 1, 0 < y < 1\}$ with Neumann boundary conditions:

$$u_x(0,y) = u_x(1,y) = 0 \text{ for } 0 < y < 1,$$

$$u_y(x,0) = 0 \text{ and } u_y(x,1) = \cos(\pi x) \text{ for } 0 < x < 1.$$

(21) Find the solution (up to an additive constant) to Laplace's equation in rectangular coordinates on the domain $\Omega = \{(x,y) \,|\, 0 < x < 2, 0 < y < 1\}$ with Neumann boundary conditions:

$$u_y(x,0) = u_y(x,1) = 0 \text{ for } 0 < x < 2,$$

$$u_x(0,y) = 0 \text{ and } u_x(2,y) = y - 1/2 \text{ for } 0 < y < 1.$$

(22) Find a bounded solution (up to an additive constant) to Laplace's equation in polar coordinates on domain $\Omega = \{(x,y) \,|\, x^2 + y^2 < 1\}$ subject to the Neumann boundary condition $u_r(1,\theta) = \sin\theta$ for $-\pi < \theta < \pi$.

(23) Find a bounded solution (up to an additive constant) to Laplace's equation in polar coordinates on domain $\Omega = \{(x,y) \,|\, x^2 + y^2 < 9\}$ subject to the Neumann boundary condition $u_r(3,\theta) = \cos(5\theta)$ for $-\pi < \theta < \pi$.

(24) Find the solution to Laplace's equation in rectangular coordinates on the domain $\Omega = \{(x,y) \,|\, 0 < x < 2, \, 0 < y < 1\}$ with mixed boundary conditions:

$$u(x,0) = u_y(x,1) = 0 \text{ for } 0 < x < 2,$$
$$u(0,y) = 0 \text{ and } u_x(2,y) = y - 1/2 \text{ for } 0 < y < 1.$$

(25) Find the solution to Laplace's equation in rectangular coordinates on the domain $\Omega = \{(x,y) \,|\, 0 < x < 3, \, 0 < y < 2\}$ with mixed boundary conditions:

$$u_y(x,0) = u(x,2) = 0 \text{ for } 0 < x < 3,$$
$$u_x(0,y) = 0 \text{ and } u(3,y) = y \text{ for } 0 < y < 2.$$

(26) Solve the following boundary value problem with $a > 0$:

$$\triangle u = 4 \text{ for } x^2 + y^2 < a^2,$$
$$u(x,y) = 0 \text{ for } x^2 + y^2 = a^2.$$

(27) Solve the following boundary value problem with $a > 0$:

$$\triangle u = xy \text{ for } x^2 + y^2 < a^2,$$
$$u(x,y) = 0 \text{ for } x^2 + y^2 = a^2.$$

(28) Solve the following boundary value problem:

$$\triangle u = 2 \text{ for } 0 < x < \pi \text{ and } 0 < y < \pi,$$
$$u(x,y) = 0 \text{ for } x = 0 \text{ or } \pi \text{ or } y = 0 \text{ or } \pi.$$

Hint: note that $v(x,y) = x^2$ is a solution to the partial differential equation $\triangle v = 2$. Define $w(x,y) = u(x,y) - v(x,y)$, then w satisfies the boundary value problem,

$$\triangle w = 0 \text{ for } 0 < x < \pi \text{ and } 0 < y < \pi,$$
$$w(x,y) = -v(x,y) \text{ for } x = 0 \text{ or } \pi \text{ or } y = 0 \text{ or } \pi.$$

Solve this boundary value problem for $w(x,y)$.

(29) Let $v(r,\theta)$ be harmonic on the open disk $D_a = \{(r,\theta) \,|\, r < a\}$, i.e., v satisfies Laplace's equation

$$v_{rr} + \frac{1}{r}v_r + \frac{1}{r^2}v_{\theta\theta} = 0 \text{ for } r < a.$$

Assume further that v is nonnegative on the circle $C_a = \{(r,\theta) \,|\, r = a\}$ and continuous on the closed disk $D_a \cup C_a = \{(r,\theta) \,|\, r \le a\}$. Prove that

$$\frac{a-r}{a+r}v(0,\theta) \le v(r,\theta) \le \frac{a+r}{a-r}v(0,\theta) \text{ for } r \le a \text{ and } 0 \le \theta < 2\pi,$$

where $v(0, \theta)$ is the value of v at the origin. This is a special case of Harnack's[1] inequality.

(30) Prove Theorem 6.4.

(31) Assume that $u(x, y)$ and $v(x, y)$ are both harmonic on a domain $\Omega \subset \mathbb{R}^2$ and both are continuous on $\Omega \cup \partial\Omega$. If $u(x, y) \leq v(x, y)$ for all $(x, y) \in \partial\Omega$, show that $u(x, y) \leq v(x, y)$ for all $(x, y) \in \Omega$.

(32) Prove Theorem 6.5.

(33) Use a geometric series and Euler's identity to show that

$$\sum_{n=1}^{\infty} \rho^n \sin(nx) = \frac{\rho \sin x}{1 - 2\rho \cos x + \rho^2},$$

if $|\rho| < 1$.

(34) Show that $\text{Re}\left(\dfrac{1 + \rho e^{i\theta}}{1 - \rho e^{i\theta}}\right) = \dfrac{1 - \rho^2}{1 - 2\rho \cos\theta + \rho^2}$.

---

[1]Carl Gustav Axel von Harnack, German mathematician (1851–1888).

Chapter 7

Sturm-Liouville Theory

A common theme runs through the process of solving most initial boundary value problems via the method of separation of variables. A solution is assumed to be a product of two of more functions each depending on a single variable. The functions which depend on the spatial variables (usually x, y, r, *etc.*) must satisfy the boundary conditions of the problem. This usually requires solving an ordinary differential equation with those associated boundary conditions, or as it has come to be called, a boundary value problem. See Eqs. (1.52)–(1.54) for a canonical example. The method of separation of variables usually introduces a constant which plays a role in the solution of the boundary value problem. Nontrivial solutions of the boundary value problem exist only for certain values of the constant. The values of the constant are called eigenvalues and the associated nontrivial solutions to the boundary value problem are called eigenfunctions. The boundary value problems encountered in the previous chapters on the heat, wave, and Laplace's equation possess eigenfunctions forming orthogonal sets which, as a consequence of the results on Fourier series described in Chap. 3, allow a sufficiently smooth function, at least formally, to be represented as an infinite series whose terms are constant multiples of eigenfunctions. In many cases, this enabled the initial condition of the initial boundary value problem to be satisfied as well as the boundary conditions.

This chapter is concerned with the development of a more comprehensive theory of boundary value problems that reduces to what was done for the heat, wave, and Laplace's equations (among others) as special cases. This is the Sturm[1]-Liouville[2] theory. In this chapter the reader will become acquainted with the **Sturm-Liouville boundary value problem**

---

[1] Jacques Charles François Sturm, French mathematician (1803–1855).
[2] Joseph Liouville, French mathematician (1809–1882).

(SLBVP) and a set of theorems which hold for a wide variety of SLB-VPs. Among these results will be statements about the type, ordering, cardinality, and asymptotic limits of the eigenvalues, results describing the types of eigenfunctions and their zeros, theorems about the orthogonality and completeness of the sets of eigenfunctions, and a relationship between an eigenvalue and its corresponding eigenfunction known as the **Rayleigh quotient**. Some of the proofs presented will depend on a small amount of complex variable theory. The motivated reader (even without a course in complex variables) should be able to follow the proofs (if necessary refer to App. A). Many of the results presented in the lemmas and theorems will be explored and extended for specific boundary value problems in Chap. 8 on "Special Functions".

7.1 Two-Point Boundary Value Problems of Second-Order Differential Equations

In the discussion of the heat equation with Dirichlet boundary conditions in Chap. 4, the ordinary differential equation,

$$X''(x) + \lambda X(x) = 0 \text{ for } 0 < x < L \tag{7.1}$$

with the boundary conditions,

$$X(0) = 0 \text{ and } X(L) = 0 \tag{7.2}$$

was solved. The boundary value problem stated in Eqs. (7.1) and (7.2) has nonzero solutions only for $\lambda \equiv \lambda_n = (n\pi/L)^2$, for $n \in \mathbb{N}$ with corresponding nontrivial solution,

$$X_n(x) = \sin\left(\frac{n\pi x}{L}\right).$$

The set $\{\lambda_n\}_{n=1}^{\infty}$ was called the set of **eigenvalues** of the boundary value problem and $\{X_n(x)\}_{n=1}^{\infty}$ is the corresponding set of **eigenfunctions**. Note that the eigenvalues are all real numbers and form a strictly increasing sequence,

$$0 < \lambda_1 < \lambda_2 < \cdots < \lambda_n < \cdots$$

which approaches infinity as $n \to \infty$. Recall that the set of eigenfunctions forms a complete, orthonormal set (after the appropriate normalization by nonzero multiplicative constants). The ordinary differential equation in Eq. (7.1) has also been solved subject to different boundary conditions such as

$$X'(0) = 0 \text{ and } X'(L) = 0 \tag{7.3}$$

or

$$X(0) = 0 \text{ and } X'(L) = 0 \tag{7.4}$$

or even the more general boundary conditions of the following form

$$X'(0) - \gamma_0 X(0) = 0 \text{ and } X'(L) + \gamma_L X(L) = 0 \tag{7.5}$$

where γ_0 and γ_L are positive constants. In each of these boundary value problems, a set of eigenvalues and corresponding eigenfunctions resulted.

One of the main objectives of this chapter is to consider a more general class of second-order ordinary differential equations with a more general set of boundary conditions. Consider the following second-order, linear homogeneous ordinary differential equation of the form

$$P(x)y'' + Q(x)y' + (R(x) + \lambda)y = 0 \tag{7.6}$$

for $a < x < b$. Suppose that $P(x)$, $Q(x)$, and $R(x)$ are continuous on interval $[a, b]$ and that $P(x) \neq 0$ for all x in $[a, b]$. Without loss of generality assume $P(x) > 0$ on $[a, b]$. It is more convenient to write the differential equation in a different form. For $x \in [a, b]$, define the function

$$p(x) = e^{\int_a^x \frac{Q(z)}{P(z)}\, dz} \tag{7.7}$$

then, multiplying both sides of Eq. (7.6) by $r(x) \equiv p(x)/P(x)$ yields

$$0 = p(x)y'' + p'(x)y' + \left(p(x)\frac{R(x)}{P(x)} + \lambda r(x)\right) y$$

$$= [p(x)y']' + \left(p(x)\frac{R(x)}{P(x)} + \lambda r(x)\right) y.$$

Setting $q(x) = p(x)R(x)/P(x)$ results in the equation,

$$[p(x)y']' + (q(x) + \lambda r(x))y = 0. \tag{7.8}$$

Throughout this chapter, it is assumed that in Eq. (7.8) the function $p(x)$ is continuously differentiable and positive on $[a, b]$, function $r(x)$ is continuous and positive on interval $[a, b]$, and that $q(x)$ is continuous on $[a, b]$ (to reduce repetition, these assumptions will not necessarily be included as assumptions in each of the lemmas, corollaries, and theorems which follow). By the assumptions on functions P and Q made earlier, the derivative $p'(x)$ is continuous on $[a, b]$ as well. An ordinary differential equation written in the form of Eq. (7.8) is said to be **self-adjoint**.

To avoid repeatedly writing out the left-hand side of Eq. (7.8), define the operator L as

$$L[y] = [p(x)y']' + q(x)y. \tag{7.9}$$

An operator such as L can be thought of as behaving like a function, except that the domain of the operator is an appropriate set of functions defined on (a, b) and the result is a function defined on (a, b). If it is necessary to refer to a value of this function, notation such as $L[y](x)$ will be employed. Equation (7.8) can then be written as

$$L[y] + \lambda r(x)y = 0 \text{ for } a < x < b. \tag{7.10}$$

At the endpoints of the interval the following homogeneous boundary conditions (sometimes called **separated boundary conditions**) are imposed:

$$\begin{aligned} \alpha_1 y(a) + \beta_1 y'(a) &= 0, \\ \alpha_2 y(b) + \beta_2 y'(b) &= 0, \end{aligned} \tag{7.11}$$

where $\alpha_1^2 + \beta_1^2 > 0$ and $\alpha_2^2 + \beta_2^2 > 0$. Another important class of boundary conditions are those known as **periodic boundary conditions**:

$$\begin{aligned} y(a) &= y(b), \\ y'(a) &= y'(b). \end{aligned} \tag{7.12}$$

Note that the periodic boundary conditions cannot be written in the form of separated boundary conditions.

Equations written in the form of Eq. (7.10) are referred to as **Sturm-Liouville equations**. A Sturm-Liouville equation is said to be **regular** on the interval $[a, b]$ if both $p(x)$ and $r(x)$ are strictly positive functions on $[a, b]$. As defined in Eq. (7.7) the function $p(x) > 0$, but $p(x)$ is not limited to functions of that form. A Sturm-Liouville equation is called **singular** if it is not regular. There are many ways in which a Sturm-Liouville equation may be singular. For example $p(x)$ or $r(x)$ may vanish somewhere in the interval $[a, b]$ or the interval itself may be unbounded. A regular (singular) Sturm-Liouville equation coupled with the separated boundary conditions as in Eq. (7.11) is called a **regular (singular) Sturm-Liouville boundary value problem**. Most of the results of this chapter will pertain to regular Sturm-Liouville problems, though similar results hold for most singular problems. Since Eq. (7.10) is homogeneous and the boundary conditions of Eq. (7.11) are homogeneous, then for any value of parameter λ the function $y(x) = 0$ is a solution to the Sturm-Liouville boundary value problem. This chapter is chiefly concerned with showing that nontrivial solutions (or eigenfunctions) to Eqs. (7.10) and (7.11) exist for infinitely many choices of λ (the eigenvalues) and cataloguing the properties of the eigenvalues and the eigenfunctions.

If λ is a value such that the Sturm-Liouville boundary value problem (comprised of Eqs. (7.10) and (7.11)), has a nontrivial solution $\phi(x)$, then λ is called an **eigenvalue** of the problem and $\phi(x)$ is called an **eigenfunction** corresponding to the eigenvalue λ. It is easy to see that Eq. (7.1) coupled with boundary conditions of the types shown in Eqs. (7.2), (7.3), (7.4), or (7.5) are all regular Sturm-Liouville boundary value problems. The eigenvalues and eigenfunctions of such problems were discussed extensively in Chap. 4. The next section explores some of the properties of the eigenvalues and eigenfunctions of the general Sturm-Liouville boundary value problem. This introduction concludes with the following example.

Example 7.1. Consider the boundary value problem:

$$x^2 y'' + x y' + \lambda y = 0 \text{ for } 1 < x < e,$$
$$y(1) = y(e) = 0.$$

Write the equation in the form of a self-adjoint Sturm-Liouville equation and solve the boundary value problem.

Solution. After dividing both sides of the ordinary differential equation by x, and defining the functions $p(x) = x$, $q(x) = 0$, and $r(x) = 1/x$, the equation can be written in self-adjoint form as

$$[x\, y']' + \frac{\lambda}{x}\, y = 0.$$

The Sturm-Liouville equation is regular since $p(x) > 0$ and $r(x) > 0$ on $[1, e]$. It is an Euler equation whose indicial equation is

$$r(r - 1) + r + \lambda = r^2 + \lambda = 0,$$

which has solutions $r = \pm\sqrt{-\lambda}$. As in previous examples there are three cases to consider when seeking nontrivial solutions to the boundary value problem.

Case $\lambda < 0$: r_1 and r_2 are real numbers and the general solution of the equation is

$$y(x) = c_1 x^{r_1} + c_2 x^{r_2}.$$

It is routine to check that, in this case, there is no nontrivial solution that satisfies the boundary conditions.

Case $\lambda = 0$: $r_1 = r_2 = 0$ and the general solution of the equation is

$$y(x) = c_1 + c_2 \ln x.$$

Again, it can be shown that there is no nontrivial solution satisfying the boundary conditions.

Case $\lambda > 0$: if $\lambda = \sigma^2$ where $\sigma > 0$, then $r_{1,2} = \pm \sigma i$ and the general solution of the equation is

$$y(x) = c_1 \cos(\sigma \ln x) + c_2 \sin(\sigma \ln x).$$

The boundary condition $y(1) = 0$ implies that $c_1 = 0$. The boundary condition $y(e) = 0$ requires that $y(e) = c_2 \sin \sigma = 0$. In order to obtain a nontrivial solution, it must be the case that $\sin \sigma = 0$. Therefore, $\lambda \equiv \lambda_n = \sigma_n^2 = n^2 \pi^2$ and the corresponding nontrivial eigenfunction is $\phi_n(x) = c_n \sin(n\pi \ln x)$ where c_n is any nonzero constant.

7.2 Properties of Eigenvalues and Eigenfunctions

This section will discuss the properties of the eigenvalues and eigenfunctions of regular Sturm-Liouville boundary value problems expressed in the form of the ordinary differential equation in Eq. (7.10) with boundary conditions in Eq. (7.11). To be clear, λ is an eigenvalue of the Sturm-Liouville boundary value problem if there exists a nontrivial function $\phi(x)$ solving Eq. (7.10) and satisfying the boundary conditions of Eq. (7.11). Let $x_0 \in (a, b)$ and suppose $y(x_0) = 1$ and $y'(x_0) = 0$, then there exists a unique solution $y_1(x)$ to Eq. (7.10) satisfying the initial conditions at x_0 under the assumptions made on $p(x)$ and $q(x)$ (see [Boyce and DiPrima (2012), Chap. 3]). This solution exists for $a < x < b$. Likewise if $y(x_0) = 0$ and $y'(x_0) = 1$ there exists a unique solution $y_2(x)$ to Eq. (7.10) satisfying this different set of initial conditions at x_0. The functions $y_1(x)$ and $y_2(x)$ are linearly independent on (a, b). The reader should note that these solutions may depend on the value of the parameter λ. If $\phi(x)$ is an eigenfunction of the Sturm-Liouville boundary value problem then there must be two constants c_1 and c_2 (not both zero) such that $\phi(x) = c_1 y_1(x) + c_2 y_2(x)$ on interval $[a, b]$ and for which

$$\alpha_1 \phi(a) + \beta_1 \phi'(a) = 0,$$
$$\alpha_2 \phi(b) + \beta_2 \phi'(b) = 0.$$

Using the linear combination of $y_1(x)$ and $y_2(x)$ in place of $\phi(x)$, this system of equations is equivalent to the matrix equation,

$$\begin{bmatrix} \alpha_1 y_1(a) + \beta_1 y_1'(a) & \alpha_1 y_2(a) + \beta_1 y_2'(a) \\ \alpha_2 y_1(b) + \beta_2 y_1'(b) & \alpha_2 y_2(b) + \beta_2 y_2'(b) \end{bmatrix} \begin{bmatrix} c_1 \\ c_2 \end{bmatrix} = \begin{bmatrix} 0 \\ 0 \end{bmatrix}.$$

This matrix equation has nontrivial solutions if and only if the matrix on the left-hand side is singular or equivalently if and only if the determinant,

$$\begin{vmatrix} \alpha_1 y_1(a) + \beta_1 y_1'(a) & \alpha_1 y_2(a) + \beta_1 y_2'(a) \\ \alpha_2 y_1(b) + \beta_2 y_1'(b) & \alpha_2 y_2(b) + \beta_2 y_2'(b) \end{vmatrix} = 0. \qquad (7.13)$$

To summarize, the Sturm-Liouville boundary value problem possesses a nontrivial solution $\phi(x)$ corresponding to parameter λ if and only if Eq. (7.13) holds. The astute reader will notice that well-known results concerning the existence and uniqueness of the solutions to *initial* value problems have been used to determine a necessary and sufficient condition for the existence of the solution to a *boundary* value problem. However, in general $y_1(x)$ and $y_2(x)$ are unknown and the existence and uniqueness of solutions to boundary value problems is more difficult to establish than for initial value problems.

The exploration of eigenvalues and eigenfunctions of the Sturm-Liouville boundary value problem continues with a convenient result known as **Lagrange's identity**.[3]

Lemma 7.1 (Lagrange's Identity). *Let L be the operator defined in Eq. (7.9) and let u and v be any twice continuously differentiable functions on (a, b), then*

$$u\,L[v] - v\,L[u] = \frac{d}{dx}\left[p(x)\left(u(x)\frac{dv}{dx} - v(x)\frac{du}{dx} \right) \right]. \qquad (7.14)$$

Proof. Note that functions u and v are not assumed to be eigenfunctions of a Sturm-Liouville boundary value problem. Applying the operator L defined in Eq. (7.9) to each function produces

$$L[u] = \frac{d}{dx}\left[p(x)\frac{du}{dx} \right] + q(x)u(x),$$

$$L[v] = \frac{d}{dx}\left[p(x)\frac{dv}{dx} \right] + q(x)v(x),$$

---

[3] Joseph-Louis Lagrange, Italian mathematician and astronomer (1736–1813).

so that

$$u L[v] - v L[u] = u(x)\frac{d}{dx}\left[p(x)\frac{dv}{dx}\right] - v(x)\frac{d}{dx}\left[p(x)\frac{du}{dx}\right]$$

$$= u(x)\frac{dp}{dx}\frac{dv}{dx} + u(x)p(x)\frac{d^2v}{dx^2} - v(x)\frac{dp}{dx}\frac{du}{dx} - v(x)p(x)\frac{d^2u}{dx^2}$$

$$= \frac{dp}{dx}\left(u(x)\frac{dv}{dx} - v(x)\frac{du}{dx}\right) + p(x)\left(u(x)\frac{d^2v}{dx^2} - v(x)\frac{d^2u}{dx^2}\right)$$

$$= \frac{d}{dx}\left[p(x)\left(u(x)\frac{dv}{dx} - v(x)\frac{du}{dx}\right)\right].$$

Thus Lagrange's identity is established. □

There is also an integral form of Lagrange's identity known as **Green's formula**[4].

Lemma 7.2 (Green's Formula). *Let L be the operator defined in Eq. (7.9) and let u and v be any twice continuously differentiable functions on* (a, b), *then*

$$\int_a^b (uL[v] - vL[u]) \, dx = \left[p(x)\left(u(x)\frac{dv}{dx} - v(x)\frac{du}{dx}\right)\right]_a^b. \tag{7.15}$$

Proof. See Exercise 11. □

The next result, known as **Picone's identity**[5] simplifies the derivation of many Sturm-Liouville results [Picone (1910)]. It is similar to Lagrange's identity.

Lemma 7.3 (Picone's Identity). *Suppose that* $u(x)$ *and* $v(x)$ *are solutions to the following ordinary differential equations:*

$$[p_1(x)u']' + q_1(x)u = 0,$$
$$[p_2(x)v']' + q_2(x)v = 0.$$

For all x such that $v(x) \neq 0$, *Picone's identity holds,*

$$\left[\frac{u}{v}(p_1u'v - p_2uv')\right]' \tag{7.16}$$

$$= (q_2 - q_1)u^2 + (p_1 - p_2)(u')^2 + \frac{p_2(u'v - uv')^2}{v^2}.$$

Proof. See Exercise 12. □

---

[4] George Green, British mathematical physicist (1793–1841).
[5] Mauro Picone, Italian mathematician (1885–1977).

The reader should note that Lagrange's identity, Green's formula, and Picone's identity hold for both regular and singular Sturm-Liouville boundary value problems.

An eigenvalue λ is said to be **simple** if there is only one linearly independent eigenfunction ϕ corresponding to λ. On the other hand, if there exist k linearly independent eigenfunctions corresponding to λ, the eigenvalue is said to be of **multiplicity** k. For Sturm-Liouville boundary value problems, the multiplicity of any eigenvalue is at most 2, since a second-order linear ordinary differential equation can have at most two linearly independent solutions. In fact, it can be shown that all the eigenvalues for the Sturm-Liouville equation with homogeneous separated boundary conditions are simple. Eigenvalues of multiplicity two exist for the case of periodic boundary conditions. The next several theorems and corollaries follow readily from Lagrange's identity and Green's formula.

Theorem 7.1. *All eigenvalues of the regular Sturm-Liouville boundary value problem of Eq. (7.10) with separated boundary conditions given in Eq. (7.11) are simple.*

Proof. Suppose $\phi(x)$ and $\psi(x)$ are eigenfunctions corresponding to the eigenvalue λ of the boundary value problem of Eqs. (7.10) and (7.11). The eigenfunctions satisfy the ordinary differential equations:

$$L[\phi](x) + \lambda r(x)\phi(x) = 0,$$
$$L[\psi](x) + \lambda r(x)\psi(x) = 0,$$

and boundary conditions at $x = a$,

$$\alpha_1 \phi(a) + \beta_1 \phi'(a) = 0,$$
$$\alpha_1 \psi(a) + \beta_1 \psi'(a) = 0.$$

The set of simultaneous boundary conditions can be written in matrix form,

$$\begin{bmatrix} \phi(a) & \phi'(a) \\ \psi(a) & \psi'(a) \end{bmatrix} \begin{bmatrix} \alpha_1 \\ \beta_1 \end{bmatrix} = \begin{bmatrix} 0 \\ 0 \end{bmatrix}.$$

Since $\alpha_1^2 + \beta_1^2 > 0$, the matrix on the left-hand side of the equation must be singular. This implies the vectors $\langle \phi(a), \phi'(a) \rangle$ and $\langle \psi(a), \psi'(a) \rangle$ are linearly dependent. There exists a scalar constant k, such that

$$\langle \phi(a), \phi'(a) \rangle = k \langle \psi(a), \psi'(a) \rangle.$$

Since Eq. (7.10) is linear and homogeneous, by the uniqueness of solutions, it must be the case that $\phi(x) = k\,\psi(x)$ for $a < x < b$ which establishes that eigenfunctions are unique up to multiplication by a scalar constant. Thus all eigenvalues of regular Sturm-Liouville boundary value problems are simple. □

In Exercise 18 the reader will be asked to show that the positive eigenvalues of Sturm-Liouville boundary value problems with periodic boundary conditions are of multiplicity two. See also Exercise 19 of Chap. 1.

The concept of function orthogonality used earlier in Chap. 3 can be generalized by introducing a **weight function**. Let $f(x)$ and $g(x)$ be two integrable functions on the interval $[a, b]$, and let $\rho(x)$ be a positive, continuous function on $[a, b]$. Define $\langle \cdot, \cdot \rangle_\rho$ as

$$\langle f, g \rangle_\rho = \int_a^b f(x)g(x)\rho(x)\, dx. \tag{7.17}$$

The reader will be asked to verify in Exercise 19 that $\langle \cdot, \cdot \rangle_\rho$ is an **inner product**. Functions f and g are said to be **orthogonal on $[a, b]$ with respect to a weight function** ρ if $\langle f, g \rangle_\rho = 0$. A set of integrable functions defined on $[a, b]$ is an **orthogonal set with respect to a weight function** ρ if the functions in the set are pairwise orthogonal, that is, any pair of functions in the set are orthogonal on $[a, b]$ with respect to the weight function $\rho(x)$. For example, the set of trigonometric functions

$$\{1, \cos(n\pi x/L), \sin(n\pi x/L)\}_{n=1}^{\infty}$$

seen in Sec. 3.2 is an orthogonal set of functions on $[-L, L]$ with respect to the weight function $\rho(x) = 1$.

The next theorem establishes the orthogonality of eigenfunctions corresponding to different eigenvalues of Sturm-Liouville boundary value problems.

Theorem 7.2. *Suppose $\phi_\lambda(x)$ and $\phi_\mu(x)$ are solutions of Eq. (7.10) corresponding to eigenvalues λ and μ respectively. If $\phi_\lambda(x)$ and $\phi_\mu(x)$ both satisfy the boundary conditions of Eq. (7.11) and if $\lambda \neq \mu$, then $\phi_\lambda(x)$ and $\phi_\mu(x)$ are orthogonal with respect to the weight function $r(x)$. That is, eigenfunctions corresponding to distinct eigenvalues of a Sturm-Liouville boundary value problem are orthogonal.*

Proof. By assumption functions $\phi_\lambda(x)$ and $\phi_\mu(x)$ satisfy the equations:

$$L[\phi_\lambda](x) = -\lambda r(x)\phi_\lambda(x),$$
$$L[\phi_\mu](x) = -\mu r(x)\phi_\mu(x).$$

Let $u(x) = \phi_\lambda(x)$ and $v(x) = \phi_\mu(x)$ in Green's formula of Eq. (7.15), then

$$\left[p(x) \left(\phi_\lambda(x) \frac{d\phi_\mu}{dx} - \phi_\mu(x) \frac{d\phi_\lambda}{dx} \right) \right]_a^b$$

$$= \int_a^b \left(\phi_\lambda(x) L[\phi_\mu](x) - \phi_\mu(x) L[\phi_\lambda](x) \right) dx$$

$$= \int_a^b \left(-\mu \phi_\lambda(x) \phi_\mu(x) r(x) + \lambda \phi_\mu(x) \phi_\lambda(x) r(x) \right) dx$$

$$= (\lambda - \mu) \int_a^b \phi_\lambda(x) \phi_\mu(x) r(x) \, dx = (\lambda - \mu) \langle \phi_\lambda, \phi_\mu \rangle_r.$$

As long as ϕ_λ and ϕ_μ satisfy the same boundary conditions the left-hand side of the equation is zero (see Exercise 13). By assumption $\lambda \neq \mu$ and thus $\langle \phi_\lambda, \phi_\mu \rangle_r = 0$. $\quad\square$

The reader should note that the orthogonality result holds for regular and singular Sturm-Liouville boundary value problems as well as for those with periodic boundary conditions.

The next result of this section reveals that the eigenvalues of Sturm-Liouville boundary value problems must be real numbers.

Theorem 7.3. *The eigenvalues of the Sturm-Liouville boundary value problem of Eqs. (7.10) and (7.11) are all real and the corresponding eigenfunctions are real-valued functions (except possibly for a complex-valued constant multiplicative factor).*

Proof. Let λ be any eigenvalue of the Sturm-Liouville boundary value problem in Eqs. (7.10) and (7.11). By definition,

$$L[\phi](x) + \lambda r(x)\phi(x) = 0.$$

Taking the complex conjugate of both sides of this equation yields

$$0 = \overline{L[\phi](x) + \lambda r(x)\phi(x)}$$
$$= \overline{[p(x)\phi'(x)]' + q(x)\phi(x) + \lambda r(x)\phi(x)}$$
$$= [p(x)\overline{\phi}'(x)]' + q(x)\overline{\phi}(x) + \overline{\lambda} r(x)\overline{\phi}(x).$$
$$= L[\overline{\phi}](x) + \overline{\lambda} r(x)\overline{\phi}(x).$$

Thus $\overline{\phi}$ is a solution of Eq. (7.10) corresponding to eigenvalue $\overline{\lambda}$. Since function ϕ satisfies the boundary conditions which have real coefficients, function $\overline{\phi}$ will as well. Hence $\overline{\phi}$ is an eigenfunction of the Sturm-Liouville boundary value problem corresponding to eigenvalue $\overline{\lambda}$. From the proof of Theorem 7.2, ϕ and $\overline{\phi}$ satisfy

$$0 = (\lambda - \overline{\lambda}) \int_a^b \phi(x)\overline{\phi}(x)r(x)\,dx = (\lambda - \overline{\lambda}) \int_a^b |\phi(x)|^2\, r(x)\,dx.$$

Thus it must be the case that $\lambda = \overline{\lambda}$ since by the assumptions the integrand must be positive. Thus λ is a real number.

Since ϕ and $\overline{\phi}$ are eigenfunctions corresponding to eigenvalue λ, the function $u(x) = (\phi(x) + \overline{\phi}(x))/2$ (which is a real-valued function) is also a solution to Eq. (7.10) satisfying the boundary conditions of Eq. (7.11). Consequently $u(x)$ is an eigenfunction of the Sturm-Liouville boundary value problem corresponding to λ. By Theorem 7.1 functions ϕ and $\overline{\phi}$ are (possibly complex) multiples of the real-valued eigenfunction u. $\square$

Since the eigenvalues of a regular Sturm-Liouville boundary value problem must be real numbers, the eigenvalues can be ordered.

7.3   Zeros of Eigenfunctions

While by definition an eigenfunction is a nontrivial function, eigenfunctions can equal zero in the interval $[a, b]$. This section will present several results which govern the placement of zeros of eigenfunctions and the relationships between zeros of different eigenfunctions. In summary, the zeros of an eigenfunction to a regular Sturm-Liouville boundary value problem are isolated, meaning that there is an open interval centered at each zero which contains no other zeros of the eigenfunction. Two linearly independent eigenfunctions to the same regular Sturm-Liouville boundary value problem must have interlaced zeros. Two eigenfunctions to different (but related) regular Sturm-Liouville boundary value problems will also possess interlaced zeros.

Theorem 7.4. *Let $\phi(x) \in C^1[a, b]$ be a nontrivial solution to Eq. (7.10). If $\phi(x_0) = 0$ for some $x_0 \in (a, b)$ then there is an open interval $(x_0 - \delta, x_0 + \delta)$ with $\delta > 0$ which contains no other zero of $\phi(x)$.*

Proof. Suppose $\phi(x)$ has consecutive zeros at $\alpha < \beta$ (both in interval (a, b)). Since $\phi(x)$ is differentiable on (a, b), then by the mean value theorem $\phi'(z) = 0$ for some $z \in (\alpha, \beta)$. Now suppose there is a sequence $\{x_n \in (a, b)\}_{n=1}^{\infty}$ such that $\phi(x_n) = 0$ for $n \in \mathbb{N}$ and $x_n \to x_0 \in (a, b)$ as $n \to \infty$. Using the mean value theorem argument there is a sequence $\{z_n \in (a, b)\}_{n=1}^{\infty}$ for which $\phi'(z_n) = 0$ for $n \in \mathbb{N}$. Note that $x_n < z_n < x_{n+1}$ for each $n \in \mathbb{N}$

which implies $z_n \to x_0$ as $n \to \infty$. By assumption $\phi(x)$ and $\phi'(x)$ are continuous on (a, b), therefore

$$0 = \lim_{n\to\infty} \phi(x_n) = \phi\left(\lim_{n\to\infty} x_n\right) = \phi(x_0),$$

$$0 = \lim_{n\to\infty} \phi'(z_n) = \phi'\left(\lim_{n\to\infty} z_n\right) = \phi'(x_0).$$

Since $\phi(x_0) = 0 = \phi'(x_0)$ then by uniqueness of solutions [Boyce and DiPrima (2012), Sec. 3.2], $\phi(x) \equiv 0$ which contradicts the assumption that $\phi(x)$ is a nontrivial function. Consequently there can be no sequence of zeros in (a, b) converging to a limit point in (a, b). Therefore around each zero of $\phi(x)$ in (a, b) there is neighborhood of radius $\delta > 0$ which contains no other zero of $\phi(x)$. □

The theorem and proof can be generalized to the cases in which the sequence of zeros $\{x_n\}_{n=1}^{\infty}$ limits on one of the endpoints either a or b. This is left to the reader.

Theorem 7.5 (Sturm Separation Theorem). *Suppose $\phi(x)$ and $\psi(x)$ are linearly independent solutions to Eq. (7.10), then $\psi(x)$ has a zero strictly between any two zeros of $\phi(x)$.*

Proof. By the assumption of linear independence of the solutions, neither $\phi(x)$ nor $\psi(x)$ is identically zero. Furthermore the two solutions cannot have a zero in common, for if they did (say at $x = x_0$), then

$$W(\phi, \psi)(x_0) = \phi(x_0)\psi'(x_0) - \phi'(x_0)\psi(x_0) = 0$$

which implies $\phi(x)$ and $\psi(x)$ are linearly dependent by Abel's[6] Theorem [Boyce and DiPrima (2012), Sec. 3.3].

Suppose $\phi(x)$ has consecutive zeros at $\alpha < \beta$ in the interval $[a, b]$. Without loss of generality assume $\phi(x) > 0$ for $x \in (\alpha, \beta)$. Clearly, $\phi'(\alpha) \geq 0$. In fact, it must be the case that $\phi'(\alpha) > 0$ since otherwise $\phi(\alpha) = \phi'(\alpha) = 0$ which implies that $\phi(x) \equiv 0$ for all $x \in [a, b]$ by uniqueness of solutions to the initial value problem. Similar reasoning implies that $\phi'(\beta) < 0$. If $\psi(x)$ has a zero in (α, β) then the proof is complete. If $\psi(x)$ has no zero in (α, β), then assume $\psi(x) > 0$ on (α, β) (or replace $\psi(x)$ by $-\psi(x)$). By assumption,

$$[p(x)\phi']' + (q(x) + \lambda r(x))\phi = 0,$$
$$[p(x)\psi']' + (q(x) + \lambda r(x))\psi = 0,$$

---

[6]Niels Henrik Abel, Norwegian mathematician (1802–1829).

which implies

$$0 = \psi \left[p(x)\phi' \right]' - \phi \left[p(x)\psi' \right]' = \left[p(x)(\psi\phi' - \phi\psi') \right]'.$$

Integrating this expression over the interval (α, β) produces

$$0 = \int_{\alpha}^{\beta} \left[p(x)(\psi\phi' - \phi\psi') \right]' \, dx = p(\beta)\psi(\beta)\phi'(\beta) - p(\alpha)\psi(\alpha)\phi'(\alpha).$$

However, the right-hand side of this equation is negative which is a contradiction. Thus $\psi(x)$ has a zero in (α, β). $\qquad\square$

An immediate corollary of Theorem 7.5 is that if two solutions, $\phi(x)$ and $\psi(x)$, to Eq. (7.10) have a zero in common, then $\phi(x)$ and $\psi(x)$ are linearly dependent. The next theorem has a conclusion which demonstrates that the zeros of solutions to Sturm-Liouville boundary value problems are interlaced just as in the Sturm Separation Theorem; however, rather than being two linearly independent solutions to the same differential equation, in the next theorem, the two solutions are to two different, but related differential equations.

Theorem 7.6 (Sturm Comparison Theorem). *Suppose $\phi(x)$ and $\psi(x)$ are nontrivial solutions to the ordinary differential equations:*

$$\left[p_1(x)\phi'(x) \right]'(x) + (q_1(x) + \lambda_1 r_1(x))\phi(x) = 0,$$
$$\left[p_2(x)\psi'(x) \right]'(x) + (q_2(x) + \lambda_2 r_2(x))\psi(x) = 0$$

on interval (a, b). Suppose further that $0 < p_2(x) \leq p_1(x)$ and $q_1(x) + \lambda_1 r_1(x) \leq q_2(x) + \lambda_2 r_2(x)$ for all $x \in [a, b]$. If $\phi(x)$ has zeros at $x = \alpha$ and $x = \beta$ with $\alpha < \beta$, then $\psi(x)$ must have a zero in $[\alpha, \beta]$.

Proof. According to Theorem 7.4 the zeros of $\phi(x)$ are isolated and without loss of generality α and β are assumed to be consecutive zeros of $\phi(x)$ with $\phi(x) > 0$ on interval (α, β). To save space let $Q_i(x) = q_i(x) + \lambda_i r_i(x)$ for $i = 1, 2$. If $p_1(x) = p_2(x)$ and $Q_1(x) = Q_2(x)$ for all $x \in [\alpha, \beta]$, then $\phi(x)$ and $\psi(x)$ satisfy the same differential equation on $[\alpha, \beta]$ and the result is true by the Sturm Separation Theorem (Theorem 7.5). In the following it is assumed that $p_1(x) = p_2(x)$ or $Q_1(x) = Q_2(x)$ does not hold for all $x \in [\alpha, \beta]$. If $\psi(x)$ has no zero in $[\alpha, \beta]$ then by Picone's identity expressed in Eq. (7.16),

$$\left[\frac{\phi}{\psi}(p_1\phi'\psi - p_2\phi\psi') \right]' = (Q_2 - Q_1)\phi^2 + (p_1 - p_2)(\phi')^2 + \frac{p_2(\phi\psi' - \phi'\psi)^2}{\psi^2}.$$

Integrating both sides of this equation over the interval $[\alpha, \beta]$ and using the fact that $\phi(\alpha) = \phi(\beta) = 0$ produce the equation,

$$0 = \int_\alpha^\beta (Q_2 - Q_1)\phi^2 \, dx + \int_\alpha^\beta (p_1 - p_2)(\phi')^2 \, dx + \int_\alpha^\beta \frac{p_2(\phi\psi' - \phi'\psi)^2}{\psi^2} \, dx,$$

which is equivalent to

$$\int_\alpha^\beta (Q_2 - Q_1)\phi^2 \, dx + \int_\alpha^\beta (p_1 - p_2)(\phi')^2 \, dx = -\int_\alpha^\beta \frac{p_2(\phi\psi' - \phi'\psi)^2}{\psi^2} \, dx.$$

The assumptions on $p_i(x)$ and $Q_i(x)$ imply the left-hand side of this equation is positive while the right-hand side is nonpositive which is a contradiction. Thus $\psi(x)$ has a zero in $[\alpha, \beta]$. $\square$

Interested readers can find more detailed proofs of the Sturm Comparison Theorem and other related results in [Hartman (1982), Chap. 3].

7.4 Generalized Fourier Series

The trigonometric functions used in Chap. 3 are just the eigenfunctions of specific Sturm-Liouville boundary value problems and any given function $f(x)$ on $[a, b]$ can be written as a series of trigonometric functions with modest assumptions on $f(x)$. In Sec. 7.6 the eigenvalues of the Sturm-Liouville boundary value problem of Eqs. (7.10) and (7.11) will be described as a sequence $\{\lambda_n\}_{n=1}^\infty$ with the functions $\{X_n(x)\}_{n=1}^\infty$ as their corresponding eigenfunctions. Can a given function f be represented as a series involving the eigenfunctions of a general Sturm-Liouville boundary value problem? That is, if $\{\lambda_n\}_{n=1}^\infty$ is the set of all the eigenvalues of a Sturm-Liouville boundary value problem can constants c_n for $n \in \mathbb{N}$, be found such that

$$f(x) = \sum_{n=1}^\infty c_n X_n(x) \tag{7.18}$$

where $X_n(x)$ is an eigenfunction corresponding to the eigenvalue λ_n? For the moment assume Eq. (7.18) is valid. Multiplying both sides of Eq. (7.18) by $X_m(x)r(x)$ and integrating over interval $[a, b]$ produce

$$\int_a^b f(x)X_m(x)r(x) \, dx = \sum_{n=1}^\infty c_n \int_a^b X_n(x)X_m(x)r(x) \, dx,$$

where it is assumed that the order of integration and summation can be interchanged on the right-hand side. By Theorem 7.2,

$$\int_a^b f(x)X_m(x)r(x) \, dx = c_m \int_a^b (X_m(x))^2 r(x) \, dx = c_m \langle X_m, X_m \rangle_r$$

and hence

$$c_m = \frac{\int_a^b f(x) X_m(x) r(x)\, dx}{\int_a^b (X_m(x))^2 r(x)\, dx} = \frac{\langle f, X_m \rangle_r}{\langle X_m, X_m \rangle_r}.$$

Define

$$\phi_n(x) = \frac{X_n(x)}{(\int_a^b (X_n(x))^2 r(x)\, dx)^{1/2}} = \frac{X_n(x)}{\sqrt{\langle X_n, X_n \rangle_r}}$$

as the normalized eigenfunction corresponding to eigenvalue λ_n. The coefficient can be chosen as

$$c_n = \int_a^b f(x) \phi_n(x) r(x)\, dx = \langle f, \phi_n \rangle_r. \tag{7.19}$$

The following theorem describes how $f(x)$ may be expressed as an infinite series in terms of the eigenfunctions $\{\phi_n(x)\}_{n=1}^\infty$.

Theorem 7.7. *Assume that $\{\lambda_n\}_{n=1}^\infty$ is the set of eigenvalues of a regular Sturm-Liouville boundary value problem of Eqs. (7.10) and (7.11). Let $\{\phi_n(x)\}_{n=1}^\infty$ be the corresponding normalized eigenfunctions. If $f(x)$ is piecewise smooth on interval $[a, b]$, the series*

$$\sum_{n=1}^\infty c_n \phi_n(x), \tag{7.20}$$

where c_n is defined in Eq. (7.19) converges pointwise to $\frac{1}{2}[f(x+) + f(x-)]$ for all $x \in (a, b)$. If $f(x)$ is continuous on $[a, b]$ and satisfies the boundary conditions of Eq. (7.11) then the series of Eq. (7.20) converges uniformly and absolutely to f for all $x \in [a, b]$.

The theorem above is a generalization of the Dirichlet Convergence Theorem (Theorem 3.1) covered in Chap. 3. Its proof is beyond the scope of this book. Interested readers can find proofs in [Atkinson (1964), Sec. 8.9] and [Courant and Hilbert (1989a), Chap. 6, Sec. 3]. The series given in Theorem 7.7 is called the **generalized Fourier series** of the function f and the constants c_n are called the **generalized Fourier coefficients** of the function f.

Example 7.2. Consider the boundary value problem:

$$X'' + \lambda X = 0 \text{ for } 0 < x < 1,$$
$$X(0) - h_0 X'(0) = 0,$$
$$X(1) + h_1 X'(1) = 0,$$

where h_0 and h_1 are positive constants. Determine the eigenvalues and normalized eigenfunctions of this Sturm-Liouville boundary value problem.

Solution. As seen in Sec. 4.1.3 on p. 129, if $h_0 h_1 \neq ((2n-1)\pi/2)^2$ for any positive integer n, then λ is an eigenvalue of the boundary value problem if and only if $\lambda > 0$ is a solution of

$$\tan \lambda = \frac{(h_0 + h_1)\lambda}{\lambda^2 - h_0 h_1}.$$

Though the equation above cannot be solved explicitly for λ, it can be shown that there are infinitely many solutions $\{\lambda_n\}_{n=1}^{\infty}$ satisfying

$$(n-1)\pi < \lambda_n < \frac{(2n-1)\pi}{2}$$

For the case when $h_0 = h_1 = 1$, Fig. 4.3 provides a graphical view of the first few solutions of the equation above and their approximate values were also given. The eigenfunction corresponding to eigenvalue λ_n is expressed as

$$X_n(x) = c \left(\cos \lambda_n x + \frac{1}{\lambda_n} \sin \lambda_n x \right), \tag{7.21}$$

where $c \neq 0$ is an arbitrary constant. The normalized eigenfunction can then be expressed as

$$\phi_n(x) = \frac{\cos \lambda_n x + \frac{1}{\lambda_n} \sin \lambda_n x}{(\int_0^1 (\cos \lambda_n x + \frac{1}{\lambda_n} \sin \lambda_n x)^2 \, dx)^{1/2}}$$

$$= \frac{2\lambda_n^{3/2} \cos \lambda_n x + 2\lambda_n^{1/2} \sin \lambda_n x}{(2\lambda_n(\lambda_n^2 + 2 - \cos 2\lambda_n) + (\lambda_n^2 - 1)\sin 2\lambda_n)^{1/2}}.$$

7.5 Estimating Eigenvalues and the Rayleigh Quotient

Before delving into the rather technical issues of establishing the existence of eigenvalues and eigenfunctions, this section reveals another relationship between eigenvalues and eigenfunctions. Suppose $\phi(x)$ is an eigenfunction corresponding to eigenvalue λ of a Sturm-Liouville boundary value problem. Multiplying both sides of Eq. (7.10) by $\phi(x)$ and integrating over the interval $[a, b]$ produce

$$0 = \int_a^b \left(\frac{d}{dx} \left[p(x) \frac{d\phi}{dx} \right] \phi(x) + (\lambda r(x) + q(x))(\phi(x))^2 \right) dx.$$

Solving for λ yields the following,

$$\lambda = \frac{- \int_a^b \left(\phi(x) \frac{d}{dx} \left[p(x) \frac{d\phi}{dx} \right] + q(x)(\phi(x))^2 \right) dx}{\int_a^b r(x)(\phi(x))^2 \, dx}. \tag{7.22}$$

Splitting the integral in the numerator of Eq. (7.22) into two definite integrals and applying integration by parts to the first definite integral result in an alternative formula,

$$\lambda = \frac{\left[-p(x)\phi(x)\frac{d\phi}{dx}\right]_a^b + \int_a^b \left[p(x)\left(\frac{d\phi}{dx}\right)^2 - q(x)(\phi(x))^2\right] dx}{\int_a^b r(x)(\phi(x))^2 \, dx}. \tag{7.23}$$

While Eq. (7.23) is not a practical means of calculating eigenvalues λ (since the eigenfunction $\phi(x)$ must be known in order to use it), this equation does provide information about the eigenvalues without solving Eq. (7.10).

Theorem 7.8. *All eigenvalues of the Sturm-Liouville boundary value problem of Eqs. (7.10) and (7.11) are nonnegative provided $q(x) \leq 0$ on $[a, b]$ and*

$$p(b)\phi(b)\frac{d\phi}{dx}(b) - p(a)\phi(a)\frac{d\phi}{dx}(a) \leq 0. \tag{7.24}$$

Proof. See Exercise 25. $\qquad\qquad\qquad\qquad\qquad\qquad\qquad\qquad\qquad\quad$ $\square$

The inequality in Eq. (7.24) is satisfied by many physically meaningful boundary conditions. For example the Dirichlet boundary conditions of the heat and wave equations require $\phi(a) = \phi(b) = 0$ and thus Eq. (7.24) is satisfied.

In the next section the existence of an infinite sequence of eigenvalues $\{\lambda_n\}_{n=1}^{\infty}$ for the Sturm-Liouville boundary value problem will be established. These eigenvalues are ordered in the sense that $\lambda_n < \lambda_{n+1}$ for all $n \in \mathbb{N}$ and hence λ_1 is the smallest eigenvalue. Motivated by Eq. (7.22), the Rayleigh quotient operator $R[\cdot]$ on a space of functions is defined as

$$R[y] = \frac{-\int_a^b y(x)\, L[y](x)\, dx}{\int_a^b r(x)(y(x))^2 \, dx}. \tag{7.25}$$

Notice that for this operator the domain is a set of appropriately differentiable functions and the range is a subset of the real numbers. Eigenfunction $\phi_1(x)$ and its corresponding eigenvalue λ_1 satisfy the relationship, $R[\phi_1] = \lambda_1$. Recall that an eigenfunction is a twice continuously differentiable solution to Eq. (7.10) satisfying the boundary conditions of Eq. (7.11). Operator $R[\cdot]$ can be applied to a broader class of functions than merely eigenfunctions. Suppose $y(x)$ is a $C^2[a, b]$ function satisfying the boundary conditions stated in Eq. (7.11), then $R[y]$ can be evaluated, though the result is not necessarily an eigenvalue of the boundary value problem of Eqs. (7.10) and (7.11). Such a function $y(x)$ is often referred to as a **trial function**. The use of trial functions provides a convenient way to estimate the value of the smallest eigenvalue λ_1.

Theorem 7.9. *Suppose $y(x)$ is piecewise smooth on $[a, b]$ such that $R[y]$ is well-defined and $y(x)$ satisfies the boundary conditions of Eq. (7.11). Then $R[y] \geq \lambda_1$ where λ_1 is the smallest eigenvalue the Sturm-Liouville boundary value problem of Eq. (7.10) with boundary conditions given in Eq. (7.11). Equality is achieved only when $y(x) = \phi_1(x)$, the eigenfunction corresponding to λ_1.*

Proof. Let $y(x)$ be an appropriately smooth function on $[a, b]$ satisfying the boundary conditions of Eq. (7.11) and use Theorem 7.7 to expand $y(x)$ as a generalized Fourier series in terms of the normalized eigenfunctions of the Sturm-Liouville boundary value problem. Applying operator $L[\cdot]$ to $y(x)$ yields

$$L[y](x) = L\left[\sum_{n=1}^{\infty} c_n \phi_n\right](x) = \sum_{n=1}^{\infty} c_n L[\phi_n](x) = -\sum_{n=1}^{\infty} c_n \lambda_n r(x)\phi_n(x)$$

where the order of application of the linear operator and summation have been interchanged and Eq. (7.10) has been used. Substituting this expression into Eq. (7.25) and integrating term by term produce

$$R[y] = \frac{\int_a^b \left(\sum_{m=1}^{\infty} c_m \phi_m(x)\right)\left(\sum_{n=1}^{\infty} c_n \lambda_n r(x)\phi_n(x)\right) dx}{\int_a^b r(x)\left(\sum_{m=1}^{\infty} c_m \phi_m(x)\right)\left(\sum_{n=1}^{\infty} c_n \phi_n(x)\right) dx}$$

$$= \frac{\sum_{m=1}^{\infty}\sum_{n=1}^{\infty} c_m c_n \lambda_n \int_a^b \phi_m(x)\phi_n(x)r(x) dx}{\sum_{m=1}^{\infty}\sum_{n=1}^{\infty} c_m c_n \int_a^b \phi_m(x)\phi_n(x)r(x) dx}.$$

For each $n \in \mathbb{N}$ there is only a single nonzero term in the double summations above due to the orthogonality of the eigenfunctions as expressed in Theorem 7.2. Hence

$$R[y] = \frac{\sum_{n=1}^{\infty} c_n^2 \lambda_n \int_a^b (\phi_n(x))^2 r(x) dx}{\sum_{n=1}^{\infty} c_n^2 \int_a^b (\phi_n(x))^2 r(x) dx} = \frac{\sum_{n=1}^{\infty} c_n^2 \lambda_n}{\sum_{n=1}^{\infty} c_n^2}.$$

Since λ_1 is assumed to be the smallest eigenvalue

$$R[y] \geq \frac{\lambda_1 \sum_{n=1}^{\infty} c_n^2}{\sum_{n=1}^{\infty} c_n^2} = \lambda_1,$$

which establishes that the smallest eigenvalue is a lower bound of the Rayleigh quotient operator. Suppose $y(x) = \phi_1(x)$. Since $\phi_1(x)$ satisfies the hypotheses of this theorem and since the generalized Fourier coefficients in the expansion of $\phi_1(x)$ are $c_1 = 1$ and $c_n = 0$ for $n \geq 2$ then

$$R[\phi_1] = \frac{\lambda_1}{1} = \lambda_1,$$

and, therefore, the first eigenvalue λ_1 is the minimum of the Rayleigh Quotient. $\qquad\qquad\square$

Before proceeding to the challenging task of establishing the existence of eigenvalues and eigenfunctions of a regular Sturm-Liouville boundary value problem, the next example illustrates Theorem 7.9.

Example 7.3. Consider the following boundary value problem:

$$y'' + \lambda y = 0 \text{ for } 0 < x < 1,$$
$$y'(0) = y'(1) = 0.$$

Use trial functions to estimate the smallest eigenvalue of this problem.

Solution. Note that $[a, b] = [0, 1]$, $p(x) = 1$, $r(x) = 1$, $q(x) = 0$, and for any function satisfying these boundary conditions the Rayleigh Quotient simplifies to

$$R[y] = \frac{\int_0^1 (y'(x))^2 \, dx}{\int_0^1 (y(x))^2 \, dx}.$$

One trial function is $y_1(x) = (x - 1)^2 + \frac{2}{3}(x - 1)^3$ (the reader may verify that this function satisfies the boundary conditions but not the ordinary differential equation).

$$R[y_1] = \frac{\int_0^1 \left(2(x - 1) + 2(x - 1)^2\right)^2 \, dx}{\int_0^1 \left((x - 1)^2 + \frac{2}{3}(x - 1)^3\right)^2 \, dx} = \frac{2/15}{13/315} = \frac{42}{13} \approx 3.23077.$$

Another trial function is $y_2(x) = \cos \pi x$. This is an eigenfunction of the boundary value problem corresponding to $\lambda = \pi^2$.

$$R[y_2] = \frac{\int_0^1 (-\pi \sin \pi x)^2 \, dx}{\int_0^1 (\cos \pi x)^2 \, dx} = \frac{\pi^2/2}{1/2} = \pi^2 \approx 9.86960.$$

Yet a third trial function is $y_3(x) = 1$ which is actually the first eigenfunction for the boundary value problem and corresponds to the smallest eigenvalue $\lambda_1 = 0$.

$$R[y_3] = R[1] = 0 = \lambda_1.$$

Readers may have noticed that two eigenfunctions were used as trial functions in the previous example. Since $\phi_1(x) = y_3(x)$, Theorem 7.9 guaranteed that $R[\phi_1] = \lambda_1$. However, use of the eigenfunction $y_2(x)$ reveals a generalization of Theorem 7.9. In general if a trial function $y(x)$ is orthogonal to eigenfunctions $\phi_k(x)$ for $k = 1, 2, \ldots, n-1$ then $R[y] \geq \lambda_n$ with equality achieved for $y(x) = \phi_n(x)$, the nth eigenfunction. Consequently $R[\phi_2] = \lambda_2$.

This observation illustrates one application of the Rayleigh quotient, the numerical estimation of the eigenvalues and eigenfunctions of a Sturm-Liouville boundary value problem. Minimizing the Rayleigh quotient over the appropriate vector space of smooth functions $y(x)$ defined on $[a, b]$ yields λ_1 and $\phi_1(x)$. Repeating the process of minimizing $R[y]$ over the functions orthogonal to $\phi_1(x)$ produces λ_2 and $\phi_2(x)$. This process can be iterated to yield numerical estimates of the eigenvalues and eigenfunctions for the Sturm-Liouville boundary value problem of Eq. (7.10) with boundary conditions given in Eq. (7.11).

7.6   Existence of Eigenfunctions and Eigenvalues

In the previous sections the existence of eigenvalues and their corresponding eigenfunctions was assumed so that their properties could be described and discussed. In this section the steps necessary to establish their existence will be taken. The reader is forewarned that some of the steps involved in proving the existence of eigenfunctions and eigenvalues are rather technical. If desired, this section can be skimmed or skipped without hampering the reader's understanding and application of the remainder of this text. In this section, the existence of eigenvalues and eigenfunctions will be rigorously established only for regular Sturm-Liouville boundary value problems of the form in Eq. (7.10) with boundary conditions as in Eq. (7.11).

The approach taken to establish the existence of eigenvalues and eigenfunctions for the Sturm-Liouville boundary value problem involves two standard conversions of the ordinary differential equation in Eq. (7.10). First, it is converted from a single second-order differential equation to a system of two first-order ordinary differential equations. Let

$$X(x) = y(x) \text{ and } Y(x) = p(x)y'(x). \tag{7.26}$$

Differentiating $Y(x)$ and using Eq. (7.10) produce the following pair of first-order ordinary differential equations,

$$X'(x) = \frac{1}{p(x)} Y(x),$$
$$Y'(x) = -(q(x) + \lambda r(x))X(x). \tag{7.27}$$

These equations can be converted into a more convenient form in polar coordinates by letting

$$X(x) = R(x)\sin\theta(x),$$
$$Y(x) = R(x)\cos\theta(x). \tag{7.28}$$

This is known as the Prüfer[7] substitution [Prüfer (1926)]. While readers may argue that the definitions of $X(x)$ and $Y(x)$ in terms of the polar coordinate functions $R(x)$ and $\theta(x)$ are interchanged, this choice is deliberate in that it results in a later, beneficial simplification. For any nontrivial solution of the Sturm-Liouville differential equation, the associated function $R(x)$ of Eq. (7.27) must be positive, since if $R(x_0) = 0$ for some $x_0 \in [a, b]$ then $y(x_0) = 0$ and $y'(x_0) = 0$, which implies that $y(x) = 0$ for all $x \in [a, b]$ by the uniqueness of solutions. Differentiating the change of variables in Eq. (7.28) yields

$$X'(x) = R'(x)\sin\theta(x) + R(x)\theta'(x)\cos\theta(x),$$
$$Y'(x) = R'(x)\cos\theta(x) - R(x)\theta'(x)\sin\theta(x).$$

Substituting these derivatives into Eq. (7.27) produces a system of equations in the polar coordinate system:

$$\frac{1}{p(x)} R(x)\cos\theta(x) = R'(x)\sin\theta(x) + R(x)\theta'(x)\cos\theta(x),$$
$$-(q(x) + \lambda r(x))R(x)\sin\theta(x) = R'(x)\cos\theta(x) - R(x)\theta'(x)\sin\theta(x).$$

Solving this set of equations for $R'(x)$ and $\theta'(x)$ produces a system of two first-order ODEs:

$$R'(x) = \frac{1}{2}\left(\frac{1}{p(x)} - [q(x) + \lambda r(x)]\right) R(x)\sin(2\theta(x)), \tag{7.29}$$

$$\theta'(x) = \frac{1}{p(x)}\cos^2\theta(x) + [q(x) + \lambda r(x)]\sin^2\theta(x). \tag{7.30}$$

Equation (7.10) is equivalent to the set of Eqs. (7.29) and (7.30) provided the change of variables is invertible, in other words, if given $X(x)$ and $Y(x)$ the quantities $R(x)$ and $\theta(x)$ can unambiguously be determined. Suppose

---

[7]Heinz Prüfer, German mathematician (1896–1934).

$y(x)$ is a nontrivial solution of Eq. (7.10), then as mentioned earlier $R(x) > 0$ for all $x \in [a, b]$. Thus the formula $R(x) = [(X(x))^2 + (Y(x))^2]^{1/2}$ and $\theta(x)$ can be determined from $\tan \theta(x) = X(x)/Y(x)$ and the signs of $X(x)$ and $Y(x)$. Consequently, the change of variables in the Prüfer substitution is invertible. As a result any solution to Eq. (7.10) implies a solution to the system in Eqs. (7.29) and (7.30) and vice versa.

Note that the right side of Eq. (7.30) is independent of R and is continuously differentiable with respect to θ. Therefore, for any initial value $\theta_0 \in [0, 2\pi)$, there exists a unique solution to Eq. (7.30), denoted as $\theta_\lambda(x)$, such that $\theta_\lambda(a) = \theta_0$. Once Eq. (7.30) is solved its solution may be substituted in Eq. (7.29) which happens to be first-order linear in R. Thus

$$R_\lambda(x) = R_\lambda(a)e^{\int_a^x \frac{1}{2}\left(\frac{1}{p(z)} - [q(z) + \lambda r(z)]\right) \sin(2\theta_\lambda(z)) \, dz}.$$

The investigation of the zeroes of a solution $y(x)$ of Eq. (7.10) plays an essential role in the remainder of this section and is made easier by studying the circumstances under which $\theta_\lambda(x)$ is an integer multiple of π for $x \in [a, b]$, for if $\theta_\lambda(x) = k\pi$ for some $k \in \mathbb{N}$ then $X(x) = y(x) = 0$. The function $\theta_\lambda(x)$ depends both on $x \in [a, b]$ and the parameter λ. The next two results explore the dependency of the behavior of this angular function θ on these two quantities. From a first glance at Eq. (7.30) it may appear that θ is a monotone increasing function of x, but this is not true in general. However, $\theta_\lambda(x)$ is "monotone increasing" at each x for which $\theta_\lambda(x)$ is an integer multiple of π. This behavior is illustrated in Fig. 7.1 and proved in the following lemma.

Lemma 7.4. *Let λ be fixed. Suppose $z \in (a, b)$ is such that $\theta_\lambda(z) = k\pi$ for some $k \in \mathbb{Z}$, then z is the unique solution to $\theta_\lambda(x) = k\pi$ in (a, b). Furthermore $\theta_\lambda(x) < k\pi$ for $x < z$ and $\theta_\lambda(x) > k\pi$ for $x > z$.*

Proof. Fix $k \in \mathbb{Z}$ such that there exists $z \in (a, b)$ for which $\theta_\lambda(z) = k\pi$. Evaluating Eq. (7.30) at such a z yields $\theta_\lambda'(z) = 1/p(z)$. Since $p(x) > 0$ and continuous on $[a, b]$, $\theta_\lambda'(x) > 0$ on some open interval containing z. Thus $\theta_\lambda(x)$ is strictly increasing in a neighborhood of every integer multiple of π. $\square$

The next theorem facilitates the comparison of the behavior of solutions of Eq. (7.10).

Theorem 7.10. *Let $p_i(x)$ be continuously differentiable on $[a, b]$ and let $q_i(x)$ be continuous on $[a, b]$ for $i = 1, 2$. Suppose $0 < p_2(x) \le p_1(x)$ and*

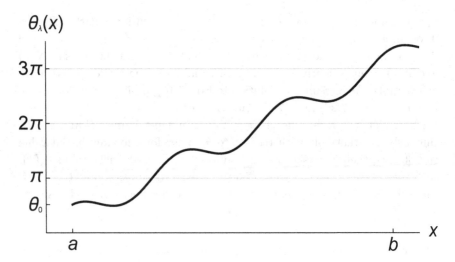

Fig. 7.1   Function $\theta_\lambda(x)$ is strictly increasing at each x for which $\theta_\lambda(x)$ is an integer multiple of π.

$q_1(x) \le q_2(x)$ on $[a, b]$. *Let y_i be a solution of*

$$[p_i(x)y'(x)]' + q_i(x)y(x) = 0,$$

and let $R_i(x)$ and $\theta_i(x)$ be the corresponding solutions of Eqs. (7.29) and (7.30). If $\theta_1(a) \le \theta_2(a)$, then $\theta_1(x) \le \theta_2(x)$ for all $x \in [a, b]$. If in addition $q_1(x) < q_2(x)$ on $[a, b]$ then $\theta_1(x) < \theta_2(x)$ on $(a, b]$.

Proof. Define the function $u(x) = \theta_2(x) - \theta_1(x)$. Using Eq. (7.30) the derivative of u may be expressed as

$$u'(x) = \frac{\cos^2 \theta_2(x)}{p_2(x)} - \frac{\cos^2 \theta_1(x)}{p_1(x)} + q_2(x)\sin^2 \theta_2(x) - q_1(x)\sin^2 \theta_1(x)$$

$$= \frac{\cos^2 \theta_2(x) - \cos^2 \theta_1(x)}{p_1(x)} + q_1(x)(\sin^2 \theta_2(x) - \sin^2 \theta_1(x)) + h(x),$$

where

$$h(x) = \left(\frac{1}{p_2(x)} - \frac{1}{p_1(x)} \right) \cos^2 \theta_2(x) + (q_2(x) - q_1(x))\sin^2 \theta_2(x).$$

By the assumptions on $p_i(x)$ and $q_i(x)$, the function $h(x) \ge 0$ on $[a, b]$. Using the Pythagorean trigonometric identity, $u'(x)$ can be written as

$$u'(x) = \left(q_1(x) - \frac{1}{p_1(x)} \right) (\sin^2 \theta_2(x) - \sin^2 \theta_1(x)) + h(x).$$

Define the function $f(x)$ as

$$f(x) = \begin{cases} 2\sin\theta_1(x)\cos\theta_1(x) & \text{if } \theta_1(x) = \theta_2(x), \\ \frac{\sin^2\theta_2(x) - \sin^2\theta_1(x)}{\theta_2(x) - \theta_1(x)} & \text{if } \theta_1(x) \neq \theta_2(x). \end{cases}$$

The function $f(x)$ can be treated as a continuous function and, if $g(x) = (q_1(x) - 1/p_1(x))f(x)$, then

$$u'(x) = g(x)u(x) + h(x),$$

with $u(a) \geq 0$. Note that the ordinary differential equation for $u(x)$ is first-order linear. The solution to this first-order ODE can be written as

$$u(x) = u(a)e^{\int_a^x g(t)\,dt} + \int_a^x h(t)e^{\int_t^x g(v)\,dv}\,dt,$$

for $a \leq x \leq b$. Consequently $u(x) \geq 0$ on $[a, b]$ which implies $\theta_2(x) \geq \theta_1(x)$ on $[a, b]$. Under the final assumption that $q_1(x) < q_2(x)$ on $[a, b]$ then $h(x) > 0$ except at an isolated set of points and thus $u(x) > 0$ on $(a, b]$ which implies $\theta_1(x) < \theta_2(x)$ on $(a, b]$. $\qquad\square$

Now consider the problem of finding nontrivial solutions to the Sturm-Liouville boundary value problem as expressed in Eqs. (7.10) and (7.11). Finding such solutions will establish the existence of eigenfunctions and eigenvalues. The boundary conditions in Eq. (7.11) can be written in a form that is more convenient to use in the corresponding Prüfer form expressed in Eqs. (7.29) and (7.30). According to the definition of X and Y given in Eq. (7.26), the boundary condition at $x = a$ can be written as $\alpha_1 X(a) + \beta_1 Y(a)/p(a) = 0$ or equivalently, as

$$\frac{\alpha_1 p(a)}{\sqrt{(\alpha_1 p(a))^2 + \beta_1^2}}X(a) + \frac{\beta_1}{\sqrt{(\alpha_1 p(a))^2 + \beta_1^2}}Y(a) = 0. \qquad (7.31)$$

Without loss of generality, assume $\beta_1 \geq 0$. Then there exists a unique $A \in [0, \pi)$ such that

$$\cos A = \frac{\alpha_1 p(a)}{\sqrt{(\alpha_1 p(a))^2 + \beta_1^2}} \text{ and } \sin A = \frac{\beta_1}{\sqrt{(\alpha_1 p(a))^2 + \beta_1^2}}.$$

The boundary condition at $x = a$ found in Eq. (7.11) becomes

$$(\cos A)X(a) + (\sin A)Y(a) = 0,$$

or in terms of $y(x)$,

$$(\cos A)y(a) + (\sin A)p(a)y'(a) = 0. \qquad (7.32)$$

By similar reasoning there is a unique constant $B \in (0, \pi]$ such that the boundary condition at $x = b$ in Eq. (7.11) is equivalent to

$$(\cos B)y(b) + (\sin B)p(b)y'(b) = 0. \tag{7.33}$$

The boundary conditions given in Eqs. (7.32) and (7.33) are more convenient to use for the solutions of the equivalent Prüfer form of Eq. (7.10).

Two lemmas will be of use in the proof of the main result about Sturm-Liouville boundary value problems. These lemmas establish the asymptotic behavior of $\theta_\lambda(x)$ of Eq. (7.30) as $\lambda \to \pm\infty$.

Lemma 7.5. *Let $\phi_\lambda(x)$ be the unique solution to the regular Sturm-Liouville differential equation of Eq. (7.10) satisfying the initial conditions*

$$\phi_\lambda(a) = \sin A \ \text{ and } \ \phi'_\lambda(a) = -\frac{\cos A}{p(a)},$$

where A is given in Eq. (7.32). If $\theta_\lambda(x)$ is the solution to Eq. (7.30) corresponding to $\phi_\lambda(x)$ then for any $z \in (a, b]$,

$$\lim_{\lambda \to \infty} \theta_\lambda(z) = \infty.$$

Proof. The solution $\phi_\lambda(x)$ satisfies the boundary condition of Eq. (7.32) and hence also satisfies the boundary condition of Eq. (7.11) at $x = a$. If $R_\lambda(x)$ and $\theta_\lambda(x)$ are the corresponding solutions to the Prüfer form of Eq. (7.10), then $\phi_\lambda(x) = R_\lambda(x) \sin \theta_\lambda(x)$ and $\phi'_\lambda(x) = p(x)R_\lambda(x)\cos\theta_\lambda(x)$. In particular, $R_\lambda(a) = 1$ and $\theta_\lambda(a) = A$.

Function $\phi_\lambda(x) = 0$ if and only if $\theta_\lambda(x) \equiv 0 \pmod{\pi}$. Theorem 7.10 and Lemma 7.4 imply that $\theta_\lambda(x)$ is monotone increasing in λ for each fixed $x \in [a, b]$ and is a strictly increasing function of x in a neighborhood of every integer multiple of π for each fixed λ. Since $\theta_\lambda(a) = A \geq 0$, then $\theta'_\lambda(a) > 0$ for sufficiently large λ by Eq. (7.30). Consequently $\theta_\lambda(x) > 0$ for all large λ and $x \in [a, b]$. By the continuity of functions p, q, and r, there exist constants P_M, Q_M, and R_m such that $p(x) \leq P_M$, $-q(x) \leq Q_M$, and $0 < R_m \leq r(x)$ on interval $(a, b]$. Let z_0 and z be fixed with $a < z_0 < z \leq b$ and λ be positive. Consider the second-order, constant coefficient, ordinary differential equation

$$P_M y'' + (\lambda R_m - Q_M)y = 0. \tag{7.34}$$

This ordinary differential equation has a corresponding Prüfer form where the angular variable, denoted as $w(x)$, satisfies the differential equation

$$w'(x) = \frac{1}{P_M}\cos^2 w(x) + (\lambda R_m - Q_M)\sin^2 w(x).$$

Let function $w_\lambda(x)$ be the solution to this differential equation with the initial value

$$w(z_0) = \tan^{-1}\left(\frac{\phi_\lambda(z_0)}{p(z_0)\phi'_\lambda(z_0)}\right) = \theta_\lambda(z_0).$$

As a result of Theorem 7.10, $\theta_\lambda(x) \geq w_\lambda(x)$ for $z_0 < x \leq z$. The choice of initial condition $w_\lambda(z_0) = \theta_\lambda(z_0)$ implies

$$\theta_\lambda(z) - \theta_\lambda(z_0) \geq w_\lambda(z) - w_\lambda(z_0). \tag{7.35}$$

Let λ be large enough that $\lambda R_m - Q_M > 0$. Any solution $y(x)$ of Eq. (7.34) is a linear combination of sine and cosine functions with common frequency $f_\lambda = \sqrt{(\lambda R_m - Q_M)/P_M}$. Therefore successive zeroes of $y(x)$ are separated by intervals of length

$$T_\lambda = \frac{\pi\sqrt{P_M}}{\sqrt{\lambda R_m - Q_M}}.$$

Note that $T_\lambda \to 0$ as $\lambda \to \infty$. If k is an integer greater than one, there exists a positive λ such that $k T_\lambda \leq z - z_0$ which implies the solution of Eq. (7.34) will have k zeroes in the interval $[z_0, z]$. Therefore, for such a value of λ it is the case that $w_\lambda(z) - w_\lambda(z_0) \geq k\pi$ and by Eq. (7.35) likewise $\theta_\lambda(z) - \theta_\lambda(z_0) \geq k\pi$. Since k is arbitrary then $\theta_\lambda(z) - \theta_\lambda(z_0) \to \infty$ as $\lambda \to \infty$. This implies that if $z \in (a, b]$ then $\theta_\lambda(z) \to \infty$ as $\lambda \to \infty$. $\square$

Lemma 7.6. *Let $\phi_\lambda(x)$ be the unique solution to the regular Sturm-Liouville differential equation of Eq. (7.10) satisfying the initial conditions*

$$\phi_\lambda(a) = \sin A \text{ and } \phi'_\lambda(a) = -\frac{\cos A}{p(a)},$$

where A is given in Eq. (7.32). If $\theta_\lambda(x)$ is the solution to Eq. (7.30) corresponding to $\phi_\lambda(x)$ then for any $z \in (a, b]$,

$$\lim_{\lambda \to -\infty} \theta_\lambda(z) = 0.$$

Proof. Fix $z \in (a, b]$ and let $0 < \epsilon < (\pi - A)/2$. The continuity assumptions on functions p, q, and r imply that there exist positive constants p_M, q_M and R_m such that $1/p(x) \leq p_M$, $|q(x)| \leq q_M$, and $0 < R_m \leq r(x)$ for $x \in [a, b]$. For brevity define $K = p_M + q_M$. Equation (7.30) and the boundedness of the sine and cosine functions imply that for all λ

$$\theta'_\lambda(x) \leq K + \lambda r(x) \sin^2 \theta_\lambda(x),$$

and $\theta'_\lambda(x) \leq K$, when $\lambda \leq 0$. Suppose $\lambda \leq 0$, then by the Mean Value Theorem,

$$\theta_\lambda(x) \leq \theta_\lambda(a) + K(x - a) = A + K(x - a).$$

When $0 \leq A < \pi$, if $x - a$ is sufficiently small, then $\theta_\lambda(x) < A + \epsilon < \pi - \epsilon$. Choose $x_0 \in (a, z]$ such that $\theta_\lambda(x_0) < A + \epsilon$. The line in the $x\theta$-plane connecting the points $(x_0, A + \epsilon)$ and (z, ϵ) can be expressed as

$$s(x) = -\frac{A}{z - x_0}(x - x_0) + A + \epsilon.$$

Define Λ as

$$\Lambda = -\frac{\frac{A}{z - x_0} + K}{R_m \sin^2 \epsilon} < 0$$

and suppose $\lambda < \Lambda$. By the choice of x_0, $\theta_\lambda(x_0) < s(x_0)$. If there exists $x \in (x_0, z]$ for which $\theta_\lambda(x) \geq s(x)$ then let x_1 be the first value of x for which $\theta_\lambda(x_1) = s(x_1)$. At this first intersection of the graphs of $s(x)$ and $\theta_\lambda(x)$, $\theta'_\lambda(x_1) \geq -A/(z - x_0)$. Note that

$$\epsilon < s(x_1) = A + \epsilon - \frac{A(x_1 - x_0)}{z - x_0} < A + \epsilon < \pi - \epsilon$$

which implies $\theta_\lambda(x_1) \in (\epsilon, \pi - \epsilon)$. Consequently $\sin \epsilon < \sin \theta_\lambda(x_1)$. Thus

$$\frac{-A}{z - x_0} \leq \theta'_\lambda(x_1) \leq K + \lambda r(x_1) \sin^2 \theta_\lambda(x_1) < K + \lambda R_m \sin^2 \epsilon < \frac{-A}{z - x_0},$$

a contradiction. Hence $\theta_\lambda(x) < s(x)$ for all $x \in (x_0, z]$. In particular $\theta_\lambda(z) < s(z) = \epsilon$. Since $z \in (a, b]$ was arbitrary, when $\lambda < \Lambda$, then $\theta_\lambda(x) < \epsilon$ for $x \in (a, b]$.   □

The asymptotic limits of $\theta_\lambda(x)$ allow the establishment of the eigenvalues, eigenfunctions, and their properties. The next theorem contains some of the fundamental results of Sturm-Liouville theory. The reader should carefully think about the generality of the following result in the context of the richness of the types of Sturm-Liouville boundary value problems which may be encountered.

Theorem 7.11. *The eigenvalues and eigenfunctions of a regular Sturm-Liouville boundary value problem of Eqs. (7.10) and (7.11) have the following properties.*

- *The boundary value problem possesses an infinite sequence of eigenvalues $\{\lambda_n\}_{n=1}^{\infty}$ that can be arranged in increasing order as*

$$\lambda_1 < \lambda_2 < \cdots < \lambda_n < \cdots$$

with $\lim_{n \to \infty} \lambda_n = \infty$.

- *Any eigenfunction $\phi_n(x)$ corresponding to eigenvalue λ_n has exactly $n-1$ zeroes in $[a, b]$.*
- *The zeroes of $\phi_{n+1}(x)$ lie between the zeroes of $\phi_n(x)$.*

Proof. As was assumed in the two previous lemmas, let $\phi_\lambda(x)$ be the unique solution to the Sturm-Liouville differential equation found in Eq. (7.10) satisfying the initial conditions,

$$\phi_\lambda(a) = \sin A \text{ and } \phi_\lambda'(a) = -\frac{\cos A}{p(a)}.$$

These values for the initial conditions of $\phi_\lambda(x)$ satisfy the boundary condition of Eq. (7.32) and hence also satisfy the boundary condition of Eq. (7.11) at $x = a$. If $R_\lambda(x)$ and $\theta_\lambda(x)$ are the corresponding solutions of Eqs. (7.29) and (7.30), the equivalent Prüfer form of Eq. (7.10), then $R_\lambda(x)$ and $\theta_\lambda(x)$ satisfy the initial conditions $R_\lambda(a) = 1$ and $\theta_\lambda(a) = A$. The parameter λ will be an eigenvalue if it assumes a value such that $\phi_\lambda(b)$ satisfies the boundary condition of Eq. (7.11) at $x = b$ or equivalently if it satisfies Eq. (7.33). From the latter equation it is evident that λ is an eigenvalue if and only if $\theta_\lambda(b) = B + n\pi$ for some integer n.

Since Lemma 7.6 establishes that $\theta_\lambda(b) \to 0$ as $\lambda \to -\infty$ and Lemma 7.5 implies $\theta_\lambda(b) \to \infty$ monotonically as $\lambda \to \infty$ then there exists a unique value $\lambda = \lambda_1$ for which $\theta_{\lambda_1}(b) = B$. Since $\theta_{\lambda_1}(a) = A \in [0, \pi)$ and $\theta_{\lambda_1}(b) = B \in (0, \pi]$, then λ_1 is an eigenvalue and the corresponding solution $\phi_1(x) = \phi_{\lambda_1}(x)$ is an eigenfunction of the regular Sturm-Liouville boundary value problem of Eqs. (7.10) and (7.11). Since $\theta_{\lambda_1}(x)$ is an increasing function of x in a neighborhood of the x values where $\theta_{\lambda_1} = 0$ and $\theta_{\lambda_1} = \pi$, then $\theta_{\lambda_1}(x) \in (0, \pi)$ for $x \in (a, b)$. Consequently the corresponding solution $\phi_1(x)$ has no zeroes in the interval (a, b). Continuing in this way, there is a unique $\lambda_2 > \lambda_1$ for which $\theta_{\lambda_2}(b) = B + \pi$. This implies that λ_2 is an eigenvalue and the corresponding solution $\phi_2(x) = \phi_{\lambda_2}(x)$ is an eigenfunction of the Sturm-Liouville boundary value problem. Since $A = \theta_{\lambda_2}(a) < \pi < \theta_{\lambda_2}(b) = B + \pi$ and since $\theta_{\lambda_2}(x)$ is monotone increasing at any x for which $\theta_{\lambda_2}(x) = \pi$ the corresponding solution $\phi_{\lambda_2}(x)$ has precisely one zero in the interval (a, b). Using similar reasoning there exists a unique $\lambda_{n+1} > \lambda_n$ for which $\theta_{\lambda_{n+1}}(b) = B + n\pi$. The corresponding solution $\phi_{n+1}(x)$ is an eigenfunction corresponding to eigenvalue λ_{n+1}. This eigenfunction will have n zeroes in the interval (a, b). The zeroes of the functions $\phi_n(x)$ and $\phi_{n+1}(x)$ are interlaced as a consequence of Theorem 7.6. □

Example 7.4. Consider the boundary value problem:

$$X'' + \lambda X = 0 \text{ for } 0 < x < 1,$$
$$X(0) - h_0 X'(0) = 0,$$
$$X(1) + h_1 X'(1) = 0,$$

where h_0 and h_1 are positive constants. For the case when $h_0 = h_1 = 1$ the non-normalized eigenfunctions are given in Eq. (7.21). An illustration of the roots of one of the eigenfunctions is shown in Fig. 7.2.

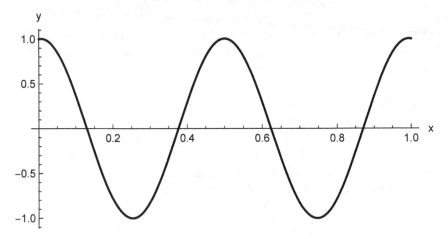

Fig. 7.2   A plot of the eigenfunction corresponding to $\lambda_5 \approx 12.7232$ showing the existence of its four roots in the interval $[0, 1]$.

7.7   Readings and Further Results

This chapter has provided a brief introduction and justification of the theory of regular Sturm-Liouville boundary value problems. The existence of eigenfunctions and the orthogonality of those eigenfunctions drives the generalized Fourier series methods used to solve many partial differential equations of interest in the physical sciences, applied mathematics, and engineering. Many theorems similar to those proved here hold for boundary value problems with periodic boundary conditions and for singular Sturm-Liouville boundary value problems though the boundary conditions and other assumptions must be chosen carefully. Some of the important eigenfunctions encountered in applications of partial differential equations arise from singular Sturm-Liouville boundary value problem. These include the

Legendre, Hermite, Laguerre, and Bessel functions studied in more detail in Chap. 8.

Readers interested in precise statements and proofs of these additional results and more general versions of the theorems presented here should consult (in increasing order of mathematical sophistication) [Al-Gwaiz (2008); González-Velasco (1995); Zettl (2005)].

7.8 Exercises

(1) Solve the boundary value problem:
$$y'' + 1 = 0 \text{ for } 0 < x < 2,$$
$$y(0) = 2,$$
$$y(2) = 3.$$

(2) Solve the boundary value problem:
$$y'' + x = 0 \text{ for } 0 < x < \pi,$$
$$y(0) = 1,$$
$$y(\pi) + y'(\pi) = 2.$$

(3) Determine if the following boundary value problem has any nontrivial solutions. If so, find them.
$$y'' - 9y = 0 \text{ for } 0 < x < 1,$$
$$y(0) + y'(0) = 0,$$
$$y(1) = 0.$$

(4) Determine if the following boundary value problem has any nontrivial solutions. If so, find them.
$$y'' + 9y = 0 \text{ for } 0 < x < \pi,$$
$$y'(0) = 0,$$
$$y(\pi) = 0.$$

(5) Suppose $1 \leq x \leq 2$ and re-write the following ordinary differential equation in self-adjoint form.
$$x^2 y'' + x y' + y = 0.$$

(6) Suppose a, b, and c are constants with $a > 0$. Re-write the following ordinary differential equation in self-adjoint form.
$$a y'' + b y' + c y = 0.$$

(7) Consider the boundary value problem:

$$[(1+x)^2 y']' + \lambda y = 0 \text{ for } 0 < x < 1,$$
$$y(0) = y(1) = 0.$$

Show that the eigenvalues and eigenfunctions of this boundary value problem are respectively:

$$\lambda_n = \frac{1}{4} + \left(\frac{n\pi}{\ln 2}\right)^2,$$

$$\phi_n(x) = \frac{1}{\sqrt{x+1}} \sin\left(\frac{n\pi \ln(x+1)}{\ln 2}\right)$$

for $n \in \mathbb{N}$.

(8) Show that the operator L defined in Eq. (7.9) is a linear operator, that is, show for all twice differentiable functions $y_1(x)$ and $y_2(x)$ and every real constant c that
   (a) $L[y_1 + y_2] = L[y_1] + L[y_2]$, and
   (b) $L[c\, y_1] = cL[y_1]$.

(9) Let $L[y] = y'' - 6y' + 9y$ and apply this operator to the following functions:
   (a) $f(x) = x^2 - 6x + 9$,
   (b) $g(x) = e^{3x}$,
   (c) $h(x) = xe^{3x}$,
   (d) $k(x) = \cosh(3x)$.

(10) Consider the second-order, linear ordinary differential equation

$$P(x)y'' + Q(x)y' + (R(x) + \lambda S(x))y = 0.$$

Re-write this equation in the form of a Sturm-Liouville boundary value problem as in Eq. (7.10).

(11) Derive Eq. (7.15).

(12) Derive Eq. (7.16).

(13) Show that if $u(x)$ and $v(x)$ satisfy Eq. (7.11) then,

$$\left[p(x)\left(u(x)\frac{dv}{dx} - v(x)\frac{du}{dx}\right)\right]_a^b = 0.$$

(14) Find the eigenvalues and corresponding eigenfunctions of the Sturm-Liouville boundary value problem:

$$y'' + \lambda y = 0 \text{ for } 0 < x < L,$$
$$y(0) = 0,$$
$$y'(L) = 0.$$

(15) Find the eigenvalues and corresponding eigenfunctions of the Sturm-Liouville boundary value problem:

$$y'' + \lambda\, y = 0 \text{ for } 0 < x < 1,$$
$$y(0) + y'(0) = 0,$$
$$y(1) + y'(1) = 0.$$

(16) Find the eigenvalues and corresponding eigenfunctions of the singular Sturm-Liouville boundary value problem:

$$x^2 y'' + x\, y' + \lambda x^2 y = 0 \text{ for } 0 < x < 1,$$
$$y'(0) = 0,$$
$$y(1) = 0.$$

(17) Show that the eigenfunctions found in Example 7.1 are pairwise orthogonal with respect to $r(x) = 1/x$ on the interval $[1, e]$.

(18) Consider the Sturm-Liouville boundary value problem:

$$y'' + \lambda\, y = 0 \text{ for } -L < y < L$$

subject to periodic boundary conditions:

$$y(-L) = y(L),$$
$$y'(-L) = y'(L).$$

Show that every positive eigenvalue possesses two corresponding linearly independent eigenfunctions.

(19) Let f, g, and h be continuous functions on $[a, b]$ and γ be a real number. Show that the function defined in Eq. (7.17) possesses the properties of an inner product:

(a) $\langle f + g, h \rangle_\rho = \langle f, h \rangle_\rho + \langle g, h \rangle_\rho$,
(b) $\langle \gamma\, f, g \rangle_\rho = \gamma \langle f, g \rangle_\rho$,
(c) $\langle f, g \rangle_\rho = \langle g, f \rangle_\rho$,
(d) $\langle f, f \rangle_\rho > 0$ if and only if $f(x) \not\equiv 0$ on (a, b).

(20) Show directly that the function $u(x)$ defined in the proof of Theorem 7.3 solves Eqs. (7.10) and (7.11).

(21) Consider Bessel's equation of order ν:

$$x^2 y'' + x\, y' + (x^2 - \nu^2) y = 0.$$

(a) Assuming $x > 0$ and defining $v = y\sqrt{x}$, show that Bessel's equation can be re-written as

$$v'' + \left(1 + \frac{1 - 4\nu^2}{4x^2}\right) v = 0.$$

(b) Suppose that $0 < \nu < 1/2$ and show that $v(x)$ has a root in each interval of the form $((n-1)\pi, n\pi)$ with $n \in \mathbb{N}$. (*Hint*: use the Sturm Comparison Theorem and the ordinary differential equation $u'' + u = 0$.)

(c) Show that $y(x)$ has a root in every interval of the form $((n-1)\pi, n\pi)$ with $n \in \mathbb{N}$ when $0 < \nu < 1/2$.

(22) Use separation of variables on the heat equation with convection:

$$u_t = k\, u_{xx} - C_0 u_x \text{ for } 0 < x < L.$$

Show that the ordinary differential equations for the spatially dependent part of the solution can be put into the Sturm-Liouville form of Eq. (7.10).

(23) Let K and L be positive constants and let f be a continuous nonnegative function on interval $[\alpha, \beta]$ satisfying the inequality

$$f(x) \leq K + \int_\alpha^x L\, f(s)\, ds.$$

Show that

$$f(t) \leq K e^{L(t-\alpha)}.$$

(24) Find all the continuous, nonnegative functions f defined on $[0, 1]$ such that

$$f(x) \leq \int_0^x f(s)\, ds,$$

for $0 \leq x \leq 1$.

(25) Provide the details of the proof to Theorem 7.8.

(26) Show that Eq. (7.24) is satisfied by the Neumann boundary conditions

$$\frac{d\phi(a)}{dx} = \frac{d\phi(b)}{dx} = 0.$$

(27) Show that Eq. (7.24) is satisfied by the Neumann boundary conditions appropriate to Newton's Law of cooling:

$$\frac{d\phi(a)}{dx} - \beta_1\phi(a) = 0,$$
$$\frac{d\phi(b)}{dx} + \beta_2\phi(b) = 0.$$

Assume $\beta_1 > 0$ and $\beta_2 > 0$.

(28) Use trial functions to estimate the smallest eigenvalue of the boundary value problem below.

$$y'' + \lambda y = 0 \text{ for } 0 < x < 1,$$
$$y(0) = 0,$$
$$y(1) = 1.$$

(29) Use trial functions to estimate the smallest eigenvalue of the boundary value problem below.

$$y'' + \lambda y = 0 \text{ for } 0 < x < 1,$$
$$y(0) = 0,$$
$$y'(1) = 0.$$

(30) Consider the ordinary differential equation:

$$u'' + g(x)\, u = 0,$$

for which $0 < A < g(x) < B$. If $u(x)$ is a solution to the ordinary differential equation and if $z_1 < z_2$ are two consecutive zeros of u, show that

$$\frac{\pi}{\sqrt{B}} < z_2 - z_1 < \frac{\pi}{\sqrt{A}}.$$

Chapter 8

Special Functions

The reader must have noticed that performing the method of separation of variables in certain common domains such as one-dimensional intervals, rectangles, disks, sectors, and annuli results in being asked to solve similar types of ordinary differential equations and boundary value problems. So far eigenfunctions involving sine, cosine, exponential, and powers of an independent variable (in the case of Euler's equation) have been common. However, a much wider variety of ordinary differential equations and boundary value problems occur frequently in applications of partial differential equations in the physical sciences and engineering. This chapter will review some of the differential equations arising in other domains and situations including Bessel's equation, Legendre's equation, and Laguerre's equation. This chapter applies the Sturm-Liouville theory of Chap. 7, though many of the examples of the current chapter are singular Sturm-Liouville boundary value problems rather than regular problems. If the reader wishes, this chapter can be skimmed or skipped until one of these special ordinary differential equations arises in the course of carrying out the method of separation of variables. This section also serves as a review of series solutions of second-order linear ordinary differential equations near ordinary and regular singular points and the **Method of Frobenius**.[1]

8.1 The Gamma Function

The **Gamma Function** is central to the definition of many of the special functions described in this chapter. It can be thought of as a generalization of the factorial function. As an aid to exploring the other special functions of this chapter, the properties and notation of the Gamma function will be

---

[1]Ferdinand Georg Frobenius, German mathematician (1849–1917).

311

elucidated. The Gamma function is denoted as $\Gamma(p)$ and is defined by the integral,

$$\Gamma(p+1) = \int_0^\infty e^{-x} x^p \, dx. \tag{8.1}$$

The improper integral converges for all $p \geq 0$. For $p < 0$ the integrand becomes unbounded as $x \to 0^+$. The integral can be shown to converge for $p > -1$ and to diverge for $p \leq -1$. Consider the case when $p = 0$,

$$\Gamma(1) = \int_0^\infty e^{-x} \, dx = \lim_{M \to \infty} \int_0^M e^{-x} \, dx = \lim_{M \to \infty} \left(-e^{-M} + 1 \right) = 1.$$

Lemma 8.1. *For $p > 0$,*

$$\Gamma(p+1) = p\,\Gamma(p). \tag{8.2}$$

Proof. By the definition given in Eq. (8.1),

$$\Gamma(p+1) = \int_0^\infty e^{-x} x^p \, dx = p \int_0^\infty e^{-x} x^{p-1} \, dx = p\,\Gamma(p)$$

using integration by parts.   □

Equation (8.2) suggests a connection between the Gamma function and the recursive nature of the factorial operation. The next lemma makes the "factorial"-like property of the Gamma function more clear.

Lemma 8.2. *If p is a positive integer,*

$$\Gamma(p+1) = p!. \tag{8.3}$$

Proof. It has been shown that $\Gamma(1) = 1$ and that $\Gamma(p+1) = p\,\Gamma(p)$, thus

$$\Gamma(2) = \Gamma(1+1) = 1\,\Gamma(1) = 1$$

and hence the claim is true for $p = 1$. Suppose the claim is true for some integer $k \geq 2$, then

$$\Gamma(k+1) = k\,\Gamma(k) = k(k-1)! = k!$$

and hence by the principle of mathematical induction the claim is true for all $p \in \mathbb{N}$.   □

Lemma 8.2 states that the Gamma function is the factorial function if p is a positive integer. One of the advantages of the Gamma function is that it is defined for non-integer values of p, yet retains many of its factorial-like properties. As an example, consider $\Gamma(1/2)$. By definition,

$$\Gamma\left(\frac{1}{2}\right) = \int_0^\infty e^{-x} x^{-1/2} \, dx.$$

Making the substitution $u^2 = x$ and $2u\,du = dx$, the integral can be re-written as

$$\Gamma\left(\frac{1}{2}\right) = \int_0^\infty e^{-u^2} u^{-1}(2u)\,du = 2\int_0^\infty e^{-u^2}\,du = \int_{-\infty}^\infty e^{-u^2}\,du = \sqrt{\pi}.$$

A similar integral was evaluated in Eq. (4.38) of Sec. 4.4 and the reader can see the details there.

Lemma 8.3. *For $p > 0$,*

$$\frac{\Gamma(p+n)}{\Gamma(p)} = p(p+1)(p+2)\cdots(p+n-1). \tag{8.4}$$

Proof. Let $p > 0$ then

$$
\begin{aligned}
\Gamma(p+n) &= \Gamma(p+n-1+1) \\
&= (p+n-1)\Gamma(p+n-1) \\
&= (p+n-1)\Gamma(p+n-2+1) \\
&= (p+n-1)(p+n-2)\Gamma(p+n-2) \\
&\ \ \vdots \\
&= (p+n-1)(p+n-2)\cdots(p+n-(n-1))\Gamma(p+n-(n-1)) \\
&= (p+n-1)(p+n-2)\cdots(p+1)\Gamma(p+1) \\
&= (p+n-1)(p+n-2)\cdots(p+1)p\,\Gamma(p) \\
\frac{\Gamma(p+n)}{\Gamma(p)} &= (p+n-1)(p+n-2)\cdots(p+1)p,
\end{aligned}
$$

since $\Gamma(p) > 0$. $\qquad\qquad\square$

For the sake of brevity an expression of the form on the right-hand side of Eq. (8.4) is denoted as

$$(p)_n = p(p+1)(p+2)\cdots(p+n-1)$$

where $(p)_n$ is called the **Pochhammer**[2] **symbol** and its evaluation is referred to as a **rising factorial**.

Lemma 8.3 makes it possible to calculate $\Gamma(p)$ for any $p > 0$ if the value of $\Gamma(p)$ is known for $0 < p < 1$. For example,

$$\Gamma\left(\frac{3}{2}\right) = \Gamma\left(\frac{1}{2}+1\right) = \frac{1}{2}\Gamma\left(\frac{1}{2}\right) = \frac{\sqrt{\pi}}{2},$$

and also

$$\Gamma\left(\frac{11}{2}\right) = \Gamma\left(5+\frac{1}{2}\right) = \Gamma\left(\frac{1}{2}\right)\left(\frac{1}{2}\right)\left(\frac{3}{2}\right)\left(\frac{5}{2}\right)\left(\frac{7}{2}\right)\left(\frac{9}{2}\right) = \frac{945\sqrt{\pi}}{32}.$$

---

[2]Leo August Pochhammer, Prussian mathematician (1841–1920).

Lemma 8.4. *The Gamma function* $\Gamma(x)$ *is continuous for all* $x > 0$.

Proof. It is sufficient to show the Gamma function is continuous for an arbitrary positive number p. Fix $p > 0$ then for $q > 0$,

$$|\Gamma(p) - \Gamma(q)| \leq \int_0^\infty e^{-x}|x^{p-1} - x^{q-1}|\, dx.$$

Consider the function $f(t) = x^{t-1}$. By the Mean Value Theorem

$$\frac{f(p) - f(q)}{p - q} = \frac{x^{p-1} - x^{q-1}}{p - q} = f'(z) = (\ln x)x^{z-1}$$

for some z between p and q. This implies

$$|\Gamma(p) - \Gamma(q)| \leq |p - q| \int_0^\infty e^{-x}x^{z-1}|\ln x|\, dx. \tag{8.5}$$

When $x \geq 1$, $0 \leq \ln x < x$ and thus

$$\int_1^\infty e^{-x}x^{z-1}|\ln x|\, dx < \int_1^\infty e^{-x}x^z\, dx < (\llbracket z \rrbracket + 1)!.$$

When $0 < x < 1$,

$$-e^{-x}(\ln x)x^{z-1} \leq -(\ln x)x^{z-1}$$

and thus

$$\int_0^1 e^{-x}x^{z-1}|\ln x|\, dx \leq -\int_0^1 (\ln x)x^{z-1}\, dx = \frac{1}{z^2} \tag{8.6}$$

(see Exercise 1). Therefore the improper integral in Eq. (8.5) converges and hence there exists $M(z) > 0$ such that

$$|\Gamma(p) - \Gamma(q)| \leq M(z)|p - q|.$$

Let a and b be constants such that $0 < a < \min\{p, q\} \leq \max\{p, q\} < b$ and let $M = \max_{a \leq z \leq b}\{M(z)\}$. Given $\epsilon > 0$, if $|p - q| < \epsilon/M$ then $|\Gamma(p) - \Gamma(q)| < \epsilon$ and consequently Γ is continuous at $p > 0$. $\qquad\square$

The Gamma function is also differentiable for $x > 0$. The derivative can be found by differentiating inside the integral.

$$\Gamma'(x) = \frac{d}{dx}\int_0^\infty e^{-t}t^{x-1}\, dt = \int_0^\infty \frac{d}{dx}[e^{-t}t^{x-1}]\, dt = \int_0^\infty e^{-t}(\ln t)t^{x-1}\, dt.$$

So far discussion of the properties of the Gamma function has been limited to the cases where $p > 0$. Lemma 8.3 allows the Gamma function to be extended to the real numbers in the set $\{x \mid x \neq 0, -1, -2, \ldots\}$. The idea behind the extension is that if n is a natural number and if $-n < x < -n+1$

then $0 < x + n < 1$ and thus $\Gamma(x + n)$ is already defined. Using Eq. (8.4) then $\Gamma(x)$ is defined to be

$$\Gamma(x) = \frac{\Gamma(x + n)}{x(x + 1) \cdots (x + n - 1)} = \frac{\Gamma(x + n)}{(x)_n}.$$

This extended form of the Gamma function is also continuous on the set $\{x \mid x \neq 0, -1, -2, \ldots\}$. The following theorem describes the behavior of the Gamma function near the non-positive integers.

Theorem 8.1. *If $n = 0, -1, -2, \ldots$ then $\lim_{x \to n} |\Gamma(x)| = \infty$.*

Proof. By Lemma 8.1, $\Gamma(x + 1) = x\Gamma(x)$ for $x > 0$, thus

$$\Gamma(x) = \frac{\Gamma(x + 1)}{x}$$

and since the Gamma function is continuous except at the non-positive integers,

$$\lim_{x \to 0} |\Gamma(x)| = \lim_{x \to 0} \frac{|\Gamma(x + 1)|}{|x|} = \frac{|\Gamma(\lim_{x \to 0}(x + 1))|}{\lim_{x \to 0} |x|} = \infty.$$

Suppose the claim has been proved for $n = -m + 1$ with $m \in \mathbb{N}$ then

$$\lim_{x \to -m} |\Gamma(x)| = \lim_{x \to -m} \frac{|\Gamma(x + 1)|}{|x|} = \frac{\lim_{x \to -m+1} |\Gamma(x)|}{m} = \infty.$$

Hence the assertion is true by the principle of mathematical induction. $\square$

Since $|\Gamma(x)| \to \infty$ as $x \to 0, -1, -2, \ldots$ then define the reciprocal of the Gamma function at $x = 0, -1, -2, \ldots$ to be 0. Thus $1/\Gamma(x)$ is continuous for all x and $1/\Gamma(x) = 0$ at $x = 0, -1, -2, \ldots$.

Define $f(x) = \ln |\Gamma(x)|$ then

$$f'(x) = \frac{\Gamma'(x)}{\Gamma(x)} \iff \Gamma'(x) = \Gamma(x)f'(x).$$

By convention $f'(x)$ is called the **digamma function** (some authors call it the **polygamma function**) and it is denoted as $\psi(x)$. Adopting this notation allows the derivative of the Gamma function to be written as $\Gamma'(x) = \psi(x)\Gamma(x)$.

The Gamma function will play a role in the description and expression of the eigenfunctions of Sturm-Liouville boundary value problems explored in the remainder of this chapter.

8.2   Bessel's Equation of Order p

Bessel's equation is an ordinary differential equation which often arises in the study of vibration in polar and cylindrical coordinates. Consider Bessel's equation of order p:

$$x^2 y'' + x y' + (x^2 - p^2)y = 0 \text{ for } x > 0, \tag{8.7}$$

where $p \geq 0$ is a constant. Dividing Eq. (8.7) by x, Bessel's equation can be written in the self-adjoint form as

$$[x\,y']' + \left(x - \frac{p^2}{x}\right) y = 0 \text{ for } x > 0.$$

This is a singular Sturm-Liouville boundary value problem. If $P(x) = x^2$, $Q(x) = x$, and $R(x) = x^2 - p^2$, then $P(0) = 0$ (and thus $x_0 = 0$ is a singular point) with

$$\lim_{x \to 0} \frac{xQ(x)}{P(x)} = \lim_{x \to 0} \frac{(x)(x)}{x^2} = \lim_{x \to 0} 1 = 1 = p_0,$$

$$\lim_{x \to 0} \frac{x^2 R(x)}{P(x)} = \lim_{x \to 0} \frac{x^2(x^2 - p^2)}{x^2} = \lim_{x \to 0}(x^2 - p^2) = -p^2 = q_0,$$

and hence $x_0 = 0$ is a regular singular point.

The indicial equation will have the form,

$$F(r) = r(r - 1) + p_0 r + q_0 = r(r - 1) + r - p^2 = r^2 - p^2.$$

Thus the exponents of singularity are $r_1 = p$ and $r_2 = -p$. Throughout this section it will be important to remember that the exponents of singularity differ by $2p \geq 0$.

The goal is to find two linearly independent solutions to Eq. (8.7). It is always possible to find a series solution corresponding to the larger of the two exponents of singularity, so there exists a solution of the form $y_1(x) = \sum_{k=0}^{\infty} a_k x^{k+p}$.

To use the Method of Frobenius, assume $y(x) = \sum_{k=0}^{\infty} a_k x^{k+r}$, differentiate formally with respect to x, substitute the result into Eq. (8.7), and re-index and combine the series to yield

$$0 = x^2 \sum_{k=0}^{\infty} (k+r)(k+r-1)a_k x^{k+r-2} + x \sum_{k=0}^{\infty} (k+r)a_k x^{k+r-1}$$

$$+ (x^2 - p^2) \sum_{k=0}^{\infty} a_k x^{k+r}$$

$$= \sum_{k=0}^{\infty} [(k+r)(k+r-1) + (k+r) - p^2]a_k x^{k+r} + \sum_{k=0}^{\infty} a_k x^{k+r+2}$$

$$= \sum_{k=0}^{\infty} [(k+r)^2 - p^2]a_k x^{k+r} + \sum_{k=0}^{\infty} a_k x^{k+r+2}$$

Using the definition of the indicial function F and re-indexing the second infinite series enable the last equation to be re-written as

$$0 = \sum_{k=0}^{\infty} F(r+k)a_k x^{k+r} + \sum_{k=2}^{\infty} a_{k-2} x^{k+r}$$

$$= F(r)a_0 x^r + F(r+1)a_1 x^{r+1} + \sum_{k=2}^{\infty} [F(r+k)a_k + a_{k-2}]x^{k+r}. \quad (8.8)$$

Since $r = -p$ or $r = p$ then $F(r) = 0$ and for arbitrary a_0 the first term above is zero. The second term has coefficient $a_1 F(r+1) = a_1((1+r)^2 - p^2)$ which must match the coefficient of x^{r+1} on the left-hand side of the equation. This can be achieved by choosing $a_1 = 0$ regardless of the values of p or r. For $k \geq 2$ the following recurrence relation holds for any r such that $F(k+r) \neq 0$,

$$a_k = -\frac{a_{k-2}}{F(k+r)}.$$

This equation is valid as long as r is the larger of the two exponents of singularity $(r = p)$. Since $a_1 = 0$ then $a_{2k-1} = 0$ for $k \in \mathbb{N}$. Since $F(k+p) \neq 0$ for $k \in \mathbb{N}$, the recurrence relation can be simplified to

$$a_k = -\frac{a_{k-2}}{k(k+2p)}.$$

Since a_0 is an arbitrary constant, then

$$a_2 = -\frac{a_0}{2(2+2p)} = \frac{(-1)^1 a_0}{2^2(1+p)}$$

$$a_4 = -\frac{a_2}{4(4+2n)} = \frac{(-1)^2 a_0}{2^4 2!(1+p)(2+p)}$$

$$\vdots$$

$$a_{2k} = \frac{(-1)^k a_0}{2^{2k} k!(1+p)(2+p)\cdots(k+p)}.$$

Thus one solution to Bessel's equation of order p has the form,

$$y_1(x) = a_0 \sum_{k=0}^{\infty} \frac{(-1)^k}{2^{2k} k! (1+p)(2+p) \cdots (k+p)} x^{2k+p}.$$

Traditionally the first solution to Bessel's equation of order p has the arbitrary constant a_0 chosen as

$$a_0 = \frac{1}{2^p \Gamma(p+1)}.$$

This defines the **Bessel function of the first kind of order** p:

$$J_p(x) = \sum_{k=0}^{\infty} \frac{(-1)^k}{2^{2k} 2^p k! \Gamma(1+p)(1+p)(2+p) \cdots (k+p)} x^{2k+p}$$

$$= \sum_{k=0}^{\infty} \frac{(-1)^k}{k! \Gamma(k+p+1)} \left(\frac{x}{2}\right)^{2k+p}. \tag{8.9}$$

Note that when $p = n \in \{0, 1, \dots\}$ then $J_n(x)$ is a power series,

$$J_n(x) = \sum_{k=0}^{\infty} \frac{(-1)^k}{k!(k+n)!} \left(\frac{x}{2}\right)^{2k+n}.$$

The functions $J_p(x)$ for $p = 0, 1/2, 1, 3/2$, and 2 are graphed in Fig. 8.1.

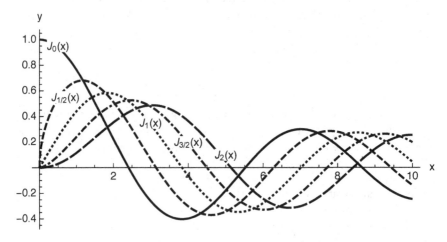

Fig. 8.1 The graphs of $J_p(x)$ for $p = 0, 1/2, 1, 3/2$, and 2. Each of these Bessel functions is bounded at $x = 0$.

A second linearly independent solution to Bessel's equation is still needed to form a fundamental set of solutions to the ordinary differential equation in Eq. (8.7). Depending on the quantity $2p$ (the difference of the exponents of singularity of Bessel's equation of order p), this can require significant effort. The simplest case is that for which $2p$ is not an integer. In this situation, replacing r in the recurrence relation by the other exponent of singularity $-p$, yields another solution

$$J_{-p}(x) = \sum_{k=0}^{\infty} \frac{(-1)^k}{k! \Gamma(k - p + 1)} \left(\frac{x}{2}\right)^{2k-p}.$$

Note that when $2p$ is not an integer, $J_p(0) = 0$ while $J_{-p}(x)$ is unbounded as $x \to 0$. Thus the two solutions are linearly independent. Once $J_p(x)$ and $J_{-p}(x)$ are found, by convention the second solution to Bessel's equation of order p, linearly independent of $J_p(x)$ is chosen to be a linear combination of $J_p(x)$ and $J_{-p}(x)$, namely

$$Y_p(x) = \frac{J_p(x) \cos(p\pi) - J_{-p}(x)}{\sin(p\pi)}. \tag{8.10}$$

The function $Y_p(x)$ is called a **Bessel function of the second kind of order** p. The denominator of this function is non-zero since $2p$ is not an integer. Since $Y_p(x)$ is a linear combination of the solutions to Bessel's equation of order p, then $Y_p(x)$ itself is a solution to Bessel's equation of order p. Since $J_{-p}(x)$ is linearly independent of $J_p(x)$ when $2p$ is not an integer, then $Y_p(x)$ is linearly independent of $J_p(x)$. Therefore the general solution of Bessel's equation of order p (when $2p$ is not an integer) is

$$y(x) = c_1 J_p(x) + c_2 Y_p(x),$$

where c_1 and c_2 are constants.

Even though finding a pair of linearly independent solutions to Bessel's equation of order p may require a great deal of effort in general when $2p$ is an integer, this is not always the case as will be shown in the next example.

Example 8.1. Find a pair of linearly independent solutions to Bessel's equation of order $p = 1/2$.

Solution. In this case the indicial function is $F(r) = r^2 - 1/4$ and the exponents of singularity are $r = \pm 1/2$. The reader should note that $F(\pm 1/2) = 0$ (as is always the case) and that $F(-\frac{1}{2} + 1) = 0$ and thus the first two terms on the right-hand side of Eq. (8.8) (reproduced below) vanish for arbitrary a_0 and a_1 when $r = -1/2$.

$$0 = F(r)a_0 x^r + F(r+1)a_1 x^{r+1} + \sum_{k=2}^{\infty} [F(r+k)a_k + a_{k-2}]x^{k+r}.$$

For $r = -1/2$ the recurrence relation can be written as

$$a_k = -\frac{a_{k-2}}{k(k-1)} \text{ for } k \geq 2.$$

Since a_1 has been chosen to be 0, all the odd-indexed coefficients in the infinite series are zero. Allowing a_0 to remain arbitrary for the moment and using the recurrence relation result in

$$a_2 = -\frac{a_0}{2!}, \quad a_4 = \frac{a_0}{4!}, \quad \ldots, \quad a_{2k} = \frac{(-1)^k a_0}{(2k)!}.$$

Thus a solution to Bessel's equation of order $1/2$ corresponding to the exponent of singularity $r = -1/2$ is

$$y_2(x) = a_0 \sum_{k=0}^{\infty} \frac{(-1)^k}{(2k)!} x^{2k-1/2} = \frac{a_0}{\sqrt{x}} \cos x.$$

If $a_0 = (2/\pi)^{1/2}$ then the function $J_{-1/2}(x)$ is defined as

$$J_{-1/2}(x) = \sqrt{\frac{2}{\pi x}} \cos x \text{ for } x > 0.$$

Another solution to Bessel's equation of order $1/2$ linearly independent from $J_{-1/2}(x)$ is, of course, $J_{1/2}(x)$ given in Eq. (8.9) when $p = 1/2$.

Next attention is given to the case in which $p = n$ is an integer but $p \neq 0$. According to Theorem 5.6.1 of [Boyce and DiPrima (2012)], a second solution of Bessel's equation of order n will have the form

$$y_2(x) = aJ_n(x)\ln x + u(x)$$

where

$$u(x) = x^{-n}\left(1 + \sum_{k=1}^{\infty} c_k x^k\right).$$

The coefficients a and c_k can be found by direct substitution of this function into Bessel's equation. Note that

$$y_2'(x) = aJ_n'(x)\ln x + \frac{a}{x}J_n(x) + u'(x),$$

$$y_2''(x) = aJ_n''(x)\ln x + \frac{2a}{x}J_n'(x) - \frac{a}{x^2}J_n(x) + u''(x).$$

Substituting these expressions into Bessel's equation of order n produces

$$\begin{aligned}
0 &= ax^2 J_n''(x)\ln x + 2axJ_n'(x) - aJ_n(x) + x^2 u''(x) \\
&\quad + axJ_n'(x)\ln x + aJ_n(x) + xu'(x) \\
&\quad + (x^2 - n^2)aJ_n(x)\ln x + (x^2 - n^2)u(x) \\
&= a\ln x\left[x^2 J_n(x) + xJ_n'(x) + (x^2 - n^2)J_n(x)\right] + 2axJ_n'(x) \\
&\quad + x^2 u''(x) + xu'(x) + (x^2 - n^2)u(x)
\end{aligned}$$

$$-2axJ_n'(x) = x^2 u''(x) + xu'(x) + (x^2 - n^2)u(x). \qquad (8.11)$$

Substituting the assumed form of function u in Eq. (8.11) produces

$$-2axJ_n'(x) = n(n+1)x^{-n}\left(1 + \sum_{k=1}^{\infty} c_k x^k\right) - 2nx^{-n+1}\sum_{k=1}^{\infty} kc_k x^{k-1}$$

$$+ x^{-n+2}\sum_{k=2}^{\infty} k(k-1)c_k x^{k-2} + x^{-n+1}\sum_{k=1}^{\infty} kc_k x^{k-1}$$

$$- nx^{-n}\left(1 + \sum_{k=1}^{\infty} c_k x^k\right) + (x^2 - n^2)x^{-n}\left(1 + \sum_{k=1}^{\infty} c_k x^k\right)$$

$$= (1-2n)\sum_{k=1}^{\infty} kc_k x^{k-n} + \sum_{k=1}^{\infty} k(k-1)c_k x^{k-n}$$

$$+ x^{2-n}\left(1 + \sum_{k=1}^{\infty} c_k x^k\right)$$

$$= \sum_{k=1}^{\infty} k(k-2n)c_k x^{k-n} + x^{2-n} + \sum_{k=1}^{\infty} c_k x^{k-n+2}.$$

Re-indexing the power series on the right-hand side of the equation yields

$$-2axJ_n'(x) = \sum_{k=0}^{\infty}(k+1)(k+1-2n)c_{k+1}x^{k-n+1} + x^{2-n} + \sum_{k=2}^{\infty} c_{k-1}x^{k-n+1}$$

$$= (1-2n)c_1 x^{1-n} + [4(1-n)c_2 + 1]x^{2-n}$$

$$+ \sum_{k=2}^{\infty}[(k+1)(k+1-2n)c_{k+1} + c_{k-1}]x^{k-n+1}.$$

If the power series form of $J_n(x)$ is used, the last equation simplifies to

$$-2a\sum_{k=0}^{\infty}\frac{(-1)^k(2k+n)}{k!(k+n)!}\left(\frac{x}{2}\right)^{2k+n}$$

$$= (1-2n)c_1 x^{1-n} + [4(1-n)c_2 + 1]x^{2-n}$$

$$+ \sum_{k=2}^{\infty}[(k+1)(k+1-2n)c_{k+1} + c_{k-1}]x^{k-n+1}.$$

The lowest power of x in the series on the left-hand side of the equation is x^n while the smallest power of x on the right-hand side of the equation is x^{1-n}. Thus it is helpful to break up the infinite series on the right-hand side of the equation and re-write the equation as

$$-2a \sum_{k=0}^{\infty} \frac{(-1)^k (2k+n)}{2^{2k+n} k! (k+n)!} x^{2k+n}$$

$$= (1-2n)c_1 x^{1-n} + [4(1-n)c_2 + 1] x^{2-n}$$

$$+ \sum_{k=2}^{2n-2} [(k+1)(k+1-2n)c_{k+1} + c_{k-1}] x^{k-n+1}$$

$$+ \sum_{k=2n-1}^{\infty} [(k+1)(k+1-2n)c_{k+1} + c_{k-1}] x^{k-n+1}. \qquad (8.12)$$

The coefficients of the first two terms and terms in the finite summation on the right-hand side of Eq. (8.12) must vanish. Thus the following system of equations must be solved:

$$(1-2n)c_1 = 0$$

$$2(2-2n)c_2 + 1 = 0$$

$$3(3-2n)c_3 + c_1 = 0$$

$$\vdots$$

$$(2n-2)(-2)c_{2n-2} + c_{2n-4} = 0$$

$$(2n-1)(-1)c_{2n-1} + c_{2n-3} = 0.$$

Therefore $c_1 = 0$ and as a result, all of the odd-indexed coefficients c_{2k+1} are zero. From the system of equations developed it can be seen that $c_2 = 1/(2(2n-2))$ and thus

$$c_{2k} = \frac{(n-k-1)!}{2^{2k} k! (n-1)!}$$

for $k = 1, 2, \ldots, n-1$.

Equating the coefficients of x^n on the left- and right-hand sides of Eq. (8.12) produces the equation,

$$-\frac{a}{2^{n-1}(n-1)!} = c_{2n-2} \iff a = \frac{-1}{2^{n-1}(n-1)!}.$$

This value of a can be used to determine the remaining coefficients in the series for y_2. Substituting the expression for a in the left-hand side of Eq. (8.12) and re-indexing the summations on the right-hand side so that only the even-indexed coefficients c_k are included produce the new equation below.

$$\sum_{k=0}^{\infty} \frac{(-1)^k (2k+n) x^{2k+n}}{2^{2(k+n-1)} k! (k+n)! (n-1)!} = \sum_{k=1}^{\infty} [2k(2k-2n)c_{2k} + c_{2k-2}] x^{2k-n}$$

$$(8.13)$$

Using the values previously determined for c_{2k} for $k = 1, 2, \ldots, n - 1$ the first $n - 1$ terms of the series on the right-hand side of Eq. (8.13) are zero. Thus this equation can be re-written as

$$\sum_{k=0}^{\infty} \frac{(-1)^k (n + 2k) x^{n+2k}}{2^{2(n+k-1)} k! (n + k)! (n - 1)!}$$

$$= \sum_{k=n}^{\infty} \left[2k(2k - 2n) c_{2k} + c_{2k-2} \right] x^{2k-n}$$

$$= \sum_{k=0}^{\infty} \left[2k(2n + 2k) c_{2n+2k} + c_{2n+2k-2} \right] x^{n+2k} \qquad (8.14)$$

after one more re-indexing of the series on the right-hand side. Equating powers of x on both sides of Eq. (8.14) produces the recurrence relation

$$c_{2n+2k} = \frac{(-1)^k \left(\frac{1}{k} + \frac{1}{n+k} \right)}{2^{2(k+n)} k! (k + n)! (n - 1)!} - \frac{c_{2n+2k-2}}{2k(2n + 2k)}, \qquad (8.15)$$

which holds for $k \in \mathbb{N}$. The coefficient c_{2n} is arbitrary but will be chosen to be

$$c_{2n} = \frac{H_n}{2^{2n} n! (n - 1)!}$$

where $H_n \equiv 1 + 1/2 + \cdots + 1/n$ is called the nth **Harmonic number**.

Lemma 8.5. *For $k \in \mathbb{N}$, the solution to the recurrence relation in Eq. (8.15) is*

$$c_{2n+2k} = \frac{(-1)^k (H_k + H_{n+k})}{2^{2n+2k} k! (n + k)! (n - 1)!}. \qquad (8.16)$$

Proof. When $k = 1$ then

$$c_{2n+2} = \frac{-\left(1 + \frac{1}{n+1} \right)}{2^{2(n+1)} 1! (n + 1)! (n - 1)!} - \frac{c_{2n}}{2(2n + 2)}$$

$$= \frac{-\left(1 + \frac{1}{n+1} \right)}{2^{2(n+1)} (n + 1)! (n - 1)!} - \frac{H_n}{2(2n + 2) 2^{2n} n! (n - 1)!}$$

$$= \frac{-\left(1 + \frac{1}{n+1} \right)}{2^{2(n+1)} (n + 1)! (n - 1)!} - \frac{H_n}{2^{2(n+1)} (n + 1)! (n - 1)!}$$

$$= \frac{-(H_1 + H_{n+1})}{2^{2n+2} 1! (n + 1)! (n - 1)!}.$$

Thus the formula holds for $k = 1$. Suppose the result is true for some $m \in \mathbb{N}$ then according to the recurrence relation

$$
\begin{aligned}
c_{2n+2(m+1)} &= \frac{(-1)^{m+1}\left(\frac{1}{m+1} + \frac{1}{n+m+1}\right)}{2^{2(n+m+1)}(m+1)!(n+m+1)!(n-1)!} \\
&\quad - \frac{c_{2n+2m}}{2(m+1)(2n+2(m+1))} \\
&= \frac{(-1)^{m+1}\left(\frac{1}{m+1} + \frac{1}{n+m+1}\right)}{2^{2(n+m+1)}(m+1)!(n+m+1)!(n-1)!} \\
&\quad - \frac{(-1)^m(H_m + H_{n+m})}{2^{2(n+m+1)}(m+1)!(n+m+1)!(n-1)!} \\
&= \frac{(-1)^{m+1}(H_{m+1} + H_{n+m+1})}{2^{2(n+m+1)}(m+1)!(n+m+1)!(n-1)!}.
\end{aligned}
$$

Thus Eq. (8.16) holds for $m+1$ as well. Using the principle of mathematical induction, the formula holds for all natural numbers k. $\qquad\square$

The above provides a second solution to Bessel's equation of order $n \in \mathbb{N}$ expressed as

$$
\begin{aligned}
y_2(x) &= \frac{-\ln x}{2^{n-1}(n-1)!}J_n(x) + x^{-n}\left(1 + \sum_{k=1}^{n-1}\frac{(n-k-1)!}{k!(n-1)!}\left[\frac{x}{2}\right]^{2k}\right) \\
&\quad + x^{-n}\left(\sum_{k=0}^{\infty}\frac{(-1)^k(H_k + H_{n+k})}{k!(n+k)!(n-1)!}\left[\frac{x}{2}\right]^{2n+2k}\right).
\end{aligned}
$$

This function is linearly independent of $J_n(x)$. Since Bessel's equation is linear and homogeneous a constant multiple of $y_2(x)$ is a solution to Bessel's equation and is also linearly independent of $J_n(x)$. By convention the second linearly independent solution to Bessel's equation of order $n \in \mathbb{N}$ is expressed as

$$
\begin{aligned}
Y_n(x) &= \frac{2}{\pi}\left((\gamma - \ln 2)J_n(x) - 2^{n-1}(n-1)!y_2(x)\right) \\
&= \frac{2}{\pi}\left(\gamma + \ln\frac{x}{2}\right)J_n(x) - \frac{1}{\pi}\sum_{k=0}^{n-1}\frac{(n-k-1)!}{k!}\left(\frac{x}{2}\right)^{2k-n} \\
&\quad - \frac{1}{\pi}\sum_{k=0}^{\infty}\frac{(-1)^k(H_k + H_{n+k})}{k!(n+k)!}\left(\frac{x}{2}\right)^{2k+n}.
\end{aligned}
$$

The constant γ is called the **Euler-Máscheroni**[3] **constant** or merely the **Euler gamma**. It is defined to be

$$\gamma \equiv \lim_{n \to \infty} (H_n - \ln n) \approx 0.577216.$$

The final case to consider in this section is that of Bessel's equation of order 0. In this case both exponents of singularity are 0. Starting as before and assuming the second solution takes the form,

$$y_2(x) = (\ln x)J_0(x) + \sum_{k=1}^{\infty} c_k x^k$$

then

$$y_2'(x) = J_0'(x) \ln x + \frac{1}{x}J_0(x) + \sum_{k=1}^{\infty} k c_k x^{k-1},$$

$$y_2''(x) = J_0''(x) \ln x + \frac{2}{x}J_0'(x) - \frac{1}{x^2}J_0(x) + \sum_{k=1}^{\infty} k(k-1) c_k x^{k-2}.$$

Substituting these expressions into Bessel's equation of order 0 yields

$$0 = x^2 \left(J_0''(x) \ln x + \frac{2}{x}J_0'(x) - \frac{1}{x^2}J_0(x) + \sum_{k=1}^{\infty} k(k-1) c_k x^{k-2} \right)$$

$$+ x \left(J_0'(x) \ln x + \frac{1}{x}J_0(x) + \sum_{k=1}^{\infty} k c_k x^{k-1} \right) + x^2 \left(J_0(x) \ln x + \sum_{k=1}^{\infty} c_k x^k \right)$$

$$= (x^2 J_0''(x) + x J_0'(x) + x^2 J_0(x)) \ln x + 2x J_0'(x)$$

$$+ \sum_{k=1}^{\infty} k(k-1) c_k x^k + \sum_{k=1}^{\infty} k c_k x^k + \sum_{k=1}^{\infty} c_k x^{k+2}.$$

Solving for the term involving $J_0'(x)$ yields

$$-2x J_0'(x) = \sum_{k=1}^{\infty} k^2 c_k x^k + \sum_{k=1}^{\infty} c_k x^{k+2}$$

$$-2\sum_{k=1}^{\infty} \frac{(-1)^k (2k)}{(k!)^2} \left(\frac{x}{2} \right)^{2k} = c_1 x + 4c_2 x^2 + \sum_{k=3}^{\infty} (k^2 c_k + c_{k-2}) x^k. \qquad (8.17)$$

Since the left-hand side of Eq. (8.17) contains only even powers of x then $c_{2k-1} = 0$ for $k \in \mathbb{N}$. Consequently the infinite series on the right-hand side of Eq. (8.17) is re-indexed to retain only the even powers of x.

$$-2\sum_{k=1}^{\infty} \frac{(-1)^k (2k)}{(k!)^2} \left(\frac{x}{2} \right)^{2k} = \sum_{k=1}^{\infty} ((2k)^2 c_{2k} + c_{2k-2}) x^{2k} \qquad (8.18)$$

---

[3]Lorenzo Máscheroni, Italian mathematician (1750–1800).

Equation (8.18) introduced the coefficient c_0 not present in the presumed solution, thus c_0 can be chosen to be 0. Equating coefficients of x^{2k} on both sides of Eq. (8.18) yields the following recurrence relation.

$$c_{2k} = \frac{(-1)^{k+1}}{2^{2k}(k!)^2 k} - \frac{c_{2k-2}}{(2k)^2}. \tag{8.19}$$

Letting $k = 1$ and using the assumption regarding $c_0 = 0$ then $c_2 = 1/4$.

Lemma 8.6. *For $k \geq 1$ the solution to the recurrence relation in Eq. (8.19) is*

$$c_{2k} = \frac{(-1)^{k+1} H_k}{2^{2k}(k!)^2}.$$

Proof. The formula has already been verified when $k = 1$. Suppose the formula holds for some $m \in \mathbb{N}$, then

$$
\begin{aligned}
c_{2(m+1)} &= \frac{(-1)^{m+1+1}}{2^{2(m+1)}((m+1)!)^2(m+1)} - \frac{c_{2(m+1)-2}}{(2(m+1))^2} \\
&= \frac{(-1)^{m+2}\frac{1}{m+1}}{2^{2(m+1)}((m+1)!)^2} - \frac{(-1)^{m+1} H_m}{2^{2m}(m!)^2(2(m+1))^2} \\
&= \frac{(-1)^{m+2} H_{m+1}}{2^{2(m+1)}((m+1)!)^2}.
\end{aligned}
$$

Thus the formula holds for $m + 1$ and by the principle of mathematical induction the lemma has been proved. $\square$

For $n = 0$ a second solution to Bessel's equation can be written as

$$y_2(x) = (\ln x) J_0(x) - \sum_{k=1}^{\infty} \frac{(-1)^k H_k}{(k!)^2} \left(\frac{x}{2}\right)^{2k}.$$

Just as was the case for $n \in \mathbb{N}$, by convention the Bessel function of the second kind of order 0 is expressed as

$$
\begin{aligned}
Y_0(x) &= \frac{2}{\pi} \left((\gamma - \ln 2) J_0(x) + y_2(x)\right) \\
&= \frac{2}{\pi} \left(\gamma + \ln \frac{x}{2}\right) J_0(x) - \frac{2}{\pi} \sum_{k=1}^{\infty} \frac{(-1)^k H_k}{(k!)^2} \left(\frac{x}{2}\right)^{2k}.
\end{aligned}
$$

The graph in Fig. 8.2 show the plots of several Bessel functions of the second kind. Note that all of the Bessel functions of the second kind are unbounded as $x \to 0^+$.

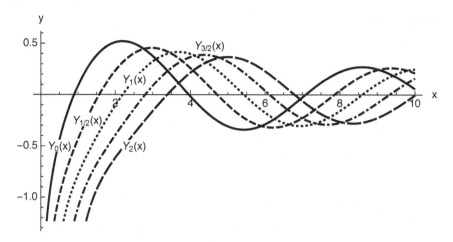

Fig. 8.2 The graphs of the Bessel functions of the second kind $Y_p(x)$ for $p = 0$, $1/2$, 1, $3/2$, and 2. The Bessel functions of the second kind are unbounded as $x \to 0^+$.

8.2.1 Properties of the Bessel Functions

Unless otherwise noted, the properties described below hold for all Bessel functions of the first kind of order p.

Theorem 8.2. *The Bessel functions of the first kind of order p satisfy the following derivative relationships:*

$$\frac{d}{dx}\left[x^{-p}J_p(x)\right] = -x^{-p}J_{p+1}(x),$$
$$\frac{d}{dx}\left[x^p J_p(x)\right] = x^p J_{p-1}(x).$$

Proof. Recall that,

$$J_p(x) = \sum_{k=0}^{\infty} \frac{(-1)^k}{k!\Gamma(k+p+1)} \left(\frac{x}{2}\right)^{2k+p}.$$

Using the product rule for derivatives and differentiating the series term by term produce

$$\frac{d}{dx}\left[x^{-p}J_p(x)\right]$$

$$= x^{-p}J_p'(x) - px^{-p-1}J_p(x)$$

$$= \frac{x^{-p}}{2}\sum_{k=0}^{\infty}\frac{(-1)^k(2k+p)}{k!\Gamma(k+p+1)}\left(\frac{x}{2}\right)^{2k+p-1} - x^{-p-1}\sum_{k=0}^{\infty}\frac{(-1)^k p}{k!\Gamma(k+p+1)}\left(\frac{x}{2}\right)^{2k+p}$$

$$= x^{-p}\sum_{k=1}^{\infty}\frac{(-1)^k k}{k!\Gamma(k+p+1)}\left(\frac{x}{2}\right)^{2k+p-1}$$

$$= -x^{-p}\sum_{k=0}^{\infty}\frac{(-1)^k}{k!\Gamma(k+p+2)}\left(\frac{x}{2}\right)^{2k+p+1} = -x^{-p}J_{p+1}(x).$$

To establish the second relationship the factor x^p is distributed over the terms of the infinite series and then the result is differentiated term by term. This avoids the use of the product rule for derivatives.

$$\frac{d}{dx}\left[x^p J_p(x)\right] = \frac{d}{dx}\left[\sum_{k=0}^{\infty}\frac{(-1)^k}{2^{2k+p}k!\Gamma(k+p+1)}x^{2k+2p}\right]$$

$$= \sum_{k=0}^{\infty}\frac{(-1)^k(2k+2p)}{2^{2k+p}k!\Gamma(k+p+1)}x^{2k+2p-1}$$

$$= x^p\sum_{k=0}^{\infty}\frac{(-1)^k}{k!\Gamma(k+p)}\left(\frac{x}{2}\right)^{2k+p-1} = x^p J_{p-1}(x).$$

Thus the derivative relationships are established.   □

Corollary 8.1. *The Bessel functions of the first kind obey the following recurrence relations.*

$$xJ_p'(x) - pJ_p(x) = -xJ_{p+1}(x) \tag{8.20}$$

$$xJ_p'(x) + pJ_p(x) = xJ_{p-1}(x) \tag{8.21}$$

$$J_{p+1}(x) + J_{p-1}(x) = \frac{2p}{x}J_p(x). \tag{8.22}$$

Proof. Theorem 8.2 implies that

$$-px^{-p-1}J_p(x) + x^{-p}J_p'(x) = -x^{-p}J_{p+1}(x).$$

Multiplying both sides of the equation by x^{p+1} yields

$$-pJ_p(x) + xJ_p'(x) = -xJ_{p+1}(x).$$

It is also the case that

$$px^{p-1}J_p(x) + x^p J_p'(x) = x^p J_{p-1}(x), \tag{8.23}$$

which when divided by x^{p-1} produces the result shown in Eq. (8.21). Adding Eqs. (8.20) and (8.21) and dividing by $2x$ yield

$$J'_p(x) = \frac{1}{2}\left(J_{p-1}(x) - J_{p+1}(x)\right).$$

Replacing $J'_p(x)$ with this expression in Eq. (8.23), dividing both sides of the resulting equation by $x^p/2$ and rearranging terms produce Eq. (8.22). $\square$

Each Bessel function has an infinite number of positive roots which will play an important role in the orthogonality of the Bessel functions. This property will be established in two steps: first, by examining the roots of $J_0(x)$ and second, by building the properties of the roots of $J_n(x)$ for $n \in \mathbb{N}$ from this foundation.

Lemma 8.7. *The positive zeros of $J_0(x)$ form a strictly increasing sequence which tends to positive infinity.*

Proof. Define the function $u(x) \equiv \sqrt{x}J_0(x)$, then

$$J_0(x) = \frac{u(x)}{\sqrt{x}},$$

$$J'_0(x) = \frac{u'(x)}{\sqrt{x}} - \frac{u(x)}{2x^{3/2}}, \text{ and}$$

$$J''_0(x) = \frac{u''(x)}{\sqrt{x}} - \frac{u'(x)}{x^{3/2}} + \frac{3u(x)}{4x^{5/2}}.$$

Substituting these expressions into Bessel's equation of order zero gives

$$0 = x^2\left(\frac{u''(x)}{\sqrt{x}} - \frac{u'(x)}{x^{3/2}} + \frac{3u(x)}{4x^{5/2}}\right) + x\left(\frac{u'(x)}{\sqrt{x}} - \frac{u(x)}{2x^{3/2}}\right) + x^2\left(\frac{u(x)}{\sqrt{x}}\right)$$

$$= x^{3/2}u''(x) + x^{3/2}u(x) + \frac{u(x)}{4\sqrt{x}}.$$

Therefore $u(x)$ is a solution to the second order linear differential equation,

$$u''(x) + \left(1 + \frac{1}{4x^2}\right)u(x) = 0.$$

If both sides of this differential equation are multiplied by $\sin x$ and the terms are re-arranged, then

$$\frac{u(x)\sin x}{4x^2} = -u''(x)\sin x - u(x)\sin x = \frac{d}{dx}\left[u(x)\cos x - u'(x)\sin x\right].$$

Integrating both sides of this equation over the interval $[2k\pi, (2k+1)\pi]$ with $k \in \{0, 1, 2, \ldots\}$ yields

$$\int_{2k\pi}^{(2k+1)\pi} \frac{u(x)\sin x}{4x^2}\,dx = -u(2k\pi) - u((2k+1)\pi).$$

Thus the function $u(x)$ must have at least one zero in every interval of the form $[2k\pi, (2k+1)\pi]$ for $k = 0, 1, \ldots$. If not then $u(x)$ strictly maintains a single sign on each such interval. If $u(x) > 0$ for $2k\pi \leq x \leq (2k+1)\pi$ the definite integral on the left-hand side of the equation is positive, while the expression on the right-hand side of the equation is negative, a contradiction. Assuming $u(x) < 0$ for $2k\pi \leq x \leq (2k+1)\pi$ results in a similar contradiction. Since $u(x) = \sqrt{x}J_0(x)$ and $\sqrt{x} > 0$ for $x > 0$ then $J_0(x)$ must have a zero in every interval of the form $[2k\pi, (2k+1)\pi]$. Therefore there exist infinitely many zeros of $J_0(x)$. Since the positive zeros of $J_0(x)$ are isolated by Theorem 7.4, they can be arranged in increasing order as

$$0 < \lambda_{0,1} < \lambda_{0,2} < \cdots < \lambda_{0,k} < \cdots,$$

where the kth positive zero of $J_0(x)$ is denoted as $\lambda_{0,k}$. Since $2k\pi \leq \lambda_{0,k}$ then certainly, $\lim_{k \to \infty} \lambda_{0,k} = \infty$. $\square$

Since the property has been established for the function $J_0(x)$, it is possible to use it to show that the zeros of $J_n(x)$ (for $n \in \mathbb{N}$) form an increasing sequence which tends to positive infinity. The zeros of $J_n(x)$ (if any) are denoted as $\lambda_{n,k}$ for $k \in \mathbb{N}$.

Theorem 8.3. *Let n be a nonnegative integer and suppose $\{\lambda_{n,k}\}_{k=1}^{\infty}$ are the positive zeros of $J_n(x)$ arranged in increasing order. Between each consecutive pair of zeros of $J_n(x)$ there exists a unique zero of $J_{n+1}(x)$.*

Proof. This proof will proceed by induction on n. It was shown in Lemma 8.7 that $J_0(x)$ has an increasing sequence of positive zeros which limits on infinity. Let $\lambda_{0,i}$ and $\lambda_{0,i+1}$ be a pair of consecutive zeros of $J_0(x)$. By Theorem 8.2,

$$\int_{\lambda_{0,i}}^{\lambda_{0,i+1}} J_1(x)\,dx = [-J_0(x)]_{\lambda_{0,i}}^{\lambda_{0,i+1}} = 0,$$

which implies $J_1(x)$ has at least one zero in $(\lambda_{0,i}, \lambda_{0,i+1})$, call it $\lambda_{1,i}$. Suppose $J_1(x)$ has another zero in $(\lambda_{0,i}, \lambda_{0,i+1})$, call it $\lambda_{1,j}$, then again by Theorem 8.2,

$$\int_{\lambda_{1,i}}^{\lambda_{1,j}} xJ_0(x)\,dx = [x\,J_1(x)]_{\lambda_{1,i}}^{\lambda_{1,j}} = 0,$$

which implies $J_0(x)$ has a zero in $(\lambda_{1,i}, \lambda_{1,j})$ contradicting the assumption that $\lambda_{0,i}$ and $\lambda_{0,i+1}$ are consecutive zeros. Thus $\lambda_{1,i}$ is the unique positive zero of $J_1(x)$ between the positive zeros, $\lambda_{0,i}$ and $\lambda_{0,i+1}$ of $J_0(x)$. As a result $\lim_{k \to \infty} \lambda_{1,k} = \infty$.

Now suppose the theorem has been proved for an integer $n > 0$, then by the principle of mathematical induction and the argument above, the theorem holds for integer $n + 1$. □

The positive zeros of $J_n(x)$ and $J_{n+1}(x)$ for $n = 0, 1, 2, \ldots$ form increasing sequences asymptotically approaching ∞ and interlaced as follows:

$$\lambda_{n,1} < \lambda_{n+1,1} < \lambda_{n,2} < \lambda_{n+1,2} < \cdots < \lambda_{n,k} < \lambda_{n+1,k} < \cdots.$$

Theorem 8.3 holds in the more general setting for $p > -1$ where p is any real number [Watson (1922), Sec. $15 \cdot 22$].

If the argument of a Bessel function of the first kind is parameterized by a zero of the function, the family of parameterized Bessel functions become orthogonal with respect to a weight function. Consider the Bessel equation of order p given in Eq. (8.7). Making the change of variable $x = \lambda z$ where $\lambda > 0$ then,

$$\frac{dy}{dx} = \frac{1}{\lambda}\frac{dy}{dz} \text{ and } \frac{d^2y}{dx^2} = \frac{1}{\lambda^2}\frac{d^2y}{dz^2}.$$

Substituting these expressions into Bessel's equation produces

$$z^2\frac{d^2y}{dz^2} + z\frac{dy}{dz} + (\lambda^2 z^2 - p^2)y(z) = 0 \tag{8.24}$$

which is called the **parametric form of Bessel's equation of order** p. To avoid confusion differentiation with respect to x will be indicated using the prime notation while differentiation with respect to z will be denoted using the Leibniz notation. The self-adjoint form of Eq. (8.24) is

$$\frac{d}{dz}\left[z\frac{dy}{dz}\right] + \frac{1}{z}(\lambda^2 z^2 - p^2)y(z) = 0. \tag{8.25}$$

The function $J_p(x)$ solves Bessel's equation of order p if and only if the function $J_p(\lambda z)$ solves the parametric form of Bessel's equation of order p. In parameterized form, the Bessel functions can be shown to be pairwise orthogonal with respect to the weighting function z on the interval $[0, 1]$.

Theorem 8.4. *Let $p \geq 0$ and let $\lambda_{p,k}$ be the kth zero of $J_p(x)$, then*

$$\int_0^1 J_p(\lambda_{p,n}z)J_p(\lambda_{p,m}z)z\,dz = \frac{1}{2}\left(J_{p+1}(\lambda_{p,n})\right)^2\delta_{mn}, \tag{8.26}$$

*for $m, n \in \mathbb{N}$, where δ_{mn} is the **Kronecker**[4] **delta function.***

---

[4]Leopold Kronecker, German mathematician (1823–1891).

The Kronecker delta is introduced to compactly express the orthogonality property. The Kronecker delta can be thought of as a discrete piecewise-defined function

$$\delta_{ij} = \begin{cases} 0 & \text{if } i \neq j, \\ 1 & \text{if } i = j. \end{cases}$$

This theorem implies the Bessel functions are orthogonal on $[0,1]$ with respect to the weighting function $w(z) = z$.

Proof. Define the linear operator $L[y]$ as

$$L[y] = \frac{d}{dz}\left[z\frac{dy}{dz}\right] - \frac{p^2}{z}y.$$

According to the comments above $J_p(\lambda_{p,n}z)$ and $J_p(\lambda_{p,m}z)$ solve respectively the parametric forms of Bessel's equation of order p:

$$L\left[J_p(\lambda_{p,n}z)\right] = -\lambda_{p,n}^2 z J_p(\lambda_{p,n}z), \tag{8.27}$$

$$L\left[J_p(\lambda_{p,m}z)\right] = -\lambda_{p,m}^2 z J_p(\lambda_{p,m}z). \tag{8.28}$$

Multiplying Eq. (8.27) by $J_p(\lambda_{p,m}z)$ and Eq. (8.28) by $J_p(\lambda_{p,n}z)$, subtracting the equations, integrating over $[0,1]$, and applying Green's formula (Lemma 7.2) with $u = J_p(\lambda_{p,m}z)$ and $v = J_p(\lambda_{p,n}z)$ result in

$$\left(\lambda_{p,m}^2 - \lambda_{p,n}^2\right)\int_0^1 J_p(\lambda_{p,m}z)J_p(\lambda_{p,n}z)z\, dz = 0.$$

Since $\lambda_{p,m}^2 \neq \lambda_{p,n}^2$ then it must be that

$$\int_0^1 J_p(\lambda_{p,m}z)J_p(\lambda_{p,n}z)z\, dz = 0,$$

which establishes Eq. (8.26) for the case in which $m \neq n$.

For the case where $m = n$ consider again Eq. (8.27). Multiplying both sides of Eq. (8.27) by $2z\frac{dJ_p}{dz}(\lambda_{p,n}z)$ produces

$$2\frac{d}{dz}\left[z\frac{dJ_p}{dz}(\lambda_{p,n}z)\right]\left(z\frac{dJ_p}{dz}(\lambda_{p,n}z)\right) = 2\left(p^2 - \lambda_{p,n}^2 z^2\right)J_p(\lambda_{p,n}z)\frac{dJ_p}{dz}(\lambda_{p,n}z)$$

$$\frac{d}{dz}\left[\left(z\frac{dJ_p}{dz}(\lambda_{p,n}z)\right)^2\right] = \left(p^2 - \lambda_{p,n}^2 z^2\right)\frac{d}{dz}[(J_p(\lambda_{p,n}z))^2].$$

Integrating both sides of the equation over $[0,1]$ yields

$$\int_0^1 \frac{d}{dz}\left[\left(z\frac{dJ_p}{dz}(\lambda_{p,n}z)\right)^2\right]dz = \int_0^1 \left(p^2 - \lambda_{p,n}^2 z^2\right)\frac{d}{dz}[(J_p(\lambda_{p,n}z))^2]\, dz$$

$$\left(\frac{dJ_p}{dz}(\lambda_{p,n})\right)^2 = -p^2\left(J_p(0)\right)^2 + 2\lambda_{p,n}^2\int_0^1 z\left(J_p(\lambda_{p,n}z)\right)^2 dz.$$

When $p = 0$ then $p^2 (J_p(0))^2 = 0$ and when $p > 0$ then $J_p(0) = 0$, thus

$$\left(\frac{dJ_p}{dz}(\lambda_{p,n})\right)^2 = 2\lambda_{p,n}^2 \int_0^1 (J_p(\lambda_{p,n}z))^2 \, z \, dz$$

Using Corollary 8.1 and the change of variable,

$$\frac{dJ_p}{dz}(\lambda_{p,n}) - pJ_p(\lambda_{p,n}) = -\lambda_{p,n}J_{p+1}(\lambda_{p,n})$$

$$\left(\frac{dJ_p}{dz}(\lambda_{p,n})\right)^2 = \lambda_{p,n}^2 (J_{p+1}(\lambda_{p,n}))^2 .$$

Hence the following equation results

$$\int_0^1 (J_p(\lambda_{p,n}z))^2 \, z \, dz = \frac{1}{2} (J_{p+1}(\lambda_{p,n}))^2 ,$$

which is Eq. (8.26) when $m = n$. $\qquad\Box$

Just as was done earlier with the trigonometric series of sine and cosine functions, a class of functions can be represented as an infinite series of Bessel functions. These series are called **Generalized Fourier Series** or more specifically **Bessel-Fourier Series**. If $f(x)$ is sufficiently smooth on $[0, 1]$, $f(x)$ can be expressed as

$$f(x) = \sum_{n=1}^{\infty} a_n J_p(\lambda_{p,n}x), \tag{8.29}$$

and Theorem 8.4 can be used to formally determine the values of a_n for $n \in \mathbb{N}$. Multiplying both sides of Eq. (8.29) by $x J_p(\lambda_{p,m}x)$ and integrating over $[0, 1]$ yield the following,

$$\int_0^1 f(x)J_p(\lambda_{p,m}x)x \, dx = \sum_{n=1}^{\infty} a_n \int_0^1 J_p(\lambda_{p,n}x)J_p(\lambda_{p,m}x)x \, dx$$

$$= a_m \int_0^1 (J_p(\lambda_{p,m}x))^2 x \, dx.$$

The orthogonality of the Bessel functions reduces the infinite series on the right-hand side to a single expression. Solving for the coefficient and using Eq. (8.26) give a formula for the nth generalized Fourier coefficient:

$$a_n = \frac{2 \int_0^1 f(x)J_p(\lambda_{p,n}x)x \, dx}{(J_{p+1}(\lambda_{p,n}))^2}. \tag{8.30}$$

Example 8.2. Formally determine the generalized Fourier coefficients for $f(x) = 1 - x$ on the interval $[0, 1]$ using the Bessel functions of the first kind of order 0. Plot $f(x)$ and the partial sums of the generalized Fourier series for $N = 1, 3$ and 5 on the same set of axes.

Solution. According to Eq. (8.30),

$$a_n = \frac{2 \int_0^1 (1 - x) J_0(\lambda_{0,n} x) x \, dx}{(J_1(\lambda_{0,n}))^2}$$

and the Nth partial sum of the can be expressed as

$$S_N(x) = \sum_{n=1}^{N} \frac{2 \int_0^1 (1 - t) J_0(\lambda_{0,n} t) t \, dt}{(J_1(\lambda_{0,n}))^2} J_0(\lambda_{0,n} x).$$

Determining the values of the coefficients requires some numerical integration. There are many resources which tabulate the zeros of the Bessel functions, for example [Weisstein (2013)]. So that the reader may compare results, the first five generalized Fourier coefficients are as follows.

n	a_n
1	0.784519
2	0.068689
3	0.053114
4	0.017363
5	0.016981

The plot of $f(x)$ and the partial sums of the generalized Fourier series are shown in Fig. 8.3.

8.2.2 Bessel Functions of Half-Integer Order

The **Bessel functions of half-integer order** are the functions $J_{n+1/2}(x)$ and $Y_{n+1/2}(x)$ where n is an integer. Such functions play a role in the definition of the spherical Bessel functions and have representations as non-power series functions. Hence the Bessel functions of half-integer order are worth studying in some detail. Using the formula for $J_p(x)$ from Eq. (8.9), letting $p = 1/2$, and simplifying produce

$$J_{1/2}(x) = \sum_{k=0}^{\infty} \frac{(-1)^k}{k! \Gamma(k + 1 + 1/2)} \left(\frac{x}{2}\right)^{2k+1/2}$$

$$= \sqrt{\frac{2}{x}} \sum_{k=0}^{\infty} \frac{(-1)^k x^{2k+1}}{2^{2k+1} k! \left(\frac{1}{2}\right) \left(\frac{3}{2}\right) \cdots \left(\frac{2k+1}{2}\right) \Gamma\left(\frac{1}{2}\right)}$$

$$= \sqrt{\frac{2}{\pi x}} \sum_{k=0}^{\infty} \frac{(-1)^k x^{2k+1}}{(2k + 1)!} = \sqrt{\frac{2}{\pi x}} \sin x.$$

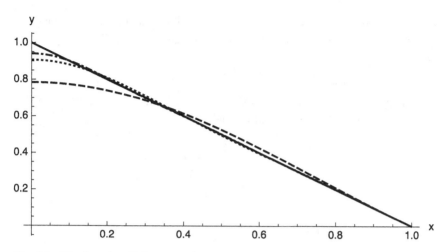

Fig. 8.3 The function $f(x) = 1 - x$ (shown as a solid line) is plotted together with the partial sums $S_N(x)$ (for $N = 1, 3, 5$) of the Bessel series. As N increases the approximation of $f(x)$ by $S_N(x)$ improves.

The recurrence relation of Corollary 8.1 defines other Bessel functions of half-integer order $n+1/2$ with $n \in \mathbb{N}$ in terms of the trigonometric functions sine and cosine. For example, by Eq. (8.20):

$$J_{3/2}(x) = \frac{1}{2x}J_{1/2}(x) - J'_{1/2}(x)$$

$$= \frac{1}{2x}\sqrt{\frac{2}{\pi x}}\sin x - \sqrt{\frac{2}{\pi x}}\cos x + \frac{1}{2x}\sqrt{\frac{2}{\pi x}}\sin x$$

$$= \sqrt{\frac{2}{\pi x}}\left(\frac{1}{x}\sin x - \cos x\right).$$

Theorem 8.5. *For* $n = 0, 1, 2, \ldots$

$$J_{n+1/2}(x) = (-1)^n x^{n+1}\sqrt{\frac{2}{\pi x}}\left(\frac{d}{x\,dx}\right)^n\left[\frac{\sin x}{x}\right]. \tag{8.31}$$

In case the reader is encountering the notation $(d/(x\,dx))^n$ for the first time, it is an operator which takes the derivative of the function to follow, multiplies that result by $1/x$ and then repeats the process $n-1$ more times. Its use may seem unconventional, but provides a compact and elegant way of expressing the Bessel functions of half-integer order.

Proof. When $n = 0$, Eq. (8.31) gives

$$(-1)^0 x^{0+1}\sqrt{\frac{2}{\pi x}}\frac{\sin x}{x} = J_{1/2}(x)$$

as derived above. To use the principle of mathematical induction, suppose the claim has been established for some integer $m \geq 0$. Then Eq. (8.20) can be re-written as

$$J_{m+3/2}(x) = \frac{m+1/2}{x} J_{m+1/2}(x) - J'_{m+1/2}(x).$$

Substituting the expression from the right-hand side of Eq. (8.31) into the right-hand side of the last equation yields

$$\begin{aligned}
J_{m+3/2}(x) &= \frac{m+1/2}{x}(-1)^m x^{m+1} \sqrt{\frac{2}{\pi x}} \left(\frac{d}{x\,dx}\right)^m \left[\frac{\sin x}{x}\right] \\
&\quad - \frac{d}{dx}\left[(-1)^m x^{m+1}\sqrt{\frac{2}{\pi x}}\left(\frac{d}{x\,dx}\right)^m\left[\frac{\sin x}{x}\right]\right] \\
&= (m+\tfrac{1}{2})(-1)^m x^m \sqrt{\frac{2}{\pi x}}\left(\frac{d}{x\,dx}\right)^m\left[\frac{\sin x}{x}\right] \\
&\quad - (-1)^m x^m \sqrt{\frac{2}{\pi x}}\left((m+\tfrac{1}{2})\left(\frac{d}{x\,dx}\right)^m\left[\frac{\sin x}{x}\right]\right. \\
&\quad \left. + x\frac{d}{dx}\left[\left(\frac{d}{x\,dx}\right)^m\left[\frac{\sin x}{x}\right]\right]\right) \\
&= (-1)^{m+1} x^{m+2}\sqrt{\frac{2}{\pi x}}\left(\frac{d}{x\,dx}\right)^{m+1}\left[\frac{\sin x}{x}\right],
\end{aligned}$$

which is Eq. (8.31) with n replaced by $m+1$. Thus the formula has been proved. $\square$

Previously in Example 8.1 it was shown that $J_{-1/2}(x) = \sqrt{2/(\pi x)}\cos x$. A result similar to that of Theorem 8.5 holds for Bessel functions of the first kind of negative half-integer order.

Theorem 8.6. *For $n = 0, 1, 2, \ldots$*

$$J_{-n-1/2}(x) = x^{n+1}\sqrt{\frac{2}{\pi x}}\left(\frac{d}{x\,dx}\right)^n\left[\frac{\cos x}{x}\right]. \tag{8.32}$$

Proof. See Exercise 15. $\square$

The Bessel functions of the second kind of half-integer order also take on special, simpler forms. Replacing p in Eq. (8.10) with $n + 1/2$ where n is an integer results in

$$\begin{aligned}
Y_{n+1/2}(x) &= \frac{J_{n+1/2}(x)\cos((n+1/2)\pi) - J_{-n-1/2}(x)}{\sin((n+1/2)\pi)} = \frac{-J_{-n-1/2}(x)}{\sin((n+1/2)\pi)} \\
&= (-1)^{n+1} J_{-n-1/2}(x). \tag{8.33}
\end{aligned}$$

In particular when $n = 0$, $Y_{1/2}(x) = -\sqrt{2/(\pi x)}\cos x$.

Example 8.3. Find expressions for $J_{-3/2}(x)$ and $Y_{3/2}(x)$ terms of the sine and cosine functions.

Solution. Substituting $n = 1$ in Eq. (8.32) produces

$$J_{-3/2}(x) = x^2 \sqrt{\frac{2}{\pi x}} \frac{d}{x\,dx} \left[\frac{\cos x}{x} \right] = x\sqrt{\frac{2}{\pi x}} \left(\frac{-x\sin x - \cos x}{x^2} \right)$$

$$= -\sqrt{\frac{2}{\pi x}} \left(\frac{1}{x}\cos x + \sin x \right).$$

Combining the expression for $J_{-3/2}(x)$ and Eq. (8.33) yields

$$Y_{3/2}(x) = -\sqrt{\frac{2}{\pi x}} \left(\frac{1}{x}\cos x + \sin x \right).$$

8.2.3 *Spherical Bessel's Equation*

A related equation to the Bessel's equation of order n is the **spherical Bessel's equation**:

$$x^2 y'' + 2x\,y' + (x^2 - n(n+1))y = 0. \tag{8.34}$$

The solution to this equation could be developed using the power series techniques and the method of Frobenius employed in the previous sections. However, there is no need to resort to this laborious procedure again. Instead set $y(x) = x^{-1/2}u(x)$ where $u(x)$ is an unknown function. Differentiating function $y(x)$ produces

$$y'(x) = -\frac{1}{2}x^{-3/2}u(x) + x^{-1/2}u'(x),$$

$$y''(x) = \frac{3}{4}x^{-5/2}u(x) - x^{-3/2}u'(x) + x^{-1/2}u''(x).$$

Substituting these derivatives into Eq. (8.34) yields

$$0 = x^2 \left(\frac{3}{4}x^{-5/2}u(x) - x^{-3/2}u'(x) + x^{-1/2}u''(x) \right)$$

$$+ 2x \left(-\frac{1}{2}x^{-3/2}u(x) + x^{-1/2}u'(x) \right) + (x^2 - n(n+1))x^{-1/2}u(x)$$

$$= x^{3/2}u''(x) + x^{1/2}u'(x) + \left(x^{3/2} - x^{-1/2}\left(n + \frac{1}{2} \right)^2 \right) u(x).$$

Multiplying both sides of the equation by $x^{1/2}$ produces

$$0 = x^2 u''(x) + xu'(x) + \left(x^2 - \left(n + \frac{1}{2} \right)^2 \right) u(x)$$

which is Bessel's equation of order $n + 1/2$ whose linearly independent solutions are $J_{n+1/2}(x)$ and $Y_{n+1/2}(x)$. Therefore denote the **spherical Bessel function of the first kind** as

$$j_n(x) = \sqrt{\frac{\pi}{2x}} J_{n+1/2}(x) \tag{8.35}$$

and the **spherical Bessel function of the second kind** as

$$y_n(x) = \sqrt{\frac{\pi}{2x}} Y_{n+1/2}(x). \tag{8.36}$$

Some authors denote the spherical Bessel function of the second kind as $n_n(x)$, but out of a desire to adopt a notation which expresses the relationship between the Bessel functions and the spherical Bessel functions, the present text will denote it by $y_n(x)$. Going forward, care should be taken when encountering a function denoted as $y_n(x)$. This common notation may sometimes refer to the nth solution to a particular equation while at other times may specify the spherical Bessel function of the second kind. The intended meaning should be clear from the context.

The following corollary states an orthogonality relationship between the spherical Bessel functions of the first kind similar to Theorem 8.4.

Corollary 8.2. *For $n = 0, 1, 2, \ldots$,*

$$\int_0^1 j_n(\lambda_{n+1/2,k} x) j_n(\lambda_{n+1/2,l} x) x^2 \, dx = \frac{1}{2} \left(j_{n+1}(\lambda_{n+1/2,k}) \right)^2 \delta_{kl} \tag{8.37}$$

where δ_{kl} is the Kronecker delta function.

Note in Eq. (8.37) the zeros of the Bessel function of the first kind of order $n + 1/2$ are used in the spherical Bessel function of the first kind of order n. Bessel's equation and its solutions will return later in Chap. 9 when the vibrating membrane and swinging chain are mathematically modeled.

8.3 The Legendre Equation

The Legendre[5] differential equation and its solution play an important role in applied mathematics and mathematical physics. For example, when using the method of separation of variables to solve Laplace's equation, $\triangle u = 0$ in spherical coordinates, the following equation arises,

$$\frac{d^2 F(\varphi)}{d\varphi^2} + \cot \varphi \frac{dF(\varphi)}{d\varphi} + \alpha(\alpha + 1) F(\varphi) = 0 \tag{8.38}$$

---

[5] Adrien-Marie Legendre, French mathematician (1752–1833).

for $0 < \varphi < \pi$, where α is a constant. Making the change of variable $x = \cos\varphi$ and defining $y = f(x) = F(\arccos x)$, then the chain rule for derivatives implies that

$$\frac{dF(\varphi)}{d\varphi} = \frac{dF(\varphi)}{dx}\frac{dx}{d\varphi} = \frac{dF(\cos^{-1}x)}{dx}(-\sin\varphi) = \frac{df}{dx}(-\sin\varphi) = (-\sin\varphi)y'.$$

Differentiating one more time produces

$$\frac{d^2F(\varphi)}{d\varphi^2} = \frac{d}{d\varphi}[(-\sin\varphi)y'] = (-\cos\varphi)y' + (\sin^2\varphi)y''.$$

Substituting these expressions into Eq. (8.38) yields

$$\begin{aligned}
0 &= (-\cos\varphi)y' + (\sin^2\varphi)y'' - (\cot\varphi)(\sin\varphi)y' + \alpha(\alpha+1)y \\
&= (\sin^2\varphi)y'' - 2(\cos\varphi)y' + \alpha(\alpha+1)y \\
&= (1 - \cos^2\varphi)y'' - 2(\cos\varphi)y' + \alpha(\alpha+1)y.
\end{aligned}$$

Reverting to the original independent variable reveals **Legendre's differential equation,**

$$(1 - x^2)y'' - 2xy' + \alpha(\alpha+1)y = 0 \text{ for } -1 < x < 1, \tag{8.39}$$

where α is a constant. This differential equation can be written in self-adjoint form as

$$[(1 - x^2)y']' + \alpha(\alpha+1)y = 0 \text{ for } -1 < x < 1. \tag{8.40}$$

The vanishing of $1 - x^2$ at $x = \pm 1$ implies this is a singular Sturm-Liouville differential equation.

The point $x_0 = 0$ is an ordinary point for Eq. (8.39). In this section power series techniques are used to determine two linearly independent solutions to this equation which converge for $|x| < 1$.

Assuming a solution of the form $y(x) = \sum_{n=0}^{\infty} a_n x^n$ and substituting this function into Eq. (8.39) yield the following,

$$\begin{aligned}
0 &= (1 - x^2)\sum_{n=2}^{\infty} n(n-1)a_n x^{n-2} - 2x\sum_{n=1}^{\infty} na_n x^{n-1} + \alpha(\alpha+1)\sum_{n=0}^{\infty} a_n x^n \\
&= \sum_{n=0}^{\infty}\left[(n+2)(n+1)a_{n+2} - (n(n+1) - \alpha(\alpha+1))a_n\right]x^n.
\end{aligned}$$

From this equation the following recurrence relation is derived:

$$a_{n+2} = \frac{(n(n+1) - \alpha(\alpha+1))a_n}{(n+2)(n+1)} = -\frac{(\alpha-n)(\alpha+n+1)a_n}{(n+2)(n+1)}. \tag{8.41}$$

If $a_0 = 1$ and $a_1 = 0$ then the even-indexed coefficients become

$$a_2 = -\frac{\alpha(\alpha+1)}{2!}$$

$$a_4 = -\frac{(\alpha+3)(\alpha-2)a_2}{(4)(3)} = \frac{\alpha(\alpha-2)(\alpha+1)(\alpha+3)}{4!}$$

$$a_6 = -\frac{(\alpha+5)(\alpha-4)a_4}{(6)(5)} = -\frac{\alpha(\alpha-2)(\alpha-4)(\alpha+1)(\alpha+3)(\alpha+5)}{6!}$$

$$\vdots$$

$$a_{2m} = (-1)^m \frac{\alpha\cdots(\alpha-2m+2)(\alpha+1)\cdots(\alpha+2m-1)}{(2m)!}. \tag{8.42}$$

The odd-indexed coefficients are all zero since a_1 was chosen to be zero. Hence one solution to Eq. (8.39) is

$$y_1(x) = 1 - \frac{\alpha(\alpha+1)}{2!}x^2 + \frac{\alpha(\alpha-2)(\alpha+1)(\alpha+3)}{4!}x^4$$

$$+ \sum_{m=3}^{\infty} \frac{\alpha\cdots(\alpha-2m+2)(\alpha+1)\cdots(\alpha+2m-1)}{(2m)!}x^{2m}.$$

The Ratio Test shows that the power series defining $y_1(x)$ converges absolutely for $|x| < 1$.

If $a_0 = 0$ and $a_1 = 1$ then the odd-indexed coefficients become

$$a_3 = -\frac{(\alpha-1)(\alpha+2)}{3!}$$

$$a_5 = -\frac{(\alpha-3)(\alpha+4)a_3}{(5)(4)} = \frac{(\alpha-1)(\alpha-3)(\alpha+2)(\alpha+4)}{5!}$$

$$a_7 = -\frac{(\alpha-5)(\alpha+6)a_5}{(7)(6)} = -\frac{(\alpha-1)(\alpha-3)(\alpha-5)(\alpha+2)(\alpha+4)(\alpha+6)}{7!}$$

$$\vdots$$

$$a_{2m+1} = (-1)^m \frac{(\alpha-1)\cdots(\alpha-2m+1)(\alpha+2)\cdots(\alpha+2m)}{(2m+1)!}. \tag{8.43}$$

In this situation the even-indexed coefficients are all zero since a_0 was chosen to be zero. Hence another solution to Eq. (8.39) is

$$y_2(x) = x - \frac{(\alpha-1)(\alpha+2)}{3!}x^3 + \frac{(\alpha-1)(\alpha-3)(\alpha+2)(\alpha+4)}{5!}x^5$$

$$+ \sum_{m=3}^{\infty} (-1)^m \frac{(\alpha-1)\cdots(\alpha-2m+1)(\alpha+2)\cdots(\alpha+2m)}{(2m+1)!}x^{2m+1}.$$

Use of the Ratio Test establishes that the power series for $y_2(x)$ converges absolutely for $|x| < 1$.

Table 8.1   A sample of $y_1(x)$ for $\alpha = 0$, 2, 4, and 6.

α	$y_1(x)$
0	1
2	$1 - 3x^2$
4	$1 - 10x^2 + 35x^4/3$
6	$1 - 21x^2 + 63x^4 - 231x^6/5$

The Wronskian determinant shows the two solutions are linearly independent.

$$W(y_1, y_2)(0) = y_1(0)y_2'(0) - y_1'(0)y_2(0) = 1 \neq 0.$$

If $\alpha = 0$ then $y_1(x) = 1$ which is a polynomial of degree zero. If $\alpha = 2n$ for some $n \in \mathbb{N}$ then the coefficient formula given in Eq. (8.42) can be re-stated as

$$a_{2m} = (-1)^m \frac{2n \cdots (2n - 2m + 2)(2n + 1) \cdots (2n + 2m - 1)}{(2m)!}.$$

Thus $a_{2m} = 0$ for all $m \geq n + 1$. This implies that

$$y_1(x) = 1 - \frac{2n(2n + 1)}{2!}x^2 + \frac{2n(2n - 2)(2n + 1)(2n + 3)}{4!}x^4$$
$$+ \sum_{m=3}^{n} \frac{2n \cdots (2n - 2m + 2)(2n + 1) \cdots (2n + 2m - 1)}{(2m)!}x^{2m},$$

is a polynomial of degree $2n$ containing only even powers of x. Table 8.1 lists $y_1(x)$ for $\alpha = 0$, 2, 4, and 6.

If $\alpha = 1$ then $y_2(x) = x$ which is a polynomial of degree 1 containing only odd powers of x. If $\alpha = 2n + 1$ for some $n \in \mathbb{N}$ then the coefficient formula of Eq. (8.43) can be rewritten as

$$a_{2m+1}$$
$$= (-1)^m \frac{(2n + 1 - 1) \cdots (2n + 1 - 2m + 1)(2n + 1 + 2) \cdots (2n + 1 + 2m)}{(2m + 1)!}.$$

Again $a_{2m+1} = 0$ for $m \geq n + 1$. This implies that

$$y_2(x) = x - \frac{(2n)(2n + 3)}{3!}x^3 + \frac{(2n)(2n - 2)(2n + 3)(2n + 5)}{5!}x^5$$
$$+ \sum_{m=3}^{n} (-1)^m \frac{(2n) \cdots (2n - 2m + 2)(2n + 3) \cdots (2n + 2m + 1)}{(2m + 1)!}$$
$$\times x^{2m + 1}$$

is a polynomial of degree $2n + 1$ containing only odd powers of x. Table 8.2 displays a sample of these odd-degree polynomial solutions to Legendre's equation.

Table 8.2   A short list of polynomial solutions of odd degree to Legendre's differential equation.

α	$y(x)$
1	x
3	$x - 5x^3/3$
5	$x - 14x^3/3 + 21x^5/5$
7	$x - 9x^3 + 99x^5/5 - 429x^7/35$

Since Legendre's equation is linear and homogeneous, if $y(x)$ solves Eq. (8.39) then $c\,y(x)$ also solves the equation for any constant c. The polynomial solutions to Legendre's equation are typically normalized by an appropriate choice for c. By convention the leading coefficient a_n of a polynomial solution of degree n to Eq. (8.39) (for either n even or n odd) is chosen to be

$$a_n = \frac{(2n)!}{2^n (n!)^2}. \tag{8.44}$$

In this case the remaining coefficients in the polynomials are found by using a backwards recurrence relation. Replacing n by the symbol m and replacing with α with $n \in \mathbb{N}$ then the original recurrence relation expressed in Eq. (8.41) can be rewritten as

$$a_{m+2} = \frac{(m(m+1) - n(n+1))a_m}{(m+2)(m+1)} \iff a_m = \frac{(m+2)(m+1)a_{m+2}}{m(m+1) - n(n+1)}.$$

Re-indexing by replacing m by $m - 2$ results in

$$a_{m-2} = -\frac{m(m-1)a_m}{(n-m+2)(n+m-1)}. \tag{8.45}$$

Equation (8.45) is the backwards recurrence relation. Using the conventional choice for a_n specified in Eq. (8.44), then

$$a_{n-2} = -\frac{n(n-1)(2n)!}{2(2n-1)2^n(n!)^2} = -\frac{n(n-1)(2n)(2n-1)(2n-2)!}{2(2n-1)2^n n^2(n-1)(n-1)!(n-2)!}$$

$$= -\frac{(2n-2)!}{2^n(n-1)!(n-2)!}.$$

This is stated as a lemma below.

Lemma 8.8. *In general,*

$$a_{n-2m} = (-1)^m \frac{(2n-2m)!}{2^n(m!)(n-m)!(n-2m)!}, \tag{8.46}$$

for $2m \le n$.

Table 8.3 Legendre polynomials $P_n(x)$
for $n = 0, 1, \ldots, 6$.

$P_0(x) = 1$
$P_1(x) = x$
$P_2(x) = (3x^2 - 1)/2$
$P_3(x) = (5x^3 - 3x)/2$
$P_4(x) = (35x^4 - 30x^2 + 3)/8$
$P_5(x) = (63x^5 - 70x^3 + 15x)/8$
$P_6(x) = (231x^6 - 315x^4 + 105x^2 - 5)/16$

Proof. Equation (8.46) has already been established for the case in which $m = 1$. Suppose it is true for $m = k$ then,

$$a_{n-2k} = (-1)^k \frac{(2n - 2k)!}{2^n (k!)(n - k)!(n - 2k)!}$$

and by the backwards recurrence relation of Eq. (8.45)

$$
\begin{aligned}
a_{n-2(k+1)} &= -\frac{(n - 2k)(n - 2k - 1)a_{n-2k}}{(n - (n - 2k) + 2)(n + (n - 2k) - 1)} \\
&= -\frac{(n - 2k)(n - 2k - 1)a_{n-2k}}{(2k + 2)(2n - 2k - 1)} \\
&= (-1)^{k+1} \frac{(n - 2k)(n - 2k - 1)(2n - 2k)!}{(2k + 2)(2n - 2k - 1)2^n (k!)(n - k)!(n - 2k)!} \\
&= (-1)^{k+1} \frac{(2n - 2(k + 1))!}{2^n (k + 1)!(n - (k + 1))!(n - 2(k + 1))!}.
\end{aligned}
$$

Thus by the principle of mathematical induction Eq. (8.46) holds. $\square$

A polynomial of degree n solving Legendre's equation and having the leading coefficient given in Eq. (8.44) is called a **Legendre polynomial** and is denoted as $P_n(x)$. Thus the general formula for $P_n(x)$ is

$$P_n(x) = \frac{1}{2^n} \sum_{k=0}^{\llbracket n/2 \rrbracket} \frac{(-1)^k (2n - 2k)!}{k!(n - k)!(n - 2k)!} x^{n-2k}, \tag{8.47}$$

where $\llbracket n/2 \rrbracket$ denotes the greatest integer less than or equal to $n/2$. Thus if α in Eq. (8.39) is zero or a positive, even integer $2n$, the series solution is a polynomial of degree $2n$ containing only even powers of x and if α is a positive, odd integer $2n + 1$, the series solution is a polynomial of degree $2n + 1$ containing only odd powers of x. To summarize Table 8.3 lists the Legendre polynomials $P_n(x)$ for $n = 0, 1, \ldots, 6$. The graphs of these Legendre polynomials are shown in Fig. 8.4.

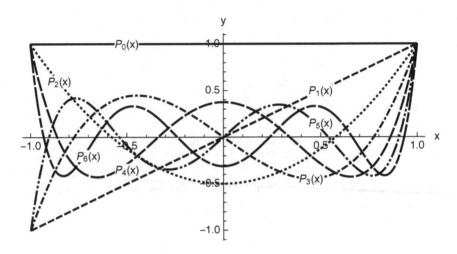

Fig. 8.4   Graphs of the Legendre polynomials $P_n(x)$ for $n = 0, 1, \ldots, 6$.

Before exploring the properties of the Legendre polynomials, a further comment on solutions to Eq. (8.39) and the value of the parameter α is in order. Although the power series $y_1(x)$ and $y_2(x)$ above converge for $-1 < x < 1$ for an arbitrary choice of α, most attention is given to the cases in which $\alpha = 0, 1, 2, \ldots$, since when α is not a nonnegative integer the series for $y_1(x)$ and $y_2(x)$ diverge at $x = \pm 1$ [Rose (2003)]. Hence the Legendre polynomials are the only solutions to Eq. (8.39) which remain finite at $x = \pm 1$. The boundedness of the Legendre polynomials is important in many physical applications.

8.3.1   *Properties of the Legendre Polynomials*

The Legendre polynomials have many interesting and useful properties. Some of these properties are more easily revealed and verified if an alternative formula for expressing the Legendre polynomials is employed. The nth Legendre polynomial can be expressed as the nth derivative of a simple polynomial. The representation of the Legendre polynomials as derivatives is known as a **Rodrigues'**[6] **formula** or **Rodrigues' form**, which has analogues for some of the other special functions to be explored in this chapter as well.

Theorem 8.7 (Rodrigues' Form of Legendre Polynomials). *For* $n =$

---

[6]Benjamin Olinde Rodrigues, French banker and mathematician (1795–1851).

$0, 1, 2, \ldots,$

$$P_n(x) = \frac{1}{2^n n!} \frac{d^n}{dx^n} \left[(x^2 - 1)^n \right]. \tag{8.48}$$

Proof. If $0 \le k \le \llbracket n/2 \rrbracket$ then

$$\frac{d^n}{dx^n} \left[x^{2n-2k} \right] = (2n - 2k)(2n - 2k - 1) \cdots (n - 2k + 1) x^{n-2k}$$

$$= \frac{(2n - 2k)!}{(n - 2k)!} x^{n-2k}.$$

If n is even and $n < 2k \le 2n$ then $n - 2k < 0 \le 2n - 2k$ which implies

$$\frac{d^n}{dx^n} [x^{2n-2k}] = 0.$$

On the other hand, if n is odd then $\llbracket n/2 \rrbracket = (n-1)/2$. If $n - 1 < 2k \le 2n$ then $n - 2k - 1 < 0 \le 2n - 2k$ and again

$$\frac{d^n}{dx^n} [x^{2n-2k}] = 0.$$

Hence starting with the definition of the nth Legendre polynomial from Eq. (8.47),

$$P_n(x) = \frac{1}{2^n} \sum_{k=0}^{\llbracket n/2 \rrbracket} \frac{(-1)^k (2n - 2k)!}{k!(n - k)!(n - 2k)!} x^{n-2k}$$

$$= \frac{1}{2^n n!} \sum_{k=0}^{\llbracket n/2 \rrbracket} \frac{(-1)^k n!}{k!(n - k)!} \frac{d^n}{dx^n} [x^{2n-2k}]$$

$$= \frac{1}{2^n n!} \sum_{k=0}^{n} \frac{(-1)^k n!}{k!(n - k)!} \frac{d^n}{dx^n} [x^{2n-2k}]$$

$$= \frac{1}{2^n n!} \frac{d^n}{dx^n} \left[\sum_{k=0}^{n} \frac{(-1)^k n!}{k!(n - k)!} (x^2)^{n-k} \right] = \frac{1}{2^n n!} \frac{d^n}{dx^n} [(x^2 - 1)^n]$$

using the binomial formula,

$$(a + b)^n = \sum_{k=0}^{n} \frac{n!}{k!(n - k)!} a^{n-k} b^k.$$

This establishes the Rodrigues' formula for the Legendre polynomials. $\square$

Since the Rodrigues' formula requires differentiating the nth power of a function n times, an extension of the product rule for derivatives may make using a Rodrigues' formula more convenient. This result is known as the **Leibniz product rule for differentiation.**

Lemma 8.9. *Let f and g be n times differentiable at x, then*

$$\frac{d^n}{dx^n}[f(x)g(x)] = \sum_{k=0}^{n} \frac{n!}{k!(n-k)!} f^{(k)}(x)g^{(n-k)}(x). \qquad (8.49)$$

Proof. See Exercise 18.   □

To illustrate the utility of the Rodrigues' formula, consider the role it plays in the proof that for $n = 0, 1, 2, \ldots$, the polynomial $P_n(1) = 1$, a conjecture the reader may make based on the graphical evidence in Fig. 8.4.

$$P_n(x) = \frac{1}{2^n n!} \frac{d^n}{dx^n}\left[(x^2-1)^n\right] = \frac{1}{2^n n!} \frac{d^n}{dx^n}\left[(x+1)^n(x-1)^n\right]$$

$$= \frac{1}{2^n n!} \sum_{k=0}^{n} \frac{n!}{k!(n-k)!} \frac{d^k}{dx^k}\left[(x+1)^n\right] \frac{d^{n-k}}{dx^{n-k}}\left[(x-1)^n\right].$$

For $k = 1, 2, \ldots, n$ the summation above contains a factor of $x - 1$ and thus will vanish when $x = 1$. Therefore,

$$P_n(1) = \frac{1}{2^n n!}\left((1+1)^n n!\right) = 1.$$

Since $P_n(x)$ is an even function when n is even and an odd function when n is odd, then

$$P_n(-x) = (-1)^n P_n(x),$$

which implies that $P_n(-1) = (-1)^n$.

The Legendre polynomial $P_n(x)$ has n distinct roots in the interval $(-1, 1)$. These roots play an important role in the numerical approximation of definite integrals via a technique called **Gaussian quadrature**.

The set of Legendre polynomials, like all of the special functions described in this chapter, possess orthogonality properties. The following lemma will help to justify the claims of orthogonality.

Lemma 8.10 (Bonnet's[7] Recurrence Relation). *For $n \in \mathbb{N}$, if $P_n(x)$ is the nth Legendre polynomial*

$$(n+1)P_{n+1}(x) + nP_{n-1}(x) = (2n+1)xP_n(x). \qquad (8.50)$$

Proof. Let $n \in \mathbb{N}$, then

$$\frac{d^2}{dx^2}\left[(x^2-1)^{n+1}\right] = 2(n+1)\left((2n+1)(x^2-1)^n + 2n(x^2-1)^{n-1}\right).$$

---

[7]Ossian Bonnet, French mathematician (1819–1892).

Therefore

$$\frac{d^{n+1}}{dx^{n+1}}\left[(x^2-1)^{n+1}\right] = \frac{d^{n-1}}{dx^{n-1}}\frac{d^2}{dx^2}[(x^2-1)^{n+1}]$$

$$= \frac{d^{n-1}}{dx^{n-1}}[2(n+1)((2n+1)(x^2-1)^n + 2n(x^2-1)^{n-1})]$$

$$2^{n+1}(n+1)!P_{n+1}(x) = 2(n+1)(2n+1)\frac{d^{n-1}}{dx^{n-1}}[(x^2-1)^n]$$

$$+ 4n(n+1)2^{n-1}(n-1)!P_{n-1}(x)$$

$$P_{n+1}(x) = \frac{2n+1}{2^n n!}\frac{d^{n-1}}{dx^{n-1}}[(x^2-1)^n] + P_{n-1}(x)$$

$$P_{n+1}(x) - P_{n-1}(x) = \frac{2n+1}{2^n n!}\frac{d^{n-1}}{dx^{n-1}}[(x^2-1)^n]. \tag{8.51}$$

The Rodrigues' formula given in Eq. (8.48) for $P_{n+1}(x)$ implies

$$P_{n+1}(x) = \frac{1}{2^{n+1}(n+1)!}\frac{d^{n+1}}{dx^{n+1}}[(x^2-1)^{n+1}]$$

$$= \frac{1}{2^n n!}\frac{d^n}{dx^n}[x(x^2-1)^n]$$

$$= \frac{1}{2^n n!}\sum_{k=0}^{n}\frac{n!}{k!(n-k)!}\frac{d^k}{dx^k}[x]\frac{d^{n-k}}{dx^{n-k}}[(x^2-1)^n]$$

$$= \frac{1}{2^n n!}\sum_{k=0}^{1}\frac{n!}{k!(n-k)!}\frac{d^k}{dx^k}[x]\frac{d^{n-k}}{dx^{n-k}}[(x^2-1)^n]$$

$$= \frac{1}{2^n n!}\left(x\frac{d^n}{dx^n}[(x^2-1)^n] + n\frac{d^{n-1}}{dx^{n-1}}[(x^2-1)^n]\right)$$

$$= xP_n(x) + \frac{n}{2^n n!}\frac{d^{n-1}}{dx^{n-1}}[(x^2-1)^n]$$

$$P_{n+1}(x) - xP_n(x) = \frac{n}{2^n n!}\frac{d^{n-1}}{dx^{n-1}}[(x^2-1)^n]. \tag{8.52}$$

Combining Eq. (8.51) and Eq. (8.52) establishes the following equation

$$\frac{P_{n+1}(x) - P_{n-1}(x)}{2n+1} = \frac{P_{n+1}(x) - xP_n(x)}{n}$$

$$n(P_{n+1}(x) - P_{n-1}(x)) = (2n+1)(P_{n+1}(x) - xP_n(x))$$

$$(2n+1)xP_n(x) = (n+1)P_{n+1}(x) + nP_{n-1}(x)$$

which is equivalent to Bonnet's recurrence relation given in Eq. (8.50). □

It is now possible to prove the orthogonality properties of the Legendre polynomials.

Theorem 8.8. *For* $n, m \in \{0, 1, 2, \ldots\}$, *if* $P_m(x)$ *and* $P_n(x)$ *are the mth and nth Legendre polynomials respectively, then*

$$\int_{-1}^{1} P_m(x) P_n(x) \, dx = \frac{2}{2n+1} \delta_{mn} \tag{8.53}$$

where δ_{mn} *is the Kronecker delta function.*

Proof. Define the linear operator $L[y] = [(1 - x^2)y']'$. Using the self-adjoint form of the Legendre equation in Eq. (8.40), Green's formula from Lemma 7.2 with $u = P_m(x)$ and $v = P_n(x)$, and integrating over $[-1, 1]$ produce

$$[m(m+1) - n(n+1)] \int_{-1}^{1} P_m(x) P_n(x) \, dx = 0.$$

Using the zero factor property, either

$$\int_{-1}^{1} P_m(x) P_n(x) \, dx = 0 \tag{8.54}$$

or

$$m(m+1) - n(n+1) = (m-n)(m+n+1) = 0.$$

Since $m, n \in \{0, 1, 2, \ldots\}$ and $m \neq n$ this last equation is not satisfied. Therefore Eq. (8.54) holds. This concludes the proof for the case in which $m \neq n$. An immediate consequence of this is that $P_n(x)$ is orthogonal to every polynomial of degree less than n.

Now consider the case in which $m = n$. Differentiating both sides of Eq. (8.52) with respect to x yields

$$P_{n+1}'(x) - xP_n'(x) - P_n(x) = \frac{n}{2^n n!} \frac{d^n}{dx^n} \left[(x^2 - 1)^n \right] = nP_n(x)$$

using the Rodrigues' formula of Eq. (8.48). Thus

$$P_{n+1}'(x) - xP_n'(x) = (n+1)P_n(x). \tag{8.55}$$

Multiplying both sides of this equation by $P_n(x)$ and integrating over $[-1, 1]$ produce

$$(n+1) \int_{-1}^{1} (P_n(x))^2 \, dx$$

$$= \int_{-1}^{1} P_{n+1}'(x) P_n(x) \, dx - \int_{-1}^{1} xP_n'(x) P_n(x) \, dx$$

$$= [P_{n+1}(x) P_n(x)]_{-1}^{1} - \int_{-1}^{1} P_{n+1}(x) P_n'(x) \, dx - \int_{-1}^{1} xP_n'(x) P_n(x) \, dx$$

$$= P_{n+1}(1) P_n(1) - P_{n+1}(-1) P_n(-1) - \int_{-1}^{1} xP_n'(x) P_n(x) \, dx$$

$$= 2 - \int_{-1}^{1} xP_n'(x) P_n(x) \, dx. \tag{8.56}$$

Using the technique of integration by parts on the right-hand side of Eq. (8.56) results in

$$(n+1) \int_{-1}^{1} (P_n(x))^2 \, dx = 2 - \left[\frac{x}{2}(P_n(x))^2\right]_{x=-1}^{x=1} + \frac{1}{2} \int_{-1}^{1} (P_n(x))^2 \, dx$$

$$\left(n + \frac{1}{2}\right) \int_{-1}^{1} (P_n(x))^2 \, dx = 2 - \frac{1}{2}(P_n(1))^2 - \frac{1}{2}(P_n(-1))^2 = 1$$

$$\int_{-1}^{1} (P_n(x))^2 \, dx = \frac{2}{2n+1}.$$

This concludes the proof of the second case of the orthogonality relationships for the Legendre polynomials. □

Given a polynomial f of degree n, it is possible to express f as a linear combination of P_0, P_1, P_2, ..., P_n:

$$f(x) = \sum_{k=0}^{n} a_k P_k(x),$$

where

$$a_k = \frac{2k+1}{2} \int_{-1}^{1} f(x) P_k(x) \, dx. \tag{8.57}$$

Example 8.4. Express $f(x) = x^5$ in terms of the Legendre polynomials of degree 5 or less.

Solution. The coefficients a_0, a_1, ..., a_5 will satisfy the equation

$$x^5 = a_0 P_0(x) + a_1 P_1(x) + \cdots + a_5 P_5(x).$$

Since $f(x)$ is an odd function and $P_0(x)$, $P_2(x)$, and $P_4(x)$ are even functions, then $a_0 = a_2 = a_4 = 0$. Calculating the odd-indexed coefficients using Eq. (8.57) produces

$$a_1 = \frac{3}{2} \int_{-1}^{1} x^5(x) \, dx = \frac{3}{7},$$

$$a_3 = \frac{7}{2} \int_{-1}^{1} x^5 \left((5x^3 - 3x)/2\right) \, dx = \frac{4}{9},$$

$$a_5 = \frac{11}{2} \int_{-1}^{1} x^5 \left((63x^5 - 70x^3 + 15x)/8\right) \, dx = \frac{8}{63}.$$

Consequently the linear combination,

$$\frac{3}{7} P_1(x) + \frac{4}{9} P_3(x) + \frac{8}{63} P_5(x) = x^5,$$

which the reader can check directly using the functions in Table 8.3.

8.3.2 Associated Legendre Polynomials

The derivatives of the Legendre polynomials, which are themselves polynomials, satisfy a second-order linear, ordinary differential equation similar to Legendre's equation. These derivative polynomials are known as the **Associated Legendre Polynomials**. This section will state and explore some of their properties.

Theorem 8.9. *Let $f(x) = P_n(x)$ then for $m \in \{0, 1, \ldots, n\}$ the function $f^{(m)}(x)$ solves the ordinary differential equation,*

$$(1 - x^2)y'' - 2x(m+1)y' + (n(n+1) - m(m+1))y = 0. \qquad (8.58)$$

Proof. When $m = 0$ then $y(x) = P_n(x)$ and the nth Legendre polynomial is a solution to the ordinary differential equation:

$$0 = (1-x^2)y'' - 2xy' + n(n+1)y = (1-x^2)y'' - 2xy' + (n(n+1) - 0(0+1))y.$$

For the purposes of induction suppose the theorem is true for $0 \le m < n$. Then $y(x) = f^{(m)}(x)$ solves the ordinary differential equation

$$
\begin{aligned}
0 &= (1 - x^2)y'' - 2x(m+1)y' + (n(n+1) - m(m+1))y \\
&= (1 - x^2)f^{(m+2)}(x) - 2x(m+1)f^{(m+1)}(x) \\
&\quad + (n(n+1) - m(m+1))f^{(m)}(x).
\end{aligned}
$$

Differentiating both sides of the equation produces

$$
\begin{aligned}
0 &= (1 - x^2)f^{(m+3)}(x) - 2x(m+2)f^{(m+2)}(x) \\
&\quad + (n(n+1) - (m+1)(m+2))f^{(m+1)}(x).
\end{aligned}
$$

which is Eq. (8.58) with m replaced by $m+1$. Hence the theorem holds by the principle of mathematical induction. $\qquad \square$

Now define the function $g(x) = (1-x^2)^{m/2}f^{(m)}(x)$. Differentiating $g(x)$ results in

$$g'(x) = (1 - x^2)^{m/2}f^{(m+1)}(x) - mx(1 - x^2)^{-1+m/2}f^{(m)}(x).$$

Multiplying $g'(x)$ by $1 - x^2$ yields the product,

$$(1 - x^2)g'(x) = (1 - x^2)^{1+m/2}f^{(m+1)}(x) - mx(1 - x^2)^{m/2}f^{(m)}(x).$$

Differentiating once more produces

$$
\begin{aligned}
[(1 - x^2)g'(x)]' &= (1 - x^2)^{m/2}[(1 - x^2)f^{(m+2)}(x) - 2(m+1)xf^{(m+1)}(x)] \\
&\quad - m(1 - x^2)^{m/2}\frac{1 - (m+1)x^2}{1 - x^2}f^{(m)}(x) \\
&= (m(m+1) - n(n+1))(1 - x^2)^{m/2}f^{(m)}(x) \\
&\quad - \frac{m - m(m+1)x^2}{1 - x^2}(1 - x^2)^{m/2}f^{(m)}(x)
\end{aligned}
$$

Table 8.4 Associated Legendre functions for $n = 0, 1, 2, 3$ and $m = 0, \ldots, n$.

	$m = 0$	$m = 1$	$m = 2$	$m = 3$
$n = 0$	1			
$n = 1$	x	$-\sqrt{1 - x^2}$		
$n = 2$	$(-1 + 3x^2)/2$	$-3x\sqrt{1 - x^2}$	$-3x^2 + 3$	
$n = 3$	$(-3x + 5x^3)/2$	$-\frac{3}{2}(-1 + 5x^2)\sqrt{1 - x^2}$	$-15x(-1 + x^2)$	$-15(1 - x^2)^{3/2}$

using Theorem 8.9. Thus

$$\left[(1 - x^2)g'(x)\right]' + \left(n(n + 1) - \frac{m^2}{1 - x^2}\right)g(x) = 0, \qquad (8.59)$$

or equivalently function $g(x)$ solves the ordinary differential equation,

$$(1 - x^2)g''(x) - 2xg'(x) + \left(n(n + 1) - \frac{m^2}{1 - x^2}\right)g(x) = 0 \qquad (8.60)$$

which is called the **associated Legendre equation**. Equation (8.59) is the self-adjoint form of the associated Legendre differential equation. The solutions are referred to as the **associated Legendre functions**, and are denoted as $P_n^m(x) = (-1)^m(1 - x^2)^{m/2}P_n^{(m)}(x)$ for $n = 0, 1, \ldots$ and $m = 0, 1, \ldots, n$ where $P_n(x)$ is the nth Legendre polynomial.[8] Table 8.4 contains a short list of associated Legendre functions.

The associated Legendre functions are useful in many applications and their definition has been extended for ease of application. The integer parameter m is called the order of the associated Legendre function. If $0 < m \le n$ the associated Legendre function for $-m < 0$ is defined as a scalar multiple of the associated Legendre function of $m > 0$,

$$P_n^{-m}(x) = (-1)^m \frac{(n - m)!}{(n + m)!}P_n^m(x) \text{ for } 0 < m \le n. \qquad (8.61)$$

Lemma 8.11. *If $n + m$ is even then $P_n^m(x)$ is an even function and if $n + m$ is odd then $P_n^m(x)$ is an odd function.*

Proof. Assuming $m \ge 0$, then

$$P_n^m(x) = (-1)^m(1 - x^2)^{m/2}P_n^{(m)}(x).$$

The factor $(1 - x^2)^{m/2}$ is even on $[-1, 1]$, thus the parity of $P_n^m(x)$ is determined by the parity of $P_n^{(m)}(x)$. There are several cases to consider.

---

[8]Some authors define $P_n^m(x) = (1 - x^2)^{m/2}P_n^{(m)}(x)$. This is merely a scalar multiple of the definition presented here and will have no effect on the results presented below.

- If n is odd then $P_n(x)$ is odd and $P_n^{(m)}$ is odd if m is even and $P_n^{(m)}(x)$ is even if m is odd. Thus $P_n^{(m)}(x)$ is even if $n + m$ is even and odd if $n + m$ is odd.
- If n is even then $P_n(x)$ is even and $P_n^{(m)}$ is odd if m is odd and $P_n^{(m)}(x)$ is even if m is even. Thus $P_n^{(m)}(x)$ is even if $n + m$ is even and odd if $n + m$ is odd.

Thus the parity of the associated Legendre functions is established.   □

The associated Legendre functions possess orthogonality properties similar to the properties of the Legendre polynomials.

Theorem 8.10. *Suppose m and n are nonnegative integers such that $m \neq n$ and suppose k is an integer such that $|k| \leq \min\{m, n\}$. Then*

$$\int_{-1}^{1} P_m^k(x) P_n^k(x) \, dx = 0. \tag{8.62}$$

Proof. Define the linear operator $L[y] = [(1-x^2)y']'$. Without loss of generality assume $m < n$. Using the self-adjoint form of the associated Legendre differential equation given in Eq. (8.59), Green's formula (Lemma 7.2) with $u = P_m^k(x)$ and $v = P_n^k(x)$, and integrating over $[-1, 1]$ produce the following,

$$(m(m + 1) - n(n + 1)) \int_{-1}^{1} P_m^k(x) P_n^k(x) \, dx = 0.$$

Since $m \neq n$ this implies $\int_{-1}^{1} P_m^k(x) P_n^k(x) \, dx = 0.$ □

Thus $P_m^k(x)$ and $P_n^k(x)$ are orthogonal on $[-1, 1]$ with respect to the unit weighting function when $m \neq n$.

Theorem 8.11. *If $|k| \leq n$, then*

$$\int_{-1}^{1} (P_n^k(x))^2 \, dx = \frac{2}{2n + 1} \frac{(n + k)!}{(n - k)!}. \tag{8.63}$$

Proof. Assume that $k \geq 0$ and recall that

$$P_n^k(x) = (-1)^k (1 - x^2)^{k/2} P_n^{(k)}(x).$$

Differentiating both sides of the equation produces

$$(-1)^k \frac{dP_n^k}{dx} = -kx(1 - x^2)^{-1+k/2} P_n^{(k)}(x) + (1 - x^2)^{k/2} P_n^{(k+1)}(x)$$

$$(1 - x^2)^{k/2} P_n^{(k+1)}(x) = kx(1 - x^2)^{-1+k/2} P_n^{(k)}(x) + (-1)^k \frac{dP_n^k}{dx}$$

$$(1 - x^2)^{\frac{k+1}{2}} P_n^{(k+1)}(x) = kx(1 - x^2)^{\frac{k-1}{2}} P_n^{(k)}(x) + (-1)^k (1 - x^2)^{\frac{1}{2}} \frac{dP_n^k}{dx}$$

$$-P_n^{k+1}(x) = kx(1 - x^2)^{-\frac{1}{2}} P_n^k(x) + (1 - x^2)^{\frac{1}{2}} \frac{dP_n^k}{dx}. \tag{8.64}$$

Squaring both sides of Eq. (8.64) produces

$$\left(P_n^{k+1}(x)\right)^2 = \frac{k^2 x^2}{1 - x^2} \left(P_n^k(x)\right)^2 + 2kx P_n^k(x) \frac{dP_n^k}{dx} + (1 - x^2) \left(\frac{dP_n^k}{dx}\right)^2. \tag{8.65}$$

Integrating both sides of Eq. (8.65) over $[-1, 1]$ yields

$$\int_{-1}^{1} \left(P_n^{k+1}(x)\right)^2 dx = \int_{-1}^{1} \frac{k^2 x^2}{1 - x^2} \left(P_n^k(x)\right)^2 dx + 2k \int_{-1}^{1} x P_n^k(x) \frac{dP_n^k}{dx} dx$$

$$+ \int_{-1}^{1} (1 - x^2) \left(\frac{dP_n^k}{dx}\right)^2 dx. \tag{8.66}$$

Integrating by parts in the second integral on the right-hand side of Eq. (8.66) yields

$$2k \int_{-1}^{1} x P_n^k(x) \frac{dP_n^k}{dx} dx = [2kx(P_n^k(x))^2]_{x=-1}^{x=1} - 2k \int_{-1}^{1} x P_n^k(x) \frac{dP_n^k}{dx} dx$$

$$- 2k \int_{-1}^{1} \left(P_n^k(x)\right)^2 dx$$

$$4k \int_{-1}^{1} x P_n^k(x) \frac{dP_n^k}{dx} dx = 2k \left(P_n^k(1)\right)^2 + 2k \left(P_n^k(-1)\right)^2 - 2k \int_{-1}^{1} \left(P_n^k(x)\right)^2 dx.$$

Note that if $k = 0$ then $2k \left(P_n^k(1)\right)^2 + 2k \left(P_n^k(-1)\right)^2 = 0$ for all n. If $k > 0$ then $P_n^k(\pm 1) = 0$ since $(1 - x^2)^{k/2} = 0$ when $x = \pm 1$ and again $2k \left(P_n^k(1)\right)^2 + 2k \left(P_n^k(-1)\right)^2 = 0$. This implies

$$2k \int_{-1}^{1} x P_n^k(x) \frac{dP_n^k}{dx} dx = -k \int_{-1}^{1} \left(P_n^k(x)\right)^2 dx. \tag{8.67}$$

Substituting the expression on the right-hand side of Eq. (8.67) into Eq. (8.66) produces

$$\int_{-1}^{1} \left(P_n^{k+1}(x)\right)^2 dx$$

$$= \int_{-1}^{1} \frac{k^2 x^2}{1 - x^2} \left(P_n^k(x) \right)^2 \, dx - k \int_{-1}^{1} \left(P_n^k(x) \right)^2 \, dx + \int_{-1}^{1} (1 - x^2) \left(\frac{dP_n^k}{dx} \right)^2 \, dx$$

$$= \int_{-1}^{1} \left(\frac{k^2 x^2}{1 - x^2} - k \right) \left(P_n^k(x) \right)^2 \, dx + \int_{-1}^{1} (1 - x^2) \left(\frac{dP_n^k}{dx} \right)^2 \, dx. \qquad (8.68)$$

Now focus on the last integral on the right-hand side of Eq. (8.68). Using integration by parts gives

$$\int_{-1}^{1} (1 - x^2) \left(\frac{dP_n^k}{dx} \right)^2 \, dx = - \int_{-1}^{1} P_n^k(x) \left[(1 - x^2) \frac{dP_n^k}{dx} \right]' \, dx,$$

and thus

$$\int_{-1}^{1} \left(P_n^{k+1}(x) \right)^2 \, dx$$

$$= \int_{-1}^{1} \left(\frac{k^2 x^2}{1 - x^2} - k \right) \left(P_n^k(x) \right)^2 \, dx - \int_{-1}^{1} P_n^k(x) \left[(1 - x^2) \frac{dP_n^k}{dx} \right]' \, dx$$

$$= \int_{-1}^{1} \left(\frac{k^2 x^2}{1 - x^2} - k \right) \left(P_n^k(x) \right)^2 \, dx + \int_{-1}^{1} \left(n(n + 1) - \frac{k^2}{1 - x^2} \right) \left(P_n^k(x) \right)^2 \, dx$$

$$= \int_{-1}^{1} \left(n(n + 1) + \frac{k^2 x^2}{1 - x^2} - k - \frac{k^2}{1 - x^2} \right) \left(P_n^k(x) \right)^2 \, dx$$

$$= (n - k)(n + k + 1) \int_{-1}^{1} \left(P_n^k(x) \right)^2 \, dx.$$

Thus there is a recurrence formula for the integral. Consequently,

$$\int_{-1}^{1} \left(P_n^k(x) \right)^2 \, dx$$

$$= (n - k + 1)(n + k) \int_{-1}^{1} \left(P_n^{k-1}(x) \right)^2 \, dx$$

$$= (n - k + 1)(n - k + 2)(n + k)(n + k - 1) \int_{-1}^{1} \left(P_n^{k-2}(x) \right)^2 \, dx$$

$$\vdots$$

$$= (n - k + 1)(n - k + 2) \cdots n(n + k)(n + k - 1) \cdots (n + 1) \int_{-1}^{1} \left(P_n(x) \right)^2 \, dx$$

$$= (n - k + 1)(n - k + 2) \cdots n(n + k)(n + k - 1) \cdots (n + 1) \frac{2}{2n + 1}$$

$$= \frac{2}{2n + 1} \frac{(n + k)!}{(n - k)!}.$$

This completes the proof of Eq. (8.63). □

The associated Legendre functions can be used to expand a function $f(x)$ as an infinite series. If $f(x)$ is sufficiently smooth on $[-1, 1]$, the associated Legendre function expansion of $f(x)$ of order k is

$$f(x) = \sum_{n=k}^{\infty} a_n P_n^k(x),$$

where

$$a_n = \frac{2n+1}{2} \frac{(n-k)!}{(n+k)!} \int_{-1}^{1} f(x) P_n^k(x)\, dx \tag{8.69}$$

for $n = k, k+1, k+2, \ldots$. If $k = 0$ this reduces to the Legendre function expansion.

Example 8.5. Consider the piecewise-defined function

$$f(x) = \begin{cases} 1 & \text{if } -1 < x < 0, \\ 0 & \text{if } 0 < x < 1. \end{cases}$$

Find the associated Legendre function expansion of $f(x)$ of order 3. Plot $f(x)$ and the Nth partial sums of the series expansion for $N = 3, 5$ and 10.

Solution. For $n \geq 3$, Eq. (8.69) implies

$$a_n = \frac{2n+1}{2} \frac{(n-3)!}{(n+3)!} \int_{-1}^{1} f(x) P_n^3(x)\, dx = \frac{2n+1}{2} \frac{(n-3)!}{(n+3)!} \int_{-1}^{0} P_n^3(x)\, dx.$$

After integrating, the coefficients are summarized in the following table.

$$a_3 = -\frac{7\pi}{12}, \qquad a_7 = -\frac{45\pi}{131072},$$
$$a_4 = \frac{3}{160}, \qquad a_8 = \frac{187}{161280},$$
$$a_5 = -\frac{11\pi}{8192}, \qquad a_9 = -\frac{133\pi}{1048576},$$
$$a_6 = \frac{13}{26880}, \qquad a_{10} = \frac{3}{112640}.$$

The plots of $f(x)$ and the partial sums of the associated Legendre function series are shown in Fig. 8.5.

8.4 Spherical Harmonics

The spherical coordinate system is often the most natural and convenient coordinate system in which to analyze a boundary value problem. Given its importance, it is worth the effort to explore the eigenfunctions and eigenvalues of Laplace's equation in spherical coordinates. The eigenfunctions

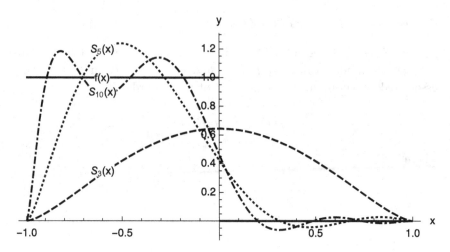

Fig. 8.5   The graph of $f(x)$ and the Nth partial sums of the associated Legendre functions of order 3 for $N = 3, 5$ and 10.

form an important class of orthogonal functions known as the **spherical harmonics**.

Consider Laplace's equation in spherical coordinates (ρ, φ, θ),[9]

$$\frac{\partial^2 u}{\partial \rho^2} + \frac{2}{\rho}\frac{\partial u}{\partial \rho} + \frac{1}{\rho^2}\left(\frac{\partial^2 u}{\partial \varphi^2} + \cot\varphi\frac{\partial u}{\partial \varphi} + \csc^2\varphi\frac{\partial^2 u}{\partial \theta^2}\right) = 0. \tag{8.70}$$

Assuming a product solution of the form $u(\rho, \varphi, \theta) = R(\rho)\Phi(\varphi)\Theta(\theta)$ and separating variables result in the following.

$$0 = R''(\rho)\Phi(\varphi)\Theta(\theta) + \frac{2}{\rho}R'(\rho)\Phi(\varphi)\Theta(\theta) + \frac{1}{\rho^2}R(\rho)\Phi''(\varphi)\Theta(\theta)$$

$$+ \frac{1}{\rho^2}\left(\cot\varphi R(\rho)\Phi'(\varphi)\Theta(\theta) + \csc^2\varphi R(\rho)\Phi(\varphi)\Theta''(\theta)\right)$$

$$-\frac{\Theta''(\theta)}{\Theta(\theta)} = \sin^2\varphi\left(\rho^2\frac{R''(\rho)}{R(\rho)} + 2\rho\frac{R'(\rho)}{R(\rho)} + \frac{\Phi''(\varphi)}{\Phi(\varphi)} + \cot\varphi\frac{\Phi'(\varphi)}{\Phi(\varphi)}\right).$$

The right-hand side of the equation depends on ρ and φ and the left-hand side depends on θ, thus both sides equal a constant, denoted as σ. Consequently one of the ordinary differential equations implied by the separation

---

[9]The reader should note that coordinate φ is the angle between the positive z-axis and the line formed by connecting a point in three-dimensional space to the origin. As such, $0 \le \varphi \le \pi$. The coordinate θ is the azimuthal coordinate as in the polar coordinate system. This is conventional notation for mathematicians though physicists and engineers often reverse the interpretations of φ and θ. The reader will find a derivation of Eq. (8.70) in [Pinsky (1998), Sec. 4.1.1].

of variables procedure is

$$\Theta''(\theta) + \sigma\Theta(\theta) = 0.$$

The solution to Eq. (8.70) should be 2π-periodic in variable θ, thus the needed constant is $\sigma = m^2$ where $m = 0, 1, 2, \ldots$ For convenience express the general solution as a linear combination of the complex exponentials $\Theta(\theta) = c_1 e^{i m \theta} + c_2 e^{-i m \theta}$ where $i = \sqrt{-1}$. During applications of the spherical harmonics the reader can select the real or complex part of the solution as needed. Turning attention to the remaining equation,

$$\sin^2\varphi\left(\rho^2\frac{R''(\rho)}{R(\rho)} + 2\rho\frac{R'(\rho)}{R(\rho)} + \frac{\Phi''(\varphi)}{\Phi(\varphi)} + \cot\varphi\frac{\Phi'(\varphi)}{\Phi(\varphi)}\right) = m^2,$$

the variables ρ and ϕ can be separated as follows

$$\rho^2\frac{R''(\rho)}{R(\rho)} + 2\rho\frac{R'(\rho)}{R(\rho)} = -\frac{\Phi''(\varphi)}{\Phi(\varphi)} - \cot\varphi\frac{\Phi'(\varphi)}{\Phi(\varphi)} + m^2\csc^2\varphi.$$

It is readily seen that the right-hand side of the equation is a function of φ only and the left-hand side is a function of ρ only, thus both sides are constant, denoted as λ. Consider the equation:

$$-\frac{\Phi''(\varphi)}{\Phi(\varphi)} - \cot\varphi\frac{\Phi'(\varphi)}{\Phi(\varphi)} + m^2\csc^2\varphi = \lambda$$

$$\Phi''(\varphi) + \cot\varphi\Phi'(\varphi) + (\lambda - m^2\csc^2\varphi)\Phi(\varphi) = 0,$$

where λ is a constant and make the change of variable $x = \cos\varphi$. Making use of the chain rule for derivatives, the equation above can be rewritten as

$$(1 - x^2)\frac{d^2\Phi}{dx^2} - 2x\frac{d\Phi}{dx} + \left(\lambda - \frac{m^2}{1 - x^2}\right)\Phi = 0.$$

If $\lambda = n(n+1)$ where $n = 0, 1, 2, \ldots$, this becomes the associated Legendre differential equation discussed in Sec. 8.3.2. The solutions are the $2n + 1$ linearly independent functions $P_n^m(x)$ for $-n \leq m \leq n$.

Combining the two factors of the product solution found so far yields

$$f(\varphi, \theta) = e^{i m \theta}P_n^m(\cos\varphi)$$

with $n = 0, 1, 2, \ldots$ and $-n \leq m \leq n$. Even though $f(\varphi, \theta)$ depends on two independent variables, these functions have orthogonality properties similar to the eigenfunctions studied earlier in this chapter.

Theorem 8.12. *Let $n = 0, 1, 2, \ldots$ and let m be an integer with $-n \leq m \leq n$. Likewise let $\hat{n} = 0, 1, 2, \ldots$ and let $\hat{m}$ be an integer with $-\hat{n} \leq \hat{m} \leq \hat{n}$ then*

$$\int_{-\pi}^{\pi}\int_{0}^{\pi} e^{i m \theta}P_n^m(\cos\varphi)e^{-i\hat{m}\theta}P_{\hat{n}}^{\hat{m}}(\cos\varphi)\sin\varphi\, d\varphi\, d\theta = 0 \qquad (8.71)$$

Table 8.5   Spherical harmonic functions for $n = 0, 1, 2, 3$ and $m = 0, \ldots, n$.

	$m = 0$	$m = 1$	$m = 2$	$m = 3$
$n = 0$	$\frac{1}{2\sqrt{\pi}}$			
$n = 1$	$\frac{1}{2}\sqrt{\frac{3}{\pi}}\cos\varphi$	$-\frac{1}{2}\sqrt{\frac{3}{2\pi}}e^{i\theta}\sin\varphi$		
$n = 2$	$\frac{1}{4}\sqrt{\frac{5}{\pi}}(3\cos^2\varphi - 1)$	$-\frac{1}{2}\sqrt{\frac{15}{2\pi}}e^{i\theta}\cos\varphi\sin\varphi$	$\frac{1}{4}\sqrt{\frac{15}{2\pi}}e^{2i\theta}\sin^2\varphi$	
$n = 3$	$\frac{1}{4}\sqrt{\frac{7}{\pi}}(5\cos^3\varphi - 3\cos\varphi)$	$-\frac{1}{8}\sqrt{\frac{21}{\pi}}e^{i\theta}(5\cos^2\varphi - 1)\sin\varphi$	$\frac{1}{4}\sqrt{\frac{105}{2\pi}}e^{2i\theta}\cos\varphi\sin^2\varphi$	$-\frac{1}{8}\sqrt{\frac{35}{\pi}}e^{3i\theta}\sin^3\varphi$

if $n \neq \hat{n}$ or $m \neq \hat{m}$. On the other hand if $n = \hat{n}$ and $m = \hat{m}$, then

$$\int_{-\pi}^{\pi}\int_{0}^{\pi} \left(P_n^m(\cos\varphi)\right)^2 \sin\varphi \, d\varphi \, d\theta = \frac{4\pi}{2n+1}\frac{(n+m)!}{(n-m)!}. \tag{8.72}$$

Proof. See Exercise 27.   □

For integers $n = 0, 1, 2, \ldots$ and $-n \leq m \leq n$ the normalized function $f(\varphi, \theta)$ is denoted as $Y_n^m(\varphi, \theta)$ where

$$Y_n^m(\varphi, \theta) = \sqrt{\frac{2n+1}{4\pi}\frac{(n-m)!}{(n+m)!}}e^{im\theta}P_n^m(\cos\varphi). \tag{8.73}$$

These functions are called the spherical harmonic functions. There are $2n + 1$ orthonormal functions corresponding to the eigenvalue $n(n + 1)$. Some authors denote the spherical harmonics as $S_{n,m}(\varphi, \theta)$, though the notation adopted here seems to be more common. The reader should be careful not to confuse the notation for the spherical harmonic functions with the notation for the Bessel functions of the second kind. Table 8.5 lists several examples of spherical harmonic functions for some small values of n.

8.5   The Laguerre Equation

The **Laguerre**[10] **differential equation** is

$$xy'' + (1 - x)y' + \lambda y = 0 \text{ for } x > 0. \tag{8.74}$$

The solutions to the Laguerre equation find applications in quantum mechanics, numerical integration, and applied mathematics. This section will explore the infinite series solution to Eq. (8.74) and show that when $\lambda \in \{0, 1, 2, \ldots\}$ one solution to the equation is a polynomial. These polynomial solutions form a family of special functions known collectively as

---

[10]Edmond Nicolas Laguerre, French mathematician (1834–1886).

the **Laguerre polynomials.** As always, Laguerre's differential equation has an expression in self-adjoint form,

$$\left[xe^{-x}y'\right]' + \lambda e^{-x}y = 0 \text{ for } x > 0. \tag{8.75}$$

This is a singular Sturm-Liouville boundary value problem. Equation (8.74) has a regular singular point at $x = 0$ since

$$\lim_{x\to 0} x\left(\frac{1-x}{x}\right) = \lim_{x\to 0}(1-x) = 1 \text{ and}$$

$$\lim_{x\to 0} x^2\left(\frac{\lambda}{x}\right) = \lim_{x\to 0}\lambda x = 0.$$

The solution to Eq. (8.74) can be found using the Method of Frobenius. Assume the series solution has the form $y(x) = \sum_{n=0}^{\infty} a_n x^{r+n}$ and substitute this solution into Eq. (8.74) to obtain

$$0 = x\sum_{n=0}^{\infty}(r+n)(r+n-1)a_n x^{r+n-2} + (1-x)\sum_{n=0}^{\infty}(r+n)a_n x^{r+n-1}$$

$$+ \lambda\sum_{n=0}^{\infty} a_n x^{r+n}$$

$$= r^2 a_0 x^{r-1} + \sum_{n=1}^{\infty}\left[(r+n)^2 a_n + (\lambda - r - n + 1)a_{n-1}\right]x^{r+n-1}.$$

Thus the indicial equation is $r^2 = 0$ which has exponents of singularity $r_1 = r_2 = 0$. The recurrence relation then is

$$a_n = \frac{(n-1-\lambda)a_{n-1}}{n^2} \text{ for } n \geq 1.$$

Now if a_0 is arbitrary, then

$$a_1 = -\frac{\lambda}{1^2}a_0 = -\frac{\lambda}{(1!)^2}a_0,$$

$$a_2 = \frac{1-\lambda}{2^2}a_1 = -\frac{\lambda(1-\lambda)}{(2!)^2}a_0,$$

$$a_3 = \frac{2-\lambda}{3^2}a_2 = -\frac{\lambda(1-\lambda)(2-\lambda)}{(3!)^2}a_0,$$

and in general

$$a_n = \frac{\prod_{k=0}^{n-1}(k-\lambda)}{(n!)^2}. \tag{8.76}$$

Thus one solution to Laguerre's equation is

$$y_1(x) = a_0\left(1 + \sum_{n=1}^{\infty}\frac{\prod_{k=0}^{n-1}(k-\lambda)}{(n!)^2}x^n\right). \tag{8.77}$$

Table 8.6   The formulas for the Laguerre polynomials $L_m(x)$ for $m = 0$, 1, 2, 3, and 4.

$L_0(x) = 1$
$L_1(x) = 1 - x$
$L_2(x) = 1 - 2x + x^2/2$
$L_3(x) = 1 - 3x + 3x^2/2 - x^3/6$
$L_4(x) = 1 - 4x + 3x^2 - 2x^3/3 + x^4/24$

When $\lambda = m \in \{0, 1, 2, \ldots\}$ then according to Eq. (8.76) the coefficients of the series solution

$$a_{m+1} = a_{m+2} = \cdots = a_{m+k} = \cdots = 0,$$

and $y_1(x)$ contains powers of x less than or equal to m, in other words, $y_1(x)$ is a polynomial of degree at most m. In this case, upon choosing $a_0 = 1$ the solution $y_1(x)$ traditionally denoted as $L_m(x)$ becomes

$$L_m(x) = 1 + \sum_{n=1}^{m} \frac{\prod_{k=0}^{n-1}(k - m)}{(n!)^2} x^n$$

$$= 1 + \sum_{n=1}^{m} \frac{(-1)^n m(m-1)(m-2)\cdots(m-(n-1))}{(n!)^2} x^n$$

$$= m! \sum_{n=0}^{m} \frac{(-1)^n}{(n!)^2(m-n)!} x^n.$$

Table 8.6 lists the first several Laguerre polynomials while Fig. 8.6 displays the graphs of these Laguerre polynomials.

8.5.1   *Properties of the Laguerre Polynomials*

The Laguerre polynomials possess many interesting relationships similar to those of the Legendre polynomials. For example there exists a Rodrigues' formula for the Laguerre polynomials.

Lemma 8.12. *For $n = 0, 1, 2, \ldots$ the Rodrigues' form of the nth Laguerre polynomial is*

$$L_n(x) = \frac{e^x}{n!} \frac{d^n}{dx^n}[e^{-x} x^n]. \tag{8.78}$$

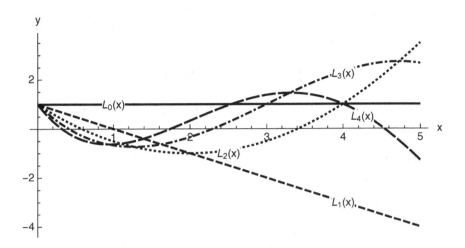

Fig. 8.6   The graphs of the Laguerre polynomials $L_m(x)$ for $m = 0, 1, 2, 3,$ and 4.

Proof. Re-writing the nth Laguerre polynomial as

$$L_n(x) = \frac{e^x}{n!} \sum_{k=0}^{n} \frac{(-1)^k (n!)^2 e^{-x}}{(k!)^2 (n-k)!} x^k$$

$$= \frac{e^x}{n!} \sum_{k=0}^{n} \left[\frac{n!}{k!(n-k)!} (-1)^k e^{-x} \right] \left[\frac{n!}{(n-(n-k))!} x^k \right]$$

$$= \frac{e^x}{n!} \sum_{k=0}^{n} \frac{n!}{k!(n-k)!} \frac{d^k}{dx^k} [e^{-x}] \frac{d^{n-k}}{dx^{n-k}} [x^n] = \frac{e^x}{n!} \frac{d^n}{dx^n} \left[e^{-x} x^n \right].$$

The last step follows from the Leibniz product rule for the nth derivative given in Lemma 8.9. $\qquad\square$

The Rodrigues' formula for the Laguerre polynomials helps to establish their orthogonality properties. Consider the integral,

$$\int_0^{\infty} (L_n(x))^2 e^{-x} \, dx.$$

Replacing one instance of the Laguerre polynomial by its Rodrigues' formula and the other instance by its summation formula produces

$$\int_0^{\infty} (L_n(x))^2 e^{-x} \, dx = \int_0^{\infty} \left(n! \sum_{k=0}^{n} \frac{(-1)^k x^k}{(k!)^2 (n-k)!} \right) \frac{e^x}{n!} \frac{d^n}{dx^n} [e^{-x} x^n] e^{-x} \, dx$$

$$= \sum_{k=0}^{n} \frac{(-1)^k}{(k!)^2 (n-k)!} \int_0^{\infty} x^k \frac{d^n}{dx^n} [e^{-x} x^n] \, dx.$$

Since $\frac{d^n}{dx^n}[x^k] = 0$ for $k = 0, 1, \ldots, n-1$, then by using integration by parts n times, it can be shown that

$$\int_0^\infty x^k \frac{d^n}{dx^n}[e^{-x}x^n]\, dx = 0$$

for $k = 0, 1, \ldots, n-1$. Thus the only nonzero term in the summation above is the $k = n$ term. Hence repeatedly using integration by parts produces

$$\int_0^\infty (L_n(x))^2 e^{-x}\, dx$$

$$= \frac{(-1)^n}{(n!)^2} \int_0^\infty x^n \frac{d^n}{dx^n}[e^{-x}x^n]\, dx$$

$$= \left[\frac{(-1)^n}{(n!)^2} x^n \frac{d^{n-1}}{dx^{n-1}}\left[e^{-x}x^n\right]\right]_0^\infty + \frac{(-1)^{n-1}}{n!(n-1)!} \int_0^\infty x^{n-1} \frac{d^{n-1}}{dx^{n-1}}\left[e^{-x}x^n\right]\, dx$$

$$= \frac{1}{n!} \int_0^\infty e^{-x}x^n\, dx = 1.$$

Define the linear operator $L[y] = [xe^{-x}y']'$. Using the self-adjoint form of the Laguerre differential equation given in Eq. (8.75), Green's formula stated in Lemma 7.2 with $u = L_m(x)$ and $v = L_n(x)$, and integrating over $[0, \infty)$ yield

$$(n - m) \int_0^\infty L_m(x)L_n(x)e^{-x}\, dx = 0.$$

Since $n \neq m$, then $\int_0^\infty L_m(x)L_n(x)e^{-x}\, dx = 0$ and the following orthogonality result for the Laguerre polynomials has been established.

Theorem 8.13. *The Laguerre polynomials are orthogonal on $[0, \infty)$ with respect to the weighting function e^{-x}. For all nonnegative integers m and n, then*

$$\int_0^\infty L_m(x)L_n(x)e^{-x}\, dx = \delta_{mn},$$

where δ_{mn} is the Kronecker delta function.

The reader should note that the Laguerre polynomials as defined form an orthonormal set with respect to the weighting function e^{-x}. If $f(x)$ is sufficiently smooth on the interval $[0, \infty)$ and if

$$\int_0^\infty (f(x))^2 e^{-x}\, dx < \infty,$$

then $f(x)$ may be represented as a generalized Fourier series (a Laguerre series) in terms of the Laguerre polynomials:

$$f(x) = \sum_{n=0}^{\infty} a_n L_n(x).$$

The coefficients a_n of the series may be found from the integral formula

$$a_n = \int_0^{\infty} f(x) L_n(x) e^{-x} \, dx \tag{8.79}$$

for $n = 0, 1, \ldots$.

Example 8.6. Find the generalized Fourier series for $f(x) = e^{-x}$ in terms of the Laguerre polynomials. Plot $f(x)$ and the Nth partial sums of the series on the same set of axes for $N = 3, 5$ and 10.

Solution. First, checking that

$$\int_0^{\infty} (e^{-x})^2 e^{-x} \, dx = \int_0^{\infty} e^{-3x} \, dx = \frac{1}{3} < \infty,$$

then Eq. (8.79) may be used to determine the Laguerre coefficients.

$$a_n = \int_0^{\infty} e^{-x} L_n(x) e^{-x} \, dx = \frac{1}{n!} \int_0^{\infty} \frac{d^n}{dx^n} \left[e^{-x} x^n \right] e^{-x} \, dx$$

$$= \frac{1}{n!} \int_0^{\infty} e^{-x} x^n e^{-x} \, dx$$

The last expression is produced by using integration by parts n times. Using integration by parts n more times yields $a_n = 1/2^{n+1}$. Therefore the Laguerre series expansion of $f(x) = e^{-x}$ is

$$e^{-x} = \sum_{n=0}^{\infty} \frac{1}{2^{n+1}} L_n(x).$$

Figure 8.7 shows the graph of $f(x)$ and the Nth partial sums of the Laguerre series for $N = 3, 5$ and 10.

8.5.2 *Associated Laguerre Equation and Polynomials*

The **associated Laguerre differential equation** is

$$xy'' + (\alpha + 1 - x)y' + \lambda y = 0 \text{ for } x \geq 0, \tag{8.80}$$

where $\alpha > -1$ is a real number and λ is also a real number. The associated Laguerre equation often appears when solving the Schrödinger[11] equation

---

[11]Erwin Schrödinger, Austrian physicist (1887–1961).

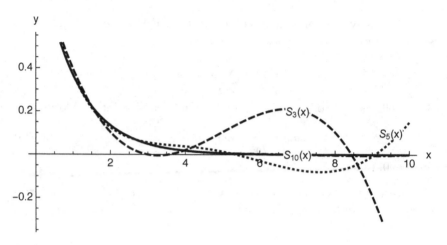

Fig. 8.7   The graph of e^{-x} and the Nth partial sums of the Laguerre series for $N = 3, 5$ and 10. The graph of e^{-x} is the solid curve. The Laguerre series approximations $S_N(x)$ approach e^{-x} as N increases.

of quantum mechanics. Much of the behavior of the associated Laguerre equation parallels that of the Laguerre equation, Eq. (8.74). Multiplying both sides of Eq. (8.80) by $x^\alpha e^{-x}$ and using the product rule for derivatives result in the self-adjoint form of the associated Laguerre differential equation,

$$\left[x^{\alpha+1}e^{-x}y'\right]' + \lambda x^\alpha e^{-x}y = 0 \text{ for } x \geq 0. \tag{8.81}$$

This is another example of a singular Sturm-Liouville boundary value problem. The value $x = 0$ is a regular singular point for Eq. (8.80) since

$$\lim_{x \to 0} x\left(\frac{\alpha + 1 - x}{x}\right) = \lim_{x \to 0}(\alpha + 1 - x) = \alpha + 1 \text{ and}$$

$$\lim_{x \to 0} x^2\left(\frac{\lambda}{x}\right) = \lim_{x \to 0} \lambda x = 0.$$

The exponents of singularity are $r = 0$ and $r = -\alpha$.

Assume the equation has a series solution of the form $y(x) = \sum_{k=0}^{\infty} a_k x^k$ (corresponding to the exponent of singularity $r = 0$) and substitute it into the associated Laguerre equation. This yields

$$0 = x\sum_{k=0}^{\infty} k(k-1)a_k x^{k-2} + (\alpha + 1 - x)\sum_{k=0}^{\infty} ka_k x^{k-1} + \lambda \sum_{k=0}^{\infty} a_k x^k$$

$$= \sum_{k=0}^{\infty} \left[(k+1)(k+1+\alpha)a_{k+1} + (\lambda - k)a_k\right] x^k.$$

The recurrence relation is

$$a_{k+1} = -\frac{(\lambda - k)a_k}{(k+1)(k+1+\alpha)} \text{ for } k \geq 0.$$

Note that the coefficients are indexed by k and the expression λ is a parameter of the associated Laguerre equation. The first term a_0, is arbitrary and by convention it is chosen to be

$$a_0 = \frac{\Gamma(1+\lambda+\alpha)}{\Gamma(1+\lambda)\Gamma(1+\alpha)}, \text{ which implies}$$

$$a_1 = -\frac{\lambda a_0}{(1)(1+\alpha)} = \frac{(-1)^1(\lambda-1+1)_1\Gamma(1+\lambda+\alpha)}{1!\,(1+\alpha)_1\Gamma(1+\lambda)\Gamma(1+\alpha)}$$

$$\vdots$$

$$a_k = \frac{(-1)^k(\lambda-k+1)_k\Gamma(1+\lambda+\alpha)}{k!\,(1+\alpha)_k\Gamma(1+\lambda)\Gamma(1+\alpha)} \text{ for } k \geq 0.$$

The Pochhammer symbol has been used to simplify the formula for the coefficients. A general infinite series solution to the associated Laguerre equation corresponding to the exponent of singularity $r = 0$ can be expressed as

$$y_1(x) = \frac{\Gamma(1+\lambda+\alpha)}{\Gamma(1+\lambda)\Gamma(1+\alpha)} \sum_{k=1}^{\infty} \frac{(-1)^k(\lambda-k+1)_k}{k!\,(1+\alpha)_k} x^k.$$

A second linearly independent solution to the associated Laguerre equation can be found using the method of Frobenius with the exponent of singularity $r = -\alpha$. This solution will not be needed and thus is omitted from the discussion. An application of the Ratio Test suggests $y_1(x)$ converges absolutely for all $x \geq 0$. The ratio $a_{k+1}/a_k = O(1/k)$ for large k and thus it can be shown by induction that $a_{k+1} \approx C/k!$ where C is a constant when k is large. Consequently for a large integer N the infinite series can be written as

$$y_1(x) = a_0 \sum_{k=1}^{N} \frac{(-1)^k(\lambda-k+1)_k}{k!\,(1+\alpha)_k} x^k + a_0 \sum_{k=N+1}^{\infty} \frac{(-1)^k(\lambda-k+1)_k}{k!\,(1+\alpha)_k} x^k$$

$$\approx a_0 \sum_{k=1}^{N} \frac{(-1)^k(\lambda-k+1)_k}{k!\,(1+\alpha)_k} x^k + C \sum_{k=N+1}^{\infty} \frac{x^k}{(k-1)!} \approx C x\, e^x,$$

for sufficiently large x. Therefore $y_1(x)$ grows like the expression $x\,e^x$ for large x. For this reason a special case of the associated Laguerre equation is preferred in applications, one in which the solutions grow like polynomials.

Table 8.7   The first four associated Laguerre polynomials $L_n^\alpha(x)$.

$L_0^\alpha(x) = 1$
$L_1^\alpha(x) = 1 + \alpha - x$
$L_2^\alpha(x) = 1 + 3\alpha/2 + \alpha^2/2 - (2 + \alpha)x + x^2/2$
$L_3^\alpha(x) = 1 + 11\alpha/6 + \alpha^2 + \alpha^3/6 - (3 + 5\alpha/2)x + (3/2 + \alpha/2)x^2 - x^3/6$

When $\lambda = n$, a nonnegative integer, then the series coefficients simplify to

$$a_k = \frac{(-1)^k \Gamma(1 + n + \alpha)}{k!(n-k)!\Gamma(1 + k + \alpha)} \text{ for } k = 0, 1, \ldots, n,$$

and $a_k = 0$ for $k \geq n + 1$. Thus when $\lambda = n$, a nonnegative integer, one solution to the associated Laguerre equation is the polynomial

$$L_n^\alpha(x) = \sum_{k=0}^{n} \frac{(-1)^k \Gamma(1 + n + \alpha)x^k}{k!(n-k)!\Gamma(1 + k + \alpha)}. \tag{8.82}$$

It should be noted that when $\alpha = 0$ the associated Laguerre equation simplifies to the original Laguerre equation and

$$L_n^0(x) = \sum_{k=0}^{n} \frac{(-1)^k \Gamma(1 + n + 0)x^k}{k!(n-k)!\Gamma(1 + k + 0)} = n! \sum_{k=0}^{n} \frac{(-1)^k x^k}{(k!)^2(n-k)!} = L_n(x).$$

The formulas for the first four associated Laguerre polynomials $L_n^\alpha(x)$ are listed in Table 8.7. For the remainder of this section, λ will be assumed to be a nonnegative integer n.

Like many of the families of special functions studied in this chapter, the associated Laguerre polynomials have a Rodrigues' formula from which other properties are easier to derive.

Lemma 8.13. *The associated Laguerre polynomial can be expressed as*

$$L_n^\alpha(x) = \frac{x^{-\alpha} e^x}{n!} \frac{d^n}{dx^n} [e^{-x} x^{n+\alpha}].$$

Proof. Rewrite the associated Laguerre polynomial as

$$L_n^\alpha(x) = \frac{x^{-\alpha}e^x}{n!} \sum_{k=0}^{n} \frac{(-1)^k n! \Gamma(n+1+\alpha)e^{-x}x^{k+\alpha}}{k!(n-k)!\Gamma(k+1+\alpha)}$$

$$= \frac{x^{-\alpha}e^x}{n!} \sum_{k=0}^{n} \left[\frac{n!}{k!(n-k)!}(-1)^k e^{-x}\right]\left[\frac{\Gamma(n+1+\alpha)}{\Gamma(k+1+\alpha)}x^{k+\alpha}\right]$$

$$= \frac{x^{-\alpha}e^x}{n!} \sum_{k=0}^{n} \frac{n!}{k!(n-k)!}\frac{d^k}{dx^k}[e^{-x}][(k+1+\alpha)_{n-k}x^{k+\alpha}]$$

$$= \frac{x^{-\alpha}e^x}{n!} \sum_{k=0}^{n} \frac{n!}{k!(n-k)!}\frac{d^k}{dx^k}[e^{-x}]\frac{d^{n-k}}{dx^{n-k}}[x^{n+\alpha}]$$

$$= \frac{x^{-\alpha}e^x}{n!}\frac{d^n}{dx^n}[e^{-x}x^{n+\alpha}].$$

The last step results from the use of the Leibniz product rule for differentiation. □

Now the orthogonality properties of the associated Laguerre polynomials may be derived in a manner similar to the orthogonality properties of the other eigenfunctions studied so far.

Theorem 8.14. *The associated Laguerre polynomials $L_n^\alpha(x)$, are orthogonal on $[0,\infty)$ with respect to the weighting function $e^{-x}x^\alpha$. For all non-negative integers m and n,*

$$\int_0^\infty L_m^\alpha(x)L_n^\alpha(x)e^{-x}x^\alpha\,dx = \frac{\Gamma(n+\alpha+1)}{n!}\delta_{mn}, \qquad (8.83)$$

where δ_{mn} is the Kronecker delta function.

Proof. Define the linear operator $L[y] = [x^{\alpha+1}e^{-x}y']'$. Assume $m \neq n$, then using the self-adjoint form of the associated Laguerre differential equation in Eq. (8.81), applying Green's formula from Lemma 7.2 with $u = L_n^\alpha(x)$ and $v = L_m^\alpha(x)$, and integrating over $[0,\infty)$ yield

$$(n-m)\int_0^\infty L_m^\alpha(x)L_n^\alpha(x)e^{-x}x^\alpha\,dx = 0.$$

Thus when $m \neq n$ the improper integral $\int_0^\infty L_m^\alpha(x)L_n^\alpha(x)e^{-x}x^\alpha\,dx = 0$. Now consider the case when $m = n$ and the improper integral

$$\int_0^\infty (L_n^\alpha(x))^2 e^{-x}x^\alpha\,dx.$$

In the integrand, one factor of $L_n^\alpha(x)$ is replaced by its Rodrigues' formula from Lemma 8.13. Consequently the integral can be expressed as

$$\int_0^\infty (L_n^\alpha(x))^2 e^{-x} x^\alpha \, dx$$

$$= \int_0^\infty L_n^\alpha(x) \left(\frac{x^{-\alpha} e^x}{n!} \frac{d^n}{dx^n} \left[e^{-x} x^{n+\alpha} \right] \right) e^{-x} x^\alpha \, dx$$

$$= \frac{1}{n!} \int_0^\infty L_n^\alpha(x) \frac{d^n}{dx^n} \left[e^{-x} x^{n+\alpha} \right] dx$$

$$= \frac{1}{n!} \left[L_n^\alpha(x) \frac{d^{n-1}}{dx^{n-1}} \left[e^{-x} x^{n+\alpha} \right] \right]_0^\infty - \frac{1}{n!} \int_0^\infty (L_n^\alpha)'(x) \frac{d^{n-1}}{dx^{n-1}} \left[e^{-x} x^{n+\alpha} \right] dx.$$

Consider the term in the square brackets above. Using Lemma 8.9 to expand the derivative produces in every term containing a factor of e^{-x}, a factor of $x^{\alpha+k}$ with $k = 1, 2, \ldots, n$ and all other factors being constants. Thus as $x \to \infty$ the expression in the brackets approaches zero asymptotically. Since $\alpha > -1$ then $x^{\alpha+k} = 0$ for $k = 1, 2, \ldots, n$ when $x = 0$. Hence the term in the square brackets above vanishes. Applying integration by parts $n - 1$ more times produces

$$\int_0^\infty (L_n^\alpha(x))^2 e^{-x} x^\alpha \, dx = \frac{(-1)^n}{n!} \int_0^\infty (L_n^\alpha)^{(n)}(x) e^{-x} x^{n+\alpha} \, dx.$$

Note that according to the form of the associated Laguerre polynomial given in Eq. (8.82), $(L_n^\alpha)^{(n)}(x) = (-1)^n$ so

$$\int_0^\infty (L_n^\alpha(x))^2 e^{-x} x^\alpha \, dx = \frac{1}{n!} \int_0^\infty e^{-x} x^{n+\alpha} \, dx = \frac{\Gamma(n + \alpha + 1)}{n!}.$$

This establishes the mutual orthogonality property of the associated Laguerre polynomials.   □

Using Theorem 8.14, suitably smooth functions can be represented as **generalized Laguerre series**, that is,

$$f(x) = \sum_{n=0}^\infty a_n L_n^\alpha(x),$$

where

$$a_n = \frac{n!}{\Gamma(n + \alpha + 1)} \int_0^\infty f(x) L_n^\alpha(x) e^{-x} x^\alpha \, dx \qquad (8.84)$$

is the nth **generalized Laguerre coefficient**. Since the coefficients depend on α, many authors choose to denote the coefficients as a_n^α.

Example 8.7. Find the generalized Laguerre series expansion for $f(x) = \sqrt{x}$ for $\alpha = 1/2$. Plot $f(x)$ and the partial sums of the series $S_N(x)$ for $N = 3, 5$, and 10.

Solution. Using the formula for the nth generalized Laguerre coefficient given in Eq. (8.84), for $n \in \mathbb{N}$,

$$
\begin{aligned}
a_n &= \frac{n!}{\Gamma(n + \frac{1}{2} + 1)} \int_0^\infty x^{1/2} L_n^{1/2}(x) e^{-x} x^{1/2}\, dx \\
&= \frac{1}{\Gamma(n + \frac{3}{2})} \int_0^\infty x^{1/2} \frac{d^n}{dx^n} \left[e^{-x} x^{n+1/2} \right] dx \\
&= \frac{-(1)(3)\cdots(2n-3)}{2^n \Gamma(n + \frac{3}{2})} \int_0^\infty x e^{-x}\, dx \\
&= \frac{-(1)(3)\cdots(2n-3)}{2^n \Gamma(n + \frac{3}{2})} = -\frac{2}{(2n+1)(2n-1)\sqrt{\pi}}.
\end{aligned}
$$

When $n = 0$, then

$$
a_0 = \frac{1}{\Gamma(3/2)} = \frac{2}{\sqrt{\pi}}.
$$

The plots of $f(x)$ and its Nth partial sums for $N = 3, 5$, and 10 are shown in Fig. 8.8.

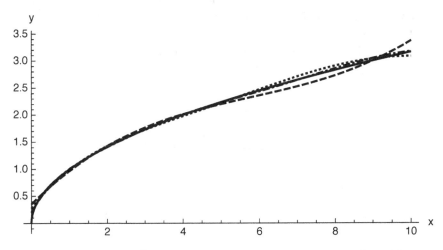

Fig. 8.8 The graphs of $\sqrt{x}$ and the Nth partial sums for $N = 3, 5$, and 10 of its generalized Laguerre series expansion. The graph of $\sqrt{x}$ is the solid curve. The generalized Laguerre series appear to converge to $\sqrt{x}$ as N increases.

Many of the results mentioned in this section depended on the assumption that the parameter $\alpha > -1$. The associated Laguerre polynomials can be extended for arbitrary (complex) α, though this extension is not needed in this text. The interested reader can consult [Szegö (1939), Sec. 5.2].

8.6 The Hermite Polynomials and Their Properties

The equation

$$y'' - 2xy' + \lambda y = 0 \text{ for } -\infty < x < \infty, \tag{8.85}$$

where λ is a constant, is known as the **Hermite**[12] **equation**. This is an important equation in mathematical physics and is often encountered in applications involving the quantum harmonic oscillator. This section will explore the solutions to Eq. (8.85) and their properties. Equation (8.85) can be expressed in self-adjoint form as

$$[e^{-x^2}y']' + \lambda e^{-x^2}y = 0 \text{ for } -\infty < x < \infty,$$

which is a singular Sturm-Liouville equation. The presence of the weighting function e^{-x^2} should signal a connection between Hermite's equation and probability. Readers with interests and backgrounds in probability may already be familiar with the functions which solve the equation.

The point $x_0 = 0$ is an ordinary point for Eq. (8.85). Power series techniques will reveal the solution to the Hermite equation. Assume the solution is a power series of the form $y(x) = \sum_{n=0}^{\infty} a_n x^n$ and substitute this series into Eq. (8.85).

$$
\begin{aligned}
0 &= \sum_{n=2}^{\infty} n(n-1)a_n x^{n-2} - 2x \sum_{n=1}^{\infty} n a_n x^{n-1} + \lambda \sum_{n=0}^{\infty} a_n x^n \\
&= \sum_{n=0}^{\infty} [(n+2)(n+1)a_{n+2} - 2na_n + \lambda a_n] x^n.
\end{aligned}
\tag{8.86}
$$

By equating coefficients of x on both sides of Eq. (8.86), the following recurrence relation results:

$$a_{n+2} = \frac{(2n - \lambda)a_n}{(n+2)(n+1)}. \tag{8.87}$$

---

[12]Charles Hermite, French mathematician (1822–1901).

If $a_0 = 1$ and $a_1 = 0$ then the following coefficients may be calculated for one solution to Eq. (8.85).

$$a_2 = -\frac{\lambda}{2!}$$

$$a_4 = \frac{(4 - \lambda)a_2}{(4)(3)} = -\frac{\lambda(4 - \lambda)}{4!}$$

$$\vdots$$

$$a_{2n} = \frac{\prod_{k=1}^{n}(4k - 4 - \lambda)}{(2n)!}.$$

Since $a_1 = 0$ then $0 = a_{2n-1}$ for $n \in \mathbb{N}$. Hence one solution to Eq. (8.85) can be written as

$$y_1(x) = 1 + \sum_{n=1}^{\infty} \frac{\prod_{k=1}^{n}(4k - 4 - \lambda)}{(2n)!} x^{2n}. \tag{8.88}$$

If $a_0 = 0$ and $a_1 = 1$ then the coefficients of another solution become:

$$a_3 = \frac{(2 - \lambda)}{3!}$$

$$a_5 = \frac{(6 - \lambda)a_3}{(5)(4)} = \frac{(2 - \lambda)(6 - \lambda)}{5!}$$

$$\vdots$$

$$a_{2n+1} = \frac{\prod_{k=1}^{n}(4k - 2 - \lambda)}{(2n + 1)!}.$$

Since $a_0 = 0$ then $a_{2n} = 0$ for $n \in \mathbb{N}$. Thus a second solution to Eq. (8.85) is

$$y_2(x) = x + \sum_{n=1}^{\infty} \frac{\prod_{k=1}^{n}(4k - 2 - \lambda)}{(2n + 1)!} x^{2n+1}. \tag{8.89}$$

The Ratio Test reveals that $y_1(x)$ and $y_2(x)$ converge absolutely for all $x \in \mathbb{R}$. The solutions are linearly independent since the Wronskian is not zero.

If $\lambda = 2m$ for some $m \in \mathbb{N}$ then one of the infinite series solutions terminates after a finite number of terms and can be written as a polynomial. In this case the ordinary differential equation in Eq. (8.85) is called the **Hermite equation of order** m. Replacing λ by $2m$ in Eq. (8.87) produces

$$a_{n+2} = \frac{(2n - 2m)a_n}{(n + 2)(n + 1)},$$

Table 8.8   The Hermite polynomials
$H_n(x)$ for $n = 0, 1, \ldots, 5$.

$H_0(x) = 1$
$H_1(x) = 2x$
$H_2(x) = 4x^2 - 2$
$H_3(x) = 8x^3 - 12x$
$H_4(x) = 16x^4 - 48x^2 + 12$
$H_5(x) = 32x^5 - 160x^3 + 120x$

which implies $a_{m+2} = 0$ and thus $a_{m+2k} = 0$ for all $k \in \mathbb{N}$. In other words, all series coefficients after the mth will be zero.

If m is even (or equivalently if $\lambda = 4N$ for some $N \in \{0, 1, 2, \ldots\}$) then

$$y_1(x) = 1 + \sum_{n=1}^{N} \frac{\prod_{k=1}^{n}(4k - 4 - 4N)}{(2n)!} x^{2n}. \tag{8.90}$$

If m is odd (or equivalently if $\lambda = 4N - 2$ for some $N \in \mathbb{N}$) then

$$y_2(x) = x + \sum_{n=1}^{N} \frac{\prod_{k=1}^{n}(4k - 4N)}{(2n + 1)!} x^{2n+1}. \tag{8.91}$$

The **Hermite polynomial** denoted as $H_n(x)$ is defined to be the polynomial solution to the Hermite equation with $\lambda = 2n$ for which the coefficient of x^n is 2^n. The even- and odd-degree Hermite polynomials can be represented by the single formula:

$$H_n(x) = n! \sum_{k=0}^{\llbracket n/2 \rrbracket} \frac{(-1)^k}{k!(n - 2k)!} (2x)^{n-2k}. \tag{8.92}$$

Table 8.8 lists $H_n(x)$ for $n = 0, 1, \ldots, 5$. When n is even, $H_n(x)$ contains only even powers of x and thus is an even function. When n is odd, $H_n(x)$ contains only odd powers of x and thus $H_n(x)$ is an odd function. The graphs of these polynomials are illustrated in Fig. 8.9.

The parity of the Hermite polynomials has already been mentioned. It should come as no surprise that there is a Rodrigues' formula for the Hermite polynomials and that this formula will be useful in determining the orthogonality properties of the Hermite polynomials. To derive the Rodrigues' formula for the Hermite polynomials, start with the weighting function $u(x) = e^{-x^2}$. Since $u'(x) = -2xe^{-x^2}$ then function u solves the ordinary differential equation

$$u'(x) + 2xu(x) = 0. \tag{8.93}$$

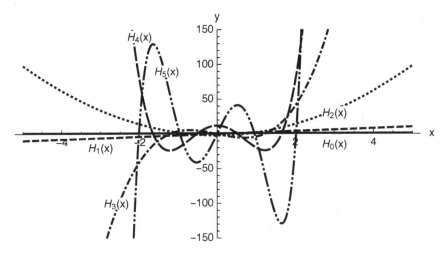

Fig. 8.9   The graphs of $H_n(x)$ for $n = 0, 1, \ldots, 5$. When n is an even (odd) integer, $H_n(x)$ is an even (odd) function.

Differentiating both sides of Eq. (8.93) $n + 1$ times produces

$$0 = u^{(n+2)}(x) + 2xu^{(n+1)}(x) + 2(n+1)u^{(n)}(x). \qquad (8.94)$$

The final step in the derivation begins with the definition of function $v(x) = (-1)^n e^{x^2} u^{(n)}(x)$. This definition is equivalent to $u^{(n)}(x) = (-1)^n e^{-x^2} v(x)$. Substituting this formulation of $u^{(n)}(x)$ into Eq. (8.94) produces

$$0 = (-1)^n [e^{-x^2} v(x)]'' + 2x(-1)^n [e^{-x^2} v(x)]' + 2(n+1)(-1)^n e^{-x^2} v(x)$$

$$= (-1)^n e^{-x^2} (v''(x) - 2xv'(x) + 2nv(x)). \qquad (8.95)$$

Thus $v(x)$ is a solution to the Hermite equation with $\lambda = 2n$. Consequently $v(x)$ is a multiple of $H_n(x)$. Since each derivative of e^{-x^2} will contain a factor of e^{-x^2}, the factor of e^{x^2} in the definition of $v(x)$ will cancel the e^{-x^2} factor and thus $v(x)$ is a polynomial. If the leading coefficient of $v(x)$ is 2^n, the same as the leading coefficient of $H_n(x)$, then $v(x) = H_n(x)$. It is easy to see that

$$(-1)^0 e^{x^2} [e^{-x^2}]^{(0)} = 1 = 2^0.$$

Thus when $n = 0$, $v(x) = H_0(x)$. Suppose the leading coefficient of $v(x)$ is 2^m for all $0 \leq m \leq n$. Consider

$$(-1)^{n+1} e^{x^2} \frac{d^{n+1}}{dx^{n+1}} [e^{-x^2}] = (-1)^{n+1} e^{x^2} ((-2x)^{n+1} e^{-x^2} + \cdots)$$

$$= 2^{n+1} x^{n+1} + \cdots$$

and note the assertion is also true for $m = n + 1$. Therefore $v(x) = H_n(x)$ for all $n \in \{0, 1, 2, \ldots\}$. Thus the following Rodrigues' formula for the Hermite polynomials is valid.

$$H_n(x) = (-1)^n e^{x^2} \frac{d^n}{dx^n}[e^{-x^2}]. \tag{8.96}$$

Equation (8.96) will be employed to prove the orthogonality properties of the Hermite polynomials.

Theorem 8.15. *The Hermite polynomials $H_n(x)$ are orthogonal on the interval $(-\infty, \infty)$ with respect to the weighting function e^{-x^2}. For $m, n \in \{0, 1, 2, \ldots\}$,*

$$\int_{-\infty}^{\infty} H_m(x) H_n(x) e^{-x^2}\, dx = 2^n n! \sqrt{\pi} \delta_{mn}, \tag{8.97}$$

where δ_{mn} is the Kronecker delta function.

Proof. Since the exponential function e^{-x^2} dominates the polynomials, improper integrals of the form shown in Eq. (8.97) will converge. If $m, n \in \{0, 1, 2, \ldots\}$ with $m \neq n$ then without loss of generality assume $m < n$. Thus,

$$\int_{-\infty}^{\infty} H_m(x) H_n(x) e^{-x^2}\, dx = (-1)^n \int_{-\infty}^{\infty} H_m(x) \frac{d^n}{dx^n}[e^{-x^2}]\, dx$$

where the $H_n(x)$ factor in the integrand has been replaced by its Rodrigues' formula. Applying integration by parts n times, keeping in mind that $H_m(x)$ is a polynomial of degree $m < n$, produces

$$\int_{-\infty}^{\infty} H_m(x) H_n(x) e^{-x^2}\, dx = 0$$

for all $m, n \in \{0, 1, 2, \ldots\}$. Now suppose $m = n$ and consider the integral

$$\int_{-\infty}^{\infty} (H_n(x))^2 e^{-x^2}\, dx = (-1)^n \int_{-\infty}^{\infty} H_n(x) \frac{d^n}{dx^n}[e^{-x^2}]\, dx.$$

Again applying integration by parts n times produces

$$\int_{-\infty}^{\infty} (H_n(x))^2 e^{-x^2}\, dx = \int_{-\infty}^{\infty} \frac{d^n}{dx^n}[H_n(x)]\, e^{-x^2}\, dx = 2^n n! \int_{-\infty}^{\infty} e^{-x^2}\, dx$$

recalling that by convention the leading coefficient of the nth Hermite polynomial is 2^n. Making use of the fact that

$$\int_{-\infty}^{\infty} e^{-x^2}\, dx = \Gamma\left(\frac{1}{2}\right) = \sqrt{\pi},$$

then

$$\int_{-\infty}^{\infty} (H_n(x))^2 e^{-x^2}\, dx = 2^n n! \sqrt{\pi}.$$

Thus the Hermite polynomials are pairwise orthogonal on $(-\infty, \infty)$ with respect to the weighting function e^{-x^2}. $\qquad \square$

Once more, if $f(x)$ is sufficiently smooth on $(-\infty, \infty)$ and if $\int_{-\infty}^{\infty} (f(x))^2 e^{-x^2} dx < \infty$, it can be represented as a series of Hermite polynomials of the form

$$f(x) = \sum_{n=0}^{\infty} a_n H_n(x).$$

Then the **Hermite coefficients** for the Hermite series expansion are given by

$$a_n = \frac{1}{2^n n! \sqrt{\pi}} \int_{-\infty}^{\infty} f(x) H_n(x) e^{-x^2} dx. \tag{8.98}$$

Example 8.8. Find the Hermite series expansion for $f(x) = \cos x$. Plot $f(x)$ and the Nth partial sums of the Hermite series for $N = 4$, 6, and 10.

Solution. Since $f(x)$ is an even function and $H_n(x)$ is an even function if and only if n is even, then without formally carrying out the integration, the odd-subscripted coefficients vanish. Thus only the Hermite coefficients of the form a_{2n} need attention. According to Eq. (8.98), by using the Rodrigues' formula for the Hermite polynomial $H_{2n}(x)$ and performing integration by parts $2n$ times,

$$
\begin{aligned}
a_{2n} &= \frac{1}{2^{2n}(2n)!\sqrt{\pi}} \int_{-\infty}^{\infty} (\cos x) H_{2n}(x) e^{-x^2} dx \\
&= \frac{1}{2^{2n}(2n)!\sqrt{\pi}} \int_{-\infty}^{\infty} (\cos x)(-1)^{2n} e^{x^2} \frac{d^{2n}}{dx^{2n}} [e^{-x^2}] e^{-x^2} dx \\
&= \frac{1}{2^{2n}(2n)!\sqrt{\pi}} \int_{-\infty}^{\infty} (\cos x) \frac{d^{2n}}{dx^{2n}} [e^{-x^2}] dx \\
&= \frac{(-1)^n}{2^{2n}(2n)!\sqrt{\pi}} \int_{-\infty}^{\infty} \cos x \, e^{-x^2} dx.
\end{aligned}
$$

To evaluate the remaining improper integral, the Euler identity in Eq. (3.40) is used. The function $\cos x$ can be written as $(e^{ix} + e^{-ix})/2$. Making this replacement in the previous integrand, using the symmetry of the integrand on the real number line, and completing the square in the exponent twice produce

$$
\begin{aligned}
a_{2n} &= \frac{(-1)^n}{2^{2n}(2n)!\sqrt{\pi}} \int_0^\infty (e^{ix} + e^{-ix}) e^{-x^2}\, dx \\
&= \frac{(-1)^n e^{-1/4}}{2^{2n}(2n)!\sqrt{\pi}} \left(\int_0^\infty e^{-(x-i/2)^2}\, dx + \int_0^\infty e^{-(x+i/2)^2}\, dx \right) \\
&= \frac{(-1)^n e^{-1/4}}{2^{2n}(2n)!\sqrt{\pi}} \left(\int_{-i/2}^{\infty-i/2} e^{-u^2}\, du + \int_{i/2}^{\infty+i/2} e^{-v^2}\, dv \right) \\
&= \frac{(-1)^n e^{-1/4}}{2^{2n}(2n)!\sqrt{\pi}} \left(\frac{\sqrt{\pi}}{2} + \frac{\sqrt{\pi}}{2} \right) = \frac{(-1)^n e^{-1/4}}{2^{2n}(2n)!}.
\end{aligned}
$$

Strictly speaking, the Cauchy-Goursat integral theorem (Theorem A.6) was used in the calculation of the integrals above. Function $f(x) = \cos x$ may be expressed as

$$
\cos x = e^{-1/4} \sum_{n=0}^\infty \frac{(-1)^n}{2^{2n}(2n)!} H_{2n}(x).
$$

Figure 8.10 shows the graph of $f(x) = \cos x$ and the Nth partial sums of the Hermite series expansion of $f(x)$ for $N = 4$, 6, and 10.

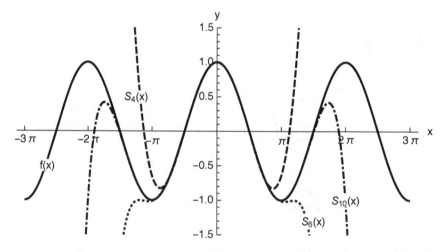

Fig. 8.10   The graph of $f(x) = \cos x$ and the Nth partial sums of the Hermite series expansion for $\cos x$ for $N = 4$, 6, and 10. The Hermite series converge to $f(x)$ as N increases.

8.7 The Chebyshev Polynomials and Their Properties

The Chebyshev[13] differential equation is

$$(1 - x^2)y'' - xy' + \alpha^2 y = 0 \text{ for } -1 < x < 1, \qquad (8.99)$$

where α is a constant. The solutions of this equation have many applications in the physical sciences and in numerical mathematics. This section will derive these solutions and many of their useful properties and relationships. The self-adjoint form of Chebyshev's differential equation is

$$[\sqrt{1 - x^2}\, y']' + \frac{\alpha^2}{\sqrt{1 - x^2}} y = 0 \text{ for } -1 < x < 1.$$

The Chebyshev equation is yet another example of a singular Sturm-Liouville boundary value problem.

The point $x_0 = 0$ is an ordinary point for Eq. (8.99) and thus solutions can be found using infinite series techniques. Assuming $y(x) = \sum_{n=0}^{\infty} a_n x^n$ and substituting this into Eq. (8.99) yield

$$0 = (1 - x^2) \sum_{n=2}^{\infty} n(n-1)a_n x^{n-2} - x \sum_{n=1}^{\infty} na_n x^{n-1} + \alpha^2 \sum_{n=0}^{\infty} a_n x^n$$

$$= \sum_{n=0}^{\infty} [(n+2)(n+1)a_{n+2} + (\alpha^2 - n^2)a_n]x^n. \qquad (8.100)$$

The uniqueness of Taylor coefficients applied to Eq. (8.100) implies the recurrence relation:

$$a_{n+2} = \frac{(n^2 - \alpha^2)a_n}{(n+1)(n+2)}. \qquad (8.101)$$

If $a_0 = 1$ and $a_1 = 0$ then the following coefficients can be calculated.

$$a_2 = -\frac{\alpha^2}{2!}$$

$$a_4 = \frac{(2^2 - \alpha^2)a_2}{(3)(4)} = -\frac{\alpha^2(2^2 - \alpha^2)}{4!}$$

$$\vdots$$

$$a_{2n} = \frac{\prod_{k=0}^{n-1}((2k)^2 - \alpha^2)}{(2n)!}.$$

Since $a_1 = 0$, all the odd-subscripted coefficients in the series solution are zero. Hence one solution of Eq. (8.99) is the series

$$y_1(x) = 1 + \sum_{n=1}^{\infty} \frac{\prod_{k=0}^{n-1}((2k)^2 - \alpha^2)}{(2n)!} x^{2n}. \qquad (8.102)$$

---

[13]Pafnuty Lvovich Chebyshev, Russian mathematician (1821–1894).

Using the Ratio Test it is seen that the power series $y_1(x)$ converges absolutely for $|x| < 1$.

If $a_0 = 0$ and $a_1 = 1$ then the coefficients are found to be

$$a_3 = \frac{(1 - \alpha^2)}{3!}$$

$$a_5 = \frac{(3^2 - \alpha^2)a_3}{(4)(5)} = \frac{(1 - \alpha^2)(3^2 - \alpha^2)}{5!}$$

$$\vdots$$

$$a_{2n+1} = \frac{\prod_{k=0}^{n-1}((2k + 1)^2 - \alpha^2)}{(2n + 1)!}.$$

All the even-subscripted coefficients in this solution are zero since a_0 has been chosen to be zero. Hence another solution to Chebyshev's differential equation is

$$y_2(x) = x + \sum_{n=1}^{\infty} \frac{\prod_{k=0}^{n-1}((2k + 1)^2 - \alpha^2)}{(2n + 1)!} x^{2n+1}. \tag{8.103}$$

Using the Ratio Test once again it can be shown that $y_2(x)$ converges absolutely for $|x| < 1$.

The Wronskian determinant verifies the solutions are linearly independent.

$$W(y_1, y_2)(0) = y_1(0)y_2'(0) - y_1'(0)y_2(0) = 1 \neq 0.$$

If α is a nonnegative integer m, then one of these series solutions terminates after a finite number of nonzero terms. In fact, suppose $\alpha = m$ for some $m \in \{0, 1, 2, \ldots\}$. According to the recurrence relation in Eq. (8.101), $a_{m+2} = 0$ which implies $a_{m+2k} = 0$ for all $k \in \mathbb{N}$. As a consequence, one solution is a polynomial of degree at most m.

If α is an even integer then $\alpha = 2N$ where N is a nonnegative integer and the solution in Eq. (8.102) reduces to the polynomial,

$$y_1(x) = 1 + \sum_{n=1}^{N} \frac{\prod_{k=0}^{n-1}((2k)^2 - (2N)^2)}{(2n)!} x^{2n}$$

$$= 1 + \sum_{n=1}^{N} (-1)^n \frac{\prod_{k=0}^{n-1}(N + k)(N - k)}{(2n)!} (2x)^{2n} \tag{8.104}$$

which is a polynomial of degree $2N$. If α is an odd integer then $\alpha = 2N + 1$ for some nonnegative integer N. The solution expressed in Eq. (8.103) becomes the polynomial,

$$y_2(x) = x + \sum_{n=1}^{N} \frac{\prod_{k=0}^{n-1}((2k+1)^2 - (2N+1)^2)}{(2n+1)!} x^{2n+1}$$

$$= x + \frac{1}{2}\sum_{n=1}^{N}(-1)^n \frac{\prod_{k=0}^{n-1}(N+k+1)(N-k)}{(2n+1)!}(2x)^{2n+1} \quad (8.105)$$

which is a polynomial of degree $2N + 1$.

Since the Chebyshev equation is linear and homogeneous, then when α is a nonnegative integer, a constant multiple of the polynomial solution in either Eq. (8.104) or Eq. (8.105) will also solve Eq. (8.99). By common convention the **Chebyshev polynomials** are denoted by $T_n(x)$ and have degree n. When $n = 2N$, $T_n(x) = (-1)^N y_1(x)$ where $y_1(x)$ is expressed in Eq. (8.104). When $n = 2N + 1$, $T_n(x) = (-1)^N(2N+1)y_2(x)$ where $y_2(x)$ is found in Eq. (8.105). To be precise these are the Chebyshev polynomials of the first kind. There are also Chebyshev polynomials of the second, third, and fourth kind, but they are not discussed here. It is easy to calculate $T_0(x) = 1$ and $T_1(x) = x$, but what are the higher degree Chebyshev polynomials? The remaining Chebyshev polynomials can be found by exploiting a remarkable relationship between the Chebyshev polynomials and the trigonometric function cosine.

Lemma 8.14. *For each nonnegative integer n,*

$$T_n(\cos\theta) = \cos(n\theta). \quad (8.106)$$

Proof. This result has already been established for $n = 0$ and $n = 1$. Let $\theta = \arccos x$ and $y(x) = \cos(n\arccos x)$. Note that

$$y'(x) = \frac{n\sin(n\arccos x)}{\sqrt{1-x^2}}$$

$$y''(x) = \frac{nx\sin(n\arccos x)}{(1-x^2)^{3/2}} - \frac{n^2\cos(n\arccos x)}{1-x^2}.$$

Substituting $y(x)$ and its derivatives into Eq. (8.99) produces

$$0 = (1-x^2)\left(\frac{nx\sin(n\arccos x)}{(1-x^2)^{3/2}} - \frac{n^2\cos(n\arccos x)}{1-x^2}\right)$$

$$- \frac{nx\sin(n\arccos x)}{\sqrt{1-x^2}} + n^2\cos(n\arccos x),$$

which shows $y(x)$ satisfies the Chebyshev differential equation. If n is an odd integer, then

$$y(0) = \cos\left(n \arccos 0\right) = \cos\left(\frac{n\pi}{2}\right) = 0 = T_n(0)$$

$$y'(0) = \frac{n \sin\left(n \arccos 0\right)}{\sqrt{1 - 0^2}} = n \sin\left(\frac{n\pi}{2}\right) = n(-1)^{(n-1)/2} = T_n'(0).$$

Likewise if n is an even integer, then

$$y(0) = \cos\left(n \arccos 0\right) = \cos\left(\frac{n\pi}{2}\right) = (-1)^{n/2} = T_n(0)$$

$$y'(0) = \frac{n \sin\left(n \arccos 0\right)}{\sqrt{1 - 0^2}} = n \sin\left(\frac{n\pi}{2}\right) = 0 = T_n'(0).$$

By uniqueness of solutions to the differential equation, $y(x) = T_n(x)$. $\square$

As a consequence of Lemma 8.14 and at the risk of slightly obscuring the polynomial character of the Chebyshev solutions, the Chebyshev polynomials may be written as

$$T_n(x) = \cos(n \arccos x). \tag{8.107}$$

The ability to express the Chebyshev polynomials in terms of the cosine function enables the development of a recurrence relation for the Chebyshev polynomials. Consider the cosine of the sum and difference of the angles $n\theta$ and θ.

$$\cos((n + 1)\theta) = \cos(n\theta)\cos(\theta) - \sin(n\theta)\sin(\theta), \tag{8.108}$$

$$\cos((n - 1)\theta) = \cos(n\theta)\cos(\theta) + \sin(n\theta)\sin(\theta). \tag{8.109}$$

Adding Eqs. (8.108) and (8.109) yields

$$\cos((n + 1)\theta) + \cos((n - 1)\theta) = 2\cos(n\theta)\cos(\theta)$$
$$T_{n+1}(\cos\theta) + T_{n-1}(\cos\theta) = 2\cos(\theta)T_n(\cos\theta)$$
$$T_{n+1}(x) = 2x\, T_n(x) - T_{n-1}(x) \tag{8.110}$$

which is the recurrence relation for the Chebyshev polynomials. By using Eq. (8.110), the higher degree Chebyshev polynomials are easily found. For example,

$$T_2(x) = 2xT_1(x) - T_0(x) = 2x^2 - 1.$$

Table 8.9 contains the formulas for the Chebyshev polynomials $T_n(x)$ for $n = 0, 1, \ldots, 5$. Graphs of the Chebyshev polynomials are shown in Fig. 8.11.

Table 8.9   The Chebyshev polynomials $T_n(x)$ for $n = 0, 1, \ldots, 5$.

$T_0(x) = 1$
$T_1(x) = x$
$T_2(x) = 2x^2 - 1$
$T_3(x) = 4x^3 - 3x$
$T_4(x) = 8x^4 - 8x^2 + 1$
$T_5(x) = 16x^5 - 20x^3 + 5x$

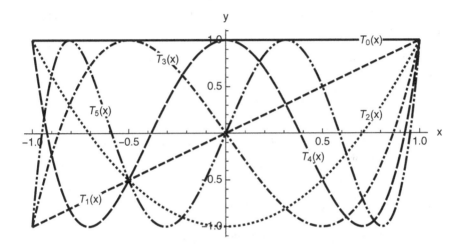

Fig. 8.11   Graphs of the Chebyshev polynomials $T_n(x)$ for $n = 0, 1, \ldots, 5$.

Most of the properties of the Chebyshev polynomials described in the remainder of this section will follow almost immediately from the cosine definition of $T_n(x)$ given in Eq. (8.107). For example, it is easy to see that for all $n \in \{0, 1, 2, \ldots\}$,

$$T_n(1) = \cos(n \arccos 1) = \cos(0) = 1.$$

Other properties are explored below.

Theorem 8.16. *The polynomial $T_n(x)$ has n distinct zeros in the interval $(-1, 1)$ located at*

$$x_k = \cos \frac{(2k-1)\pi}{2n} \text{ for } k = 1, 2, \ldots, n.$$

Proof. This assertion follows from Eq. (8.107). □

Theorem 8.17. *For $n \in \mathbb{N}$ the leading coefficient of $T_n(x)$ is 2^{n-1}.*

Proof. This is certainly true for $T_1(x) = x$. Suppose the claim is true for all $1 \leq n \leq m$. Using the recurrence formula of Eq. (8.110)

$$T_{m+1}(x) = 2x\, T_m(x) - T_{m-1}(x).$$

Since $T_{m-1}(x)$ has degree $m - 1$, the leading coefficient of $T_{m+1}(x)$ is the same as the leading coefficient of $(2x)T_m(x)$. By the induction hypothesis regarding the leading coefficient of $T_m(x)$, the leading coefficient of $T_{m+1}(x)$ will be $(2)2^{m-1} = 2^m$. Thus by mathematical induction the theorem is proved. $\square$

There are many alternative formulas for the Chebyshev polynomials in addition to the recursive formula given by Eq. (8.110).

Theorem 8.18. *For $n \in \mathbb{N}$ the nth Chebyshev polynomial is given by the formula*

$$T_n(x) = 2^{n-1} \prod_{k=1}^{n} \left(x - \cos \frac{(2k-1)\pi}{2n} \right). \tag{8.111}$$

Proof. Since $T_n(x)$ is a polynomial of degree n with n distinct roots described in Theorem 8.16, $T_n(x)$ must be divisible by

$$x - \cos \frac{(2k-1)\pi}{2n}$$

for $k = 1, 2, \ldots, n$. By Theorem 8.17 the leading coefficient is 2^{n-1}. $\square$

Finally the Chebyshev polynomials satisfy an orthogonality relationship on the interval $[-1, 1]$.

Theorem 8.19. *For $m, n \in \{0, 1, 2, \ldots\}$ the Chebyshev polynomials are orthogonal on $[-1, 1]$ with respect to the weighting function $(1 - x^2)^{-1/2}$ and*

$$\int_{-1}^{1} \frac{T_m(x)T_n(x)}{\sqrt{1 - x^2}}\, dx = \begin{cases} \pi & \text{if } m = n = 0, \\ \dfrac{\pi}{2}\delta_{mn} & \text{otherwise,} \end{cases} \tag{8.112}$$

where δ_{mn} is the Kronecker delta function.

Proof. Making use of the change of variable $x = \cos\theta$, the integral of Eq. (8.112) can be written as

$$\int_{-1}^{1} \frac{T_m(x)T_n(x)}{\sqrt{1 - x^2}}\, dx = -\int_{\pi}^{0} T_m(\cos\theta)T_n(\cos\theta)\, d\theta$$

$$= \int_{0}^{\pi} \cos(m\theta)\cos(n\theta)\, d\theta = \begin{cases} \pi & \text{if } m = n = 0, \\ \dfrac{\pi}{2}\delta_{mn} & \text{otherwise.} \end{cases}$$

Hence Eq. (8.112) holds. $\square$

A consequence of this theorem is that the Chebyshev polynomial of degree n is orthogonal on $[-1, 1]$ to every polynomial of smaller degree with respect to the weighting function $(1 - x^2)^{-1/2}$. This observation is important in the derivation of the Rodrigues' formula for the Chebyshev polynomials.

Theorem 8.20. *For $n \in \{0, 1, 2, \ldots\}$ the Chebyshev polynomial of degree n, $T_n(x)$ can be expressed as*

$$T_n(x) = \frac{(-2)^n n!}{(2n)!}(1 - x^2)^{1/2}\frac{d^n}{dx^n}[(1 - x^2)^{n-1/2}]. \tag{8.113}$$

Proof. The truth of Eq. (8.113) is evident when $n = 0$ or $n = 1$, so suppose n is an integer greater than 1. Define the function $f(x) = \frac{d^n}{dx^n}\left[(1 - x^2)^{n-1/2}\right]$. Let m be a nonnegative integer with $m < n$, then

$$\int_{-1}^{1} x^m f(x)\, dx = \int_{-1}^{1} x^m \frac{d^n}{dx^n}[(1 - x^2)^{n-1/2}]\, dx$$

$$= -m \int_{-1}^{1} x^{m-1} \frac{d^{n-1}}{dx^{n-1}}[(1 - x^2)^{n-1/2}]\, dx$$

$$= (-1)^m m! \left[\frac{d^{n-m-1}}{dx^{n-m-1}}[(1 - x^2)^{n-1/2}]\right]_{-1}^{1} = 0.$$

Consequently $f(x)(1 - x^2)^{1/2}$ is orthogonal on $[-1, 1]$ to x^m with respect to the weighting function $(1 - x^2)^{-1/2}$ for $m = 0, 1, \ldots, n - 1$. Differentiation shows that $f(x)(1 - x^2)^{1/2}$ is a polynomial of degree n and that

$$\left[(1 - x^2)^{1/2}\frac{d^n}{dx^n}[(1 - x^2)^{n-1/2}]\right]_{x=1} = (-1)^n(1)(3)\cdots(2n - 1).$$

Therefore, $f(x)(1 - x^2)^{1/2}$ must be a multiple of $T_n(x)$, see Exercise 50. Since $T_n(1) = 1$, then

$$(1 - x^2)^{1/2}\frac{d^n}{dx^n}[(1 - x^2)^{n-1/2}] = (-1)^n(1)(3)\cdots(2n - 1)T_n(x)$$

and the proof is complete. $\square$

To express a function as a **Chebyshev series**, the orthogonality of the Chebyshev polynomials is employed. If $f(x)$ is sufficiently smooth on $(-1, 1)$ then

$$f(x) = \frac{a_0}{2} + \sum_{n=1}^{\infty} a_n T_n(x),$$

where

$$a_n = \frac{2}{\pi} \int_{-1}^{1} \frac{f(x)T_n(x)}{\sqrt{1-x^2}} \, dx, \qquad (8.114)$$

for $n = 0, 1, 2, \ldots$.

Example 8.9. Express $f(x) = \arccos x$ as a Chebyshev series. Plot $f(x)$ and the Nth partial sums of the series for $N = 3$, 5, and 9.

Solution. Making use of the Chebyshev coefficient formula in Eq. (8.114) for the case of $n = 0$,

$$a_0 = \frac{2}{\pi} \int_{-1}^{1} \frac{T_0(x) \arccos x}{\sqrt{1-x^2}} \, dx = \left[-\frac{1}{\pi} (\arccos x)^2 \right]_{-1}^{1} = -\frac{1}{\pi}(0 - \pi^2) = \pi.$$

Now consider the even-subscripted Chebyshev coefficients.

$$a_{2n} = \frac{2}{\pi} \int_{-1}^{1} \frac{T_{2n}(x) \arccos x}{\sqrt{1-x^2}} \, dx$$

$$= \frac{2}{\pi} \left(\int_{-1}^{0} \frac{T_{2n}(x) \arccos x}{\sqrt{1-x^2}} \, dx + \int_{0}^{1} \frac{T_{2n}(x) \arccos x}{\sqrt{1-x^2}} \, dx \right)$$

Making the substitution $u = -x$ in the first integral produces

$$a_{2n} = \frac{2}{\pi} \left(-\int_{1}^{0} \frac{T_{2n}(-u) \arccos(-u)}{\sqrt{1-(-u)^2}} \, du + \int_{0}^{1} \frac{T_{2n}(x) \arccos x}{\sqrt{1-x^2}} \, dx \right)$$

$$= \frac{2}{\pi} \left(\int_{0}^{1} \frac{T_{2n}(u)(\pi - \arccos u)}{\sqrt{1-u^2}} \, du + \int_{0}^{1} \frac{T_{2n}(x) \arccos x}{\sqrt{1-x^2}} \, dx \right)$$

$$= 2 \int_{0}^{1} \frac{T_{2n}(x)}{\sqrt{1-x^2}} \, dx = 0.$$

All that remains to be calculated are the odd-subscripted Chebyshev coefficients.

$$a_{2n-1} = \frac{2}{\pi} \int_{-1}^{1} \frac{T_{2n-1}(x) \arccos x}{\sqrt{1-x^2}} \, dx$$

$$= \frac{2}{\pi} \int_{-1}^{1} \frac{\cos((2n-1)\arccos x) \arccos x}{\sqrt{1-x^2}} \, dx$$

$$= \frac{2}{\pi} \int_{0}^{\pi} u \cos((2n-1)u) \, du$$

$$= \frac{2}{\pi} \left(\left[\frac{u \sin((2n-1)u)}{2n-1} \right]_{0}^{\pi} - \int_{0}^{\pi} \frac{\sin((2n-1)u)}{2n-1} \, du \right)$$

$$= \left[\frac{2}{(2n-1)^2 \pi} \cos((2n-1)u) \right]_{0}^{\pi} = \frac{-4}{(2n-1)^2 \pi}.$$

Thus the Chebyshev series expansion for $f(x) = \arccos x$ is

$$\arccos x = \frac{\pi}{2} - \frac{4}{\pi} \sum_{n=1}^{\infty} \frac{1}{(2n-1)^2} T_{2n-1}(x).$$

The graph of $\arccos x$ and the Nth partial sums of the series for $N = 3, 5,$ and 9 are shown in Fig. 8.12.

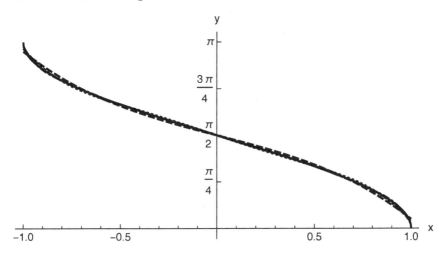

Fig. 8.12 The graph of $f(x) = \arccos x$ and the Nth partial sums of the Chebyshev series for $f(x)$ for $N = 3, 5,$ and 9. The graph of $f(x)$ is the solid curve. Even for small values of N the Chebyshev series approximation to $f(x)$ is accurate on the interval $[-1, 1]$.

The special functions described in this chapter will play roles in the solution of several important (initial) boundary value problems which arise in the physical sciences. These applications will appear in Chap. 9.

8.8 Exercises

(1) Verify the value of the improper integral in Eq. (8.6).

(2) Show that an equivalent definition of the Gamma function is

$$\Gamma(p+1) = 2 \int_{0}^{\infty} e^{-x^2} x^{2p+1} \, dx.$$

(3) Evaluate $\Gamma(-1/2)$.

(4) Show that for $n = 0, 1, 2, \ldots,$

$$\Gamma\left(n + \frac{1}{2}\right) = \frac{(2n)!\sqrt{\pi}}{2^{2n} n!}.$$

(5) Show that the nth derivative of the Gamma function is

$$\Gamma^{(n)}(x) = \int_0^\infty e^{-t}(\ln t)^n t^{x-1}\, dt.$$

(6) A very useful identity involving the Gamma function is the following.

$$\frac{\Gamma(p)\Gamma(q)}{\Gamma(p+q)} = 2\int_0^{\pi/2} \cos^{2p-1} t \sin^{2q-1} t\, dt. \tag{8.115}$$

Follow the steps outlined in the exercises below to establish Eq. (8.115).

(a) For $p \geq 0$ and $q \geq 0$ use the result of Exercise 2 to show that

$$\Gamma(p)\Gamma(q) = 4\int_0^\infty \int_0^\infty e^{-u^2-v^2} u^{2p-1} v^{2q-1}\, du\, dv.$$

(b) Convert the double integral to polar coordinates to show that

$$\Gamma(p)\Gamma(q) = 2\Gamma(p+q)\int_0^{\pi/2} \cos^{2p-1} t \sin^{2q-1} t\, dt.$$

(7) Use the main result of Exercise 6 to evaluate the following definite integrals:

(a) $\int_0^{\pi/2} \cos^2 t \sin t\, dt.$

(b) $\int_0^{\pi/2} \cos^3 t \sin t\, dt.$

(c) $\int_0^{\pi/2} \cos^2 t \sin^2 t\, dt.$

(d) $\int_0^{\pi/2} \cos^3 t \sin^2 t\, dt.$

(8) Use the main result of Exercise 6 to derive the **Wallis[14] formulas.**

(a) $\int_0^{\pi/2} \sin^{2n} t\, dt = \frac{\pi}{2}\frac{(2n)!}{2^{2n}(n!)^2} = \int_0^{\pi/2} \cos^{2n} t\, dt$ for $n = 0, 1, 2, \ldots$.

(b) $\int_0^{\pi/2} \sin^{2n+1} t\, dt = \frac{2^{2n}(n!)^2}{(2n+1)!} = \int_0^{\pi/2} \cos^{2n+1} t\, dt$ for $n = 0, 1, 2, \ldots$.

(9) Express the arc length of the lemniscate $r^2 = 2\cos 2\theta$ in terms of the Gamma function and other constants.

(10) The Bessel function of the first kind of order 0 can be expressed as an integral. Show that

$$J_0(x) = \frac{1}{\pi}\int_0^\pi \cos(x \sin \theta)\, d\theta.$$

---

[14] John Wallis, English mathematician (1616–1703).

(11) The integral representation of the Bessel function of the first kind established in Exercise 10 can be generalized. For $n = 1, 2, \ldots$,

$$J_n(x) = \frac{1}{\pi} \int_0^\pi \cos(x \sin\theta - n\theta) \, d\theta.$$

Show that

$$J_{2n-1}(x) = \frac{2}{\pi} \int_0^{\pi/2} \sin(x \sin\theta) \sin((2n-1)\theta) \, d\theta,$$

and

$$J_{2n}(x) = \frac{2}{\pi} \int_0^{\pi/2} \cos(x \sin\theta) \cos(2n\theta) \, d\theta.$$

(12) Prove the "Pythagorean Theorem" for Bessel functions:

$$J_{1/2}^2(x) + J_{-1/2}^2(x) = \frac{2}{\pi x}.$$

(13) Find the generalized Fourier (Bessel-Fourier) series of the indicated order p for the following functions defined on $[0, 1]$.

(a) $f(x) = \begin{cases} 1 & \text{if } 0 < x < 1/2, \\ 0 & \text{if } 1/2 < x < 1, \end{cases}$ with $p = 0$.

(b) $f(x) = x^2$ with $p = 2$.

(c) $f(x) = x^3$ with $p = 3$.

(d) $f(x) = x^4$ with $p = 4$.

(14) Using the explicit form given for $J_{3/2}(x)$, numerically approximate its first five positive zeroes.

(15) Prove Theorem 8.6.

(16) Use Eq. (8.35) to find the explicit form of $j_n(x)$ for $n = 0, 1$ and 2.

(17) Use Eq. (8.35) to establish the following formulas for the spherical Bessel functions of the first kind.

(a) $\frac{d}{dx}\left[x^{n+1} j_n(x)\right] = x^{n+1} j_{n-1}(x)$.

(b) $\frac{d}{dx}\left[x^{-n} j_n(x)\right] = -x^{-n} j_{n+1}(x)$.

(c) $j_{n-1}(x) + j_{n+1}(x) = \frac{2n+1}{x} j_n(x)$.

(d) $n j_{n-1}(x) - (n+1) j_{n+1}(x) = (2n+1) j_n'(x)$.

(18) Prove Lemma 8.9.

(19) Establish the following identity for $n \in \{0, 1, 2, \ldots\}$.

$$\int_{-1}^1 f(x) P_n(x) \, dx = \frac{(-1)^n}{2^n n!} \int_{-1}^1 f^{(n)}(x)(x^2 - 1)^n \, dx.$$

Use integration by parts and Eq. (8.48).

(20) Suppose $f(x)$ is a polynomial of degree n. Show that

$$\int_{-1}^{1} f(x) P_m(x) \, dx = 0,$$

for all $m > n$.

(21) Evaluate the following definite integrals.

(a) $\int_{-1}^{1} (1 - x) P_2(x) \, dx$.

(b) $\int_{-1}^{1} (1 - x^2) P_2(x) \, dx$.

(c) $\int_{-1}^{1} x^n P_n(x) \, dx$.

(d) $\int_{-1}^{1} x^n P_{n-1}(x) \, dx$.

(22) Use Lemma 8.9 to help evaluate the following definite integrals.

(a) $\int_{-1}^{1} x \ln(1 - x) P_n(x) \, dx$.

(b) $\int_{-1}^{1} x \ln(1 + x) P_n(x) \, dx$.

(23) Find the generalized Fourier series in terms of the Legendre polynomials for the following functions defined on $[-1, 1]$.

(a) $f(x) = \begin{cases} -1 & \text{if } -1 < x < 0, \\ 1 & \text{if } 0 < x < 1. \end{cases}$

(b) $f(x) = \begin{cases} 0 & \text{if } -1 < x < 0, \\ x & \text{if } 0 < x < 1. \end{cases}$

(c) $f(x) = |x|$.

(d) $f(x) = \sin \dfrac{\pi x}{2}$.

(24) Find the associated Legendre function $P_n^m(x)$ for the following values of n and m.

(a) $n = 2$ and $m = -1$.

(b) $n = 3$ and $m = 0$.

(c) $n = 4$ and $m = 2$.

(d) $n = 4$ and $m = -2$.

(25) Find the series expansion for $f(x) = \cos \pi x$ on $[-1, 1]$ in terms of the associated Legendre functions of order 2. Calculate at least the first 5 nonzero coefficients.

(26) Find the series expansion for $f(x) = 1 - x^2$ on $[-1, 1]$ in terms of the associated Legendre functions of order 1. Calculate at least the first 5 nonzero coefficients.

(27) Prove Theorem 8.12.

(28) Suppose $f(x)$ is n times continuously differentiable on $(0, \infty)$ and that $\lim_{x\to\infty} f^{(k)}(x)e^{-x} = 0$ for $k = 0, 1, \ldots, n$. Show that

$$\int_0^\infty f(x)L_n(x)e^{-x}\,dx = \frac{(-1)^n}{n!}\int_0^\infty f^{(n)}(x)x^n e^{-x}\,dx.$$

(29) Find the Laguerre series expansion for the constant function $f(x) = 1$.

(30) Find the Laguerre series expansion for $f(x) = J_0(2\sqrt{x})$.

(31) Express the generalized Laguerre differential equation of Eq. (8.80) in self-adjoint form.

(32) Express $L_4^\alpha(x)$ as a polynomial in x.

(33) Use the Rodrigues' formula for the generalized Laguerre polynomials to show that

$$\frac{d}{dx}[L_n^\alpha(x)] = -L_{n-1}^{\alpha+1}(x).$$

(34) Show that the generalized Laguerre polynomials satisfy the following identity.

$$L_n^\alpha(x) = \frac{x}{n}\frac{d}{dx}\left[L_{n-1}^\alpha(x)\right] + \left(1 + \frac{\alpha}{n} - \frac{x}{n}\right)L_{n-1}^\alpha(x).$$

(35) Show that the generalized Laguerre polynomials satisfy the following identity.

$$\frac{d}{dx}[L_n^\alpha(x)] = \frac{d}{dx}\left[L_{n-1}^\alpha(x)\right] - L_{n-1}^\alpha(x).$$

(36) Show that the generalized Laguerre polynomials satisfy the following recurrence relation.

$$(n+1)L_{n+1}^\alpha(x) - (2n + \alpha + 1 - x)L_n^\alpha(x) + (n+\alpha)L_{n-1}^\alpha(x) = 0.$$

(37) Suppose $f(x)$ is n times continuously differentiable on $(0, \infty)$ and that $\lim_{x\to\infty} f^{(k)}(x)x^m e^{-x} = 0$ for all non-negative integers k and m. Show that

$$\int_0^\infty f(x)L_n^\alpha(x)x^\alpha e^{-x}\,dx = \frac{(-1)^n}{n!}\int_0^\infty f^{(n)}(x)x^{n+\alpha}e^{-x}\,dx.$$

(38) Find the generalized Laguerre series expansion for $f(x) = x$.

(39) Suppose function $f(x)$ and its derivatives are integrable and that

$$\lim_{x \to \pm\infty} f^{(k)}(x) x^m e^{-x^2} = 0$$

for all non-negative integers k and m. Show that

$$\int_{-\infty}^{\infty} f(x) H_n(x) e^{-x^2} \, dx = \int_{-\infty}^{\infty} f^{(n)}(x) e^{-x^2} \, dx.$$

(40) Show that the Hermite polynomials satisfy the recurrence formula

$$H_{n+1}(x) = 2x \, H_n(x) - 2n \, H_{n-1}(x),$$

for $n \in \mathbb{N}$.

(41) Show that the Hermite polynomials satisfy the derivative relationship

$$H_n'(x) = 2n \, H_{n-1}(x),$$

for $n \in \mathbb{N}$.

(42) Find the Hermite series expansion for $f(x) = x^2 + x + 1$.

(43) Find the Hermite series expansion for $f(x) = \sin x$.

(44) Find the Hermite series expansion for $f(x) = e^x$.

(45) Establish the following identity relating the generalized Laguerre polynomial of order $\alpha = -1/2$ and the Hermite polynomials.

$$L_n^{-1/2}(x) = \frac{(-1)^n}{2^{2n} n!} H_{2n}(\sqrt{x}).$$

(46) Establish the following identity relating the generalized Laguerre polynomial of order $\alpha = 1/2$ and the Hermite polynomials.

$$L_n^{1/2}(x) = \frac{(-1)^n}{2^{2n+1} n! \sqrt{x}} H_{2n+1}(\sqrt{x}).$$

(47) Establish the following formulas for the Chebyshev polynomials.

(a) $T_{2N}(x) = N \sum_{n=0}^{N} (-1)^{N+n} \dfrac{(N+n-1)!}{(N-n)!(2n)!} (2x)^{2n}$ for $N \in \mathbb{N}$.

(b) $T_{2N+1}(x) = \dfrac{2N+1}{2} \sum_{n=0}^{N} (-1)^{N+n} \dfrac{(N+n)!}{(N-n)!(2n+1)!} (2x)^{2n+1}$ for $N = 0, 1, \ldots$.

(c) $T_m(x) = \dfrac{1}{2} \sum_{n=0}^{\llbracket m/2 \rrbracket} (-1)^n \dfrac{m}{m-n} \binom{m-n}{n} (2x)^{m-2n}$ for $m \in \mathbb{N}$.

(48) Show that the Chebyshev polynomials satisfy the following recurrence relations for $n \in \mathbb{N}$.

(a) $(x - 1)[T_{2n+1}(x) - 1] = [T_{n+1}(x) - T_n(x)]^2$.

(b) $2(x^2 - 1)[T_{2n}(x) - 1] = [T_{n+1}(x) - T_{n-1}(x)]^2$.

(49) Show that for $n \in \mathbb{N}$, the Chebyshev polynomial $T_n(x)$ can be expressed as the determinant of the $n \times n$ matrix

$$
\begin{bmatrix}
x & 1 & 0 & 0 & \cdots & 0 & 0 & 0 \\
1 & 2x & 1 & 0 & \cdots & 0 & 0 & 0 \\
0 & 1 & 2x & 1 & \cdots & 0 & 0 & 0 \\
0 & 0 & 1 & 2x & \cdots & 0 & 0 & 0 \\
0 & 0 & 0 & 1 & \cdots & 0 & 0 & 0 \\
\vdots & \vdots & \vdots & \vdots & & \vdots & \vdots & \vdots \\
0 & 0 & 0 & 0 & \cdots & 1 & 2x & 1 \\
0 & 0 & 0 & 0 & \cdots & 0 & 1 & 2x
\end{bmatrix}.
$$

(50) Suppose $p(x)$ is a polynomial of degree n which is orthogonal on $[-1, 1]$ to every polynomial of smaller degree with respect to the weighting function $(1 - x^2)^{-1/2}$ for $m = 0, 1, \ldots, n - 1$. Show $p(x)$ is a multiple of $T_n(x)$.

(51) Find the Chebyshev series expansions for the following functions on $[-1, 1]$.

(a) x^2.

(b) x^3.

(c) x^4.

(d) $\arcsin x$.

Chapter 9

Applications of PDEs in the Physical Sciences

This chapter will explore some mathematical models of physical phenomena and will make use of the mathematical methods and solution techniques developed in earlier chapters. In particular the method of separation of variables and Fourier analysis will be used in several new situations. The applications of partial differential equation techniques to be presented here have been selected to illustrate the utility of some of the special functions introduced in Chap. 8. Many other applications could be explored as well and the interested reader is encouraged to pursue readings in electricity and magnetism, classical and quantum mechanics, and optics. The interested reader can refer to [Garrity (2015); Griffiths (1999); Rauch (2012)].

9.1 Swinging Chain

The mathematical model of the swinging or vibrating chain is related to two other situations which the reader may have studied, the simple pendulum [Awrejcewicz (2012), Chap. 2] and the vibrating string (see Chap. 5). Consider a chain or string with one end fixed at the origin of a rectangular coordinate system, with the positive x-axis oriented in the downward direction. The length of the chain is L and the linear density of the chain is $\rho(x)$. The chain is subject to gravitational acceleration g (again in the downward, positive x-direction). If the chain is unperturbed by other forces, it will remain still occupying the interval $[0, L]$ along the x-axis. If the chain is moved by a small amount from this equilibrium position, let $u(x, t)$ denote the displacement of the chain from the x-axis at position x and time t.

    The derivation of the initial boundary value problem modeling the swinging chain is similar to the development of the wave equation in Sec. 1.4. The main differences are the end of the chain at $x = L$ is not fixed and the tension force in the chain is not constant. At position x along the chain, the tension is the weight of the segment of chain occupying the interval $[x, L]$. Using the density function, the tension at position x is thus

$$T(x) = g \int_x^L \rho(s) \, ds.$$

Assuming as in Sec. 1.4 that the displacements of the chain from equilibrium are small, substituting the expression for the tension into Eq. (1.35) results in

$$\rho(x)u_{tt} = -g\rho(x)u_x + T(x)u_{xx} + f(x, t) \tag{9.1}$$

where $f(x, t)$ represents any external forces on the chain. For the sake of simplicity in this initial analysis of the model, assume $f(x, t) = 0$. Note that Eq. (9.1) can then be written as

$$\rho(x)\frac{\partial^2 u}{\partial t^2} = \frac{\partial}{\partial x}\left[T(x)\frac{\partial u}{\partial x}\right]. \tag{9.2}$$

If the initial displacement and initial velocity of the chain are $u(x, 0) = f(x)$ and $u_t(x, 0) = g(x)$ respectively, then the boundary and initial conditions imposed on the mathematical model in Eq. (9.2) are

$$u(0, t) = 0 \text{ for } t > 0,$$

$$u(x, 0) = f(x) \text{ and } u_t(x, 0) = g(x) \text{ for } 0 < x < L.$$

Either the initial displacement or the initial velocity could be zero corresponding to the struck and plucked string situations explored in Chap. 5.

The remainder of this section will explore a special case of Eq. (9.2) for which the linear density of the chain is assumed constant so that $T(x) = g\rho(L - x)$ where ρ is the constant linear density. This simplifies the analysis permitting a product solution to be found. Using this linear expression for the tension in Eq. (9.2) produces the equation

$$\rho\frac{\partial^2 u}{\partial t^2} = \frac{\partial}{\partial x}\left[g\rho(L - x)\frac{\partial u}{\partial x}\right]$$

which simplifies to

$$u_{tt} = -gu_x + g(L - x)u_{xx}. \tag{9.3}$$

Assuming a product solution of the form $u(x, t) = X(x)T(t)$ the reader is asked to show in Exercise 1 that

$$\frac{T''(t)}{gT(t)} = -\frac{X'(x)}{X(x)} + (L - x)\frac{X''(x)}{X(x)} = c, \tag{9.4}$$

where c is a constant. Thus the variables can be separated. The resulting boundary value problem for $X(x)$ can be expressed as

$$(L - x)X''(x) - X'(x) - c\,X(x) = 0 \text{ for } 0 < x < L,$$
$$X(0) = 0. \tag{9.5}$$

While not stated explicitly as a boundary condition, any physically meaningful solution to the initial boundary value problem of the swinging chain must be bounded. The boundedness condition is used to determine that $c < 0$ (see Exercise 2). This ordinary differential equation does not match exactly any of the equations solved in Chap. 8, but can be transformed into a familiar equation via a change of variable. Let $L - x = \alpha \xi^\beta$ where α and β are constants to be chosen in order to simplify the ordinary differential equation. Making this change of variable results in the following ordinary differential equation (see Exercise 4),

$$\xi^{2-\beta} \frac{d^2 X}{d\xi^2} + \xi^{1-\beta} \frac{dX}{d\xi} - \alpha \beta^2 c X = 0. \tag{9.6}$$

If $\beta = 2$, $\alpha = -1/(4c)$, and both sides of Eq. (9.6) are multiplied by ξ^2 then Bessel's equation of order zero results (see Sec. 8.2):

$$\xi^2 \frac{d^2 X}{d\xi^2} + \xi \frac{dX}{d\xi} + \xi^2 X = 0.$$

Consequently the general solution can be expressed as $X(\xi) = c_1 J_0(\xi) + c_2 Y_0(\xi)$. Since the solution must be bounded as $x \to L^-$ which is equivalent to $\xi \to 0^+$, then $c_2 = 0$. In terms of the original independent variable $X(x) = c_1 J_0(2\sqrt{-c(L - x)})$. Choosing $c = -\lambda^2$ with $\lambda > 0$ and imposing the boundary condition at $u(0,t) = 0$ for all $t > 0$ implies $J_0(2\lambda\sqrt{L}) = 0$ which in turn implies that $2\lambda\sqrt{L}$ is a zero of the Bessel function of the first kind of order zero. Solving the time-dependent portion of Eq. (9.4) yields $T(t) = c_3 \cos\left(\sqrt{g}\lambda t\right) + c_4 \sin\left(\sqrt{g}\lambda t\right)$. Hence the product solutions indexed by $\lambda > 0$ take the form

$$u_\lambda(x,t) = [A_\lambda \cos\left(\sqrt{g}\lambda t\right) + B_\lambda \sin\left(\sqrt{g}\lambda t\right)] J_0(2\lambda\sqrt{L - x}),$$

where A_λ and B_λ are constants. Thus if $\lambda_{0,n}$ denotes the nth zero of the Bessel function of the first kind of order zero, then

$$\lambda \equiv \lambda_n = \frac{\lambda_{0,n}}{2\sqrt{L}}.$$

Motivated by the Principle of Superposition of solutions, the formal solution to the swinging chain can be written as the Fourier series:

$$u(x,t) \tag{9.7}$$
$$= \sum_{n=1}^{\infty} \left[A_n \cos\left(\frac{\lambda_{0,n}}{2}\sqrt{\frac{g}{L}}t\right) + B_n \sin\left(\frac{\lambda_{0,n}}{2}\sqrt{\frac{g}{L}}t\right) \right] J_0\left(\lambda_{0,n}\sqrt{1 - \frac{x}{L}}\right).$$

Example 9.1. Determine the formal solution to the swinging chain initial boundary value problem for initial conditions $u(x,0) = x$ and $u_t(x,0) = 0$.

Solution. Since the initial velocity of the chain is zero then $B_n = 0$ for $n \in \mathbb{N}$. Setting $t = 0$ in Eq. (9.7) produces

$$x = \sum_{n=1}^{\infty} A_n J_0 \left(\lambda_{0,n} \sqrt{1 - \frac{x}{L}} \right).$$

Equation (8.30) with $p = 0$ can be used to determine A_n; however, a change of variable is necessary. Let

$$\xi = \sqrt{1 - \frac{x}{L}} \iff x = L(1 - \xi^2)$$

then the initial displacement of the chain can be expressed as

$$L(1 - \xi^2) = \sum_{n=1}^{\infty} A_n J_0 (\lambda_{0,n}\xi)$$

and by Eq. (8.30),

$$
\begin{aligned}
A_n &= \frac{2 \int_0^1 L(1 - \xi^2) J_0(\lambda_{0,n}\xi)\xi \, d\xi}{(J_1(\lambda_{0,n}))^2} \\
&= \frac{2L}{(J_1(\lambda_{0,n}))^2} \left(\frac{1}{\lambda_{0,n}^2} \int_0^{\lambda_{0,n}} u J_0(u) \, du - \frac{1}{\lambda_{0,n}^4} \int_0^{\lambda_{0,n}} u^2(u J_0(u)) \, du \right) \\
&= \frac{2L}{(J_1(\lambda_{0,n}))^2} \left(\frac{2}{\lambda_{0,n}^4} \int_0^{\lambda_{0,n}} u^2 J_1(u) \, du \right) \quad \text{(see Exercise 5)} \\
&= \frac{4L}{\lambda_{0,n}^4 (J_1(\lambda_{0,n}))^2} \left[u^2 J_2(u) \right]_0^{\lambda_{0,n}} = \frac{4L J_2(\lambda_{0,n})}{\lambda_{0,n}^2 (J_1(\lambda_{0,n}))^2} \\
&= \frac{8L}{\lambda_{0,n}^3 J_1(\lambda_{0,n})} \quad \text{(by Eq. (8.22))}.
\end{aligned}
$$

Numerically approximating the definite integrals for the first several coefficients produces $A_1 \approx 1.10802L$, $A_2 \approx -0.139778L$, $A_3 \approx 0.0454765L$, $A_4 \approx -0.0209909L$, and $A_5 \approx 0.0116362L$. For this example the position of the swinging chain can be expressed as

$$u(x,t) = 8L \sum_{n=1}^{\infty} \frac{1}{\lambda_{0,n}^3 J_1(\lambda_{0,n})} J_0 \left(\lambda_{0,n} \sqrt{1 - \frac{x}{L}} \right).$$

Figure 9.1 shows the solution to the swinging chain problem for several different values of t when $L = 1$.

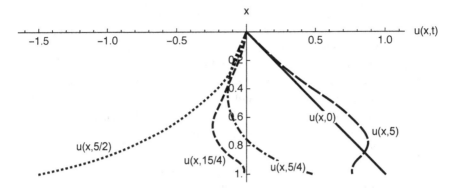

Fig. 9.1   The position of the swinging chain for several values of t. For convenience L is chosen to be 1.

In addition to the similar processes used to develop the mathematical models for the vibrating string and the swinging chain, there are some similar features of the general solutions. Equation (5.15) defined the nth normal mode of vibration for the vibrating string. Likewise the nth term of Eq. (9.7),

$$u_n(x,t) = \left[A_n \cos \left(\frac{\lambda_{0,n}}{2} \sqrt{\frac{g}{L}} t \right) + B_n \sin \left(\frac{\lambda_{0,n}}{2} \sqrt{\frac{g}{L}} t \right) \right] J_0 \left(\lambda_{0,n} \sqrt{1 - \frac{x}{L}} \right)$$

is called the nth **normal mode** of vibration for the swinging chain. The expression $(\lambda_{0,n}/2)\sqrt{g/L}$ is the frequency of the nth normal mode of vibration and corresponding to $n = 1$, $(\lambda_{0,1}/2)\sqrt{g/L}$ is the **fundamental frequency** for the swinging chain. An important difference between the frequencies of the vibrating string and the swinging chain is the fact that the frequencies of two different normal modes of vibration for the vibrating string are rationally related, which enables the blending of different modes of vibration to form sounds which are pleasing to the human ear. For the case of the swinging chain, the ratio of the frequencies of the nth and mth modes is $\lambda_{0,n}/\lambda_{0,m}$ which in general is not a rational number. Hence swinging chains make unusual musical instruments.

9.2   Vibration of a Circular Drum Head

Continuing the comments just made about musical instruments, the reader should consider a typical percussion instrument, the drum. Most drums consist of a rigid frame across which is stretched a flexible membrane or "skin". While the cross section of a drum can take many shapes, one of

the most common is that of a circle. In this section, a mathematical model
of a vibrating, circular drum head will be presented as an initial boundary
value problem. The partial differential equation will be solved using the
familiar method of separation of variables and the solution will involve
eigenfunctions satisfying an appropriate set of boundary conditions. Finally
a solution satisfying the initial conditions can be expressed by means of a
generalized Fourier series. By now the reader should begin to appreciate
the wide applicability of the solution techniques developed in the earlier
chapters of this text.

Suppose a flexible circular membrane of radius $r_0 > 0$ is centered at
the origin and has its circular edge fixed so that it cannot be displaced
from the xy-plane. Suppose the initial displacement of the membrane for
$x^2 + y^2 < r_0^2$ is described by the function $f(x, y)$ and the initial velocity of
the membrane is 0. According to Sec. 1.4.2 (see page 24) the displacement u
of the membrane from the xy-plane can be described by the initial boundary
value problem:

$$u_{tt} = c^2(u_{xx} + u_{yy}) \text{ for } x^2 + y^2 < r_0^2 \text{ and } t > 0,$$
$$u(x, y, t) = 0 \text{ for } x^2 + y^2 = r_0^2 \text{ and } t > 0,$$
$$u(x, y, 0) = f(x, y) \text{ for } x^2 + y^2 < r_0^2,$$
$$u_t(x, y, 0) = 0 \text{ for } x^2 + y^2 < r_0^2.$$

The constant $c > 0$ is the wave speed of vibrations in the membrane and
is directly proportional to the square root of the tension present in the
membrane and inversely proportional to the square root of the density
(mass per unit area) of the membrane.

The circular geometry of the membrane suggests that the solution of the
initial boundary value problem may be more easily found using the polar
coordinate system. Re-writing the problem (and, for now, dropping the initial condition for the initial velocity of the membrane) in polar coordinates
produces

$$u_{tt} = c^2 \left(u_{rr} + \frac{1}{r}u_r + \frac{1}{r^2}u_{\theta\theta} \right) \text{ for } r < r_0, \ \theta \in \mathbb{R}, \ t > 0, \quad (9.8)$$
$$u(r_0, \theta, t) = 0 \text{ for } \theta \in \mathbb{R} \text{ and } t > 0, \quad (9.9)$$
$$u(r, \theta, 0) = f(r, \theta) \text{ for } r < r_0 \text{ and } \theta \in \mathbb{R}. \quad (9.10)$$

The 2π-periodicity of the solution in the variable θ will also be helpful in
determining the proper solution. Using separation of variables and assuming a product solution of the form $u(r, \theta, t) = R(r)\Theta(\theta)T(t)$ to Eq. (9.8)

result in the following equation:

$$R(r)\Theta(\theta)T''(t) = c^2 \left(R''(r)\Theta(\theta)T(t) + \frac{R'(r)\Theta(\theta)T(t)}{r} + \frac{R(r)\Theta''(\theta)T(t)}{r^2} \right)$$

$$\frac{T''(t)}{c^2 T(t)} = \frac{R''(r)}{R(r)} + \frac{1}{r}\frac{R'(r)}{R(r)} + \frac{1}{r^2}\frac{\Theta''(\theta)}{\Theta(\theta)}. \tag{9.11}$$

Since the left-hand side of the equation depends on t alone while the right-hand side depends on (r, θ), then both sides must be constant. Denote the constant $-\lambda^2 \leq 0$. If the constant is positive, the function $T(t)$ is unbounded. Consider the right-hand side of Eq. (9.11):

$$\frac{R''(r)}{R(r)} + \frac{1}{r}\frac{R'(r)}{R(r)} + \frac{1}{r^2}\frac{\Theta''(\theta)}{\Theta(\theta)} = -\lambda^2$$

$$r^2\frac{R''(r)}{R(r)} + r\frac{R'(r)}{R(r)} + \lambda^2 r^2 = -\frac{\Theta''(\theta)}{\Theta(\theta)}.$$

Once again separation of variables has been achieved. Since the right-hand side of the last equation is a function of θ only and the solution to the initial boundary value problem should be 2π-periodic in θ then both sides of the equation are equal to the constant m^2 where $m = 0, 1, 2, \ldots$ (see Exercise 19 of Chap. 1). Consequently

$$\Theta(\theta) \equiv \Theta_m(\theta) = c_1 \cos(m\theta) + c_2 \sin(m\theta),$$

where c_1 and c_2 are constant coefficients. The ordinary differential equation implied by the left-hand side of the separated variables equation can be written as

$$r^2 R''(r) + r R'(r) + (\lambda^2 r^2 - m^2) R(r) = 0.$$

Equation (8.24) referred to this as the parametric form of Bessel's equation of order m. Thus the general solution is $R(r) = c_3 J_m(\lambda r) + c_4 Y_m(\lambda r)$. However, since the solution should be bounded as $r \to 0^+$, the radial part of the product solution will be expressed as $R_m(r) = J_m(\lambda r)$. Readers will recall that $Y_m(x) \to -\infty$ as $x \to 0^+$ for $m = 0, 1, 2, \ldots$. The boundary condition $u(r_0, \theta, t) = 0$ implies $J_m(\lambda r_0) = 0$ for each m, and hence $\lambda \equiv \lambda_{m,n}/r_0$ where $\lambda_{m,n}$ is the nth zero of the Bessel function of the first kind of order m.

Finally, the ordinary differential equation implied by the time dependent portion of the PDE is

$$T''(t) + \left(\frac{c\lambda_{m,n}}{r_0} \right)^2 T(t) = 0.$$

As a result, the time dependent portion of the solution can be written as

$$T(t) = c_5 \cos \left(\frac{c\lambda_{m,n}t}{r_0} \right) + c_6 \sin \left(\frac{c\lambda_{m,n}t}{r_0} \right).$$

Thus a product solution can be expressed in the form,

$$u_{m,n}(r,\theta,t) = J_m \left(\frac{\lambda_{m,n}r}{r_0} \right) (c_1 \cos(m\theta) + c_2 \sin(m\theta))$$

$$\left(c_5 \cos \left(\frac{c\lambda_{m,n}t}{r_0} \right) + c_6 \sin \left(\frac{c\lambda_{m,n}t}{r_0} \right) \right)$$

for $m = 0, 1, 2, \ldots$ and $n \in \mathbb{N}$. The Principle of Superposition suggests that a double infinite sum of terms of the form above also formally solves the boundary value problem. Assuming the double summation can be differentiated term by term with respect to t, the assumption that the initial velocity of the membrane is 0 implies $c_6 = 0$. Consequently the solution to the initial boundary value product is written formally as a double infinite series of the form

$$u(r,\theta,t) \tag{9.12}$$

$$= \sum_{m=0}^{\infty} \sum_{n=1}^{\infty} \left[J_m \left(\frac{\lambda_{m,n}r}{r_0} \right) (A_{m,n} \cos(m\theta) + B_{m,n} \sin(m\theta)) \cos \left(\frac{c\lambda_{m,n}t}{r_0} \right) \right]$$

where $A_{m,n}$ and $B_{m,n}$ are coefficients which will be determined from the initial displacement of the membrane.

In general the coefficients can be found by first setting $t = 0$ in which case,

$$f(r,\theta) \tag{9.13}$$

$$= \sum_{m=0}^{\infty} \sum_{n=1}^{\infty} \left[A_{m,n} J_m \left(\frac{\lambda_{m,n}r}{r_0} \right) \cos(m\theta) + B_{m,n} J_m \left(\frac{\lambda_{m,n}r}{r_0} \right) \sin(m\theta) \right].$$

Multiply both sides of Eq. (9.13) by $\pi^{-1} \cos(k\theta)$ and integrate over $[-\pi, \pi]$. The only nonzero term in the summation over index m will be the term for which $m = k$, thus

$$\frac{1}{\pi} \int_{-\pi}^{\pi} f(r,\theta) \cos(k\theta) \, d\theta = \sum_{n=1}^{\infty} A_{k,n} J_k \left(\frac{\lambda_{k,n}r}{r_0} \right). \tag{9.14}$$

Similarly if both sides of Eq. (9.13) are multiplied by $\pi^{-1} \sin(k\theta)$ and integrated over $[-\pi, \pi]$,

$$\frac{1}{\pi} \int_{-\pi}^{\pi} f(r,\theta) \sin(k\theta) \, d\theta = \sum_{n=1}^{\infty} B_{k,n} J_k \left(\frac{\lambda_{k,n}r}{r_0} \right). \tag{9.15}$$

The coefficients can now be found individually using Bessel series coefficient formulas. Make the substitution $z = r/r_0$. In the special case where $k = 0$, Eq. (9.14) simplifies to

$$\frac{1}{\pi} \int_{-\pi}^{\pi} f(r_0 z, \theta)\, d\theta = \sum_{n=1}^{\infty} A_{0,n} J_0(\lambda_{0,n} z). \tag{9.16}$$

Using the Bessel series coefficient formula given in Eq. (8.30),

$$A_{0,n} = \frac{2}{\pi (J_1(\lambda_{0,n}))^2} \int_0^1 z\, J_0(\lambda_{0,n} z) \left(\int_{-\pi}^{\pi} f(r_0 z, \theta)\, d\theta \right) dz. \tag{9.17}$$

For $k \in \mathbb{N}$,

$$A_{k,n} = \frac{2 \int_0^1 z\, J_k(\lambda_{k,n} z) \left(\int_{-\pi}^{\pi} f(r_0 z, \theta) \cos(k\theta)\, d\theta \right) dz}{\pi (J_{k+1}(\lambda_{k,n}))^2}, \tag{9.18}$$

$$B_{k,n} = \frac{2 \int_0^1 z\, J_k(\lambda_{k,n} z) \left(\int_{-\pi}^{\pi} f(r_0 z, \theta) \sin(k\theta)\, d\theta \right) dz}{\pi (J_{k+1}(\lambda_{k,n}))^2}. \tag{9.19}$$

As was the case for the vibrating string, the motion of the vibrating membrane can be thought of as the superposition of a number of modes of vibration. The (m, n)th mode of vibration is given by the expression,

$$J_m\left(\frac{\lambda_{m,n} r}{r_0} \right) (A_{m,n} \cos(m\theta) + B_{m,n} \sin(m\theta)) \cos\left(\frac{c \lambda_{m,n} t}{r_0} \right).$$

The frequency associated with this mode is

$$f_{m,n} = \frac{c \lambda_{m,n}}{2\pi r_0}$$

and therefore the fundamental frequency corresponding to $m = 0$ and $n = 1$ is $f_{0,1} = c\lambda_{0,1}/(2\pi r_0)$.

Example 9.2. Find the solution to the initial boundary value problem for the vibrating circular drumhead given in Eqs. (9.8)–(9.10) for the case where $c = 1$, $r_0 = 1$, and $f(r, \theta) = r(1 - r^2) \sin \theta$.

Solution. Since $f(r, \theta)$ is an odd function in θ, then by Eqs. (9.17) and (9.18) the coefficients $A_{k,n} = 0$ for $k = 0, 1, 2, \ldots$ and $n \in \mathbb{N}$. The orthogonality of the trigonometric functions on $[-\pi, \pi]$ used in Eq. (9.19) implies $B_{k,n} = 0$ for $k = 2, 3, \ldots$ and $n \in \mathbb{N}$. Thus the infinite series solution simplifies considerably to the form,

$$u(r, \theta, t) = \sin(\theta) \sum_{n=1}^{\infty} B_{1,n} J_1(\lambda_{1,n} r) \cos(\lambda_{1,n} t).$$

The coefficients $B_{1,n}$ for $n \in \mathbb{N}$ can be found by integration

$$B_{1,n} = \frac{2}{\pi(J_2(\lambda_{1,n}))^2} \int_0^1 z\, J_1(\lambda_{1,n}z) \left(\int_{-\pi}^{\pi} z(1-z^2) \sin^2(\theta)\, d\theta \right) dz$$

$$= \frac{2}{(J_2(\lambda_{1,n}))^2} \int_0^1 (z^2 - z^4)\, J_1(\lambda_{1,n}z)\, dz = \frac{16}{\lambda_{1,n}^3 J_2(\lambda_{1,n})}.$$

The evaluation of this type of definite integral depends on properties of the Bessel functions and on integration by parts. The reader will find help in evaluating this integral exactly in Exercise 5. The formal solution can then be expressed as

$$u(r, \theta, t) = 16 \sin(\theta) \sum_{n=1}^{\infty} \frac{J_1(\lambda_{1,n}r)}{\lambda_{1,n}^3 J_2(\lambda_{1,n})} \cos(\lambda_{1,n}t).$$

Contour plots of the displacement of the vibrating membrane are shown in Fig. 9.2. The displacement is depicted at $t = 0$, $1/4$, $1/2$, and $3/4$.

9.3 Steady-State Temperature in a Sphere

This application topic makes use of the spherical coordinate system for $\mathbb{R}^3$ which has been briefly mentioned earlier. Despite the increasing number of dimensions to the mathematical model, the familiar techniques, particularly separation of variables and generalized Fourier series, continue to provide solutions.

Consider a solid spherical shell of radius 1 whose center lies at the origin of a three-dimensional coordinate system and for which the temperature of the surface of the northern hemisphere is kept at $100°$C and the temperature of the surface of the southern hemisphere is $0°$C. In this section an expression for the steady-state temperature of the sphere will be found using a three-dimensional version of the heat equation. The steady-state temperature distribution u satisfies Laplace's equation, $\triangle u = 0$. To conveniently accommodate the geometry of the problem, the Laplacian operator will be expressed in terms of the spherical coordinates (ρ, φ, θ),

$$\triangle u = \frac{1}{\rho^2} \left[\rho(\rho\, u)_{\rho\rho} + \csc\varphi(\sin\varphi\, u_{\varphi})_{\varphi} + \csc^2\varphi\, u_{\theta\theta} \right] = 0.$$

The reader may want to verify this is equivalent to the form given in Eq. (8.70). Since the temperature distribution must be defined throughout the interior of the sphere, it is natural to require that the temperature

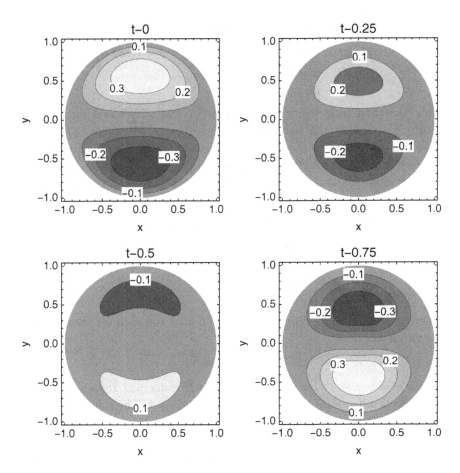

Fig. 9.2   The displacement of the circular drumhead for $t = 0$, $1/4$, $1/2$, and $3/4$.

distribution be bounded as $\rho \to 0^+$ in addition to the boundary conditions imposed on the surface of the sphere. These boundary conditions are the given temperature distribution on the top and bottom surfaces of the sphere. Using spherical coordinates the temperature on the surface of the sphere can be expressed as $u(\rho, \varphi, \theta) = 100$ for $\rho = 1$, $0 \leq \theta \leq 2\pi$, $0 \leq \varphi \leq \pi/2$ and $u(\rho, \varphi, \theta) = 0$ for $\rho = 1$, $0 \leq \theta \leq 2\pi$, $\pi/2 < \varphi \leq \pi$. Finally since the hemisphere is invariant in the θ coordinate, the temperature distribution should be independent of θ. Consequently the temperature distribution must satisfy the boundary value problem,

$$0 = \rho(\rho\,u)_{\rho\rho} + \csc\varphi(\sin\varphi\,u_\varphi)_\varphi \text{ for } 0 < \rho < 1 \text{ and } 0 < \varphi < \pi,$$

$$u(1,\varphi) = \begin{cases} 100 & \text{for } 0 \le \varphi \le \pi/2, \\ 0 & \text{for } \pi/2 < \varphi \le \pi. \end{cases}$$

Note that assuming the solution is independent of θ based on physical intuition simplifies the boundary value problem. The heat equation in spherical coordinates without this simplification will be revisited in Sec. 9.5.

As usual, the starting point is the assumption that $u(\rho, \varphi)$ can be written as a product solution $u(\rho, \varphi) = R(\rho)\Phi(\varphi)$. Differentiating the product solution and substituting it into Laplace's equation produce

$$\rho(\rho\,R(\rho))_{\rho\rho}\Phi(\varphi) + \csc\varphi(\sin\varphi\,\Phi'(\varphi))_\varphi R(\rho) = 0,$$

$$\frac{(2\rho R'(\rho) + \rho^2 R''(\rho))\Phi(\varphi) + (\cot\varphi\,\Phi'(\varphi) + \Phi''(\varphi))R(\rho)}{R(\rho)\Phi(\varphi)} = 0,$$

$$\frac{\rho^2 R''(\rho) + 2\rho R'(\rho)}{R(\rho)} = \frac{-\Phi''(\varphi) - \cot\varphi\,\Phi'(\varphi)}{\Phi(\varphi)} = c$$

where c is a constant. The ordinary differential equation for $\Phi(\varphi)$ can be rewritten as

$$\Phi''(\varphi) + \cot\varphi\,\Phi'(\varphi) + c\,\Phi(\varphi) = 0.$$

When $c = n(n+1)$ for $n = 0, 1, 2, \ldots$ the reader will recognize this as Legendre's differential equation explored in Sec. 8.3. The solutions will be denoted as $P_n(\cos\varphi)$ for $n = 0, 1, 2, \ldots$. If $c \ne n(n+1)$ where n is a nonnegative integer, then $\Phi(\varphi)$ is not finite when $\varphi = 0$ or $\varphi = \pi$ (corresponding to the north and south poles of the sphere respectively). Hence this case is not considered further. The ordinary differential equation for $R(\rho)$ can be written as

$$\rho^2 R''(\rho) + 2\rho R'(\rho) - n(n+1)R(\rho) = 0.$$

This is Euler's equation with general solution,

$$R_n(\rho) = a_n\rho^n + b_n\rho^{-1-n}.$$

To ensure the boundedness of solutions as $\rho \to 0^+$ the coefficients b_n are chosen to be zero, thus a product solution takes the form $u_n(\rho, \varphi) = a_n\rho^n P_n(\cos\varphi)$. The Principle of Superposition suggests that the temperature distribution can be written formally as the infinite series,

$$u(\rho, \varphi) = \sum_{n=0}^{\infty} a_n\rho^n P_n(\cos\varphi).$$

The boundary condition $u(1, \varphi)$ determines the values of the series coefficients a_n,

$$u(1, \varphi) = \sum_{n=0}^{\infty} a_n P_n(\cos \varphi) = \begin{cases} 100 & \text{for } 0 \le \varphi \le \pi/2, \\ 0 & \text{for } \pi/2 < \varphi \le \pi. \end{cases}$$

Multiply both sides of the equation by $P_m(\cos \varphi) \sin \varphi$ and integrate from $\varphi = 0$ to $\varphi = \pi$.

$$\sum_{n=0}^{\infty} a_n \int_0^{\pi} P_n(\cos \varphi) P_m(\cos \varphi) \sin \varphi \, d\varphi = 100 \int_0^{\pi/2} P_m(\cos \varphi) \sin \varphi \, d\varphi$$

$$\sum_{n=0}^{\infty} a_n \int_{-1}^{1} P_n(x) P_m(x) \, dx = 100 \int_0^1 P_m(x) \, dx.$$

By Theorem 8.8 the only nonzero term in the infinite series on the left-hand side of the equation occurs when $m = n$, thus

$$\frac{2a_n}{2n+1} = 100 \int_0^1 P_n(x) \, dx.$$

The reader will be asked to evaluate the integral on the right-hand side of the equation in Exercises 8 and 9 and to obtain

$$a_n = \begin{cases} 50 & \text{if } n = 0, \\ 0 & \text{if } n = 2, 4, 6, \ldots, \\ \dfrac{50(2n+1)(-1)^{(n+3)/2}(n+1)!}{n \, 2^{n+1} \left(\frac{n+1}{2}\right)! \left(\frac{n+1}{2}\right)!} & \text{if } n = 1, 3, 5, \ldots. \end{cases}$$

Therefore the formal solution describing the temperature distribution on the sphere can be expressed as

$$u(\rho, \varphi) = 50 - 50 \sum_{k=1}^{\infty} \frac{(-1)^k (2k)!(4k-1)}{2^{2k}(2k-1)(k!)^2} \rho^{2k-1} P_{2k-1}(\cos \varphi).$$

Figure 9.3 illustrates the solution as a contour plot. Rather than depicting the entire sphere, a "slice" through the north and south poles is shown since the solution is independent of the azimuthal coordinate θ.

9.4   Vibrating Sphere

In this text the wave equation has been studied in the context of a vibrating string, swinging chain, and a vibrating circular drumhead. In these explorations the Cartesian and polar coordinate systems have been used respectively. In this section another application of the wave equation is

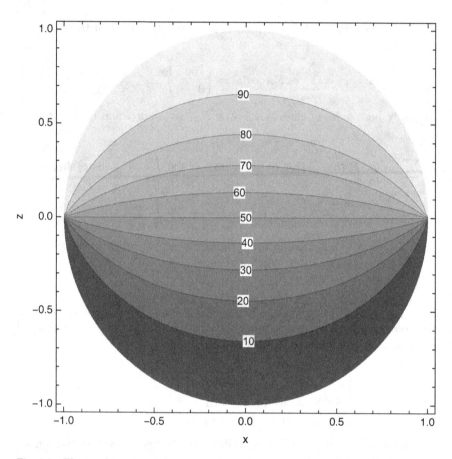

Fig. 9.3   The steady-state temperature distribution within a sphere for which the surface of the northern hemisphere is kept at 100C and the southern hemisphere is kept at 0C. The temperature distribution is independent of the azimuthal coordinate θ.

studied which necessitates the use of the spherical coordinate system. In particular, consider a flexible membrane in the shape of a hollow sphere of radius ρ_0. In the physical world this may be thought of as a rubber ball or balloon. If a point on the surface of the sphere is displaced radially from the equilibrium radius ρ_0, the tension in the surrounding area will attempt to restore the point to the equilibrium radius. In this way the reader can imagine the surface of the sphere "vibrating". In the spherical coordinate system the mathematical model of this situation is described by the following initial boundary value problem. Let $u(\varphi, \theta, t)$ be the radial displacement of the point $(\rho_0, \varphi, \theta)$ from equilibrium at time t.

$$u_{tt} = c^2 \triangle u - \omega^2 u \text{ for } 0 < \varphi < \pi, \ -\pi < \theta < \pi, \ t > 0, \quad (9.20)$$

$$u(\varphi, \theta, 0) = f(\varphi, \theta) \text{ for } 0 < \varphi < \pi, \ -\pi < \theta < \pi,$$

$$u_t(\varphi, \theta, 0) = g(\varphi, \theta) \text{ for } 0 < \varphi < \pi, \ -\pi < \theta < \pi.$$

The constant c is the wave speed and ω is also a constant. Boundary conditions are implied by the 2π-periodicity of the solution in the θ variable and by the assumption that the solution is bounded at $\varphi = 0$ and $\varphi = \pi$.

As expected, the first step in solving the initial boundary value problem is to use the method of separation of variables to determine the eigenvalues and eigenfunctions of the model. If $u(\varphi, \theta, t) = F(\varphi, \theta)T(t)$ is a product solution to the partial differential equation in Eq. (9.20), then

$$F(\varphi, \theta)T''(t) = c^2 T(t)\left(\triangle F(\varphi, \theta)\right) - \omega^2 F(\varphi, \theta)T(t)$$

$$\frac{1}{c^2}\left(\frac{T''(t)}{T(t)} + \omega^2\right) = \frac{\triangle F(\varphi, \theta)}{F(\varphi, \theta)}.$$

Since the left-hand side of the equation depends only on t and the right-hand side depends only on φ and θ, this use of separation of variables has been successful. Both sides of the equation are equal to a constant denoted as $-\lambda$.

The right-hand side of the equation implies the differential equation

$$\triangle F(\varphi, \theta) + \lambda F(\varphi, \theta) = 0.$$

The eigenvalues and eigenfunctions to this equation were found in Sec. 8.4. The eigenvalues are $\lambda \equiv \lambda_n = n(n+1)$ for $n = 0, 1, 2, \ldots$ and the eigenfunctions are the spherical harmonics $Y_n^m(\varphi, \theta)$ for $m = -n, -n+1, \ldots, n$. Any other choice of λ results in solutions which are unbounded at $\varphi = 0$ and $\varphi = \pi$ and hence are not suitable as a factor in the product solution to this initial boundary value problem.

Using the eigenvalues found, the t-dependent portion of the product solution satisfies the ordinary differential equation,

$$T''(t) + (\omega^2 + c^2 n(n+1))T(t) = 0$$

which has the general solution,

$$T_n(t) = c_1 \cos\left(\sqrt{\omega^2 + c^2 n(n+1)}t\right) + c_2 \sin\left(\sqrt{\omega^2 + c^2 n(n+1)}t\right).$$

Thus a product solution to Eq. (9.20) has the form

$$u_{n,m}(\varphi, \theta, t)$$

$$= Y_n^m(\varphi, \theta)\left(a_{n,m}\cos\left(\sqrt{\omega^2 + c^2\lambda_n}t\right) + b_{n,m}\sin\left(\sqrt{\omega^2 + c^2\lambda_n}t\right)\right)$$

for $n = 0, 1, 2, \ldots$ and $m = -n, -n+1, \ldots, n$. Note that if $\omega = 0$, then
$$u_{0,0}(\varphi, \theta, t) = a_{0,0} Y_0^0(\varphi, \theta).$$
The radial displacement of the sphere from equilibrium can be expressed formally as the infinite series,
$$u(\varphi, \theta, t) = \sum_{n=0}^{\infty} \sum_{m=-n}^{n} u_{n,m}(\varphi, \theta, t).$$

The initial conditions and the orthogonality of the spherical harmonics can be used to determine the coefficients $a_{n,m}$ and $b_{n,m}$. For $t = 0$ the infinite series takes the form,
$$u(\varphi, \theta, 0) = f(\varphi, \theta) = \sum_{n=0}^{\infty} \sum_{m=-n}^{n} a_{n,m} Y_n^m(\varphi, \theta).$$

Multiply both sides of the equation by the complex conjugate of $Y_{\hat{n}}^{\hat{m}}(\varphi, \theta) \sin \varphi$ and integrate with respect to θ over interval $[-\pi, \pi]$ and with respect to φ over $[0, \pi]$. Using the definition of the spherical harmonic function in Eq. (8.73) and the orthogonality of the spherical harmonics as described in Theorem 8.12, the only nonzero term in the infinite series occurs when $m = \hat{m}$ and $n = \hat{n}$. In this case,
$$a_{n,m} = \int_{-\pi}^{\pi} \int_0^{\pi} f(\varphi, \theta) \overline{Y_n^m(\varphi, \theta)} \sin \varphi \, d\varphi \, d\theta$$
$$= \sqrt{\frac{2n+1}{4\pi} \frac{(n-m)!}{(n+m)!}} \int_{-\pi}^{\pi} \int_0^{\pi} f(\varphi, \theta) e^{-im\theta} P_n^m(\cos \varphi) \sin \varphi \, d\varphi \, d\theta.$$

Differentiating the formal solution with respect to t and setting $t = 0$ produce the equation
$$u_t(\varphi, \theta, 0) = g(\varphi, \theta) = \sum_{n=0}^{\infty} \sum_{m=-n}^{n} b_{n,m} \sqrt{\omega^2 + c^2 n(n+1)} Y_n^m(\varphi, \theta).$$

Multiplying both sides of the equation by the complex conjugate of $Y_{\hat{n}}^{\hat{m}}(\varphi, \theta) \sin \varphi$ and integrating with respect to θ over interval $[-\pi, \pi]$ and with respect to φ over $[0, \pi]$ generate
$$b_{n,m} = \frac{1}{\sqrt{\omega^2 + c^2 n(n+1)}} \int_{-\pi}^{\pi} \int_0^{\pi} g(\varphi, \theta) \overline{Y_n^m(\varphi, \theta)} \sin \varphi \, d\varphi \, d\theta.$$

Note that if $\omega = 0$, then $b_{0,0} = 0$. Scant attention has been paid to the conditions under which these double integrals exist and the infinite series converges and is differentiable. Generally if $f(\varphi, \theta)$ and $g(\varphi, \theta)$ are sufficiently smooth functions on their domains, the formal solution can be shown to be a valid solution to the initial boundary value problem.

Example 9.3. Find a formal series solution describing the radial displacement from equilibrium of a point on a flexible spherical membrane of equilibrium radius $\rho_0 = 1$ subject to an initial displacement of $f(\varphi, \theta) = 32^{-1} \sin^2(2\varphi) \sin(2\theta)$ with zero initial velocity. Assume the wave speed $c = 1$ and the constant $\omega = 0$ in Eq. (9.20).

Solution. Since the initial velocity is zero, the coefficients $b_{n,m} = 0$ for all n and m. From the discussion above,

$$a_{n,m} = \frac{1}{32} \int_{-\pi}^{\pi} \int_0^{\pi} \sin^2(2\varphi) \sin(2\theta) \overline{Y_n^m(\varphi, \theta)} \sin\varphi \, d\varphi \, d\theta$$

$$= \frac{1}{32} \int_0^{\pi} \sin^2(2\varphi) \sin\varphi \int_{-\pi}^{\pi} \sin(2\theta) \overline{Y_n^m(\varphi, \theta)} \, d\theta \, d\varphi.$$

Recall the spherical harmonic function $Y_n^m(\varphi, \theta)$ can be expressed as

$$Y_n^m(\varphi, \theta) = \sqrt{\frac{2n+1}{4\pi} \frac{(n-m)!}{(n+m)!}} e^{im\theta} P_n^m(\cos\varphi)$$

for $m \geq 0$. Thus when $m \geq 0$ the coefficient,

$$a_{n,m} = \frac{1}{32} \sqrt{\frac{2n+1}{4\pi} \frac{(n-m)!}{(n+m)!}} \int_0^{\pi} \tag{9.21}$$

$$\times \sin^2(2\varphi) P_n^m(\cos\varphi) \sin\varphi \, d\varphi \int_{-\pi}^{\pi} e^{-im\theta} \sin(2\theta) \, d\theta.$$

In Exercise 14 the reader will be asked to verify the following equation for $m > 0$,

$$a_{n,-m} = \frac{1}{32} \int_{-\pi}^{\pi} \int_0^{\pi} \sin^2(2\varphi) \sin(2\theta) \overline{Y_n^{-m}(\varphi, \theta)} \sin\varphi \, d\varphi \, d\theta$$

which can be written as:

$$32(-1)^m a_{n,-m} \tag{9.22}$$

$$= \sqrt{\frac{2n+1}{4\pi} \frac{(n-m)!}{(n+m)!}} \int_0^{\pi} \sin^2(2\varphi) P_n^m(\cos\varphi) \sin\varphi \, d\varphi \int_{-\pi}^{\pi} e^{im\theta} \sin(2\theta) \, d\theta.$$

Suppose $|m| \neq 2$, then

$$\int_{-\pi}^{\pi} e^{im\theta} \sin(2\theta) \, d\theta = \left[\frac{e^{im\theta}}{m^2 - 4} \left(2\cos(2\theta) - im\sin(2\theta) \right) \right]_{-\pi}^{\pi} = 0.$$

This implies $a_{n,m} = 0$ for $m \neq \pm 2$. Thus the solution to the vibrating sphere problem takes on the form

$$u(\varphi, \theta, t) = \sum_{n=2}^{\infty} \left[\cos(\sqrt{n(n+1)}t) \left(a_{n,-2} Y_n^{-2}(\varphi, \theta) + a_{n,2} Y_n^2(\varphi, \theta) \right) \right]$$

$$= \sum_{n=2}^{\infty} \left[\cos(\sqrt{n(n+1)}t) \left(a_{n,-2} Y_n^2(\varphi, -\theta) + a_{n,2} Y_n^2(\varphi, \theta) \right) \right]$$

where the last equality is a direct result of the definition of the spherical harmonics $Y_n^m(\varphi, \theta)$ given in Eq. (8.73). Using integration by parts the reader can show that (Exercise 15),

$$\int_{-\pi}^{\pi} e^{\pm 2i\theta} \sin(2\theta)\, d\theta = \pm \pi i \tag{9.23}$$

where $i = \sqrt{-1}$. Letting $m = 2$ and substituting the result of Eq. (9.23) in Eqs. (9.21) and (9.22) reveal that

$$a_{n,2} = -a_{n,-2} = -\frac{\pi i}{32} \sqrt{\frac{2n+1}{4\pi} \frac{(n-2)!}{(n+2)!}} \int_0^{\pi} \sin^2(2\varphi) P_n^2(\cos \varphi) \sin \varphi \, d\varphi.$$

Making the substitution $x = \cos \phi$ produces

$$
\begin{aligned}
a_{n,2} = -a_{n,-2} &= -\frac{\pi i}{8} \sqrt{\frac{2n+1}{4\pi} \frac{(n-2)!}{(n+2)!}} \int_{-1}^{1} x^2(1-x^2) P_n^2(x)\, dx \\
&= \frac{\pi i}{8} \sqrt{\frac{2n+1}{4\pi} \frac{(n-2)!}{(n+2)!}} \int_{-1}^{1} x^2(1-x^2)^2 \frac{d^2}{dx^2}[P_n(x)]\, dx \\
&= \begin{cases} \frac{-i\sqrt{\pi}}{14\sqrt{30}} & \text{if } n = 2, \\ \frac{-i\sqrt{\pi}}{21\sqrt{10}} & \text{if } n = 4, \\ 0 & \text{otherwise.} \end{cases}
\end{aligned} \tag{9.24}
$$

The reader is asked to provide the justification for this result in Exercise 16. This result further reduces the infinite series solution to the vibrating sphere problem to an exact solution containing only a finite number of terms,

$$
\begin{aligned}
u(\varphi, \theta, t) &= \frac{i\sqrt{\pi}}{14\sqrt{30}} \cos(\sqrt{6}t) \left(Y_2^2(\varphi, -\theta) - Y_2^2(\varphi, \theta) \right) \\
&\quad + \frac{i\sqrt{\pi}}{21\sqrt{10}} \cos(\sqrt{20}t) \left(Y_4^2(\varphi, -\theta) - Y_4^2(\varphi, \theta) \right) \\
&= \left(\frac{1}{168} \cos(\sqrt{6}t) P_2^2(\cos \varphi) + \frac{1}{420} \cos(2\sqrt{5}t) P_4^2(\cos \varphi) \right) \sin(2\theta).
\end{aligned}
$$

Figure 9.4 illustrates the displacement of the spherical membrane at $t = \pi/(3\sqrt{6})$. Since the displacement of the membrane from equilibrium is small, the membrane retains a roughly spherical shape.

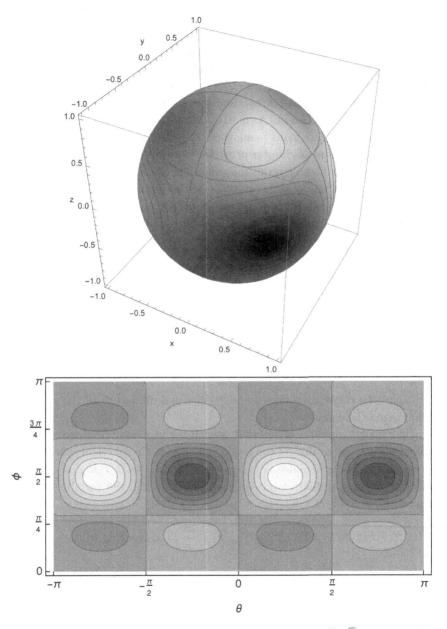

Fig. 9.4   The displacement of the spherical membrane at $t = \pi/(3\sqrt{6})$. The top plot depicts the displacement as a spherical plot. Since the displacements are small, the membrane still appears nearly spherical. The bottom plot is a cylindrical projection of the vibrating membrane. Lighter areas in both plots show regions where the radius of the membrane is greater than its equilibrium radius (here chosen to be $\rho_0 = 1$) while darker areas show regions where the radius of the membrane is smaller than its equilibrium radius.

9.5 Heat Equation in a Sphere

This section will explore the solution to the time-dependent heat equation on a solid sphere. Once again the reader will have the opportunity to use the method of separation of variables on a mathematical model set in a three-dimensional spatial domain and will explore the eigenfunctions and eigenvalues of this situation. A small addition to the initial boundary value problem described here leads to another interesting problem which is left for the reader in the end-of-chapter projects. Consider a solid sphere of radius a constructed of a homogeneous material. The surface of the sphere is kept at temperature 0 and at time $t = 0$ the initial distribution of temperature within the sphere is described by the function $f(\rho, \varphi, \theta)$ where (ρ, φ, θ) are the spherical coordinates. If $u(\rho, \varphi, \theta, t)$ denotes the temperature in the sphere at time t, then an initial boundary value problem which models this situation can be written as,

$$u_t = k \triangle u \tag{9.25}$$

$$\text{for } 0 \leq \rho < a, \ 0 < \varphi < \pi, \ -\pi < \theta < \pi, \text{ and } t > 0,$$

$$u(a, \varphi, \theta, t) = 0 \text{ for } 0 < \varphi < \pi, \ -\pi < \theta < \pi, \text{ and } t > 0, \tag{9.26}$$

$$u(\rho, \varphi, \theta, 0) = f(\rho, \varphi, \theta) \tag{9.27}$$

$$\text{for } 0 \leq \rho < a, \ 0 < \varphi < \pi, \text{ and } -\pi < \theta < \pi.$$

As usual the constant k represents the thermal diffusivity of the material comprising the sphere.

The reader should now be familiar with the process of finding the eigenfunctions and eigenvalues of the Laplacian operator in spherical coordinates. Assuming a product solution of the form $u(\rho, \varphi, \theta, t) = T(t)R(\rho)F(\varphi, \theta)$ exists for the partial differential equation in Eq. (9.25), the variables can be separated as follows:

$$T'(t)R(\rho)F(\varphi, \theta) = kT(t)R''(\rho)F(\varphi, \theta) + \frac{2k}{\rho}T(t)R'(\rho)F(\varphi, \theta)$$

$$+ \frac{k}{\rho^2}\left(TRF_{\varphi\varphi} + \cot\varphi TRF_{\varphi} + \csc^2\varphi TRF_{\theta\theta}\right)$$

$$\frac{T'(t)}{kT(t)} = \frac{R''(\rho)}{R(\rho)} + \frac{2}{\rho}\frac{R'(\rho)}{R(\rho)} + \frac{1}{\rho^2 F}\left(F_{\varphi\varphi} + \cot\varphi F_{\varphi} + \csc^2\varphi F_{\theta\theta}\right).$$

The left-hand side of the equation depends only on t while the right-hand side depends only on the spatial variables (ρ, φ, θ). Hence both sides are equal to a constant denoted as $-\lambda$. Treating just the right-hand side of the last equation, the variables can be further separated as

$$-\lambda = \frac{R''(\rho)}{R(\rho)} + \frac{2}{\rho}\frac{R'(\rho)}{R(\rho)} + \frac{1}{\rho^2 F(\varphi,\theta)}\left(F_{\varphi\varphi} + \cot\varphi F_\varphi + \csc^2\varphi F_{\theta\theta}\right)$$

$$\frac{\triangle F(\varphi,\theta)}{F(\varphi,\theta)} = -\lambda\rho^2 - \frac{\rho^2 R''(\rho)}{R(\rho)} - \frac{2\rho R'(\rho)}{R(\rho)}.$$

The numerator on left-hand side of the last equation involves the Laplacian operator applied to a function of (φ,θ). The right-hand side depends only on ρ. Again both sides must equal the same constant. This situation has arisen before and $F(\varphi,\theta)$ must be a spherical harmonic function $Y_n^m(\varphi,\theta)$ and the constant must be $-n(n+1)$ for $n = 0, 1, 2, \dots$. Moving to the radial coordinate, the implied ordinary differential equation is

$$\rho^2 R''(\rho) + 2\rho R'(\rho) + (\lambda\rho^2 - n(n+1))R(\rho) = 0.$$

Thinking ahead to the time-dependent portion of the product solution, it is expected for λ to be positive. Based on prior exposure to the heat equation with homogeneous Dirichlet boundary conditions and the physical intuition that, in the absence of sources of heat energy within the sphere, the asymptotic temperature distribution within the sphere should be zero. Thus $T(t)$ should decay asymptotically to zero. To clarify the notation, then replace λ with ν^2 where $\nu > 0$. Alternatively the reader can consider the cases in which $\lambda \leq 0$ and show that no nontrivial solutions to the ordinary differential equation above satisfy the Dirichlet boundary conditions at $\rho = a$ and the implicit assumption that the solution is bounded at $\rho = 0$.

Consider the differential equation now written as

$$\rho^2 R''(\rho) + 2\rho R'(\rho) + (\nu^2\rho^2 - n(n+1))R(\rho) = 0.$$

Make the change of variable $x = \nu\rho$, then

$$x^2 \frac{d^2 R}{dx^2} + 2x \frac{dR}{dx} + (x^2 - n(n+1))R(x) = 0$$

which the reader may recognize from Eq. (8.34) as the spherical Bessel's equation. According to Eqs. (8.35) and (8.36) the general solution to the spherical Bessel's equation is

$$R_n(x) = A_n j_n(x) + B_n y_n(x).$$

However, since $y_n(x)$ is unbounded as $x \to 0^+$, the coefficient B_n must be chosen to be 0. In Exercise 19 the reader will be asked to confirm that the spherical Bessel functions of the first kind are bounded as x approaches 0 for $n = 0, 1, 2, \dots$. Hence $R_n(\rho) = A_n j_n(\nu\rho)$. The homogeneous Dirichlet boundary condition requires $R_n(a) = 0$ and thus $\nu a = \lambda_{n+1/2,q}$, the qth zero of the Bessel function of the first kind of order $n + 1/2$. Finally the radial portion of the product solution may be expressed as, omitting the dependence on an arbitrary multiplicative constant, $R_{n,m}(\rho) = j_n(\lambda_{n+1/2,q}\rho/a)$.

With $\lambda = \nu^2 = \lambda_{n+1/2,q}^2$, the time-dependent ordinary differential equation can be expressed as

$$\frac{T'(t)}{kT(t)} = -\left(\frac{\lambda_{n+1/2,q}}{a}\right)^2 \implies T_{n,q}(t) = e^{-k\lambda_{n+1/2,q}^2 t/a^2}.$$

A product solution can be assembled from its factors,

$$u_{n,m,q}(\rho, \varphi, \theta, t) = j_n\left(\frac{\lambda_{n+1/2,q}\rho}{a}\right) Y_n^m(\varphi, \theta) e^{-k\lambda_{n+1/2,q}^2 t/a^2}.$$

By the Principle of Superposition, a formal solution to the heat equation on the sphere can be expressed as a triple infinite series,

$$u(\rho, \varphi, \theta, t) = \sum_{n=0}^{\infty} \sum_{q=1}^{\infty} \sum_{m=-n}^{n} A_{n,m,q} j_n\left(\frac{\lambda_{n+1/2,q}\rho}{a}\right) Y_n^m(\varphi, \theta) e^{-k\lambda_{n+1/2,q}^2 t/a^2}.$$

The coefficients $A_{n,m,q}$ are determined by the initial temperature distribution within the sphere. In order to calculate these coefficients the eigenfunctions must be shown to be orthogonal over the sphere.

For the sake of concise notation, let

$$U_{n,m,q}(\rho, \varphi, \theta) = j_n(\lambda_{n+1/2,q}\rho/a)Y_n^m(\varphi, \theta).$$

Consider the triple integral,

$$\int_0^a \int_0^\pi \int_{-\pi}^\pi U_{n,m,q}(\rho, \varphi, \theta)\overline{U_{\hat{n},\hat{m},\hat{q}}(\rho, \varphi, \theta)}\rho^2 \sin\varphi \, d\theta \, d\phi \, d\rho$$

$$= \int_0^\pi \int_{-\pi}^\pi Y_n^m(\varphi, \theta)\overline{Y_{\hat{n}}^{\hat{m}}(\varphi, \theta)} \sin\varphi \, d\theta \, d\varphi$$

$$\times \int_0^a j_n\left(\frac{\lambda_{n+\frac{1}{2},q}\rho}{a}\right) j_{\hat{n}}\left(\frac{\lambda_{\hat{n}+\frac{1}{2},\hat{q}}\rho}{a}\right)\rho^2 \, d\rho$$

$$= \frac{a^3}{2}\left(j_{n+1}(\lambda_{n+1/2,q})\right)^2 \delta_{n\hat{n}}\delta_{m\hat{m}}\delta_{q\hat{q}}.$$

This result follows from Corollary 8.2 by way of Exercise 20 and Theorem 8.12. Hence when $t = 0$,

$$f(\rho, \varphi, \theta) = \sum_{n=0}^{\infty} \sum_{q=1}^{\infty} \sum_{m=-n}^{n} A_{n,m,q} j_n\left(\frac{\lambda_{n+1/2,q}\rho}{a}\right) Y_n^m(\varphi, \theta).$$

If both sides of the equation are multiplied by $\overline{U_{n,m,q}(\rho, \varphi, \theta)}\rho^2 \sin\varphi$ and integrated over the sphere of radius a, then

$$A_{n,m,q}$$

$$= \frac{2}{a^3\left(j_{n+1}(\lambda_{n+1/2,q})\right)^2} \int_0^a \int_0^\pi \int_{-\pi}^\pi f(\rho, \varphi, \theta)\overline{U_{n,m,q}(\rho, \varphi, \theta)}\rho^2 \sin\varphi \, d\theta \, d\varphi \, d\rho.$$

Example 9.4. Find the temperature distribution within a sphere as described in the initial boundary value problem in Eqs. (9.25)–(9.27) where the initial heat distribution is

$$f(\rho, \varphi, \theta) = j_2 \left(\frac{\lambda_{5/2,1}\rho}{a} \right) \left(P_2(\cos\varphi) + \sin(2\theta) P_2^2(\cos\varphi) \right).$$

Solution. Quite a great deal of work is saved if the initial condition is written in the equivalent form

$$f(\rho, \varphi, \theta) = j_2 \left(\frac{\lambda_{5/2,1}\rho}{a} \right) \left(\sqrt{\frac{4\pi}{5}} Y_2^0(\varphi, \theta) + 2i\sqrt{\frac{6\pi}{5}} (Y_2^{-2}(\varphi, \theta) - Y_2^2(\varphi, \theta)) \right).$$

Once the initial condition is expressed in terms of eigenfunctions, the solution can be expressed as

$$\begin{aligned}
u(\rho, \varphi, \theta, t) &= \sqrt{\frac{4\pi}{5}} Y_2^0(\varphi, \theta) A_{2,0,1} j_2 \left(\frac{\lambda_{5/2,1}\rho}{a} \right) e^{-k\lambda_{5/2,1}^2 t/a^2} \\
&\quad + 2i\sqrt{\frac{6\pi}{5}} Y_2^{-2}(\varphi, \theta) A_{2,-2,1} j_2 \left(\frac{\lambda_{5/2,1}\rho}{a} \right) e^{-k\lambda_{5/2,1}^2 t/a^2} \\
&\quad - 2i\sqrt{\frac{6\pi}{5}} Y_2^2(\varphi, \theta) A_{2,2,1} j_2 \left(\frac{\lambda_{5/2,1}\rho}{a} \right) e^{-k\lambda_{5/2,1}^2 t/a^2} \\
&= e^{-k\lambda_{5/2,1}^2 t/a^2} f(\rho, \phi, \theta).
\end{aligned}$$

9.6   Quantum Harmonic Oscillator

In this section and the next, the solution methods for partial differential equations will be applied to examples from **quantum mechanics**. While the detailed study of quantum mechanics would fill dozens of books and likely require years of study to adequately comprehend, the mathematical principles required to analyze simple examples have already been presented. During the 17th, 18th, and 19th centuries the Newtonian or classical view of mechanics was deemed to accurately describe the physical and engineering phenomena of interest. While clearly an oversimplification, Newton's second law of motion, force equals mass times acceleration or $F = ma$, is certainly a good starting point in the study of many physical situations. Thus physicists, mathematicians, and engineers found a great deal of utility in the solution of initial value problems consisting of an ordinary differential equation together with initial position and initial velocity conditions. As the 20th century dawned, physicists were beginning to focus on developing an understanding of the atom. At first atoms were imagined to be tiny solar systems with the atomic nucleus playing the role of the star and the electrons moving around the nucleus like planets orbiting a star. Since the field of celestial mechanics had been so successful at describing the motions of the sun and planets, this was a natural place to start. Unfortunately classical mechanics models of the atom behaved in very different ways from actual observations of atoms. Gradually a new approach came into vogue. While classical mechanics is useful for modeling most macroscopic phenomena (human scale or even celestial scale), microscopic phenomena (molecular or atomic scale) require a quantum theoretical framework. Part of this quantum theory involved the assumption that energy was apportioned in discrete amounts (called quanta) and that the position and velocity of objects at the atomic scale should be treated probabilistically rather than deterministically. In short, the mathematical models shifted from specifying the deterministic position and velocity of an object to specifying a probability distribution for the position and velocity which are now thought of as random variables.

In this section the differences between the two approaches will be illustrated by the study of the harmonic oscillator. In most courses on the topic of ordinary differential equations, students study the simple harmonic oscillator, often as a mathematical model of the displacement $u(t)$ of a mass attached to a linear spring. As the mass is displaced from its equilibrium position, the mass begins to move in a periodic fashion which can be determined from knowledge of the mass m, the spring constant k, the initial position $u(t_0) = u_0$, and the initial velocity $u'(t_0) = v_0$. The classical spring-mass problem (sometimes called the simple harmonic oscillator) will be compared to the quantum harmonic oscillator. The reader may think of this as a model of a molecule consisting of two bound atoms. As one atom is moved from its equilibrium position relative to the other atom, it begins to oscillate as well. Since the masses and distances involved are so small and since there is no physical spring connecting the two atoms, a quantum mechanical analysis must be undertaken. However, some similarities between the two physical problems can be noted. In both cases, the total energy can be described as the sum of the kinetic and potential energies. Expressions can be developed describing the probability of observing the mass or the atom in a small interval of length Δu during its motion.

Before introducing the new quantum mechanical ideas necessary to describe the molecular phenomena, the more familiar spring-mass system will be reviewed. The mathematical model of the classical spring-mass system (without damping) can be expressed as,

$$m\,u'' + k\,u = 0,$$
$$u(0) = u_0,$$
$$u'(0) = v_0.$$

The reader will provide some of the details of the solution to a simple harmonic motion problem in the exercises. In the mean time the displacement of the mass can be described by the function,

$$u(t) = u_0 \cos\left(\sqrt{\frac{k}{m}}\,t\right) + v_0\sqrt{\frac{m}{k}}\sin\left(\sqrt{\frac{k}{m}}\,t\right). \qquad (9.28)$$

Let the quantity $\sqrt{k/m} = \omega$, then the period of the oscillation is $T = 2\pi/\omega$ and the frequency of the oscillation is $f = \omega/(2\pi) = 1/T$. The amplitude of the oscillation is $A = \sqrt{u_0^2 + v_0^2/\omega^2}$. The total energy of the spring-mass system is the sum of the kinetic and potential energies:

$$E = \frac{1}{2}m\left(u'(t)\right)^2 + \frac{1}{2}k\left(u(t)\right)^2. \qquad (9.29)$$

Since there are no other forces present in the spring-mass system the total energy is constant, and thus

$$E = \frac{1}{2}k\left(u_0^2 + \frac{v_0^2}{\omega^2}\right). \qquad (9.30)$$

When the mass is at maximum displacement $u(t) = A$, its velocity is zero and by Eq. (9.29), $A = \sqrt{2E/k}$. Substituting this expression into Eq. (9.29) and rearranging terms produce the following expression for the velocity of the mass:

$$u'(t) = \pm\omega\sqrt{A^2 - \left(u(t)\right)^2}. \qquad (9.31)$$

For the classical spring-mass system the mass oscillates in the interval $[-A, A]$. Suppose an observer examines a subinterval of $[-A, A]$ of length Δu. What is the probability of observing the mass in this subinterval? Let the probability distribution for the displacement u defined for $u \in [-A, A]$ be denoted as $P(u)$. For any fixed $u \in (-A, A)$, if Δu is small, the probability of observing the mass in the subinterval $[u, u + \Delta u]$ is approximately $P(u)\Delta u$. Let Δt be the amount of time required for the mass to travel a distance Δu. During a complete period of oscillation the mass will traverse the subinterval of length Δu twice and hence,

$$P(u)\Delta u = \frac{2\Delta t}{T} \implies P(u)\frac{\Delta u}{\Delta t} = \frac{2}{T}.$$

As Δu and Δt approach zero, $P(u) = 2/(|u'(t)|T)$ where $|u'(t)| = \omega\sqrt{A^2 - (u(t))^2}$ is the speed of the mass. Thus for the classical simple harmonic oscillator, the probability density function for the displacement of the mass is

$$P(u) = \frac{2}{\omega\sqrt{A^2 - (u(t))^2}\,\frac{2\pi}{\omega}} = \frac{1}{\pi\sqrt{A^2 - u^2}}. \qquad (9.32)$$

Now attention is shifted to the quantum harmonic oscillator. Since classical Newtonian mechanics performed poorly when describing atomic scale phenomena, physicists suggested that very small objects had the properties of waves rather than the behavior of point masses. In 1926, Louis de Broglie[1] stated that all matter has wave-like properties. This assertion has since become known as the **de Broglie hypothesis**. The fundamental property of an object is its **wave function** $\Psi(x,t)$, from which quantities such as the position and velocity of the object can be derived. The wave function is a complex-valued function. Max Born[2] developed the statistical interpretation of the wave function which implies the probability of finding an object in the interval $[x, x+dx]$ at time t is $P(x,t)\,dx = \Psi(x,t)\overline{\Psi(x,t)}\,dx$ where $\overline{\Psi(x,t)}$ is the complex conjugate of the wave function. The wave function of an object solves the linear partial differential equation known as the linear **Schrödinger equation**. The one-dimensional version of the linear Schrödinger equation can be written as

$$i\hbar\frac{\partial\Psi}{\partial t} = -\frac{\hbar^2}{2\mu}\frac{\partial^2\Psi}{\partial x^2} + V(x)\Psi(x,t).$$

This equation was developed by Erwin Schrödinger in 1926 and should be treated as an axiom of quantum mechanics the same way that $F = m\,a$ is treated as an axiom of classical mechanics. As usual $i = \sqrt{-1}$, x denotes position, and $V(x)$ is the potential energy of the object. The symbol $\hbar$ is the reduced Planck[3] constant and μ denotes the reduced mass of the object. The concept of reduced mass is used in mechanics in situations where one smaller mass interacts with one larger mass. This helps to simplify the "two-body" problem to a "one-body" problem. The precise meaning of the term is not important to the mathematical analysis carried out below for the harmonic oscillator.

---

[1] Louis de Broglie, French physicist (1892–1987).
[2] Max Born, German physicist and mathematician (1882–1970).
[3] Max Karl Ernst Ludwig Planck, German physicist (1858–1947).

If the potential energy of the quantum harmonic oscillator is $V(x) = kx^2/2$ (the same as for the classical harmonic oscillator), Schrödinger's equation can be written as

$$i\hbar \frac{\partial \Psi}{\partial t} = -\frac{\hbar^2}{2\mu} \frac{\partial^2 \Psi}{\partial x^2} + \frac{1}{2} kx^2 \Psi(x, t). \tag{9.33}$$

Assume this equation is defined for $-\infty < x < \infty$ and the solutions asymptotically approach 0 as $|x| \to \infty$. Let the initial wave function be $\Psi(x, 0) = f(x)$. The method of separation of variables can be applied to Eq. (9.33). Assuming $\Psi(x, t) = X(x)T(t)$ then

$$i\hbar X(x)T'(t) = -\frac{\hbar^2}{2\mu} X''(x)T(t) + \frac{1}{2} kx^2 X(x)T(t)$$

$$i\hbar \frac{T'(t)}{T(t)} = -\frac{\hbar^2}{2\mu} \frac{X''(x)}{X(x)} + \frac{1}{2} kx^2.$$

Once again, the left-hand side of the equation depends on t while the right-hand side depends on x and thus both sides are equal to a constant. In this case the constant is denoted by E for reasons to be explained shortly. Thus the ordinary differential equation with independent variable x implied by the last equation is

$$-\frac{\hbar^2}{2\mu} X''(x) + \left(\frac{1}{2} kx^2 - E \right) X(x) = 0. \tag{9.34}$$

Let ω be a constant such that $k = \mu\omega^2$ and make the substitution $x = \xi\sqrt{\hbar/(\mu\omega)}$. Equation (9.34) can be re-written as

$$\frac{d^2 X}{d\xi^2} + \left(\frac{2E}{\hbar\omega} - \xi^2 \right) X = 0. \tag{9.35}$$

Solutions to Eq. (9.35) should have the properties that the solutions asymptotically approach 0 as $|x| \to \infty$ and also that the wave function when multiplied by its conjugate is a valid probability density function on the real line. While the connection is not immediately obvious, consider the Hermite equation of Eq. (8.85). If $\phi(x) = e^{-x^2/2} y(x)$ where $y(x)$ is a solution to Hermite's ordinary differential equation then

$$0 = [e^{x^2/2} \phi(x)]'' - 2x[e^{x^2/2} \phi(x)]' + \lambda e^{x^2/2} \phi(x)$$
$$= \left(\phi''(x) + (\lambda + 1 - x^2)\phi(x) \right) e^{x^2/2}. \tag{9.36}$$

After dividing both sides of Eq. (9.36) by $e^{x^2/2}$ note that Eq. (9.35) is of the form of Eq. (9.36) if $2E/(\hbar\omega) = \lambda + 1$. If $\lambda = 2n$ where $n = 0, 1, 2, \ldots$, then $y(x) = H_n(x)$, the nth Hermite polynomial. Thus for $E \equiv E_n =$

$\hbar\omega(n+1/2)$ the function $X_n(\xi) = e^{-\xi^2/2}H_n(\xi)$ is a solution to Eq. (9.35). Consequently

$$X_n(x) = e^{-\mu\omega x^2/(2\hbar)} H_n\left(\sqrt{\frac{\mu\omega}{\hbar}}x\right) \text{ for } n = 0, 1, 2, \ldots . \qquad (9.37)$$

Functions of the form seen in Eq. (9.37) are known as **Hermite functions**. The Hermite functions are solutions to Eq. (9.34) and vanish as $|x| \to \infty$ as can be seen from Eq. (9.37) due to the presence of the decaying exponential factor. See Fig. 9.5 for an illustration.

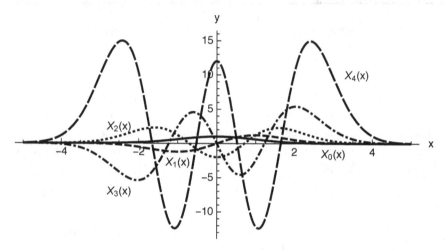

Fig. 9.5   The graphs of the first five Hermite functions $X_n(x)$. These functions quickly decay to 0 as $|x| \to \infty$.

If $\lambda \neq 2n$ where n is a nonnegative integer, then $y(x)$ is an infinite series of the form given in Eq. (8.88) or Eq. (8.89). Using the recurrence relation for the series coefficients (see Eq. (8.87)), for large n, $a_{n+2}/a_n \approx 2/n$. Thus when n is large $a_{2n} \approx C_0/n!$ and $a_{2n+1} \approx C_1/n!$ where C_0 and C_1 are constants. The infinite series solution $y_1(x)$ given in Eq. (8.88) can be thought of as the sum of a polynomial of degree $2N - 2$ and an infinite series,

$$y_1(x) = 1 + \sum_{n=1}^{N-1} \frac{\prod_{k=1}^{n}(4k - 4 - \lambda)}{(2n)!}x^{2n} + \sum_{n=N}^{\infty} a_{2n}x^{2n}$$

$$\approx 1 + \sum_{n=1}^{N-1} \frac{\prod_{k=1}^{n}(4k - 4 - \lambda)}{(2n)!}x^{2n} + C_0\sum_{n=N}^{\infty} \frac{x^{2n}}{n!}.$$

For large $|x|$ then $y_1(x) \approx C_0 e^{x^2}$. A similar conclusion holds for solution $y_2(x)$ given in Eq. (8.89). As a result, the eigenfunctions $X_\lambda(\xi) \approx C e^{\xi^2/2}$ for some constant C when $|\xi|$ is large and the squares of such expressions are not integrable over the real line. Therefore solutions to Eq. (9.36) for $\lambda \neq 2n$ for all nonnegative integers n are not appropriate for consideration.

According to Theorem 8.15, the Hermite functions are pairwise orthogonal on the real line. Using the eigenvalues E_n in the time-dependent ordinary differential equation produces $T_n(t) = e^{-i\omega(n+1/2)t}$. Thus product solutions to the one-dimensional linear Schrödinger equation for the quantum harmonic oscillator have the form,

$$\Psi_n(x,t) = e^{-i\omega(n+1/2)t} e^{-\mu\omega x^2/(2\hbar)} H_n\left(\sqrt{\frac{\mu\omega}{\hbar}} x\right) \text{ for } n = 0, 1, 2, \ldots$$

The symbol E_n was chosen for the eigenvalues, since these are typically called **energy eigenvalues**. The implication is that energies of the product solutions to Eq. (9.33) for the quantum harmonic oscillator take on values from an infinite, but discrete set of values. The energy levels are separated by the finite positive quantity $\hbar\omega$. The least energy the quantum harmonic oscillator can possess is $E_0 = \hbar\omega/2 > 0$ which is sometimes called the **zero point energy**.

The Principle of Superposition suggests the formal solution to the quantum harmonic oscillator is an infinite series of the form,

$$\Psi(x,t) = e^{-\mu\omega x^2/(2\hbar)} \sum_{n=0}^{\infty} A_n e^{-i\omega(n+1/2)t} H_n\left(\sqrt{\frac{\mu\omega}{\hbar}} x\right), \tag{9.38}$$

where the expressions A_n are constants. If the initial form of the wave function is $f(x)$, then

$$f(x) = e^{-\mu\omega x^2/(2\hbar)} \sum_{n=0}^{\infty} A_n H_n\left(\sqrt{\frac{\mu\omega}{\hbar}} x\right).$$

Multiply both sides of this equation by

$$H_m\left(\sqrt{\frac{\mu\omega}{\hbar}} x\right) e^{-\mu\omega x^2/(2\hbar)}$$

and then integrate over the real number line. Using the orthogonality of the Hermite polynomials, the only nonzero term in the infinite series is the one for which $m = n$. Therefore by Eq. (8.98) the series coefficient,

$$A_n = \frac{1}{2^n n!} \sqrt{\frac{\mu\omega}{\hbar\pi}} \int_{-\infty}^{\infty} f(x) H_n\left(\sqrt{\frac{\mu\omega}{\hbar}} x\right) e^{-\mu\omega x^2/(2\hbar)} \, dx. \tag{9.39}$$

As a simple example, suppose the probability distribution function for the position of the mass is initially Gaussian in shape with $f(x)\overline{f(x)} = \frac{1}{\sqrt{2\pi}}e^{-x^2/2}$. This implies $f(x) = \frac{1}{\sqrt[4]{2\pi}}e^{-x^2/4}$. The coefficients of the infinite series solution are then

$$A_n = \frac{1}{2^n n!}\sqrt{\frac{\mu\omega}{\hbar\pi}}\int_{-\infty}^{\infty}\frac{1}{\sqrt[4]{2\pi}}e^{-x^2/4}H_n\left(\sqrt{\frac{\mu\omega}{\hbar}}x\right)e^{-\mu\omega x^2/(2\hbar)}\,dx.$$

Since $H_n(x)$ is an odd function when n is odd, $A_n = 0$ for n odd. The calculation of the remaining coefficients is made easier if it is assumed that $\mu\omega/\hbar = 1$. In this case,

$$A_{2n} = \frac{1}{2^{2n+1/4}\pi^{3/4}(2n)!}\int_{-\infty}^{\infty}e^{-3x^2/4}H_{2n}(x)\,dx = \frac{2}{12^n(n!)\sqrt[4]{18\pi}}.$$

An approach to evaluating the improper integral above is outlined in Exercise 25.

A natural question to ask is whether the classical mechanical approach and the quantum mechanical approach are compatible with one another. It has been seen that the differences between energy levels of the quantum harmonic oscillator are integer multiples of $\hbar\omega$. As n increases the energy E_n increases and the relative change in energy vanishes,

$$\lim_{n\to\infty}\frac{\hbar\omega}{E_n} = \lim_{n\to\infty}\frac{\hbar\omega}{\left(n+\frac{1}{2}\right)\hbar\omega} = 0.$$

Thus for energetic oscillators the discrete differences between energy levels is less noticeable. Another comparison between the classical harmonic oscillator and the quantum harmonic oscillator can be made through the probability density functions governing the probabilities of observing the mass in a particular interval. Recall that the amplitude of the classical harmonic oscillator is $A = \sqrt{2E/k}$. If the classical harmonic oscillator has energy equal to one of the discrete energy levels of the quantum harmonic oscillator, then

$$A = \sqrt{\frac{2(n+1/2)\hbar\omega}{k}} = \sqrt{\frac{(2n+1)\hbar}{\mu\omega}}.$$

Plotting the probability distribution for the location of the mass of the classical harmonic oscillator from Eq. (9.32) and $\Psi_n(x,0)\overline{\Psi_n(x,0)}$ together on the same set of axes as in Fig. 9.6 shows the relationship between the probability density functions for low values of n. As n (and hence the energy of the oscillator) increases, the quantum harmonic oscillator's probability density function oscillates more rapidly about the classical harmonic

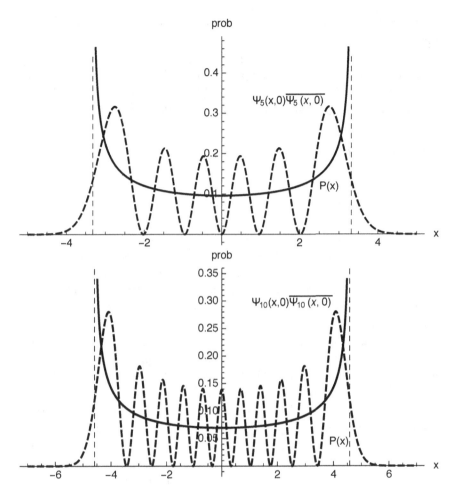

Fig. 9.6 The graph of the probability density function $P(x)$ for the classical harmonic oscillator (solid curve) and the graph of $\Psi_n(x,0)\overline{\Psi_n(x,0)}$ (dashed curve) together on the same set of axes. In the upper graph $n = 5$ while in the lower graph $n = 10$. The quantity $\mu\omega/\hbar$ was chosen to be 1. The vertical asymptotes represent the maximum displacements of the classical harmonic oscillator.

oscillator's probability density function. Thus at higher energy levels the quantum features of the harmonic oscillator are more difficult to measurably resolve. One perhaps surprising difference is apparent between the two treatments of the oscillator depicted in Fig. 9.6. The mass of the classical harmonic oscillator must be found in the bounded interval $[-A, A]$ and has probability 0 of being found outside of this interval. The mass of the quantum harmonic oscillator may be found anywhere on the real number line, though the figure strongly suggests that the probability of the mass being outside the interval $[-A, A]$ is small, but not zero!

In the next section another application of the solution techniques for quantum mechanically-derived partial differential equations will be explored. The next application will involve three-dimensional space.

9.7   Model of the Hydrogen Atom

The second application of the linear Schrödinger equation is to that of a model of the hydrogen atom, the simplest atom in the sense that it contains only one proton and one electron (ignoring isotopes which may contain one or more neutrons). The proton makes up the nucleus of the hydrogen atom and the electron "orbits" the nucleus. For reference the mass of a proton is approximately $m_p = 1.67262 \times 10^{-27}$ kg and that of an electron is $m_e = 9.10938 \times 10^{-31}$ kg [Mohr *et al.* (2015)]. Note that the proton's mass is about 1836 times that of the electron, thus the center of mass of a hydrogen atom is located nearly at the position of the proton in the nucleus. The proton and electron possess charges of $\epsilon = 1.60218 \times 10^{-19}$ Coulomb[4] and $-\epsilon$ respectively. Having opposite charges, the proton and electron are attracted to one another through electrostatic attraction. The potential for this attractive force can be written as $V(\rho) = -\epsilon^2/\rho$ where ρ is the distance separating the proton and electron. In the early part of the 20th century, physicists believed that the electron orbited the proton like a planet orbits a star. A problem with this theory is that accelerating charged particles give off energy and thus hydrogen atoms (in fact all atoms) should be radiating energy continuously and would therefore collapse. Ernest Rutherford[5] observed that the hydrogen atom is stable and emits energy corresponding to a discrete set of wavelengths. Soon thereafter Niels Bohr[6] put forth a theory that the electron in the hydrogen atom can exist in a discrete set of stable stationary states. While in these stationary states, the atom does not emit radiation. The hydrogen atom will emit or absorb radiation when the electron transitions from one stationary state to another.

In the remainder of this section, the permissible energy levels and

---

[4] The unit of charge is named for Charles-Augustin de Coulomb, French physicist (1736–1806). Many authors use the symbol e for the charge of a proton. To avoid confusion with the natural exponential function, ϵ will be used here.

[5] Ernest Rutherford, New Zealand physicist (1871–1937).

[6] Niels Henrik David Bohr, Danish physicist (1885–1962).

stationary states (or orbits) of the hydrogen atom will be determined from an analysis of the linear Schrödinger equation for the hydrogen atom,

$$i\hbar\frac{\partial\Psi}{\partial t} = -\frac{\hbar^2}{2\mu}\triangle\Psi - \frac{\epsilon^2}{\rho}\Psi. \tag{9.40}$$

In this model the origin of the spherical coordinate system is assumed to lie at the center of the proton. As in the earlier exploration of the quantum harmonic oscillator, the symbol μ represents the reduced mass of the electron,

$$\mu = \frac{m_e m_p}{m_e + m_p} \approx m_e.$$

The expression $\Psi(\rho, \varphi, \theta, t)$ is the wave function of the electron. Solutions to Eq. (9.40) should have the property that

$$\int_0^\infty \int_0^\pi \int_{-\pi}^\pi \Psi(\rho, \varphi, \theta, t)\overline{\Psi(\rho, \varphi, \theta, t)}\rho^2 \sin\varphi \, d\theta \, d\varphi \, d\rho = 1. \tag{9.41}$$

This condition imposed on the solution is sometimes referred to as the solution being **normalizable**. The wave function should also be 2π-periodic in θ, should vanish as $\rho \to \infty$, and should be bounded as $\rho \to 0^+$.

By now the steps to take to study this model should be familiar and natural to the reader. The starting point is the assumption that a product solution of the form $\Psi(\rho, \varphi, \theta, t) = T(t)R(\rho)F(\varphi, \theta)$ solves Eq. (9.40). Substituting this into the partial differential equation and rearranging terms produce

$$i\hbar\frac{T'(t)}{T(t)} = -\frac{\hbar^2}{2\mu}\frac{\triangle[R(\rho)F(\varphi, \theta)]}{R(\rho)F(\varphi, \theta)} - \frac{\epsilon^2}{\rho}. \tag{9.42}$$

The left-hand side of this equation depends on t while the right-hand side depends on the spherical coordinate system variables, thus each side is equal to a constant denoted as E. Setting the right-hand side of Eq. (9.42) equal to E and expanding the Laplacian in spherical coordinates result in

$$E = -\frac{\hbar^2}{2\mu}\frac{\triangle[R(\rho)F(\varphi, \theta)]}{R(\rho)F(\varphi, \theta)} - \frac{\epsilon^2}{\rho} \tag{9.43}$$

$$\left(E + \frac{\epsilon^2}{\rho}\right)RF = -\frac{\hbar^2}{2\mu}\left(R''F + \frac{2}{\rho}R'F\right)$$

$$-\frac{\hbar^2}{2\mu\rho^2}\left(RF_{\varphi\varphi} + \cot\varphi RF_\varphi + \csc^2\varphi RF_{\theta\theta}\right).$$

Multiplying both sides of the last equation by $2\mu\rho^2/(\hbar^2 RF)$ and moving terms depending on ρ to the left-hand side yield the following equation.

$$\rho^2\left(\frac{R''}{R} + \frac{2}{\rho}\frac{R'}{R} + \frac{2\mu E}{\hbar^2}\right) + \frac{2\mu\epsilon^2}{\hbar^2}\rho = -\left(\frac{F_{\varphi\varphi}}{F} + \cot\varphi\frac{F_\varphi}{F} + \csc^2\varphi\frac{F_{\theta\theta}}{F}\right)$$

The radial variable ρ has been successfully separated from the angular variables (φ, θ). If both sides of the equation are equal to a constant λ then the partial differential equation for the angular variables becomes

$$F_{\varphi\varphi} + \cot \varphi F_{\varphi} + \csc^2 \varphi F_{\theta\theta} + \lambda F = 0.$$

This equation has been encountered before. If $\lambda = l(l+1)$ for $l = 0, 1, 2, \ldots$, the solutions are the spherical harmonic functions $Y_l^m(\varphi, \theta)$.

Setting the ρ-dependent side of the equation equal to $l(l + 1)$ produces an ordinary differential equation for $R(\rho)$.

$$\rho^2 R''(\rho) + 2\rho R'(\rho) + \left(\frac{2\mu E}{\hbar^2}\rho^2 + \frac{2\mu\epsilon^2}{\hbar^2}\rho - l(l+1)\right) R(\rho) = 0. \qquad (9.44)$$

By a change of variable Eq. (9.44) can be re-written as

$$xy''(x) + ((2l+1) + 1 - x)y'(x) + \left(\epsilon^2\sqrt{\frac{-\mu}{2E\hbar^2}} - l - 1\right) y(x) = 0 \qquad (9.45)$$

(see Exercise 26). If the coefficient of $y(x)$ in Eq. (9.45) is a nonnegative integer, then Eq. (9.45) is an example of the associated Laguerre differential equation studied in Sec. 8.5.2 on page 363. For the sake of compact notation let $n = \epsilon^2\sqrt{-\mu/(2E\hbar^2)}$, and thus when n is a positive integer such that $n \geq l + 1$, there is a polynomial solution to Eq. (9.45), namely $y(x) = L_{n-l-1}^{2l+1}(x)$, the associated Laguerre polynomial of degree $n - l - 1$ and order $2l + 1$. If the coefficient of $y(x)$ in Eq. (9.45) is not a nonnegative integer, the resulting solution will not yield a normalizable wave function, and this case is not appropriate for consideration here.

Using the change of variable introduced in Exercise 26 and defining $\alpha = 2\sqrt{-2\mu E/\hbar^2}$, the solution to Eq. (9.44) can be written as

$$R_{n,l}(\rho) = e^{-\alpha\rho/2}(\alpha\rho)^l L_{n-l-1}^{2l+1}(\alpha\rho) \qquad (9.46)$$

for $l = 0, 1, 2, \ldots$ and n an integer for which $n \geq l + 1$.

Multiplying the factors of the product solution determined thus far yields a time-independent solution to the boundary value problem

$$R(\rho)F(\varphi, \theta) = e^{-\alpha\rho/2}(\alpha\rho)^l L_{n-l-1}^{2l+1}(\alpha\rho)Y_l^m(\varphi, \theta),$$

where l, m, and n are integers satisfying the inequality $|m| \leq l < n$ and $n \in \mathbb{N}$. To aid with notation, define $u_{l,m,n}(\rho, \varphi, \theta)$ to be the expression on the right-hand side of the last equation. The assumption that n is a positive integer implies the possible values of E, the energy of the electron, are given by

$$E \equiv E_n = -\frac{\mu\epsilon^4}{2\hbar^2 n^2}. \qquad (9.47)$$

These energies are called the energies of the bound (or stationary) states of the hydrogen atom. For a given $n \in \mathbb{N}$ there are n^2 stationary states (see Exercise 27). This in turn implies the constant $\alpha = 2\mu\epsilon^2/(\hbar^2 n)$. The reader may wish to confirm directly by differentiation that $u_{l,m,n}(\rho, \varphi, \theta)$ solves Eq. (9.43).

Returning to the left-hand side of Eq. (9.42), the implied ordinary differential equation for the time-dependent factor $T(t)$ of the product solution is

$$T'(t) = \frac{-iE_n}{\hbar} T(t),$$

and thus $T_n(t) = e^{-iE_n t/\hbar}$ for $n \in \mathbb{N}$. Since the magnitude of $T_n(t)$ is one, the wave functions are normalized when $u_{l,m,n}(\rho, \varphi, \theta)$ is normalized. By Theorem 8.12 and the definition of spherical harmonics, the triple integral

$$\int_0^\infty \int_0^\pi \int_{-\pi}^\pi u_{l,m,n}(\rho, \varphi, \theta)\overline{u_{l,m,n}(\rho, \varphi, \theta)}\rho^2 \sin \varphi \, d\theta \, d\varphi \, d\rho$$

$$= \int_0^\infty e^{-\alpha\rho}(\alpha\rho)^{2l} \left(L_{n-l-1}^{2l+1}(\alpha\rho)\right)^2 \rho^2 \, d\rho \int_0^\pi \int_{-\pi}^\pi Y_l^m(\varphi, \theta)\overline{Y_l^m(\varphi, \theta)} \sin \varphi \, d\theta \, d\varphi$$

$$= \int_0^\infty e^{-\alpha\rho}(\alpha\rho)^{2l} \left(L_{n-l-1}^{2l+1}(\alpha\rho)\right)^2 \rho^2 \, d\rho.$$

Making the substitution $r = \alpha\rho$ allows the remaining integral to be rewritten as

$$\int_0^\infty e^{-\alpha\rho}(\alpha\rho)^{2l} \left(L_{n-l-1}^{2l+1}(\alpha\rho)\right)^2 \rho^2 \, d\rho = \frac{1}{\alpha^3} \int_0^\infty r^{2l+2} \left(L_{n-l-1}^{2l+1}(r)\right)^2 e^{-r} \, dr.$$

Unfortunately the result of Theorem 8.14 does not apply here. In Exercise 28 the reader will show that

$$\int_0^\infty r^{2l+2} \left(L_{n-l-1}^{2l+1}(r)\right)^2 e^{-r} \, dr = \frac{(2n)\Gamma(n+l+1)}{(n-l-1)!}. \tag{9.48}$$

Finally the normalized product solutions to Schrödinger's equation for the hydrogen atom are

$$\Psi_{l,m,n}(\rho, \varphi, \theta, t) \tag{9.49}$$

$$= \sqrt{\frac{\alpha^3(n-l-1)!}{(2n)\Gamma(n+l+1)}} e^{-iE_n t/\hbar} e^{-\alpha\rho/2}(\alpha\rho)^l L_{n-l-1}^{2l+1}(\alpha\rho)Y_l^m(\varphi, \theta),$$

for $|m| \leq l < n$ and $n \in \mathbb{N}$. Physicists have given names to the indices of the wave function. Index l is often called the **angular momentum quantum number**, index m is the **magnetic quantum number**, and n is the **principal** or **total quantum number**. When the electron of the hydrogen atom is in energy state E_n, the wave function of the electron will be one of the n^2 functions given in Eq. (9.49). The probability density function of the electron is thus

$$
\begin{aligned}
P_{l,m,n}(\rho, \varphi, \theta) \\
= \Psi_{l,m,n}(\rho, \varphi, \theta, t)\overline{\Psi_{l,m,n}(\rho, \varphi, \theta, t)} \\
= \frac{\alpha^3(n - l - 1)!}{(2n)\Gamma(n + l + 1)} e^{-\alpha\rho}(\alpha\rho)^{2l} \left(L_{n-l-1}^{2l+1}(\alpha\rho)\right)^2 Y_l^m(\varphi, \theta)\overline{Y_l^m(\varphi, \theta)}.
\end{aligned}
\tag{9.50}
$$

As an example a density plot of the $P_{2,0,3}(\rho, \varphi, \theta)$ is shown in Fig. 9.7. If the linear Schrödinger equation stated in Eq. (9.40) is supplemented with an initial condition for the wave function $\Psi(\rho, \varphi, \theta, 0)$, then a formal solution to the initial value problem would be a sum (or infinite series) of product solutions of the type given in Eq. (9.49).

9.8 Exercises

(1) Assuming a product solution of the form $u(x, t) = X(x)T(t)$ show that Eq. (9.3) can be written as Eq. (9.4).

(2) Show that to have a bounded product solution to Eq. (9.3), the separation of variables constant c in Eq. (9.4) must be negative.

(3) Suppose that $L - x = \alpha\xi^\beta$ and show that

$$
\frac{dX}{dx} = -\frac{\xi^{1-\beta}}{\alpha\beta}\frac{dX}{d\xi}
$$

$$
\frac{d^2X}{dx^2} = \left(\frac{\xi^{1-\beta}}{\alpha\beta}\right)^2\frac{d^2X}{d\xi^2} + \frac{(1-\beta)\xi^{1-2\beta}}{\alpha^2\beta^2}\frac{dX}{d\xi}.
$$

(4) Using the change of variable described in Exercise 3 derive Eq. (9.6).

(5) Verify the following integral formulas. (*Hint*: use Theorem 8.2.)

    (a) $\int_0^{\lambda_{n,k}} x^{n+1} J_n(x)\,dx = \lambda_{n,k}^{n+1} J_{n+1}(\lambda_{n,k})$.

    (b) $\int_0^{\lambda_{n,k}} x^{n+3} J_n(x)\,dx = \lambda_{n,k}^{n+1}(\lambda_{n,k}^2 - 4(n+1))J_{n+1}(\lambda_{n,k})$.

(6) Find the solution to the swinging chain problem if the initial displacement of the chain is 0 and the initial velocity of the chain is $g(x) = x$ for $0 \leq x \leq L$.

(7) Find the formal solution to the swinging chain problem if the chain is subject to a damping force proportional to the velocity of the chain. Assume the proportionality constant γ for the damping force satisfies the inequality $0 < \gamma < \lambda_{0,0}\sqrt{g/L}$.

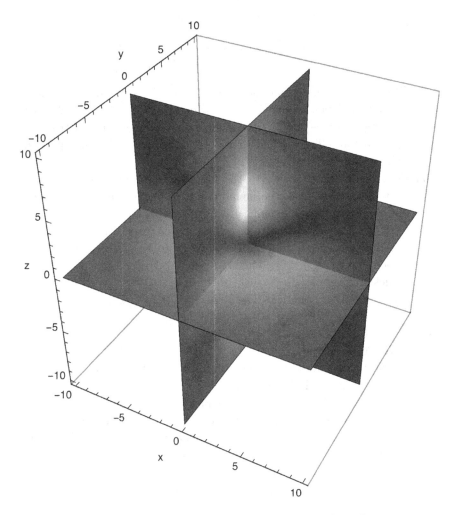

Fig. 9.7 The graph of the probability density function $P_{2,0,3}(\rho,\varphi,\theta)$ for the bound state of the electron in the hydrogen atom. Lighter areas of the plot indicate higher probabilities. To ease the process of sketching the probability density function, the quantity α was set to 1. Thus the scale of the density plot is exaggerated.

(8) Use Eq. (8.55) to show that

$$n \int_0^1 P_n(x)\, dx = -P_{n+1}(0).$$

(9) Show that for $n = 2, 4, 6, \ldots$, $\int_0^1 P_n(x)\, dx = 0$ and that for

$n = 1, 3, 5, \ldots$

$$\int_0^1 P_n(x)\,dx = \frac{(-1)^{(n+3)/2}(n+1)!}{n\,2^{n+1}\left(\frac{n+1}{2}\right)!\left(\frac{n+1}{2}\right)!}.$$

(10) Generalize the analysis done in Sec. 9.3 to the situation of a spherical shell of inner radius $r_1 > 0$ and outer radius $r_2 > r_1$. Suppose the inner surface has a time-independent temperature distribution $T_1(\varphi) = 100 - 25\cos\varphi$ and the outer surface has a time-independent temperature distribution $T_2(\varphi) = 30 + 25\cos\varphi$.

(11) Find the formal solution to the vibrating circular drumhead if the initial displacement of the membrane from the xy-plane is 0 and the initial velocity of the drumhead is given by $g(r, \theta)$.

(12) Find the formal solution to the vibrating circular drumhead with an initial displacement from the xy-plane given by $f(r, \theta)$ and an initial velocity given by $g(r, \theta)$. The reader may be helped by considering a "plucked" drum and a "struck" drum.

(13) Show for the spherical harmonic functions that $Y_n^{-m}(\varphi, \theta) = (-1)^m Y_n^m(\varphi, -\theta)$ for $m > 0$.

(14) Derive Eq. (9.22).

(15) Derive Eq. (9.23).

(16) Confirm the results of the definite integral in Eq. (9.24). (*Hint*: use Lemma 8.11 and Theorem 8.10.)

(17) Suppose the vibrating spherical membrane of Example 9.3 has an initial displacement from equilibrium described by the function $f(\varphi, \theta) = \frac{1}{16}\sin(2\varphi)\sin(3\theta)$. Find the displacement of the membrane from equilibrium as a function of (φ, θ, t).

(18) Suppose the vibrating spherical membrane of Example 9.3 is released from its equilibrium position and given an initial velocity described by the function

$$g(\varphi, \theta) = \begin{cases} \varphi\,\theta\left(\frac{\pi}{2} - \varphi\right)\left(\frac{\pi}{2} - \theta\right) & \text{for } 0 \leq \varphi,\, \theta \leq \pi/2 \\ 0 & \text{otherwise.} \end{cases}$$

Find the displacement of the membrane from equilibrium as a function of (φ, θ, t).

(19) Show that the spherical Bessel functions $j_n(x)$ are bounded as x approaches 0 for $n = 0, 1, 2, \ldots$.

(20) Use Corollary 8.2 to show that

$$\int_0^a j_n\left(\frac{\lambda_{n+1/2,k}\rho}{a}\right) j_n\left(\frac{\lambda_{n+1/2,l}\rho}{a}\right)\rho^2\,d\rho = \frac{a^3}{2}\left(j_{n+1}(\lambda_{n+1/2,k})\right)^2\delta_{kl}.$$

(21) Suppose the solid sphere of Sec. 9.5 is insulated and the initial temperature distribution within the sphere is

$$f(\rho, \varphi, \theta) = j_2 \left(\frac{\lambda_{5/2,1}\rho}{a} \right) \left(P_2(\cos \varphi) + \sin(2\theta) P_2^2(\cos \varphi) \right).$$

What is the steady-state temperature distribution within the sphere?

(22) Derive the displacement of the classical simple harmonic oscillator in Eq. (9.28) by solving the initial value problem:

$$m\, u''(t) + k\, u(t) = 0,$$
$$u(0) = u_0,$$
$$u'(0) = v_0.$$

(23) Show that the function $P(u)$ given in Eq. (9.32) is a probability density function by integrating $P(u)$ over the interval $[-A, A]$.

(24) Derive Eq. (9.39).

(25) The steps outlined below can be used to carry out the integration of

$$\int_{-\infty}^{\infty} e^{-3x^2/4} H_{2n}(x)\, dx.$$

(a) Show that

$$\int_{-\infty}^{\infty} e^{-3x^2/4} H_{2n}(x)\, dx = 2 \int_{-\infty}^{\infty} x e^{-3x^2/4} H_{2n-1}(x)\, dx$$
$$- 2(2n-1) \int_{-\infty}^{\infty} e^{-3x^2/4} H_{2n-2}(x)\, dx.$$

(*Hint*: use Exercise 40 of Chap. 8.)

(b) Show that

$$2 \int_{-\infty}^{\infty} x e^{-3x^2/4} H_{2n-1}(x)\, dx = \frac{8}{3}(2n-1) \int_{-\infty}^{\infty} e^{-3x^2/4} H_{2n-2}(x)\, dx.$$

(*Hint*: use Exercise 41 of Chap. 8.)

(c) Derive the reduction formula,

$$\int_{-\infty}^{\infty} e^{-3x^2/4} H_{2n}(x)\, dx = \frac{2}{3}(2n-1) \int_{-\infty}^{\infty} e^{-3x^2/4} H_{2n-2}(x)\, dx,$$

for $n \in \mathbb{N}$.

(d) Use induction to show

$$\int_{-\infty}^{\infty} e^{-3x^2/4} H_{2n}(x)\, dx = 2\sqrt{\frac{\pi}{3}} \frac{(2n)!}{3^n (n!)},$$

for $n = 0, 1, 2, \ldots$.

(26) Derive Eq. (9.45) from Eq. (9.44) in the following two steps.

(a) Define $\alpha = 2\sqrt{\frac{-2\mu E}{\hbar^2}}$, $x = \alpha\rho$, and $w(x) = R(x/\alpha)$ and re-write Eq. (9.44) as

$$x^2 w''(x) + 2xw'(x) + \left(\epsilon^2 \sqrt{\frac{-\mu}{2E\hbar^2}} x - \frac{1}{4}x^2 - l(l+1) \right) w(x) = 0.$$

(b) Define $y(x) = e^{x/2}x^{-l}w(x)$ and derive Eq. (9.45).

(27) Fix n as a positive integer and determine the number of stationary states of the hydrogen atom which correspond to the energy E_n. (*Hint*: for $u_{l,m,n}(\rho, \varphi, \theta)$ the index l must obey the inequality $0 \le l \le n-1$ and $|m| \le l$.)

(28) Evaluate the integral in Eq. (9.48). (*Hint*: multiply the recurrence relation for associated Laguerre polynomials established in Exercise 36 of Chap. 8 by $L_n^\alpha(x)x^\alpha e^{-x}$ and integrate over the interval $[0, \infty)$.)

(29) Physicists often measure the distance of the electron in the hydrogen atom from the nucleus in units called the **Bohr radius** defined as $a = \hbar^2/(\mu\epsilon^2)$ [Asmar (2016)]. Show that the ρ-dependent factor of the product solution for the wave function of the electron in the hydrogen atom can be written as

$$R_{n,l}(\rho) = e^{-\rho/(an)} \left(\frac{2\rho}{an} \right)^l L_{n-l-1}^{2l+1} \left(\frac{2\rho}{an} \right).$$

(30) Find the probability that the electron of the hydrogen atom is within one Bohr radius of the nucleus when the electron is in its so called **ground state** where $l = m = 0$ and $n = 1$.

Projects

(1) Nuclear bombs containing uranium as the core operate on the principle that when an atom of uranium is bombarded with a neutron, sometimes the neutron is captured. When capture occurs the uranium atom splits (or undergoes fission) which releases more neutrons which in turn bombard more nearby uranium atoms with neutrons setting off a cascade of fission which releases a great deal of energy. In this project the reader will explore the neutron density in a spherical core of uranium. The mathematical model used here will be similar to the heat equation with a source term. Let $u(\rho, \varphi, \theta, t)$ denote the density of neutrons in a sphere of uranium of radius R. To keep the discussion simple it will be

assumed that the rate of production of new neutrons will be proportional to the neutron density. Thus the neutron density in the sphere can be modeled by the partial differential equation,

$$u_t = k \triangle u + \gamma u.$$

The constant $k > 0$ is the neutron diffusivity of the nuclear core while $\gamma > 0$ is the constant of proportionality for the release of neutrons. Appropriate boundary conditions on u are that the neutron density on the surface of the sphere is 0, in other words $u(R, \varphi, \theta, t) = 0$ for $0 \le \varphi \le \pi$, $-\pi \le \theta \le \pi$, and $t > 0$. The neutron density should be bounded at the center of the sphere and for any fixed t, function $u(\rho, \varphi, \theta, t)$ should be bounded.

(a) Assuming the distribution of neutrons depends only on ρ, show the partial differential equation simplifies to

$$u_t = k \left(u_{\rho\rho} + \frac{2}{\rho} u_\rho \right) + \gamma u.$$

(b) Use the method of separation of variables to find the formal series solution to the partial differential equation above. The solution should depend only on (ρ, t).

(c) Critical mass for the nuclear weapon occurs when the core of uranium is large enough that the neutron density $u(\rho, t)$ is an increasing function of t. Show that critical mass for a spherical core is

$$m = \frac{4\pi^4}{3} \left(\frac{k}{\gamma} \right)^{3/2} \rho$$

where ρ is the material density (mass per unit volume).

(2) Consider a mass m attached to one end of a rigid mass-less rod of length l. The other end of the rod is allowed to rotate in a plane about a fixed point. The gravitational force of magnitude g attracts the mass in the downward direction. Let the angle the rod makes with the pure downward direction be ϕ. In classical mechanics this is known as a simple pendulum. This problem will explore the quantum version of the pendulum. The kinetic and potential energies of the pendulum are respectively

$$T = \frac{ml^2}{2} \left(\frac{d\phi}{dt} \right)^2 \text{ and } V = mgl(1 - \cos\phi).$$

The time-dependent linear Schrödinger equation for the quantum pendulum is

$$i\hbar \frac{\partial \Psi}{\partial t} = -\frac{\hbar^2}{2ml^2} \frac{\partial^2 \Psi}{\partial \phi^2} + mgl(1 - \cos\phi)\Psi. \tag{9.51}$$

(a) Make a change of variable by substituting $\xi = \phi + \pi$ and re-write the time-dependent linear Schrödinger equation for the quantum pendulum.

(b) Assume a product solution $\Psi(\xi, t) = X(\xi)T(t)$ and use separation of variables to show

$$\frac{d^2 X}{d\xi^2} + \frac{2ml^2}{\hbar^2}\left(E - mgl - mgl\cos\xi\right)X(\xi) \qquad (9.52)$$

where E is a constant.

(c) Equation (9.52) is related to the **Mathieu[7] equation**. Find 2π-periodic functions (called Mathieu functions) which solve Eq. (9.52).

(d) Find the energy levels of the quantum pendulum.

---

[7]Émile Léonard Mathieu, French mathematician (1835–1890).

Chapter 10

Nonhomogeneous Initial Boundary Value Problems

In previous chapters, particularly those exploring solution techniques for the heat and wave equations, some examples and exercises have included nonhomogeneous boundary conditions and *ad hoc* methods for finding the solutions to these initial boundary value problems were presented. The purpose of this chapter is to present a generally applicable method for solving nonhomogeneous initial boundary value problems. This method will systematically solve nonhomogeneous problems by breaking them into a collection of related initial boundary value problems in which the nonhomogeneous terms are isolated and separated in the various partial differential equations, boundary conditions, or initial conditions. The main result is driven by **Duhamel's**[1] **Principle** which is closely related to a method used to find particular solutions to linear, nonhomogeneous ordinary differential equations called variation of parameters.

Consider the nonhomogeneous initial boundary value problem related to the heat equation,

$$u_t = k\, u_{xx} + g(x,t) \text{ for } 0 < x < L \text{ and } t > 0,$$
$$u(0,t) = a(t) \text{ and } u(L,t) = b(t) \text{ for } t > 0, \qquad (10.1)$$
$$u(x,0) = f(x) \text{ for } 0 < x < L.$$

The partial differential equation contains a source term $g(x,t)$ and the boundary conditions, though of Dirichlet type, are nonhomogeneous. For simplicity it is assumed that $a(t)$ and $b(t)$ are continuous for $t \geq 0$, $f(x)$ is continuous on $[0, L]$, and $g(x,t)$ is continuous on the strip $[0, L] \times [0, \infty)$. The general approach to this type of nonhomogeneous problem is to assume the solution $u(x,t)$ can be written as the sum of three functions $u(x,t) =$

---

[1] Jean-Marie Constant Duhamel, French mathematician and physicist (1797–1872).

$v_1(x,t) + v_2(x,t) + v_3(x,t)$. The function $v_1(x,t)$ is chosen to satisfy the nonhomogeneous boundary conditions and is defined as

$$v_1(x,t) = (b(t) - a(t))\frac{x}{L} + a(t). \qquad (10.2)$$

Now let $u(x,t) = w(x,t) + v_1(x,t)$, then $u_t = w_t + (v_1)_t$ and $u_{xx} = w_{xx}$. Substituting these into Eq. (10.1) produces the following initial boundary value problem for $w(x,t)$:

$$w_t = k\,w_{xx} + \hat{g}(x,t) \text{ for } 0 < x < L \text{ and } t > 0,$$

$$w(0,t) = 0 \text{ and } w(L,t) = 0 \text{ for } t > 0,$$

$$w(x,0) = f(x) - (b(0) - a(0))\frac{x}{L} - a(0) \text{ for } 0 < x < L,$$

where $\hat{g}(x,t) = g(x,t) - (b'(t) + a'(t))x/L - a'(t)$. Note that while the partial differential equation satisfied by $w(x,t)$ remains nonhomogeneous, the boundary conditions are now homogeneous. Function $v_1(x,t)$ was designed to handle the nonhomogeneous boundary conditions of Eq. (10.1) so that an initial boundary value problem with homogeneous boundary conditions could be derived. Now treat $w(x,t)$ as the sum $w(x,t) = v_2(x,t) + v_3(x,t)$ where $v_2(x,t)$ solves the initial boundary value problem:

$$(v_2)_t = k(v_2)_{xx} \text{ for } 0 < x < L \text{ and } t > 0,$$

$$v_2(0,t) = 0 \text{ and } v_2(L,t) = 0 \text{ for } t > 0,$$

$$v_2(x,0) = f(x) - (b(0) - a(0))\frac{x}{L} - a(0) \text{ for } 0 < x < L.$$

The initial boundary value problem solved by $v_2(x,t)$ is homogeneous and is the type of problem readily solved by the methods covered in Chap. 4. Function $v_3(x,t)$ solves the initial boundary value problem:

$$(v_3)_t = k(v_3)_{xx} + \hat{g}(x,t) \text{ for } 0 < x < L \text{ and } t > 0,$$

$$v_3(0,t) = 0 \text{ and } v_3(L,t) = 0 \text{ for } t > 0, \qquad (10.3)$$

$$v_3(x,0) = 0 \text{ for } 0 < x < L,$$

which has homogeneous Dirichlet boundary conditions and a zero initial condition, while the partial differential equation contains the source term $\hat{g}(x,t)$. The zero boundary and initial conditions of Eq. (10.3) are compatible. The main topic of this chapter is Duhamel's principle which will be motivated using a physical argument in Sec. 10.1 along the lines of [Bleecker and Csordas (1996), Sec. 3.4]. In Sec. 10.2, Duhamel's principle will present a systematic approach to solving nonhomogeneous initial boundary value problems of the form shown in Eq. (10.3). The nonhomogeneous wave equation will be treated in a similar manner in Sec. 10.3.

10.1 Duhamel's Principle and Its Motivation

Consider the nonhomogeneous initial boundary value problem of the type derived above in Eq. (10.3) and restated below for clarity,

$$u_t = k\,u_{xx} + g(x,t) \text{ for } 0 < x < L \text{ and } t > 0,$$
$$u(0,t) = 0 \text{ and } u(L,t) = 0 \text{ for } t > 0, \qquad (10.4)$$
$$u(x,0) = 0 \text{ for } 0 < x < L.$$

To motivate the up-coming idea, note that the function $g(x,t)$ represents the rate of temperature change in the medium caused by an internal heat source. Fix a $T > 0$ and suppose the interval $[0,T]$ is partitioned into subintervals $\{[s_{i-1}, s_i]\}_{i=1}^n$ where

$$0 = s_0 \leq s_1 \leq \cdots \leq s_{i-1} \leq s_i \leq \cdots \leq s_n = T.$$

For $i = 1, \ldots, n$ define $\Delta s_i = s_i - s_{i-1}$ and define the functions $\mathbb{1}_i(t)$ as

$$\mathbb{1}_i(t) = \begin{cases} 1 & \text{if } s_{i-1} \leq t < s_i, \\ 0 & \text{otherwise.} \end{cases}$$

A function such as $\mathbb{1}_i$ is sometimes called a support or indicator function. Consider an initial boundary value problem related to Eq. (10.4) in which the heat source is turned on only during the interval $[s_{i-1}, s_i)$ and that Δs_i is small. Otherwise, the heat source term is 0. Heat from the source may flow in the medium during this short time interval. Thus the initial boundary value problem for the "switched" heat source is expressed as:

$$u_t = k\,u_{xx} + \mathbb{1}_i(t)g(x,t) \text{ for } 0 < x < L \text{ and } t > 0,$$
$$u(0,t) = 0 \text{ and } u(L,t) = 0 \text{ for } t > 0, \qquad (10.5)$$
$$u(x,0) = 0 \text{ for } 0 < x < L.$$

For $i = 1, \ldots, n$ let $u_i(x,t)$ be the solution to the corresponding initial boundary value problem of the form shown in Eq. (10.5). Note that for t in the interval $[0,T]$, $\sum_{i=1}^n \mathbb{1}_i(t)g(x,t) = g(x,t)$ and thus by the Principle of Superposition (Theorem 1.1), $\sum_{i=1}^n u_i(x,t) = u(x,t)$ where $u(x,t)$ solves the initial boundary value problem shown in Eq. (10.4) for $0 \leq t \leq T$.

Now deconstruct the initial boundary value problem in Eq. (10.5) by considering the three intervals of time, $[0, s_{i-1})$, $[s_{i-1}, s_i)$ and $[s_i, T]$. Since the boundary and initial conditions are zero in Eq. (10.5), then $u_i(x,t) = 0$ for $0 \leq t < s_{i-1}$ and $0 \leq x \leq L$. Under the usual assumptions on the material properties of the medium (such as the cross sectional area of the medium being constant and the material properties remaining constant) at $t = s_i$ the temperature distribution in the medium is approximately $g(x, s_{i-1})\Delta s_i$. For $t \geq s_i$ the heat source is turned off and a new initial boundary value problem can be considered. For the new problem, there is no heat source, but there is a nonzero, initial temperature distribution approximated as $g(x, s_{i-1})\Delta s_i$. The initial boundary value problem is expressed as:

$$u_t = k\,u_{xx} \text{ for } 0 < x < L \text{ and } t > s_i,$$
$$u(0,t) = 0 \text{ and } u(L,t) = 0 \text{ for } t > s_i, \tag{10.6}$$
$$u(x, s_i) = g(x, s_{i-1})\Delta s_i \text{ for } 0 < x < L.$$

Thus for $t \geq s_i$ the contribution to the temperature distribution in the rod from the switching on of the heat source during the time interval $[s_{i-1}, s_i)$ is approximately $v(x,t)\Delta s_i$ where $v(x,t)$ is the solution to the following initial value problem (see Exercise 1):

$$v_t = k\,v_{xx} \text{ for } 0 < x < L \text{ and } t > s_i,$$
$$v(0,t) = 0 \text{ and } v(L,t) = 0 \text{ for } t > s_i, \tag{10.7}$$
$$v(x, s_i) = g(x, s_{i-1}) \text{ for } 0 < x < L.$$

To acknowledge the dependence of v on the time at which the heat source is turned on, express v as a function of (x,t) with a parameter s as $v(x,t;s)$. Using this new notation Eq. (10.7) may be rewritten as

$$v_t = k\,v_{xx} \text{ for } 0 < x < L \text{ and } t > s_i,$$
$$v(0,t;s_{i-1}) = v(L,t;s_{i-1}) = 0 \text{ for } t > s_i, \tag{10.8}$$
$$v(x, s_i; s_{i-1}) = g(x, s_{i-1}) \text{ for } 0 < x < L.$$

The solution to Eq. (10.6) is then $u_i(x,t) = v(x,t;s_{i-1})\Delta s_i$ where $v(x,t;s_{i-1})$ solves Eq. (10.8). Therefore at $t = T$ the solution u to the initial boundary value problem in Eq. (10.4) is

$$u(x,T) \approx \sum_{i=1}^{n} v(x,T;s_{i-1})\Delta s_i.$$

In the limit as the norm of the partition approaches zero, the limit of the summation defines a definite integral and

$$u(x,T) = \lim_{n\to\infty} \sum_{i=1}^{n} v(x,T;s_{i-1})\Delta s_i = \int_0^T v(x,T;s)\,ds.$$

Since T is arbitrary, the variable can be replaced by t and $v(x,t;s)$ is a solution to the following initial boundary value problem:

$$v_t = k\,v_{xx} \text{ for } 0 < x < L \text{ and } t > s,$$
$$v(0,t;s) = v(L,t;s) = 0 \text{ for } t > s, \tag{10.9}$$
$$v(x, s; s) = g(x, s) \text{ for } 0 < x < L.$$

Readers may be uncomfortable with the choice of initial time in Eq. (10.9) as s rather than 0. This is handled by a change of variable. Define $\hat{v}(x, t - s; s) = v(x, t; s)$, then $v_t = \hat{v}_t$ and $v_{xx} = \hat{v}_{xx}$ and hence $\hat{v}(x, t; s)$ solves the initial boundary value problem:

$$\hat{v}_t = k\,\hat{v}_{xx} \text{ for } 0 < x < L \text{ and } t > 0,$$
$$\hat{v}(0, t; s) = \hat{v}(L, t; s) = 0 \text{ for } t > 0, \tag{10.10}$$
$$\hat{v}(x, 0; s) = g(x, s) \text{ for } 0 < x < L.$$

Using this newly defined $\hat{v}(x, t; s)$, the solution $u(x, t)$ to Eq. (10.4) becomes

$$u(x, t) = \int_0^t v(x, t; s)\,ds = \int_0^t \hat{v}(x, t - s; s)\,ds. \tag{10.11}$$

The initial boundary value problems stated in Eq. (10.9) or Eq. (10.10) are often referred to as **auxiliary problems** to the original initial boundary value problem in Eq. (10.4). Since the two auxiliary problems are equivalent to each other, the reader may work with either one when attempting to solve a nonhomogeneous initial boundary value problem.

10.2 Nonhomogeneous Heat Equation

With the definitions and notation introduced in the previous section, Duhamel's principle for the nonhomogeneous heat equation is stated.

Theorem 10.1. *Let $\overline{R} = \{(x, t) \,|\, 0 \leq x \leq L \text{ and } t \geq 0\}$ and suppose $g \in \mathcal{C}^2(\overline{R})$. If $\hat{v}(x, t; s)$ is a solution to the auxiliary problem given in Eq. (10.10), then $u(x, t)$ as defined in Eq. (10.11) solves the initial boundary value problem in Eq. (10.4).*

Proof. The proof proceeds by demonstrating that $u(x, t)$ satisfies the initial and boundary conditions of Eq. (10.4) and finally that the partial differential equation is satisfied as well. When $t = 0$,

$$u(x, 0) = \int_0^0 \hat{v}(x, -s; s)\,ds = 0,$$

and thus the initial condition is met. When $x = 0$,

$$u(0, t) = \int_0^t \hat{v}(0, t - s; s)\,ds = \int_0^t 0\,ds = 0,$$

and thus the boundary condition at $x = 0$ is satisfied. A similar calculation verifies that the boundary condition at $x = L$ is met. Under the assumptions of the theorem, application of Lemma 1.1 yields

$$u_t(x,t) = \frac{d}{dt}\int_0^t \hat{v}(x,t-s;s)\,ds = \hat{v}(x,t-t;t) + \int_0^t \hat{v}_t(x,t-s;s)\,ds$$

$$= g(x,t) + \int_0^t k\,\hat{v}_{xx}(x,t-s;s)\,ds$$

$$= g(x,t) + k\,u_{xx}(x,t).$$

This derivation uses the assumption that $\hat{v}(x,t;s)$ solves Eq. (10.10) and Lemma 1.1 to differentiate twice with respect to x in the final definite integral. Hence $u(x,t)$ solves the partial differential equation found in Eq. (10.4).   □

The remainder of this section contains several examples to illustrate the use of Duhamel's principle for solving nonhomogeneous initial boundary value problems.

Example 10.1. Use Duhamel's principle to find a solution to the following initial boundary value problem:

$$u_t = u_{xx} + t^2 x(1 - x/\pi) \text{ for } 0 < x < \pi \text{ and } t > 0,$$

$$u(0,t) = u(\pi,t) = 0 \text{ for } t > 0,$$

$$u(x,0) = 0 \text{ for } 0 < x < \pi.$$

Solution. The solution is a direct result of applying Theorem 10.1. The first step is setting up the auxiliary problem of the form shown in Eq. (10.10):

$$\hat{v}_t = \hat{v}_{xx} \text{ for } 0 < x < \pi \text{ and } t > 0,$$

$$\hat{v}(0,t;s) = \hat{v}(\pi,t;s) = 0 \text{ for } t > 0,$$

$$\hat{v}(x,0;s) = s^2 x(1 - x/\pi) \text{ for } 0 < x < \pi.$$

The solution to the auxiliary problem has the form:

$$\hat{v}(x,t;s) = \sum_{n=1}^{\infty} a_n e^{-n^2 t} \sin(nx),$$

where

$$a_n = \frac{2}{\pi}\int_0^\pi s^2 x(1 - x/\pi)\sin(nx)\,dx = \begin{cases} 0 & \text{if } n \text{ is even,} \\ 8s^2/(n^3\pi^2) & \text{if } n \text{ is odd.} \end{cases}$$

The solution to the original nonhomogeneous initial boundary value problem can be written formally as

$$u(x,t) = \int_0^t \hat{v}(x,t-s;s)\,ds = \sum_{n=1}^{\infty}\int_0^t a_n e^{-n^2(t-s)}\sin(nx)\,ds.$$

Note that a typical odd-indexed term in the infinite series is

$$u_n(x,t) = \int_0^t \frac{8s^2}{n^3\pi^2} e^{-n^2(t-s)} \sin(nx)\, ds = \frac{8}{n^3\pi^2} e^{-n^2 t} \sin(nx) \int_0^t s^2 e^{n^2 s}\, ds$$

$$= \frac{8}{n^3\pi^2} e^{-n^2 t} \sin(nx) \left(\frac{e^{n^2 t}(n^4 t^2 - 2n^2 t + 2) - 2}{n^6} \right)$$

$$= \frac{8}{\pi^2} \frac{n^4 t^2 - 2n^2 t + 2 - 2e^{-n^2 t}}{n^9} \sin(nx).$$

Thus the solution to the original initial boundary value problem is

$$u(x,t) = \frac{8}{\pi^2} \sum_{n=1,3,\ldots}^{\infty} \frac{n^4 t^2 - 2n^2 t + 2 - 2e^{-n^2 t}}{n^9} \sin(nx).$$

The next example builds on the features of the previous example.

Example 10.2. Use Duhamel's principle to find a solution to the following initial boundary value problem:

$$u_t = u_{xx} + t^2 x(1 - x/\pi) \text{ for } 0 < x < \pi \text{ and } t > 0,$$

$$u(0,t) = u(\pi,t) = 0 \text{ for } t > 0,$$

$$u(x,0) = 1 - \cos(2x) \text{ for } 0 < x < \pi.$$

Solution. The first step in finding the solution is to write $u(x,t) = v(x,t) + w(x,t)$ where $v(x,t)$ is a solution to the homogeneous part of the initial boundary value problem:

$$v_t = v_{xx} \text{ for } 0 < x < \pi \text{ and } t > 0,$$

$$v(0,t) = v(\pi,t) = 0 \text{ for } t > 0,$$

$$v(x,0) = 1 - \cos(2x) \text{ for } 0 < x < \pi,$$

and $w(x,t)$ is a particular solution to the nonhomogeneous equation:

$$w_t = w_{xx} + t^2 x(1 - x/\pi) \text{ for } 0 < x < \pi \text{ and } t > 0,$$

$$w(0,t) = w(\pi,t) = 0 \text{ for } t > 0,$$

$$w(x,0) = 0 \text{ for } 0 < x < \pi.$$

The function $v(x,t)$ is found by expressing the initial condition as a Fourier series as was done in Chap. 4,

$$v(x,t) = \frac{16}{3\pi} e^{-t} \sin x - \frac{16}{\pi} \sum_{n=3,5,\ldots}^{\infty} \frac{e^{-n^2 t}}{n(n^2 - 4)} \sin(nx).$$

The solution to the nonhomogeneous initial boundary value problem was found in Example 10.1 and thus the solution to the original initial boundary value problem is

$$u(x,t) = \frac{16}{3\pi} e^{-t} \sin x - \frac{16}{\pi} \sum_{n=3,5,\dots}^{\infty} \frac{e^{-n^2 t}}{n(n^2 - 4)} \sin(nx)$$

$$+ \frac{8}{\pi^2} \sum_{n=1,3,\dots}^{\infty} \frac{n^4 t^2 - 2n^2 t + 2 - 2e^{-n^2 t}}{n^9} \sin(nx).$$

The next example demonstrates the use of Duhamel's principle for solving an initial boundary value problem containing nonhomogeneous terms in the partial differential equation and in the Dirichlet boundary conditions.

Example 10.3. Use Duhamel's principle to find a solution to the following initial boundary value problem:

$$u_t = u_{xx} + e^{-t} \sin(2x) \text{ for } 0 < x < \pi \text{ and } t > 0,$$
$$u(0,t) = 1 - e^{-t} \text{ and } u(\pi,t) = 1 - \cos t \text{ for } t > 0,$$
$$u(x,0) = \sin x \text{ for } 0 < x < \pi.$$

Solution. The first step calls for finding a function which converts the boundary conditions to homogeneous form. Let

$$v_1(x,t) = (e^{-t} - \cos t)\frac{x}{\pi} + 1 - e^{-t},$$

then if $u(x,t) = v_1(x,t) + w(x,t)$,

$$u_t = (v_1)_t + w_t = (\sin t - e^{-t})\frac{x}{\pi} + e^{-t} + w_t,$$
$$u_{xx} = (v_1)_{xx} + w_{xx} = w_{xx}.$$

Thus $w(x,t)$ must solve the following initial boundary value problem:

$$w_t = w_{xx} + e^{-t}(\sin(2x) - 1) - (\sin t - e^{-t})\frac{x}{\pi} \text{ for } 0 < x < \pi \text{ and } t > 0,$$
$$w(0,t) = w(\pi,t) = 0 \text{ for } t > 0,$$
$$w(x,0) = \sin x \text{ for } 0 < x < \pi.$$

The second step is to solve the homogeneous part of the initial boundary value problem:

$$(v_2)_t = (v_2)_{xx} \text{ for } 0 < x < \pi \text{ and } t > 0,$$
$$(v_2)(0,t) = (v_2)(\pi,t) = 0 \text{ for } t > 0,$$
$$(v_2)(x,0) = \sin x \text{ for } 0 < x < \pi.$$

The reader can verify directly via differentiation that $v_2(x,t) = e^{-t}\sin x$. The third step involves the application of Duhamel's principle in the form of Theorem 10.1 to solve the initial boundary value problem containing the nonhomogeneous partial differential equation:

$$(v_3)_t = (v_3)_{xx} + e^{-t}(\sin(2x) - 1) - (\sin t - e^{-t})\frac{x}{\pi} \text{ for } 0 < x < \pi, \, t > 0,$$

$$(v_3)(0,t) = (v_3)(\pi,t) = 0 \text{ for } t > 0, \qquad\qquad (10.12)$$

$$(v_3)(x,0) = 0 \text{ for } 0 < x < \pi.$$

The solution is $v_3(x,t) = \int_0^t \hat{v}(x,t-s;s)\,ds$ where $\hat{v}$ solves the homogeneous initial boundary value problem:

$$\hat{v}_t = \hat{v}_{xx} \text{ for } 0 < x < \pi \text{ and } t > 0,$$

$$\hat{v}(0,t;s) = \hat{v}(\pi,t;s) = 0 \text{ for } t > 0,$$

$$\hat{v}(x,0;s) = e^{-s}(\sin(2x) - 1) - (\sin s - e^{-s})\frac{x}{\pi} \text{ for } 0 < x < \pi.$$

The function $\hat{v}(x,t;s)$ can be expressed formally as a Fourier series,

$$\hat{v}(x,t;s) = \sum_{n=1}^{\infty} a_n e^{-n^2 t} \sin(nx)$$

where the Fourier sine series coefficients are

$$a_n = \frac{2}{\pi} \int_0^{\pi} \left(e^{-s}(\sin(2x) - 1) - (\sin s - e^{-s})\frac{x}{\pi} \right) \sin(nx)\,dx$$

$$= \frac{2}{n\pi}((-1)^n \sin s + e^{-s}(\delta_{2n}\pi - 1)),$$

and δ_{ij} is the Kronecker delta. The solution to the nonhomogeneous initial boundary value problem given in Eq. (10.12) can be written formally as

$$u(x,t) = \int_0^t \hat{v}(x,t-s;s)\,ds = \sum_{n=1}^{\infty} \int_0^t a_n e^{-n^2(t-s)} \sin(nx)\,ds.$$

For the case of $n = 1$,

$$\frac{-2}{\pi} \int_0^t (\sin s + e^{-s})e^{-(t-s)} \sin(x)\,ds = \frac{-2e^{-t}\sin(x)}{\pi} \int_0^t (e^s \sin s + 1)\,ds$$

$$= \frac{\sin x}{\pi}(\cos t - \sin t - e^{-t}(2t + 1)).$$

When $n = 2$,

$$\frac{1}{\pi} \int_0^t (\sin s + e^{-s}(\pi - 1))e^{-4(t-s)} \sin(2x)\,ds$$

$$= \frac{\sin(2x)}{\pi} \left[\frac{4\sin t}{17} + \frac{(20 - 17\pi)e^{-4t}}{51} + \frac{(\pi - 1)e^{-t}}{3} - \frac{\cos t}{17} \right],$$

and when $n > 2$,

$$\frac{2}{n\pi} \int_0^t \left((-1)^n \sin s - e^{-s}\right) e^{-n^2(t-s)} \sin(nx) \, ds$$

$$= \frac{2e^{-n^2 t} \sin(nx)}{n\pi} \int_0^t \left((-1)^n e^{n^2 s} \sin s - e^{(n^2-1)s}\right) ds$$

$$= \frac{2\sin(nx)}{n\pi} \left[\frac{(-1)^n (e^{-n^2 t} - \cos t + n^2 \sin t)}{n^4 + 1} - \frac{e^{-t} - e^{-n^2 t}}{n^2 - 1}\right].$$

Thus the solution takes the form,

$$v_3(x,t) = \frac{\sin x}{\pi}(\cos t - \sin t - e^{-t}(2t + 1))$$

$$+ \frac{\sin(2x)}{\pi}\left[\frac{4\sin t}{17} + \frac{(20 - 17\pi)e^{-4t}}{51} + \frac{(\pi - 1)e^{-t}}{3} - \frac{\cos t}{17}\right]$$

$$+ \frac{2}{\pi}\sum_{n=3}^{\infty} \frac{\sin(nx)}{n}\left[\frac{(-1)^n(e^{-n^2 t} - \cos t + n^2 \sin t)}{n^4 + 1} - \frac{e^{-t} - e^{-n^2 t}}{n^2 - 1}\right].$$

The solution to the original initial boundary value problem is then $u(x,t) = v_1(x,t) + v_2(x,t) + v_3(x,t)$.

An alternative approach can be taken to determine the solution to the nonhomogeneous initial boundary value problem in Eq. (10.12). This approach was briefly explored when finding the solution to the nonhomogeneous heat equation in Sec. 4.2. Suppose the product solutions to the initial boundary value problem have the form $v_n(x,t) = T_n(t)\sin(nx)$, where $T_n(t)$ is an (as yet) unknown function satisfying the initial condition $T_n(0) = 0$ for $n \in \mathbb{N}$. The reader may readily check that $v_n(x,t)$ satisfies the boundary and initial conditions of Eq. (10.12). Assuming a solution of the form $v(x,t) = \sum_{n=1}^{\infty} T_n(t)\sin(nx)$ can be differentiated term by term, substituting into the partial differential equation yields the equation below:

$$\sum_{n=1}^{\infty} T_n'(t)\sin(nx) = -\sum_{n=1}^{\infty} n^2 T_n(t)\sin(nx) + e^{-t}(\sin(2x) - 1) - (\sin t - e^{-t})\frac{x}{\pi}.$$

Multiplying both sides of the equation by $\sin(mx)$ and integrating over the interval $[0, \pi]$, assuming the order of integration and summation can be interchanged, result in

$$\frac{\pi}{2}T_n'(t) = -\frac{n^2\pi}{2}T_n(t) + \frac{1}{n}((-1)^n \sin t - e^{-t}) + \frac{\pi}{2}e^{-t}\delta_{2n},$$

where δ_{ij} is the Kronecker delta. Thus $T_n(t)$ can be found by solving a system of first-order, linear initial value problems. For $n \neq 2$ the initial value problem takes the form,

$$T_n'(t) + n^2 T_n(t) = \frac{2}{n\pi}((-1)^n \sin t - e^{-t}),$$

$$T_n(0) = 0.$$

Multiplying both sides of the ordinary differential equation by the integrating factor $e^{n^2 t}$ and integrating over the interval $[0, t]$ result in

$$\int_0^t \frac{d}{ds}[e^{n^2 s} T_n(s)]\, ds = \frac{2}{n\pi} \int_0^t ((-1)^n e^{n^2 s} \sin s - e^{(n^2-1)s})\, ds.$$

There are two cases to consider when carrying out the integration. If $n = 1$ then

$$T_1(t) = \frac{1}{\pi}(\cos t - \sin t - e^{-t}(2t + 1)),$$

while if $n \geq 3$,

$$T_n(t) = \frac{2}{n\pi} \left[\frac{(-1)^n(e^{-n^2 t} - \cos t + n^2 \sin t)}{n^4 + 1} - \frac{e^{-t} - e^{-n^2 t}}{n^2 - 1} \right].$$

Readers will be asked to show in Exercise 10 that

$$T_2(t) = \frac{4}{17\pi} \sin t + \frac{(20 - 17\pi)e^{-4t}}{51\pi} + \frac{(\pi - 1)e^{-t}}{3\pi} - \frac{1}{17\pi} \cos t.$$

Therefore the solution $v_3(x, t)$ can be written formally as the infinite series

$$v_3(x, t) = \frac{\sin(x)}{\pi}(\cos t - \sin t - e^{-t}(2t + 1))$$

$$+ \frac{\sin(2x)}{\pi} \left(\frac{4 \sin t}{17} + \frac{(20 - 17\pi)e^{-4t}}{51} + \frac{(\pi - 1)e^{-t}}{3} - \frac{\cos t}{17} \right)$$

$$+ \frac{2}{\pi} \sum_{n=3}^{\infty} \frac{\sin(nx)}{n} \left[\frac{(-1)^n(e^{-n^2 t} - \cos t + n^2 \sin t)}{n^4 + 1} - \frac{e^{-t} - e^{-n^2 t}}{n^2 - 1} \right].$$

This is the same $v_3(x, t)$ as found previously via Duhamel's principle in Example 10.3.

10.2.1 *Other Boundary Conditions*

Duhamel's principle can be applied to initial boundary value problems arising from the heat equation involving other types of boundary conditions. Naturally some adjustments to Theorem 10.1 must be made when the Dirichlet boundary conditions are changed to other boundary conditions such as those of Neumann type. Consider the following nonhomogeneous initial boundary value problem:

$$u_t = k\,u_{xx} + g(x,t) \text{ for } 0 < x < L \text{ and } t > 0,$$

$$u_x(0,t) = u_x(L,t) = 0 \text{ for } t > 0, \qquad\qquad (10.13)$$

$$u(x,0) = 0 \text{ for } 0 < x < L.$$

The initial boundary value problem in Eq. (10.13) is related to the auxiliary problem:

$$\hat{v}_t = k\,\hat{v}_{xx} \text{ for } 0 < x < L \text{ and } t > 0,$$

$$\hat{v}_x(0,t;s) = \hat{v}_x(L,t;s) = 0 \text{ for } t > 0, \qquad\qquad (10.14)$$

$$\hat{v}(x,0;s) = g(x,s) \text{ for } 0 < x < L,$$

for each fixed value of the parameter $s \geq 0$. Duhamel's principle for the heat equation with Neumann boundary conditions is stated in the next theorem.

Theorem 10.2. *Let* $\overline{R} = \{(x,t)\,|\,0 \leq x \leq L \text{ and } t \geq 0\}$ *and suppose* $g \in C^2(\overline{R})$. *If* $\hat{v}(x,t;s)$ *is a solution to the auxiliary problem in Eq.* (10.14), *then* $u(x,t)$ *as defined in Eq.* (10.11) *solves the initial boundary value problem given in Eq.* (10.13).

The proof follows the same steps as for the case of the Dirichlet boundary conditions.

Example 10.4. Use Duhamel's principle to find the solution to the following initial boundary value problem:

$$u_t = u_{xx} + e^{-t}\sin(\pi x) \text{ for } 0 < x < 1 \text{ and } t > 0,$$

$$u_x(0,t) = 1 - e^{-t} \text{ and } u_x(1,t) = \sin t \text{ for } t > 0,$$

$$u(x,0) = \cos(\pi x) \text{ for } 0 < x < 1.$$

Solution. The solution is found using the same approach as used for the nonhomogeneous problem with Dirichlet boundary conditions. The solution $u(x,t)$ is decomposed as the sum of three functions $v_1(x,t)$, $v_2(x,t)$, and $v_3(x,t)$. The first function is chosen to satisfy the nonhomogeneous boundary conditions. The reader can verify by differentiation that

$$v_1(x,t) = \frac{1}{2}(\sin t + e^{-t} - 1)x^2 + (1 - e^{-t})x$$

satisfies the nonhomogeneous boundary conditions in the initial boundary value problem given above. For convenience, let $w(x,t) = u(x,t) - v_1(x,t)$, then $w(x,t)$ is the solution to the following initial boundary value problem:

$$w_t = w_{xx} + e^{-t}\sin(\pi x) + \frac{1}{2}(e^{-t} - \cos t)x^2 - e^{-t}x + e^{-t} - 1 + \sin t$$

$$\text{for } 0 < x < 1 \text{ and } t > 0,$$

$w_x(0,t) = w_x(1,t) = 0$ for $t > 0$,

$w(x,0) = \cos(\pi x)$ for $0 < x < 1$.

The solution $w(x,t)$ is further written as $w(x,t) = v_2(x,t) + v_3(x,t)$ where $v_2(x,t)$ will be chosen to solve the homogeneous part of the partial differential equation, the homogeneous Neumann boundary conditions, and the nonzero initial condition. By inspection, the reader can see that $v_2(x,t) = e^{-\pi^2 t}\cos(\pi x)$ is the desired solution. Finally $v_3(x,t)$ must solve the nonhomogeneous heat equation with zero boundary and initial conditions expressed as,

$$(v_3)_t = (v_3)_{xx} + e^{-t}\sin(\pi x) + \frac{1}{2}(e^{-t} - \cos t)x^2 - e^{-t}x + e^{-t} - 1 + \sin t$$
$$\text{for } 0 < x < 1 \text{ and } t > 0,$$

$(v_3)_x(0,t) = (v_3)_x(1,t) = 0$ for $t > 0$,

$v_3(x,0) = 0$ for $0 < x < 1$.

Using Duhamel's principle the solution of this initial boundary value problem is $v_3(x,t) = \int_0^t \hat{v}(x,t-s;s)\,ds$ where $\hat{v}(x,t;s)$ solves the following initial boundary value problem:

$$\hat{v}_t = \hat{v}_{xx} \text{ for } 0 < x < 1 \text{ and } t > 0,$$

$$\hat{v}_x(0,t;s) = \hat{v}_x(1,t;s) = 0 \text{ for } t > 0,$$

$$\hat{v}(x,0;s) = e^{-s}\sin(\pi x) + \frac{1}{2}(e^{-s} - \cos s)x^2 - e^{-s}x + e^{-s} - 1 + \sin s$$
$$\text{for } 0 < x < 1.$$

The solution can be written formally as an infinite series,

$$\hat{v}(x,t;s) = \frac{a_0}{2} + \sum_{n=1}^{\infty} a_n e^{-n^2\pi^2 t}\cos(n\pi x),$$

where

$$a_n = 2\int_0^1 \left[e^{-s}\sin(\pi x) + \frac{x^2}{2}(e^{-s} - \cos s) - e^{-s}x + e^{-s} - 1 + \sin s \right]$$
$$\times \cos(n\pi x)\,dx$$

$$= \begin{cases} 2\left(\sin s + 2e^{-s}\left(\dfrac{1}{3} + \dfrac{1}{\pi}\right) - 1 - \dfrac{1}{6}\cos s\right) & \text{if } n = 0, \\[2mm] \dfrac{2}{\pi^2}(e^{-s} + \cos s) & \text{if } n = 1, \\[2mm] \dfrac{-2(1 + n^2(\pi(1 + (-1)^n) - 1))e^{-s}}{n^2(n^2 - 1)\pi^2} - \dfrac{2(-1)^n\cos s}{n^2\pi^2} & \text{if } n > 1. \end{cases}$$

To find $v_3(x,t)$, assuming that $\hat{v}(x,t;s)$ can be integrated term by term, consider the case when $n = 0$,

$$\int_0^t \left(\sin s + 2e^{-s}\left(\frac{1}{3}+\frac{1}{\pi}\right) - 1 - \frac{1}{6}\cos s \right) ds$$

$$= 1 - \cos t + 2\left(\frac{1}{3}+\frac{1}{\pi}\right)(1-e^{-t}) - t - \frac{1}{6}\sin t.$$

When $n = 1$,

$$\int_0^t \left(\frac{2}{\pi^2}(e^{-s}+\cos s)\right) e^{-\pi^2(t-s)}\cos(\pi x)\,ds$$

$$= \frac{2e^{-\pi^2 t}}{\pi^2}\cos(\pi x)\int_0^t e^{(\pi^2-1)s}\,ds + \frac{2e^{-\pi^2 t}}{\pi^2}\cos(\pi x)\int_0^t e^{\pi^2 s}\cos s\,ds$$

$$= \frac{2}{\pi^2}\left(\frac{e^{-t}-e^{-\pi^2 t}}{\pi^2-1} + \frac{\sin t + \pi^2(\cos t - e^{-\pi^2 t})}{\pi^4+1}\right)\cos(\pi x).$$

Finally for $n = 2, 3, \ldots$,

$$-2\int_0^t \left(\frac{(1+n^2(\pi(1+(-1)^n)-1))e^{-s}}{n^2(n^2-1)\pi^2} + \frac{(-1)^n\cos s}{n^2\pi^2}\right)$$

$$\times e^{-n^2\pi^2(t-s)}\cos(n\pi x)\,ds$$

$$= \frac{-2(1+n^2(\pi(1+(-1)^n)-1))}{n^2(n^2-1)\pi^2}e^{-n^2\pi^2 t}\cos(n\pi x)\int_0^t e^{(n^2\pi^2-1)s}\,ds$$

$$- \frac{2(-1)^n}{n^2\pi^2}e^{-n^2\pi^2 t}\cos(n\pi x)\int_0^t e^{n^2\pi^2 s}\cos s\,ds$$

$$= \frac{-2(1+n^2(\pi(1+(-1)^n)-1))}{n^2(n^2-1)(n^2\pi^2-1)\pi^2}(e^{-t}-e^{-n^2\pi^2 t})\cos(n\pi x)$$

$$- \frac{2(-1)^n}{n^2\pi^2(n^4\pi^4+1)}(\sin t + n^2\pi^2\cos t - n^2\pi^2 e^{-n^2\pi^2 t})\cos(n\pi x).$$

Therefore the third solution,

$$v_3(x,t) = 1 - \cos t + 2\left(\frac{1}{3}+\frac{1}{\pi}\right)(1-e^{-t}) - t - \frac{1}{6}\sin t$$

$$+ \frac{2}{\pi^2}\left(\frac{e^{-t}-e^{-\pi^2 t}}{\pi^2-1} + \frac{\sin t + \pi^2(\cos t - e^{-\pi^2 t})}{\pi^4+1}\right)\cos(\pi x)$$

$$- \frac{2}{\pi^2}\sum_{n=2}^{\infty}\frac{(1+n^2(\pi(1+(-1)^n)-1))}{n^2(n^2-1)(n^2\pi^2-1)}(e^{-t}-e^{-n^2\pi^2 t})\cos(n\pi x)$$

$$- \frac{2}{\pi^2}\sum_{n=2}^{\infty}\frac{(-1)^n}{n^2(n^4\pi^4+1)}(\sin t + n^2\pi^2\cos t - n^2\pi^2 e^{-n^2\pi^2 t})\cos(n\pi x).$$

10.2.2 *Unbounded Intervals*

Duhamel's principle can also be applied to the heat equation model of an unbounded, infinitely long medium. As was the case in the previous section, the nonhomogeneous initial value problem,

$$u_t = k\, u_{xx} + g(x,t) \text{ for } -\infty < x < \infty \text{ and } t > 0,$$
$$u(x,0+) = 0 \text{ for } -\infty < x < \infty \tag{10.15}$$

is associated with the following parameterized auxiliary homogeneous initial value problem,

$$\hat{v}_t = k\, \hat{v}_{xx} \text{ for } -\infty < x < \infty \text{ and } t > 0,$$
$$\hat{v}(x,0+;s) = g(x,s) \text{ for } -\infty < x < \infty. \tag{10.16}$$

According to Theorem 4.3 the solution to Eq. (10.16) can be expressed as

$$\hat{v}(x,t;s) = \int_{-\infty}^{\infty} U(x-y,t)g(y,s)\,dy \tag{10.17}$$

for $-\infty < x < \infty$ and $t > 0$, where U is the fundamental solution to the heat equation given in Eq. (4.34). Note that this solution is parameterized by $s \geq 0$. A version of Duhamel's principle for infinite media is stated in the next theorem which can be proved in a manner similar to that found in [Evans (2010), p. 50].

Theorem 10.3. *Let* $\overline{R} = \{(x,t)\,|\,-\infty < x < \infty \text{ and } t \geq 0\}$, *suppose* $g \in \mathcal{C}^2(\overline{R})$, *and that g and its first and second partial derivatives are bounded on* $\overline{R}$. *A solution to Eq. (10.15) for* $-\infty < x < \infty$ *and* $t > 0$ *is given by*

$$u(x,t) = \int_0^t \hat{v}(x,t-s;s)\,ds = \int_0^t \int_{-\infty}^{\infty} U(x-y,t-s)g(y,s)\,dy\,ds. \tag{10.18}$$

Example 10.5. Find a solution to the following initial value problem:

$$u_t = u_{xx} + \sin(x-t) \text{ for } -\infty < x < \infty \text{ and } t > 0,$$
$$u(x,0+) = e^{-x^2} \text{ for } -\infty < x < \infty.$$

Solution. The strategy for finding the solution to this problem is similar to that followed in the previous section, except that no boundary conditions are present. Thus the solution can be decomposed into the sum $u(x,t) = u_1(x,t) + u_2(x,t)$ where $u_1(x,t)$ solves the homogeneous heat equation with nonzero initial condition:

$$(u_1)_t = (u_1)_{xx} \text{ for } -\infty < x < \infty \text{ and } t > 0,$$
$$u_1(x,0+) = e^{-x^2} \text{ for } -\infty < x < \infty.$$

By Theorem 4.3 (or Exercise 17 of Chap. 4),

$$u_1(x,t) = \frac{e^{-x^2/(1+4t)}}{\sqrt{1+4t}}.$$

The other function $u_2(x,t)$ must satisfy the nonhomogeneous, unbounded heat equation with zero initial condition:

$$(u_2)_t = (u_2)_{xx} + \sin(x-t) \text{ for } -\infty < x < \infty \text{ and } t > 0,$$
$$u_2(x,0+) = 0 \text{ for } -\infty < x < \infty.$$

The associated auxiliary initial value problem is

$$\hat{v}_t(x,t;s) = \hat{v}_{xx}(x,t;s) \text{ for } -\infty < x < \infty \text{ and } t > 0,$$
$$\hat{v}(x,0;s) = \sin(x-s) \text{ for } -\infty < x < \infty.$$

The version of Duhamel's principle for infinite media given in Theorem 10.3 can be used to find $u_2(x,t)$. The double integral in Eq. (10.18) will be evaluated in two steps.

$$
\begin{aligned}
\hat{v}(x,t;s) &= \int_{-\infty}^{\infty} \frac{1}{\sqrt{4\pi t}} e^{-(x-y)^2/(4t)} \sin(y-s)\,dy \\
&= \frac{1}{\sqrt{4\pi t}} \int_{-\infty}^{\infty} e^{-(x-y)^2/(4t)} \operatorname{Im}\left(e^{i(y-s)}\right)\,dy \\
&= \frac{e^{-t}}{\sqrt{4\pi t}} \operatorname{Im}\left(e^{-i(s-x)} \int_{-\infty}^{\infty} e^{-(y-x-2it)^2/(4t)}\,dy\right) \\
&= e^{-t} \operatorname{Im}\left(\frac{e^{-i(s-x)}}{\sqrt{2\pi}} \int_{-\infty}^{\infty} e^{-z^2/2}\,dz\right) = -e^{-t}\sin(s-x).
\end{aligned}
$$

A contour integral in complex variables has been used to carry out this integration step. See App. A if additional background information is needed. One more integration produces $u_2(x,t)$,

$$
\begin{aligned}
u_2(x,t) &= \int_0^t -e^{s-t}\sin(s-x)\,ds = -e^{x-t}\int_{-x}^{t-x} e^z \sin(z)\,dz \\
&= \frac{1}{2}(\cos(x-t) + \sin(x-t) - e^{-t}(\cos x + \sin x)).
\end{aligned}
$$

Therefore a solution to the original initial value problem is

$$u(x,t) = \frac{e^{-x^2/(1+4t)}}{\sqrt{1+4t}} + \frac{1}{2}(\cos(x-t) + \sin(x-t) - e^{-t}(\cos x + \sin x)).$$

10.3   Nonhomogeneous Wave Equation

In the previous section treating the heat equation, Duhamel's principle generated a solution to a nonhomogeneous partial differential equation (with zero boundary and initial conditions) by solving an auxiliary problem in which the heat equation and boundary conditions were homogeneous and the nonhomogeneous term became the initial condition. This will also be the approach taken for nonhomogeneous versions of the wave equation. Duhamel's principle finds use when solving the initial boundary value problems involving the wave equation for which the partial differential equation contains nonhomogeneous terms and/or the boundary conditions are nonhomogeneous. The simplest place to begin a discussion of the nonhomogeneous wave equation is with the case of the infinitely long string, which eliminates the need to consider boundary conditions. Later the nonhomogeneous wave equation on a bounded, finite string will be treated.

Consider the wave equation modeling an infinitely long string subject to an external force,

$$u_{tt} = c^2 u_{xx} + g(x, t).$$

The nonhomogeneous term $g(x, t)$ can be interpreted as the acceleration of the string at position x at time t and thus is sometimes referred to as the **applied acceleration**. If the vibrating string is initially at rest in its equilibrium position, then the initial value problem can be summarized as

$$u_{tt} = c^2 u_{xx} + g(x, t) \text{ for } -\infty < x < \infty \text{ and } t > 0,$$
$$u(x, 0) = u_t(x, 0) = 0 \text{ for } -\infty < x < \infty. \tag{10.19}$$

The derivation of the auxiliary equation to this nonhomogeneous initial value problem follows a similar heuristic argument as was made for the heat equation. Fix a $T > 0$ and let $\{s_i\}_{i=0}^n$ be a partition of the interval $[0, T]$ such that

$$0 = s_0 \le s_1 \le \cdots \le s_{i-1} \le s_i \le \cdots \le s_n = T.$$

For $i = 1, \ldots, n$ define $\Delta s_i = s_i - s_{i-1}$ and define the support functions $\mathbb{1}_i(t)$ as before. Suppose at time $t = s_{i-1}$ the applied acceleration is turned on and it is turned off at $t = s_i$. The initial boundary value problem for the infinitely long string with external acceleration applied only during $[s_{i-1}, s_i)$ is

$$u_{tt} = c^2 u_{xx} + \mathbb{1}_i(t) g(x, t) \text{ for } -\infty < x < \infty \text{ and } t > 0,$$
$$u(x, 0) = u_t(x, 0) = 0 \text{ for } -\infty < x < \infty. \tag{10.20}$$

If $u_i(x,t)$ is the solution to the corresponding initial value problem as shown in Eq. (10.20) for $i = 1, \ldots, n$, then as before $\sum_{i=1}^{n} u_i(x,t) = u(x,t)$, the solution to the initial value problem in Eq. (10.19) for $0 \le t \le T$.

Breaking down the initial value problem shown in Eq. (10.20) by intervals in time, then $u_i(x,t) = 0$ for $0 \le t < s_{i-1}$. If Δs_i is small then during $[s_{i-1}, s_i)$ the external force accelerates the string to a velocity of approximately $g(x, s_{i-1})\Delta s_i$. The external force displaces the string by an approximate amount $g(x, s_{i-1})(\Delta s_i)^2/2$ which can be ignored if Δs_i is small. Thus for $t \ge s_i$ the contribution to the displacement of the string $u_i(x,t)$ can be approximated by the solution to the following initial value problem:

$$u_{tt} = c^2 u_{xx} \text{ for } -\infty < x < \infty \text{ and } t > s_i,$$

$$u(x, s_i) = 0 \text{ for } -\infty < x < \infty,$$

$$u_t(x, s_i) = g(x, s_{i-1})\Delta s_i \text{ for } -\infty < x < \infty.$$

For $t > s_i$ the displacement $u_i(x,t) \approx v(x,t)\Delta s_i$, where $v(x,t)$ solves the initial value problem:

$$v_{tt} = c^2 v_{xx} \text{ for } -\infty < x < \infty \text{ and } t > s_i,$$

$$v(x, s_i; s_{i-1}) = 0 \text{ for } -\infty < x < \infty, \qquad (10.21)$$

$$v_t(x, s_i; s_{i-1}) = g(x, s_{i-1}) \text{ for } -\infty < x < \infty.$$

To indicate the dependence of $v(x,t)$ on s, the solution to Eq. (10.21) will be denoted as $v(x, s_i; s_{i-1})$. Using the Riemann sum argument as in the case of the nonhomogeneous heat equation yields

$$u(x,T) = \int_0^T v(x,T;s)\,ds.$$

Since T is arbitrary it can be replaced by any $t > 0$. As the norm of the partition approaches zero, $v(x,t;s)$ solves the following initial value problem:

$$v_{tt} = c^2 v_{xx} \text{ for } -\infty < x < \infty \text{ and } t > s,$$

$$v(x, s; s) = 0 \text{ for } -\infty < x < \infty, \qquad (10.22)$$

$$v_t(x, s; s) = g(x, s) \text{ for } -\infty < x < \infty.$$

Just as in the case of the heat equation, a change of variable places the initial condition at the customary time of $t = 0$. Define $\hat{v}(x, t - s; s) = v(x, t; s)$, where $\hat{v}$ solves the initial value problem:

$$\hat{v}_{tt} = c^2 \hat{v}_{xx} \text{ for } -\infty < x < \infty \text{ and } t > 0,$$

$$\hat{v}(x, 0; s) = 0 \text{ for } -\infty < x < \infty, \qquad (10.23)$$

$$\hat{v}_t(x, 0; s) = g(x, s) \text{ for } -\infty < x < \infty.$$

Equivalent initial value problems in Eqs. (10.22) and (10.23) are referred to as the auxiliary initial value problems corresponding to Eq. (10.19). With these preliminary remarks in place, Duhamel's principle for the wave equation can be stated and proved.

Theorem 10.4. *Suppose* $g(x, t) \in C^1(\mathbb{R} \times (0, \infty))$, *then*

$$u(x, t) = \frac{1}{2c} \int_0^t \int_{x-c(t-s)}^{x+c(t-s)} g(r, s) \, dr \, ds \qquad (10.24)$$

solves the initial value problem expressed in Eq. (10.19).

Proof. If $t = 0$ in Eq. (10.24) then $u(x, 0) = 0$ since $g(x, t)$ is C^1. Therefore the initial displacement of the infinite string is zero. The auxiliary problem to Eq. (10.19) is given in Eq. (10.23). According to Eq. (5.20) the d'Alembertian solution to Eq. (10.23) is

$$\hat{v}(x, t; s) = \frac{1}{2c} \int_{x-ct}^{x+ct} g(r, s) \, dr.$$

Thus the reader can see that $u(x, t)$ given in Eq. (10.24) is

$$u(x, t) = \int_0^t \hat{v}(x, t - s; s) \, ds.$$

Since $g(x, t)$ is assumed to be C^1, then $g_t(x, t)$ is continuous and $\hat{v}(x, t-s; s)$ is C^2. Using Lemma 1.1,

$$u_t(x, t) = \frac{\partial}{\partial t} \int_0^t \hat{v}(x, t - s; s) \, ds = \hat{v}(x, 0; t) + \int_0^t \hat{v}_t(x, t - s; s) \, ds$$

$$= \int_0^t \hat{v}_t(x, t - s; s) \, ds,$$

using the initial conditions of Eq. (10.23). Setting $t = 0$ again reveals $u_t(x, 0) = 0$ since $\hat{v}_t(x, t - s; s)$ is C^1 and thus the initial velocity of the infinite string is zero. Employing Lemma 1.1 once again,

$$u_{tt}(x, t) = \hat{v}_t(x, 0; t) + \int_0^t \hat{v}_{tt}(x, t - s; s) \, ds$$

$$= g(x, t) + \int_0^t c^2 \hat{v}_{xx}(x, t - s; s) \, ds = g(x, t) + c^2 u_{xx}(x, t).$$

The last equation is made possible by two uses of Leibniz's rule (Lemma 1.1). Hence it has been shown that $u(x, t)$ given in Eq. (10.24) is a C^2 solution to the nonhomogeneous wave equation in Eq. (10.19). □

Example 10.6. Use Duhamel's principle to find the solution to the following initial value problem:

$$u_{tt} = c^2 u_{xx} + e^{-(x-t)} \text{ for } -\infty < x < \infty \text{ and } t > 0,$$
$$u(x,0) = \sin x \text{ for } -\infty < x < \infty,$$
$$u_t(x,0) = \cos x \text{ for } -\infty < x < \infty.$$

Solution. The approach taken to solve this initial value problem is the familiar one of separating the solution into the sum of the solutions to two related problems, the first being the homogeneous version of the partial differential equations with the given initial conditions imposed and the second being an application of Duhamel's principle to solve the given nonhomogeneous partial differential equation with zero initial displacement and velocity. Using d'Alembert's approach, the solution to

$$u_{tt} = c^2 u_{xx} \text{ for } -\infty < x < \infty \text{ and } t > 0,$$
$$u(x,0) = \sin x \text{ for } -\infty < x < \infty,$$
$$u_t(x,0) = \cos x \text{ for } -\infty < x < \infty,$$

can be expressed as

$$u_1(x,t) = \frac{1}{2}\left(\sin(x+ct) + \sin(x-ct)\right) + \frac{1}{2c}\int_{x-ct}^{x+ct} \cos s \, ds$$
$$= \frac{1}{2}\left(\sin(x+ct) + \sin(x-ct)\right) + \frac{1}{2c}\left(\sin(x+ct) - \sin(x-ct)\right).$$

According to Theorem 10.4, the solution to the nonhomogeneous initial value problem:

$$u_{tt} = c^2 u_{xx} + e^{-(x-t)} \text{ for } -\infty < x < \infty \text{ and } t > 0,$$
$$u(x,0) = 0 \text{ for } -\infty < x < \infty,$$
$$u_t(x,0) = 0 \text{ for } -\infty < x < \infty,$$

can be written as the double integral,

$$u_2(x,t) = \frac{1}{2c}\int_0^t \int_{x-c(t-s)}^{x+c(t-s)} e^{-(r-s)} \, dr \, ds$$
$$= \frac{1}{2c}\int_0^t \left(e^{-(x-ct)}e^{(1-c)s} - e^{-(x+ct)}e^{(c+1)s}\right) ds$$
$$= \frac{1}{2c}\left(e^{-(x-ct)}\frac{e^{(1-c)t} - 1}{1 - c} - e^{-(x+ct)}\frac{e^{(1+c)t} - 1}{1 + c}\right).$$

The solution to the original initial value problem is thus $u(x,t) = u_1(x,t) + u_2(x,t)$.

This chapter concludes by illustrating the application of Duhamel's principle to a vibrating string of finite length. This topic was briefly touched upon in Sec. 5.4. For the sake of convenience the general setting of the nonhomogeneous wave equation for a finite length string will be repeated,

$$u_{tt} = c^2 u_{xx} + F(x,t) \text{ for } 0 < x < L \text{ and } t > 0,$$
$$u(0,t) = \phi(t) \text{ and } u(L,t) = \psi(t) \text{ for } t > 0, \tag{10.25}$$
$$u(x,0) = f(x) \text{ and } u_t(x,0) = g(x) \text{ for } 0 < x < L.$$

The solution to the initial boundary value problem in Eq. (10.25) can be found by solving simpler initial boundary value problems which separate the nonhomogeneous term in the partial differential equation from the nonhomogeneous boundary conditions. A reference function defined as

$$u_1(x,t) = (\psi(t) - \phi(t)) \frac{x}{L} + \phi(t)$$

agrees with the nonhomogeneous boundary conditions of Eq. (10.25) at $x = 0$ and $x = L$. Thus if the solution to Eq. (10.25) is thought of as $u(x,t) = u_1(x,t) + v(x,t)$ then $v(x,t)$ satisfies the following initial boundary value problem:

$$v_{tt} = c^2 v_{xx} + F(x,t) - (\psi''(t) - \phi''(t)) \frac{x}{L} - \phi''(t) \text{ for } 0 < x < L, t > 0,$$
$$v(0,t) = v(L,t) = 0 \text{ for } t > 0, \tag{10.26}$$
$$v(x,0) = f(x) - (\psi(0) - \phi(0)) \frac{x}{L} - \phi(0) \text{ for } 0 < x < L,$$
$$v_t(x,0) = g(x) - (\psi'(0) - \phi'(0)) \frac{x}{L} - \phi'(0) \text{ for } 0 < x < L.$$

This initial boundary value problem will be solved by splitting Eq. (10.26) into two problems, one containing the nonzero initial conditions and the other containing the nonhomogeneous partial differential equation. The initial boundary value problem:

$$(v_1)_{tt} = c^2 (v_1)_{xx} \text{ for } 0 < x < L \text{ and } t > 0,$$
$$v_1(0,t) = v_1(L,t) = 0 \text{ for } t > 0,$$
$$v_1(x,0) = f(x) - (\psi(0) - \phi(0)) \frac{x}{L} - \phi(0) \text{ for } 0 < x < L, \tag{10.27}$$
$$(v_1)_t(x,0) = g(x) - (\psi'(0) - \phi'(0)) \frac{x}{L} - \phi'(0) \text{ for } 0 < x < L,$$

can be solved using the techniques developed in Chap. 5 and the solution written as a Fourier series or in d'Alembertian form. The nonhomogeneity present in the partial differential equation of Eq. (10.26) can be addressed in the solution of the initial boundary value problem:

$$(v_2)_{tt} = c^2(v_2)_{xx} + F(x,t) - (\psi''(t) - \phi''(t))\frac{x}{L}$$

$$-\phi''(t) \text{ for } x \in (0,L), \ t > 0,$$

$$v_2(0,t) = v_2(L,t) = 0 \text{ for } t > 0, \tag{10.28}$$

$$v_2(x,0) = (v_2)_t(x,0) = 0 \text{ for } 0 < x < L.$$

The solution to Eq. (10.28) is $v_2(x,t) = \int_0^t \hat{v}(x,t-s;s)\,ds$ where $\hat{v}(x,t;s)$ is the solution to the auxiliary problem:

$$\hat{v}_{tt} = c^2\hat{v}_{xx} \text{ for } 0 < x < L \text{ and } t > 0,$$

$$\hat{v}(0,t;s) = \hat{v}(L,t;s) = 0 \text{ for } t > 0,$$

$$\hat{v}(x,0;s) = 0 \text{ for } 0 < x < L, \tag{10.29}$$

$$\hat{v}_t(x,0;s) = F(x,s) - (\psi''(s) - \phi''(s))\frac{x}{L} - \phi''(s) \text{ for } 0 < x < L.$$

Example 10.7. Use Duhamel's principle to find the solution to the initial boundary value problem stated in Example 5.7. Compare the work done to find the solution here with the work done in Chap. 5.

Solution. The reference function dispensing with the boundary conditions is $u_1(x,t) = (x/L)\sin(\omega t)$ and thus if $u(x,t) = v(x,t) + u_1(x,t)$, then $v(x,t)$ solves the following initial boundary value problem:

$$v_{tt} = c^2 v_{xx} + \frac{\omega^2 x}{L}\sin(\omega t) \text{ for } 0 < x < L \text{ and } t > 0,$$

$$v(0,t) = v(L,t) = 0 \text{ for } t > 0,$$

$$v(x,0) = 0 \text{ and } v_t(x,0) = -\frac{\omega x}{L} \text{ for } 0 < x < L.$$

The solution to the homogeneous portion of the initial boundary value problem was found to be

$$v_1(x,t) = \frac{2\omega L}{c\pi^2}\sum_{n=1}^{\infty}\frac{(-1)^n}{n^2}\sin\frac{n\pi c t}{L}\sin\frac{n\pi x}{L}.$$

All that remains is to find the particular solution to the nonhomogeneous partial differential equation with zero boundary and initial conditions. This is done using Duhamel's principle after solving the auxiliary problem:

$$\hat{v}_{tt} = c^2\hat{v}_{xx} \text{ for } 0 < x < L \text{ and } t > 0,$$

$$\hat{v}(0,t;s) = \hat{v}(L,t;s) = 0 \text{ for } t > 0,$$

$$\hat{v}(x,0;s) = 0 \text{ for } 0 < x < L,$$

$$\hat{v}_t(x,0;s) = \frac{\omega^2 x}{L}\sin(\omega s) \text{ for } 0 < x < L.$$

The solution can be written as the Fourier series,

$$\hat{v}(x,t;s) = \sum_{n=1}^{\infty} a_n \sin \frac{n\pi ct}{L} \sin \frac{n\pi x}{L}$$

where

$$\hat{v}_t(x,0;s) = \frac{\omega^2 x}{L} \sin(\omega s) = \sum_{n=1}^{\infty} \frac{a_n n\pi c}{L} \sin \frac{n\pi x}{L}.$$

Multiplying both sides of this equation by $\sin(m\pi x/L)$ and integrating over the interval $[0, L]$ result in

$$a_n = \frac{2(-1)^{n+1} L\omega^2}{n^2 \pi^2 c} \sin(\omega s).$$

The reader will be asked to provide the details of this derivation in Exercise 17. According to Duhamel's principle the solution $v_2(x,t) = \int_0^t \hat{v}(x, t-s; s)\, ds$. Assuming the Fourier series can be integrated term by term, then

$$\int_0^t a_n \sin \frac{n\pi c(t-s)}{L} \sin \frac{n\pi x}{L}\, ds = b_n \sin \frac{n\pi x}{L} \int_0^t \sin(\omega s) \sin \frac{n\pi c(t-s)}{L}\, ds$$

where for $n \in \mathbb{N}$, $b_n = 2(-1)^{n+1} L\omega^2/(n^2\pi^2 c)$. Employing the product-to-sum formula of Eq. (3.9),

$$\int_0^t \sin(\omega s) \sin \frac{n\pi c(t-s)}{L}\, ds$$

$$= \frac{1}{2} \int_0^t \cos \left[\left(\omega + \frac{n\pi c}{L} \right) s - \frac{n\pi ct}{L} \right] - \cos \left[\left(\omega - \frac{n\pi c}{L} \right) s + \frac{n\pi ct}{L} \right] ds$$

$$= \begin{cases} \dfrac{L^2\omega \sin(cn\pi\, t/L) - cLn\pi \sin(\omega t)}{(L\omega)^2 - (cn\pi)^2} & \text{if } n \neq \omega L/(c\pi), \\[2ex] \dfrac{\sin(\omega t)}{2\omega} - \dfrac{t}{2} \cos(\omega t) & \text{if } n = \omega L/(c\pi). \end{cases}$$

If there exists no $N \in \mathbb{N}$ such that $N = \omega L/(c\pi)$ then

$$v_2(x,t) = \sum_{n=1}^{\infty} b_n \frac{L^2\omega \sin(cn\pi\, t/L) - cLn\pi \sin(\omega t)}{(L\omega)^2 - (cn\pi)^2} \sin \frac{n\pi x}{L}.$$

On the other hand if there exists $N \in \mathbb{N}$ for which $N = \omega L/(c\pi)$ then

$$v_2(x,t) = b_N \left(\frac{\sin(\omega t)}{2\omega} - \frac{t}{2} \cos(\omega t) \right) \sin \frac{\omega x}{c}$$

$$+ \sum_{n=1, n \neq N}^{\infty} b_n \frac{L^2\omega \sin(cn\pi\, t/L) - cLn\pi \sin(\omega t)}{(L\omega)^2 - (cn\pi)^2} \sin \frac{n\pi x}{L}.$$

The solution to the origin initial boundary value problem can be expressed as $u(x,t) = u_1(x,t) + v_1(x,t) + v_2(x,t)$. Readers can verify that the answer derived here is the same as the solution found in Example 5.7.

10.4 Exercises

(1) Show that if $u(x,t)$ is a solution to Eq. (10.6), then $v(x,t) = u(x,t)/\Delta s$ is a solution to Eq. (10.7).

(2) Find the derivative with respect to t of the following function.
$$\int_0^t e^{st} \cos t \, ds.$$

(3) Find the derivative with respect to t of the following function.
$$\int_0^t \tan^{-1}(s^2 t)(1 + st)^2 \, ds.$$

(4) Use Duhamel's principle to find a solution to the following initial boundary value problem:
$$u_t = u_{xx} + \sin(3x) \text{ for } 0 < x < \pi \text{ and } t > 0,$$
$$u(0,t) = u(\pi,t) = 0 \text{ for } t > 0,$$
$$u(x,0) = 0 \text{ for } 0 < x < \pi.$$

(5) Use Duhamel's principle to find a solution to the following initial boundary value problem:
$$u_t = u_{xx} + t^2 \sin(3x) \text{ for } 0 < x < \pi \text{ and } t > 0,$$
$$u(0,t) = u(\pi,t) = 0 \text{ for } t > 0,$$
$$u(x,0) = 0 \text{ for } 0 < x < \pi.$$

(6) Use Duhamel's principle to find a solution to the following initial boundary value problem:
$$u_t = u_{xx} + \sin(3x) \text{ for } 0 < x < \pi \text{ and } t > 0,$$
$$u(0,t) = u(\pi,t) = 0 \text{ for } t > 0,$$
$$u(x,0) = x(1 - x/\pi) \text{ for } 0 < x < \pi.$$

(7) Use Duhamel's principle to find a solution to the following initial boundary value problem:
$$u_t = u_{xx} + t^2 \sin(3x) \text{ for } 0 < x < \pi \text{ and } t > 0,$$
$$u(0,t) = u(\pi,t) = 0 \text{ for } t > 0,$$
$$u(x,0) = e^{-x} \sin x \text{ for } 0 < x < \pi.$$

(8) Use Duhamel's principle to find a solution to the following initial boundary value problem:
$$u_t = u_{xx} + \sin(3x) \text{ for } 0 < x < \pi \text{ and } t > 0,$$
$$u(0,t) = 1 \text{ and } u(\pi,t) = e^{-t} \text{ for } t > 0,$$
$$u(x,0) = \cos(2x) \text{ for } 0 < x < \pi.$$

(9) Use Duhamel's principle to find a solution to the following initial boundary value problem:
$$u_t = u_{xx} + t^2 \sin(3x) \text{ for } 0 < x < \pi \text{ and } t > 0,$$
$$u(0,t) = \sin t \text{ and } u(\pi,t) = 1 - \cos t \text{ for } t > 0,$$
$$u(x,0) = e^{-x} \sin x \text{ for } 0 < x < \pi.$$

(10) Solve the following initial value problem:
$$T_2'(t) + 4T_2(t) = \frac{1}{\pi} \sin t + e^{-t}\left(1 - \frac{1}{\pi}\right),$$
$$T_2(0) = 0.$$

(11) Use Duhamel's principle to find a solution to the following initial boundary value problem:
$$u_t = u_{xx} + t\sin(x) \text{ for } 0 < x < \pi \text{ and } t > 0,$$
$$u_x(0,t) = 0 \text{ and } u_x(\pi,t) = 1 - \cos t \text{ for } t > 0,$$
$$u(x,0) = 1 - \cos x \text{ for } 0 < x < \pi.$$

(12) Use Duhamel's principle to find a solution to the following initial boundary value problem:
$$u_t = u_{xx} + \cos(x) \text{ for } 0 < x < \pi \text{ and } t > 0,$$
$$u_x(0,t) = 0 \text{ and } u(\pi,t) = \sin t \text{ for } t > 0,$$
$$u(x,0) = 0 \text{ for } 0 < x < \pi.$$

(13) Use Duhamel's principle to find a solution to the following initial value problem:
$$u_t = u_{xx} + \cos(x+t) \text{ for } -\infty < x < \infty \text{ and } t > 0,$$
$$u(x,0+) = e^{-|x|} \text{ for } -\infty < x < \infty.$$

(14) Use Duhamel's principle to find a solution to the following initial boundary value problem:
$$u_t = u_{xx} + e^{-(x+t)} \text{ for } 0 < x < \infty \text{ and } t > 0,$$
$$u_x(0+,t) = 0 \text{ for } t > 0,$$
$$u(x,0+) = e^{-x} \text{ for } 0 < x < \infty.$$

(15) Use Duhamel's principle to find the solution to the following initial value problem:
$$u_{tt} = u_{xx} + \sin(x-t) \text{ for } -\infty < x < \infty \text{ and } t > 0,$$
$$u(x,0) = 0 \text{ for } -\infty < x < \infty,$$
$$u_t(x,0) = \sin x \text{ for } -\infty < x < \infty.$$

(16) Use Duhamel's principle to find the solution to the following initial value problem:

$$u_{tt} = u_{xx} + e^{-x}\sin(x - t) \text{ for } -\infty < x < \infty \text{ and } t > 0,$$
$$u(x,0) = \cos x \text{ for } -\infty < x < \infty,$$
$$u_t(x,0) = x^2 \text{ for } -\infty < x < \infty.$$

(17) Use the Fourier coefficient formulas to show that if

$$\frac{\omega^2 x}{L}\sin(\omega s) = \sum_{n=1}^{\infty} \frac{a_n(s)n\pi c}{L}\sin\frac{n\pi x}{L},$$

then

$$a_n(s) = \frac{2(-1)^{n+1}L\omega^2}{n^2\pi^2 c}\sin(\omega s).$$

(18) Find the solution to the following initial boundary value problem:

$$u_{tt} = u_{xx} + \sin(x - t) \text{ for } 0 < x < \pi \text{ and } t > 0,$$
$$u(0,t) = u(\pi,t) = 0 \text{ for } t > 0,$$
$$u(x,0) = u_t(x,0) = 0 \text{ for } 0 < x < \pi.$$

(19) Find the solution to the following initial boundary value problem:

$$u_{tt} = u_{xx} + \sin(x - t) \text{ for } 0 < x < \pi \text{ and } t > 0,$$
$$u(0,t) = 0 \text{ and } u(\pi,t) = \sin t - t \text{ for } t > 0,$$
$$u(x,0) = u_t(x,0) = 0 \text{ for } 0 < x < \pi.$$

(20) Find the solution to the following initial boundary value problem:

$$u_{tt} = u_{xx} + \sin(x - t) \text{ for } 0 < x < \pi \text{ and } t > 0,$$
$$u(0,t) = 0 \text{ and } u(\pi,t) = \sin t - t \text{ for } t > 0,$$
$$u(x,0) = \sin(2x) \text{ and } u_t(x,0) = 0 \text{ for } 0 < x < \pi.$$

Chapter 11

Nonlinear Partial Differential Equations

So far this textbook has concentrated on linear partial differential equations except in Chap. 2 where the solutions to first order, quasilinear partial differential equations were found using the method of characteristics. In the real world, many physical phenomena are the result of complex interactions of many physical processes such as convection, diffusion, and reaction. To model such interactions, nonlinearity is unavoidable, and as a result, nonlinear partial differential equations arise. The subject of nonlinear partial differential equations is much more complicated and challenging than its linear counterpart. As seen in earlier chapters, the Principle of Superposition of solutions and the method of separation of variables play crucial roles in the process of finding solutions of boundary value problems for linear partial differential equations. However, these fundamental principles and methods are no longer applicable to most nonlinear partial differential equations. In fact, there is no general method that can be used to solve a large class of nonlinear partial differential equations. As Lawrence C. Evans [Lindenstrauss *et al.* (1994)] once put it:

> Keep in mind that there is in truth no central core theory of nonlinear PDE, nor can there be. The sources of PDE are so many — physical, probabilistic, geometric, etc. — that the subject is a confederation of diverse subareas, each studying different phenomena for different nonlinear PDE by utterly different methods.

The lack of a general core theory makes the study of nonlinear partial differential equations much less systematic and much more challenging. Almost all the methods or techniques for nonlinear differential equations are on an *ad hoc* basis: different nonlinear partial differential equations often call for different methods. Furthermore, the method developed for studying a particular nonlinear partial differential equation is usually closely related to the physical application from which the equation arises. The physical application often motivates the expectation of the existence of certain types of solutions and the specific properties of such solutions. For example, many nonlinear wave equations originated from wave or signal propagation and, as a result, have traveling wave solutions, which will be introduced in Sec. 11.1.

In addition to finding special types of solutions to nonlinear partial differential equations, there are methods that apply to certain types of nonlinear partial differential equations. For example, through an appropriate change of variable, Burgers'[1] equation can be transformed to the familiar homogeneous heat equation. This transformation is detailed in Sec. 11.2.

The goal of this chapter is to introduce readers to a few important nonlinear partial differential equations and several important techniques in their study. No attempt will be made for a comprehensive discussion or for going into depth in the subject of nonlinear partial differential equations. In Sec. 11.1, some introductory concepts related to traveling wave and solitary wave solutions are described. Section 11.2 is devoted to the discussion of Burgers' equation focusing on the existence of traveling wave solutions and the Cole[2]-Hopf[3] transformation which transforms the Burgers' equation to a linear homogenous heat equation. This is followed by a discussion of the existence and properties of soliton solutions to the Korteweg[4]-de Vries[5] (KdV) and nonlinear Schrödinger equations in Sec. 11.3 and Sec. 11.4 respectively. The chapter ends with some suggestions and comments for further study and a set of exercises for understanding and extending the discussions begun in this chapter.

11.1   Waves and Traveling Waves

Traveling wave solutions of nonlinear partial differential equations have drawn much attention and are one of the main topics of this chapter. In this context a wave refers to any disturbance or signal that travels through a medium over time, carrying energy with it as it propagates. It could refer to a water wave, a sound wave, or an electromagnetic wave, even a signal passing through a fiber optical cable. In the physical world, wave propagation is a commonly observed phenomenon. In general, if $u(x,t)$ represents the disturbance or signal strength at the location x at time t, it

---

[1] Johannes Martinus Burgers, Dutch physicist (1895–1981).
[2] Julian D. Cole, American mathematician (1925–1999).
[3] Eberhard Hopf, German-American mathematician and astronomer (1902–1983).
[4] Diederik Johannes Korteweg, Dutch mathematician (1848–1941).
[5] Gustav de Vries, Dutch mathematician (1866–1934).

is a **traveling wave** if u has the form,

$$u(x,t) = U(x - ct) \tag{11.1}$$

for some single-variable function U and a fixed real constant c. For a wave $u(x,t)$ given by Eq. (11.1), the initial wave or profile $u(x,0) = U(x)$ is propagated at speed $|c|$. For many nonlinear partial differential equations, bounded traveling wave solutions are of chief interest.

The reader may recall that the idea of traveling wave solutions occurred in the discussion of solutions to the one-dimensional wave equation found using d'Alembert's approach. In Chap. 5, the solution of the standard wave equation on the real number line with the initial conditions

$$u(x,0) = f(x) \text{ and } u_t(x,0) = 0,$$

(the plucked string) is given by

$$u(x,t) = \frac{1}{2}(f(x - ct) + f(x + ct)).$$

In this case, $u(x,t)$ is the sum of two simple traveling waves, that is, the initial disturbance $f(x)$ of the string is propagated to the right and left, respectively, with speed $|c|$, each at half of the magnitude. Therefore, strictly speaking, $u(x,t)$ is not a traveling wave itself, rather it is the sum of two simple traveling waves.

In general, to find the traveling wave solution to a partial differential equation, the solution $u(x,t)$ is assumed to have the form given by Eq. (11.1) and substituted into the partial differential equation to identify the constant c and the function U. The function U will satisfy an ordinary differential equation and, possibly, some boundary conditions. The following example demonstrates this basic idea.

Example 11.1. Find all bounded, traveling wave solutions of the (linear) partial differential equation,

$$u_t - u_{xxx} = 0. \tag{11.2}$$

Solution. Let c be any constant and assume that $u(x,t) = U(x - ct)$ is a solution of Eq. (11.2). Setting $\xi = x - ct$ and substituting $u(x,t) = U(\xi)$ in the equation yield

$$-cU'(\xi) - U'''(\xi) = 0.$$

Integrating the above equation once produces

$$U''(\xi) + cU(\xi) = A, \tag{11.3}$$

where A is an arbitrary constant of integration. If $c < 0$, the general solution of Eq. (11.3) is

$$U(\xi) = A_1 e^{\sqrt{-c}\,\xi} + A_2 e^{-\sqrt{-c}\,\xi} + \frac{A}{c}$$

where A_1 and A_2 are arbitrary constants. Unless A_1 and A_2 are both zero, $U(\xi)$ is unbounded. If $A_1 = A_2 = 0$, then solution $U(\xi) = A/c$ is a constant which solves the original partial differential equation. If $c = 0$, Eq. (11.3) produces no bounded solution other than constant solutions. The last case to consider is that of $c > 0$. In this case the general solution of Eq. (11.3) is

$$U(\xi) = A_1 \cos(\sqrt{c}\,\xi) + A_2 \sin(\sqrt{c}\,\xi) + \frac{A}{c}$$

where A_1 and A_2 are again arbitrary constants. The function $U(\xi)$ is bounded and leads to the following bounded traveling wave solution of Eq. (11.2),

$$u(x - ct) = A_1 \cos(\sqrt{c}(x - ct)) + A_2 \sin(\sqrt{c}(x - ct)) + \frac{A}{c} \qquad (11.4)$$

for any choices of constants A_1, A_2, and A.

Equation (11.2) is a linear, dissipative partial differential equation. The nonconstant, bounded traveling wave solutions occur only for $c > 0$. This implies that all traveling wave solutions travel to the right only. Furthermore, disregarding the constant A/c, all nonconstant traveling wave solutions are linear combinations of

$$u_1(x, t) = \cos(\sqrt{c}(x - ct)) \text{ and } u_2(x, t) = \sin(\sqrt{c}(x - ct)).$$

Setting $k = \sqrt{c}$ and $\omega = k^3$, these two solutions can be written as

$$u_1(x, t) = \cos(kx - \omega t) \text{ and } u_2(x, t) = \sin(kx - \omega t).$$

Such traveling wave solutions are often called one-dimensional **plane wave solutions** and are often conveniently represented in the form of complex exponentials as $u(x, t) = e^{i(kx - \omega t)}$ and its complex conjugate. The constant k is called the **wave number** and ω **the angular frequency**. The equality $\omega = k^3$ describes the relationship between the angular frequency and the wave number and is often referred to as the **dispersion relation** for the partial differential equation.

As another simple example, consider the partial differential equation

$$\psi_t = i\,\psi_{xx} \qquad (11.5)$$

Equation (11.5) is often referred to as the free Schrödinger equation for free particles. Free particles are particles which are not bound by any external force or equivalently exist in a region of space where the potential energy is constant (or more specifically zero). If $\psi(x,t) = e^{i(kx - \omega t)}$ is a solution to Eq. (11.5), the following dispersion relation holds,

$$-i\,\omega = -i\,k^2 \iff \omega = k^2.$$

The traveling wave solutions found in the examples above are periodic. Another class of traveling wave solution, of particular importance for nonlinear partial differential equations, is the **solitary wave** or **soliton**, solution which is a special type of traveling wave solution. In general, a solitary wave solution has two important features. First, it must have a permanent form, meaning that its profile will not change as it propagates with time. Next it is localized, meaning that it decays to 0 or approaches a constant at infinity. A soliton solution, or simply a soliton, is a solitary wave solution that is stable in the sense that it maintains its wave profile even after colliding with another soliton. Therefore, strictly speaking, solitons and solitary wave solutions are different concepts though some authors choose not to distinguish between them. The exact definition or classification is not of importance for the discussion here. Readers interested in the details of these concepts may see [Newell (1985)] or [Drazin and Johnson (1989)] for more information.

Traveling wave solutions do not exist for every nonlinear partial differential equation. For the equations for which traveling wave solutions do exist, only certain initial profiles (initial conditions) lead to traveling wave solutions. Of course, similar remarks hold for solitary wave solutions. However, traveling (solitary) wave solutions are of importance since they often reveal the interaction between different physical processes, especially between nonlinearity and the effects of reaction and/or dispersion. Note that when seeking a traveling wave solution, the wave speed c must be determined so that the function defined by Eq. (11.1) is a solution to the underlying partial differential equation. When the function defined by Eq. (11.1) is substituted into the partial differential equation, an ordinary differential equation is produced with, possibly, some specified behavior at infinity. Therefore, in many cases, to find the traveling wave solutions of a partial differential equation is essentially a problem of finding the solution or solutions to a boundary value problem of an ordinary differential equation.

The existence and properties of travelling wave solutions for a few nonlinear partial differential equations are discussed in later sections. The discussion will be restricted to one spatial dimension only, although many concepts can be adapted to higher-dimensional cases.

11.2   Burgers' Equation

The discussion of nonlinear partial differential equations starts with the well-known **Burgers' equation**. This is one of the simplest nonlinear partial differential equations. In Sec. 2.3 a simple traffic flow model for traffic on a straight stretch of highway was developed. The discussion there leads naturally to Burgers' equation. Recall that $\rho(x,t)$ was used to denote the vehicle density at position x at time t and $q(x,t)$ the traffic flux, the number of vehicles passing location x per unit time at time t. The conservation of vehicles leads to Eq. (2.41) (repeated here for convenience),

$$\frac{\partial \rho}{\partial t}(x,t) + \frac{\partial q}{\partial x}(x,t) = 0.$$

The vehicle velocity is assumed to be a linear function of the traffic density. This leads to the assumption that the traffic flux is proportional to the vehicle density and produced the model in Eq. (2.43). In reality, the traffic flux can be related to the vehicle density in a more complicated fashion or can even depend on other factors. For example, the flux $q(x,t)$ may depend on the rate of change of ρ with respect to x. Suppose $q(x,t) = Q(\rho(x,t)) - \mu\rho_x$ for some function $Q(\rho)$ and constant $\mu > 0$. If $Q(\rho)$ is taken to be the simple quadratic function $Q(\rho) = u_{max}\rho^2/\rho_{max}$, then the traffic flux is given by

$$q \equiv q(\rho) = \frac{u_{max}}{\rho_{max}}\rho^2 - \mu\frac{\partial \rho}{\partial x}$$

where u_{max} is the maximum speed limit of vehicles (for example, the posted speed limit) and ρ_{max} is the maximum vehicle density. Under these assumptions, the traffic density $\rho(x,t)$ can be modeled by the following nonlinear partial differential equation,

$$\frac{\partial \rho}{\partial t} + 2\frac{u_{max}}{\rho_{max}}\rho\frac{\partial \rho}{\partial x} - \mu\frac{\partial^2 \rho}{\partial x^2} = 0. \tag{11.6}$$

By setting $\xi = \rho_{max}x/(2u_{max})$ and $u(\xi,t) = \rho(x,t)$ and rewriting ξ as x thereafter, Eq. (11.6) becomes the following

$$\frac{\partial u}{\partial t} + u\frac{\partial u}{\partial x} - \nu\frac{\partial^2 u}{\partial x^2} = 0, \tag{11.7}$$

where the diffusion coefficient $\nu > 0$ is a constant. This equation is commonly referred to as the standard form of Burgers' equation, which is nonlinear due to the presence of the term $u\,\partial u/\partial x$.

Although Burgers' equation is derived from the traffic flow model, its applications go far beyond that context. Historically Burgers' equation was first proposed to model the motion of turbulent flow [Burgers (1948)]. In fact, Burgers' equation is commonly considered a simplified version of the Navier[6]-Stokes[7] equations, the fundamental equations of viscous fluid dynamics. In the context of fluid flow, the nonlinear term $u\,u_x$ models a convection process and the term $\nu\,u_{xx}$ models the effect of viscosity or the diffusion process. For this reason the constant ν is often referred to as the kinetic viscosity or simply, viscosity. Some authors refer to Eq. (11.7) as the viscous form of Burgers' equation. Burgers' equation arises from many other physical situations as well, such as the propagation of sound waves in a viscous medium and the study of magnetohydrodynamics. For more details, see [Hamilton and Blackstock (1998)] and [Olesen (2003)]. If the viscosity $\nu = 0$, Burgers' equation becomes one of the simplest nonlinear conservation laws (often referred to as the inviscid Burgers' equation).

Mathematically, Burgers' equation describes the balance between the effects of nonlinearity and diffusion. If u and/or u_x are so small that the nonlinear term of Eq. (11.7) can be ignored, the solutions of Burgers' equation should be similar to those of the heat equation. On the other hand, if $|u|$ and/or $|u_x|$ are large, the nonlinear term plays a more dominant role in the equation, and the balance of the nonlinear term and the diffusion term results in nonlinear waves — traveling wave solutions. It is also worthwhile to point out that Burgers' equation is parabolic if $\nu > 0$ and it is hyperbolic if $\nu = 0$. If $\nu > 0$, under relatively mild conditions on the initial data, the solution of Burgers' equation is smooth and is defined for all $t \geq 0$. However, if $\nu = 0$, the solution is not smooth and a shock wave can develop as indicated in the discussion of the traffic flow model in Sec. 2.3 and later in this section.

Before solving Eq. (11.7) a brief discussion of conservation laws will be presented. Burgers' equation is one example of a conservation law.

11.2.1 *Conservation Laws and Burgers' Equation*

A careful approach to the solution of the inviscid Burgers' equation will make use of the assumption that the viscous Burgers' equation can be interpreted as a mathematical model of fluid flow. As such, physical quantities like mass should be conserved. In fact Burgers' equation (and other similar mathematical models) are often categorized as **conservation laws**. In some cases the solution to Burgers' equation is not unique or is multi-

---

[6]Claude-Louis Navier, French engineer and physicist (1785–1836).
[7]George Gabriel Stokes, Irish mathematician (1819–1903).

valued. Once the quantity that must be conserved in Burgers' equation is determined, that quantity can be used to determine which of the multiple values (if any) of u is the "correct" solution to Burgers' equation. This will be important later in Sec. 11.2.4 when the solution to the inviscid Burgers' equation admits a shock.

To determine the conserved quantity, integrate the viscous Burgers' equation in Eq. (11.7) with respect to x over an arbitrary interval $[a, b]$.

$$\int_a^b u_t \, dx + \int_a^b u \, u_x \, dx - \nu \int_a^b u_{xx} \, dx = 0$$

$$\int_a^b u_t \, dx + \frac{1}{2}(u(b,t))^2 - \frac{1}{2}(u(a,t))^2 - \nu u_x(b,t) + \nu u_x(a,t) = 0. \quad (11.8)$$

Note that by Leibniz's rule,

$$\frac{d}{dt}\int_a^b u(x,t)\, dx = \int_a^b u_t \, dx + u(b,t)\frac{db}{dt} - u(a,t)\frac{da}{dt}. \quad (11.9)$$

Using Eq. (11.8) to make a replacement in Eq. (11.9) produces

$$\frac{d}{dt}\int_a^b u(x,t)\, dx = \left(u(b,t)\frac{db}{dt} - \frac{1}{2}(u(b,t))^2 + \nu u_x(b,t)\right)$$
$$- \left(u(a,t)\frac{da}{dt} - \frac{1}{2}(u(a,t))^2 + \nu u_x(a,t)\right). \quad (11.10)$$

Hence the rate of change of $\int_a^b u(x,t)\, dx$ is the quantity,

$$\left[u(x,t)\frac{dx}{dt} - \frac{1}{2}(u(x,t))^2 + \nu \, u_x(x,t)\right]_{x=a}^{x=b}. \quad (11.11)$$

Since the rate of change of the integral of u over $[a, b]$ is dependent only on the values at the endpoints of the integral (namely a and b) then the integral of u is said to be *conserved*.

In the following sections, attention will be focused on the existence and properties of traveling wave solutions of the viscous Burgers' equation and the Cole-Hopf transformation technique for determining solutions to the viscous Burgers' equation before turning to the inviscid Burgers' equation.

11.2.2 *Traveling Wave Solutions of Burgers' Equation*

As mentioned earlier, a traveling wave solution is a solution of the form $U(x - ct)$ for some constant c. Setting $\xi = x - ct$ and substituting $u(x,t) = U(\xi)$ into Eq. (11.7) produce the equation,

$$-c\frac{dU}{d\xi} + u\frac{dU}{d\xi} - \nu\frac{d^2U}{d\xi^2} = 0.$$

Integrating both sides of the equation with respect to ξ yields

$$-cU + \frac{1}{2}U^2 - \nu \frac{dU}{d\xi} = A, \tag{11.12}$$

where A is a constant. Therefore,

$$\frac{dU}{d\xi} = \frac{1}{2\nu}(U^2 - 2cU - 2A). \tag{11.13}$$

Equation (11.13) can be solved using the separation of variables technique for first-order ordinary differential equations. The solutions of interest in this case are the bounded solutions. It can be shown that there is no bounded solution of Eq. (11.13) if the quadratic equation $U^2 - 2cU - 2A = 0$ has only one real root (other than the constant solution) or two complex roots, see Exercise 6. Assume that the equation $U^2 - 2cU - 2A = 0$ has two distinct real roots, that is, the inequality $c^2 + 2A > 0$ is satisfied, and denote the two real roots as $U_1 = c - \sqrt{c^2 + 2A}$ and $U_2 = c + \sqrt{c^2 + 2A}$. The constant functions $U(\xi) = U_1$ and $U(\xi) = U_2$ are equilibrium solutions to Eq. (11.13). If $U(\xi)$ is a solution to Eq. (11.13) with initial value between U_1 and U_2, then by the uniqueness of solutions to the ordinary differential equation, $U(\xi)$ will be bounded and stay between U_1 and U_2. Separating the variables in Eq. (11.13) and rewriting the equation as

$$\frac{1}{(U - U_1)(U - U_2)} \frac{dU}{d\xi} = \frac{1}{2\nu}$$

allow both sides to be integrated with respect to ξ. This results in

$$\ln \left| \frac{U - U_1}{U - U_2} \right| = \frac{(U_1 - U_2)}{2\nu}\xi + B,$$

where B is an arbitrary constant of integration. Therefore using the fact that $U_1 < U(\xi) < U_2$,

$$\frac{U - U_1}{U_2 - U} = e^{\alpha\xi + B},$$

where $\alpha = (U_1 - U_2)/(2\nu) < 0$. Solving this equation for U yields

$$U(\xi) = \frac{U_1 + U_2 e^{\alpha\xi + B}}{1 + e^{\alpha\xi + B}}. \tag{11.14}$$

As expected, the solution U satisfies

$$\lim_{\xi \to -\infty} U(\xi) = U_2 \text{ and } \lim_{\xi \to \infty} U(\xi) = U_1.$$

This leads to the following traveling wave solution for Burgers' equation

$$u(x,t) = U(x - ct) = \frac{U_1 + U_2 e^{\alpha(x-ct)+B}}{1 + e^{\alpha(x-ct)+B}}, \tag{11.15}$$

A First Course in Partial Differential Equations

which satisfies the following limits for each fixed t,

$$\lim_{x \to -\infty} u(x,t) = U_2 \text{ and } \lim_{x \to \infty} u(x,t) = U_1.$$

Note that if $B = 0$, the corresponding traveling wave solution becomes

$$u(x,t) = U(x - ct) = \frac{U_1 + U_2 e^{\alpha(x-ct)}}{1 + e^{\alpha(x-ct)}}$$

and the initial wave profile is obtained by setting $t = 0$,

$$u(x,0) = U(x) = \frac{U_1 + U_2 e^{\alpha x}}{1 + e^{\alpha x}}.$$

Only when the initial condition is given by the function above or one of its horizontal translations does this lead to a traveling wave solution. The traveling wave solutions have the following properties:

(1) For any fixed t, a traveling wave solution decreases from U_2 to U_1 as x increases from $-\infty$ to ∞. Note that U_1 and U_2 are equilibrium solutions of Eq. (11.13). In the language of dynamical systems, the solution $U(\xi)$ given by Eq. (11.14) is a homoclinic orbit of the ordinary differential equation in Eq. (11.13) [Guckenheimer and Holmes (1983)].

(2) The traveling waves travel from the left to right with a speed $c = (U_1 + U_2)/2$ since U_1 and U_2 are solutions of the quadratic equation $U^2 - 2cU - 2A = (U - U_1)(U - U_2) = 0$, see Exercise 8.

(3) As the viscosity ν approaches 0, for fixed t and x, if $x < ct$, the solution $u(x,t) \to U_2$ and if $x > ct$, $u(x,t) \to U_1$. In the limit, a shock wave develops.

Figure 11.1 illustrates the traveling wave solution to Burgers' equation. In the upper plot the viscosity $\nu = 1$ and the surface plot appears smooth. In the lower plot the viscosity has been decreased to $\nu = 1/4$ and the surface appears to drop more abruptly.

11.2.3   *The Cole-Hopf Transformation*

The existence of traveling wave solutions of Burgers' equation was established in the last section. A natural question to ask then, is whether it is possible to find all solutions to Burgers' equation, not merely the traveling wave solutions. More specifically, motivated by the analysis of the heat equation and wave equation, does Burgers' equation have a solution for any given initial condition,

$$u(x,0) = f(x) \text{ for } -\infty < x < \infty. \tag{11.16}$$

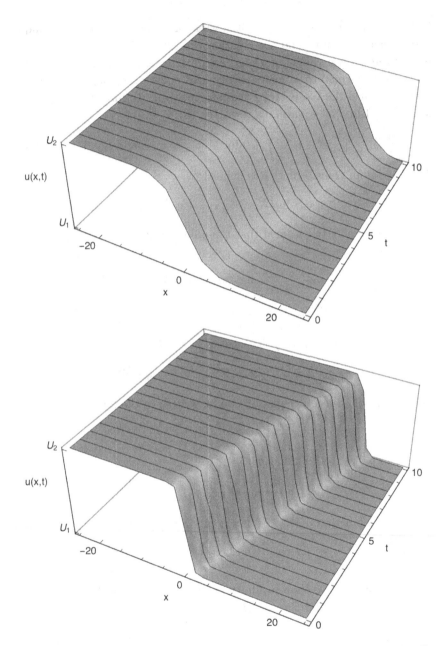

Fig. 11.1   Surface plots of the traveling wave solutions to Burgers' equation (Eq. (11.7)). In the upper plot the viscosity parameter $\nu = 1$. When the viscosity is lowered (to $\nu = 1/4$, as in the lower plot) the solution drops more abruptly. In the limit as $\nu \to 0$ it may be expected that the solution develops a discontinuity (called a shock).

If there exists a solution to the initial value problem, is the solution unique and is there a way to find the solution explicitly? This section attempts to partially answer these questions. As a matter of fact, Burgers' equation can be transformed to a homogeneous heat equation through a change of variable, called the **Cole-Hopf transformation**. To introduce the transformation, Burgers' equation is rewritten as

$$\frac{\partial u}{\partial t} + \frac{\partial}{\partial x}\left(\frac{1}{2}u^2 - \nu\frac{\partial u}{\partial x}\right) = 0 \text{ for } -\infty < x < \infty \text{ and } t > 0. \tag{11.17}$$

The Cole-Hopf transformation is defined by

$$u = -2\nu\frac{v_x}{v} \iff u = -2\nu\frac{\partial}{\partial x}\left[\ln v\right], \tag{11.18}$$

where it is implicitly assumed that the unknown function $v(x,t) > 0$. The chain rule implies

$$\frac{\partial u}{\partial t} = -2\nu\frac{v\,v_{xt} - v_x v_t}{v^2} = -2\nu\frac{v\,v_{tx} - v_t v_x}{v^2} = -2\nu\frac{\partial}{\partial x}\left[\frac{v_t}{v}\right],$$

$$\frac{\partial u}{\partial x} = -2\nu\frac{v\,v_{xx} - (v_x)^2}{v^2},$$

and therefore in Eq. (11.17) the term $u^2/2 - \nu\,\partial u/\partial x$ can be written as

$$2\nu^2\frac{(v_x)^2}{v^2} + 2\nu^2\frac{v\,v_{xx} - (v_x)^2}{v^2} = 2\nu^2\frac{v_{xx}}{v},$$

and as a result Burgers' equation can be written in the form

$$-2\nu\frac{\partial}{\partial x}\left[\frac{v_t}{v}\right] + \frac{\partial}{\partial x}\left[2\nu^2\frac{v_{xx}}{v}\right] = 0,$$

which simplifies to the following:

$$\frac{\partial}{\partial x}\left[\frac{v_t - \nu\,v_{xx}}{v}\right] = 0. \tag{11.19}$$

Clearly, if a function $v(x,t) > 0$ is a solution the linear, homogeneous heat equation

$$v_t = \nu\,v_{xx}, \tag{11.20}$$

the numerator of the term inside the partial derivative in Eq. (11.19) will be identically zero and thus $v(x,t)$ is a solution to Eq. (11.19). Hence the function $u(x,t)$ defined by Eq. (11.18) is a solution to Burgers' equation. On the other hand, if $u(x,t)$ is any solution to Eq. (11.7), then by integrating Eq. (11.18) and solving for v, the function

$$v(x,t) = e^{-\frac{1}{2\nu}\int_0^x u(y,t)\,dy} \tag{11.21}$$

is a solution of Eq. (11.19). Of course, the main interest is to obtain solutions of Burgers' equation from solutions to the heat equation. Since many solutions of the heat equation are known, by using the Cole-Hopf transformation, many solutions of Burgers' equation can be found. For example, since $v(x,t) = A + Be^{-\nu t}\sin x$ is a solution of the heat equation for any constants A and B, if A is large enough to guarantee that $v(x,t)$ is positive everywhere,

$$u(x,t) = \frac{-2\nu Be^{-\nu t}\cos x}{A + Be^{-\nu t}\sin x}$$

is a solution to Burgers's equation.

Next consider the initial value problem of Burgers' equation. Assume that Eq. (11.7) is given an initial condition defined by Eq. (11.16), then the function v will have a corresponding initial condition given by Eq. (11.21),

$$v(x,0) = g(x) = e^{-\frac{1}{2\nu}\int_0^x f(y)\,dy}. \tag{11.22}$$

By Theorem 4.3, if $g(x)$ satisfies the condition specified there, the solution to Eq. (11.20) is

$$v(x,t) = \frac{1}{\sqrt{4\pi\nu t}} \int_{-\infty}^{\infty} g(y)e^{-\frac{(x-y)^2}{4\nu t}}\,dy \text{ for } t > 0. \tag{11.23}$$

This implies

$$v_x(x,t) = \frac{-1}{\sqrt{4\pi\nu t}} \int_{-\infty}^{\infty} g(y)\left(\frac{x-y}{2\nu t}\right)e^{-\frac{(x-y)^2}{4\nu t}}\,dy \text{ for } t > 0. \tag{11.24}$$

Substituting Eqs. (11.23) and (11.24) in Eq. (11.18) yields a solution of the initial value problem of the Burgers' equation:

$$
\begin{aligned}
u(x,t) &= \frac{2\nu \int_{-\infty}^{\infty} g(y)\left(\frac{x-y}{2\nu t}\right)e^{-\frac{(x-y)^2}{4\nu t}}\,dy}{\int_{-\infty}^{\infty} g(y)e^{-\frac{(x-y)^2}{4\nu t}}\,dy} \\
&= \frac{\int_{-\infty}^{\infty}(x-y)e^{-\frac{1}{2\nu}\left(\int_0^y f(s)\,ds + \frac{(x-y)^2}{2t}\right)}\,dy}{t\int_{-\infty}^{\infty} e^{-\frac{1}{2\nu}\left(\int_0^y f(s)\,ds + \frac{(x-y)^2}{2t}\right)}\,dy}.
\end{aligned} \tag{11.25}
$$

According to Theorem 4.5 the solution to the initial value problem consisting of Eqs. (11.7) and (11.16) is unique provided f is piecewise continuous and bounded.

Example 11.2. Consider the initial value problem consisting of the viscous Burgers' equation and the initial condition:

$$u_t + u\,u_x - \nu\,u_{xx} = 0,$$

$$u(x,0) = \begin{cases} 0 & \text{if } x < 0, \\ 1 & \text{if } x \geq 0. \end{cases}$$

Find a solution to this initial value problem using the Cole-Hopf transformation.

Solution. According the formula given in Eq. (11.22),

$$g(x) = \begin{cases} 1 & \text{if } x < 0, \\ e^{-x/(2\nu)} & \text{if } x \geq 0. \end{cases}$$

Using this in Eq. (11.25), the solution to the initial value problem is

$$u(x,t) = \frac{\int_{-\infty}^{\infty} g(y)(x-y)e^{-\frac{(x-y)^2}{4\nu t}} \, dy}{t \int_{-\infty}^{\infty} g(y)e^{-\frac{(x-y)^2}{4\nu t}} \, dy}$$

$$= \frac{\int_{-\infty}^{0} (x-y)e^{-\frac{(x-y)^2}{4\nu t}} \, dy + \int_{0}^{\infty} (x-y)e^{-\frac{y}{2\nu}-\frac{(x-y)^2}{4\nu t}} \, dy}{t \int_{-\infty}^{0} e^{-\frac{(x-y)^2}{4\nu t}} \, dy + t \int_{0}^{\infty} e^{-\frac{y}{2\nu}-\frac{(x-y)^2}{4\nu t}} \, dy}$$

$$= \frac{4\nu t \int_{\frac{x}{\sqrt{4\nu t}}}^{\infty} ze^{-z^2} \, dz + te^{\frac{t-2x}{4\nu}} \int_{\frac{t-x}{\sqrt{4\nu t}}}^{\infty} (\sqrt{4\nu t} - 4\nu z)e^{-z^2} \, dz}{t\sqrt{4\nu t} \int_{\frac{x}{\sqrt{4\nu t}}}^{\infty} e^{-z^2} \, dz + te^{\frac{t-2x}{4\nu}} \sqrt{4\nu t} \int_{\frac{t-x}{\sqrt{4\nu t}}}^{\infty} e^{-z^2} \, dz}$$

$$= \frac{e^{\frac{t-2x}{4\nu}} \left(1 - \operatorname{erf}\left(\frac{t-x}{\sqrt{4\nu t}}\right)\right)}{e^{\frac{t-2x}{4\nu}} \left(1 - \operatorname{erf}\left(\frac{t-x}{\sqrt{4\nu t}}\right)\right) + 1 - \operatorname{erf}\left(\frac{x}{\sqrt{4\nu t}}\right)}$$

for $t > 0$. In order to evaluate the integrals involved in the solution, the technique of completing the square and the substitutions $z = (x-y)/\sqrt{4\nu t}$ and $z = (y+t-x)/\sqrt{4\nu t}$ were used. The error function introduced in Eq. (4.43) is used to express the values of the integrals. Figure 11.2 illustrates the solution to this example.

11.2.4 *Inviscid Burgers' Equation and Shock Waves*

If the viscosity $\nu = 0$, then Burgers' equation (Eq. (11.7)) becomes

$$\frac{\partial u}{\partial t} + u \frac{\partial u}{\partial x} = 0, \tag{11.26}$$

known as the inviscid Burgers' equation. This equation is a first order partial differential equation of hyperbolic type as opposed to the viscous Burgers' equation which is parabolic. As a result, the solutions of the inviscid Burgers' equation have dramatically different properties.

The dependence on ν is apparent in the traveling wave solution to Eq. (11.15) for the viscous Burgers' equation. As ν approaches 0, the traveling wave solutions becomes steeper for x near the value ct as can be seen in the surface plots of the solutions in Fig. 11.1. In the following, the

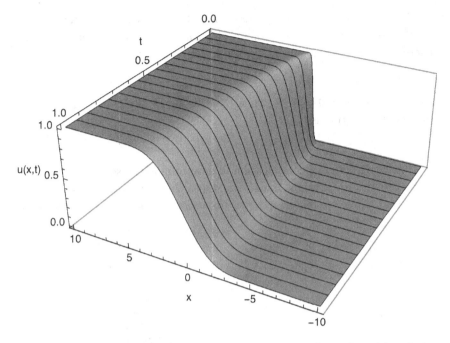

Fig. 11.2   Surface plot of the solution to the viscous Burgers' equation with unit step function as the initial condition. For the purposes of plotting the viscosity parameter is $\nu = 1$. Note that as t increases the surface becomes smoother.

traveling wave solution given by Eq. (11.15) will be denoted as $u_\nu(x,t)$. To solve the inviscid Burgers' equation in general requires the method of characteristics described in Sec. 2.2. The characteristic system for Eq. (11.26) can be expressed as:

$$\frac{dx}{ds} = u, \quad \frac{dt}{ds} = 1, \text{ and } \frac{du}{ds} = 0.$$

The characteristics are parameterized by s. The third characteristic equation indicates that u is constant along characteristics. The second characteristic equation allows the parameter s to be identified with the independent variable t. Hence the first characteristic equation may be rewritten as $dx/dt = u$ and thus the solution to the characteristic system may be expressed as

$$x = u\,t + C_1,$$

$$u = C_2,$$

where C_1 and C_2 are arbitrary constants which may be determined from initial conditions imposed on the inviscid Burgers' equation. If the initial condition is given as $u(x, 0) = f(x)$, then along the characteristic through $(x_0, 0)$ it is the case that $C_1 = x_0$ and thus $x_0 = x - u\,t$ on the characteristic. Therefore the solution can be written in implicit form as

$$u(x, t) = f(x_0) = f(x - u(x, t)t). \tag{11.27}$$

For example if $u(x, 0) = e^{-x^2}$, then by Exercise 14 of Chap. 2, the implicit form of the solution is

$$u(x, t) = e^{-(x - u(x, t)t)^2}.$$

However, there is a complication in determining the solution. Since the characteristics are straight lines with slopes that depend on u, unless u is a constant, these characteristics may intersect for some $t > 0$. If the t-axis

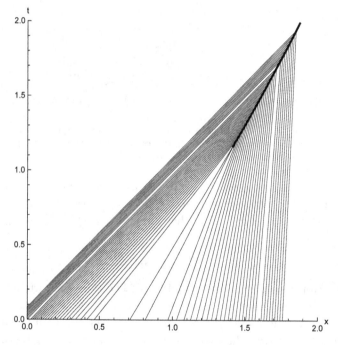

Fig. 11.3   The intersection of characteristics for the inviscid Burgers' equation (with initial condition $u(x, 0) = e^{-x^2}$) indicate that a more careful approach to the solution must be followed in the vicinity of the intersections. The thick curve is the locus of points where characteristics intersect and is called a shock. The shock is the location of a discontinuity in the solution to the inviscid Burgers' equation.

is oriented vertically and the x-axis is oriented horizontally as in Fig. 11.3, the slope of the characteristics is $1/u(x, 0)$, thus if $u(x, 0)$ is a decreasing function of x in some interval, the slopes of the characteristics will increase, leading to the intersection of the characteristics at some $t > 0$. Figure 11.3 illustrates characteristics which intersect at some time t, with $1 \leq t \leq 2$. At the location and time of intersection of two characteristics with different values of u, a phenomenon known as a **shock** occurs. If the inviscid Burgers' equation is thought of as a model of fluid flow, then the downstream fluid is moving slower than the upstream fluid, or equivalently, the upstream fluid is overtaking the downstream fluid. Since the dependent variable u is constant along each characteristic, the intersection of characteristics implies u is multivalued, which means the solution is not well-defined. Figure 11.4 illustrates a typical multivalued situation for the solution to the inviscid Burgers' equation.

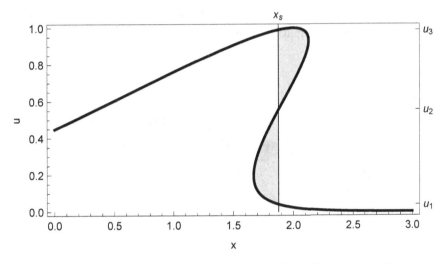

Fig. 11.4 When characteristics of the inviscid Burgers' equation intersect, the continuous solution to the equation become undefined due to it being multivalued. Hence the solution in the region where characteristics intersect must be discontinuous. In this figure $t = 2$ and the initial condition for the inviscid Burgers' equation is $u(x, 0) = e^{-x^2}$.

The shock will form at the earliest time the surface $u(x, t)$ has a vertical tangent. Using implicit differentiation on the general solution found in Eq. (11.27) to the inviscid Burgers' equation, the slope of the solution in the x-direction is

$$u_x = \frac{f'(z)}{1 + t\, f'(z)}.$$

A vertical tangent occurs when u_x is undefined, or equivalently when $t = -1/f'(z)$. Thus the earliest time this occurs is the time for which $f'(z)$ takes on its negative minimum. The reader will be asked to show in Exercise 11 that the shock forms at $t_0 = \sqrt{e/2}$ when $f(x) = e^{-x^2}$. The x-coordinate at which the shock forms is found by solving the equation $1 + t_0 f'(x_0) = 0$ which makes u_x undefined and gives the coordinate on the $t = 0$ line where the characteristic passing through the beginning of the shock passes. When $f(x) = e^{-x^2}$, solving the equation,

$$0 = 1 - 2\sqrt{\frac{e}{2}}x_0 e^{-x_0^2} \implies x_0 = \frac{1}{\sqrt{2}}.$$

Hence the shock begins at the point with coordinates:

$$(x,t) = (x_0 + t_0 f(x_0), t_0) = \left(\sqrt{2}, \sqrt{\frac{e}{2}}\right)$$

in the specific example explored here. Figure 11.5 depicts the solution to the inviscid Burgers' equation with initial condition $u(x,0) = e^{-x^2}$. Notice

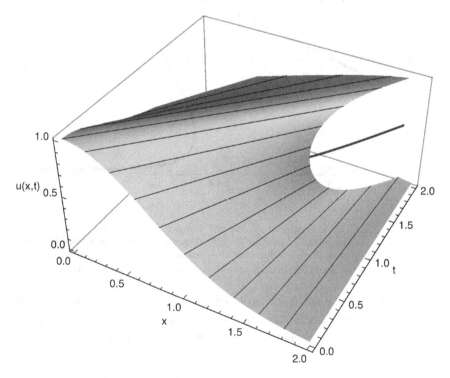

Fig. 11.5 A plot of the solution surface to the inviscid Burgers' equation with initial condition $u(x,0) = e^{-x^2}$. A shock forms at $t = \sqrt{e/2}$ and to avoid becoming multivalued the solution surface must develop a discontinuity across the shock.

the discontinuity across the shock, after the shock forms at $t = \sqrt{e/2}$. The type of solution presented in Fig. 11.5 is known as an **entropy solution**. The term "entropy" comes from the field of thermodynamics and implies that solutions may be lost along shocks.

Suppose the shock is located at position x_s (the s-subscript denoting shock) at time t. The integral of u is conserved even when passing through the shock and since the shock has infinitesimal width, the definite integral must be zero. This observation yields a method for determining the value of the discontinuous solution to the inviscid Burgers' equation along a shock. Note that in Fig. 11.4 the function $u(x,t)$ is multivalued for an interval on the x-axis. The correct value of u can be found by placing the shock (the vertical line labeled x_s in Fig. 11.4) at such as position that the shaded area to the right of x_s and left of the curve $u(x,t) = f(x - u(x,t)t)$ equals the shaded area to the left of x_s and right of the curve. Since one region is on either side of the vertical line and the regions are of equal area, their sum is 0. The value of u (labeled u_2 in Fig. 11.4) where x_s intersects the curve $u(x,t) = f(x - u(x,t)t)$ and which separates the left and right regions is then the "correct" value of $u(x,t)$.

More features of the shock can be determined as well. The position of the shock is a function of time. Thus by choosing $a < x_s$ and $b > x_s$ in Eq. (11.10) and setting $\nu = 0$, in the limit as $a \to x_s^-$ and $b \to x_s^+$,

$$u(x_s+,t)\frac{dx_s}{dt} - \frac{1}{2}(u(x_s+,t))^2 = u(x_s-,t)\frac{dx_s}{dt} - \frac{1}{2}(u(x_s-,t))^2$$

$$\frac{dx_s}{dt} = \frac{1}{2}(u(x_s+,t) + u(x_s-,t))$$

where $u(x_s+,t)$ and $u(x_s-,t)$ denote the values of the solution. Therefore the shock moves at a velocity which is the average of the values of u immediately on either side of the shock.

This section concludes with an examination of the relationship between the solution to the viscous Burgers' equation, the formula for which was given in Eq. (11.25), and the solution to the inviscid Burgers' equation. More details relevant to this discussion and more examples may be found in [Whitham (1974)]. To see the connection between the entropy solution to Eq. (11.26) and the solution to Eq. (11.7) as $\nu \to 0^+$, let $u_\nu(x,t)$ be the solution to Eq. (11.7) and let $u(x,t)$ be the entropy solution to Eq. (11.26) with the same initial conditions, $u(x,0) = f(x)$. During this discussion (x,t) will remain fixed. The solution to the viscous Burgers' equation given by Eq. (11.25) is single-valued and continuous for all t while the solution to Eq. (11.26) may be discontinuous and contain shocks depending on the initial condition $f(x)$. The limit $\nu \to 0^+$ should produce the same shock structure and values as the viscous solution.

The solution $u_\nu(x,t)$ is given by Eq. (11.25). To study the limit as $\nu \to 0^+$, consider the dominant contributions to the improper integrals in the numerator and denominator of Eq. (11.25). Since these integrals converge, the presence of the term,

$$e^{-\frac{1}{2\nu}\left(\int_0^y f(s)\,ds + (x-y)^2/(2t)\right)}$$

in both integrals suggests the dominant contributions occur in intervals where the function $h(y) \equiv \int_0^y f(s)\,ds + (x-y)^2/(2t)$ reaches a minimum. For a fixed x, $h(y)$ reaches a minimum when

$$h'(y) = f(y) - \frac{x-y}{t} = 0, \tag{11.28}$$

and $h''(y) = f'(y) + 1/t > 0$. For fixed (x,t), let ξ be a solution to Eq. (11.28) and call ξ a **stationary point**. The contribution to the improper integral from a neighborhood of the stationary point can be found by Laplace's method (sometimes called the method of steepest descent). Using Taylor's[8] theorem to expand about ξ implies

$$h(y) \approx h(\xi) + \frac{1}{2}\left(f'(\xi) + \frac{1}{t}\right)(y - \xi)^2,$$

where the truncation error is $O((y-\xi)^3)$. Thus the contribution to the improper integral from a neighborhood of a stationary point is

$$\int_{-\infty}^{\infty} e^{-h(y)/(2\nu)}\,dy \approx e^{-h(\xi)/(2\nu)} \int_{-\infty}^{\infty} e^{-\frac{1}{2\nu}\left(f'(\xi)+\frac{1}{t}\right)(y-\xi)^2/2}\,dy$$

$$= \sqrt{\frac{4\pi\nu t}{1 + tf'(\xi)}}\, e^{-h(\xi)/(2\nu)} = \sqrt{\frac{4\pi\nu}{h''(\xi)}}\, e^{-h(\xi)/(2\nu)}.$$

Likewise, in a neighborhood of a stationary point,

$$\int_{-\infty}^{\infty} \frac{x-y}{t}\, e^{-h(y)/(2\nu)}\,dy \approx \frac{x-\xi}{t}\sqrt{\frac{4\pi\nu}{h''(\xi)}}\, e^{-h(\xi)/(2\nu)}.$$

If there is only one stationary point then substituting these two approximations into Eq. (11.25) gives the approximation

$$u(x,t) \approx \frac{x-\xi}{t} = f(\xi) \tag{11.29}$$

---

[8]Brook Taylor, English mathematician (1685–1731).

using Eq. (11.28). Treating the approximation in Eq. (11.29) as an equation and solving for x yield the pair of equations:

$$x = \xi + f(\xi)t,$$
$$u = f(\xi).$$

Note that this pair of equations is merely the parametric representation of the characteristics for the inviscid Burgers' equation. The stationary point is the parameter in this representation.

Now suppose there are two stationary points, ξ_3 and ξ_1 with $\xi_3 < \xi_1$. The dominant contributions to the improper integrals in Eq. (11.25) come from both ξ_3 and ξ_1, thus

$$u(x,t) \approx \frac{(x - \xi_3)\frac{e^{-h(\xi_3)/(2\nu)}}{\sqrt{h''(\xi_3)}} + (x - \xi_1)\frac{e^{-h(\xi_1)/(2\nu)}}{\sqrt{h''(\xi_1)}}}{t\left(\frac{e^{-h(\xi_3)/(2\nu)}}{\sqrt{h''(\xi_3)}} + \frac{e^{-h(\xi_1)/(2\nu)}}{\sqrt{h''(\xi_1)}}\right)}. \tag{11.30}$$

In the case where $h(\xi_3) < h(\xi_1)$ then $h(\xi_1)/(2\nu)$ is very much larger than $h(\xi_3)/(2\nu)$ as $\nu \to 0^+$. The argument above shows that as ν approaches zero from the right $u(x,t) \approx (x - \xi_3)/t$. When $h(\xi_3) > h(\xi_1)$ then $u(x,t) \approx (x - \xi_1)/t$ as $\nu \to 0^+$. The case of $h(\xi_3) = h(\xi_1)$ implies

$$\int_0^{\xi_3} f(s)\, ds + \frac{(x - \xi_3)^2}{2t} = \int_0^{\xi_1} f(s)\, ds + \frac{(x - \xi_1)^2}{2t}.$$

By Eq. (11.28), which defines a stationary point, $t f(\xi_i) = x - \xi_i$ for $i = 1, 2$ and thus,

$$\int_0^{\xi_1} f(s)\, ds - \int_0^{\xi_3} f(s)\, ds$$
$$= \frac{(x - \xi_3)^2}{2t} - \frac{(x - \xi_1)^2}{2t}$$
$$= \frac{f(\xi_3)}{2}(x - \xi_3) - \frac{f(\xi_1)}{2}(x - \xi_1)$$
$$= \frac{f(\xi_3)}{2}(tf(\xi_1) + \xi_1 - \xi_3) - \frac{f(\xi_1)}{2}(tf(\xi_3) + \xi_3 - \xi_1)$$
$$\int_{\xi_3}^{\xi_1} f(s)\, ds = \frac{1}{2}\left(f(\xi_3) + f(\xi_1)\right)(\xi_1 - \xi_3). \tag{11.31}$$

Readers will recognize that right-hand side of Eq. (11.31) as the trapezoidal rule for approximating a definite integral. See Fig. 11.6. Hence the area under the chord connecting (ξ_3, u_3) to (ξ_1, u_1) is the value of the definite integral in Eq. (11.31). This implies the area under the curve $u = f(x)$

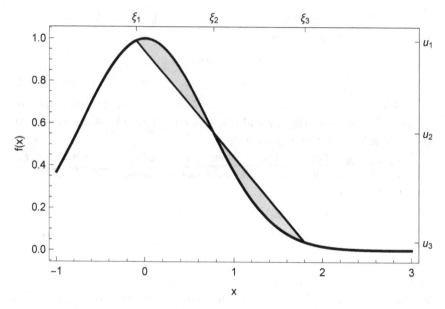

Fig. 11.6   The points on the solution curve $u(x, t) = f(x - u(x, t)t)$ of the inviscid Burgers' equation are shifted horizontally to the left by an amount $u(x, t)t$ to produce the initial data. Thus the breaking wave of Fig. 11.4 is shifted into the initial condition. The vertical line which denoted the position of the shock becomes the chord seen in this figure. The shaded areas are equal. For the purpose of drawing this curve, the function $f(x) = e^{-x^2}$.

and above the chord over the interval $[\xi_3, \xi_2]$ equals the area under the chord and above the curve in the interval $[\xi_2, \xi_1]$. This equal area notion was encountered earlier when determining the value of the solution to the inviscid Burgers' equation along the shock. Equation (11.31) provides a way to determine the value of the solution to the inviscid Burgers' equation along a shock. If $\xi_3 < \xi_1$ are values of x such that Eq. (11.31) is satisfied then the point (ξ_2, u_2) where the chord connecting (ξ_3, u_3) to (ξ_1, u_1) intersects the graph of $u = f(x)$ provides the value u_2 of the solution to the inviscid Burgers' equation on the shock. Since the stationary points ξ_1 and ξ_3 are functions of (x, t) then the direction of the inequality between $h(\xi_1)$ and $h(\xi_3)$ causes the value of $u(x, t)$ to abruptly "switch" when crossing the shock. This analysis has assumed that the initial condition is a "one-humped" function, but the idea can be generalized.

11.3 The Korteweg-de Vries Equation

The **Korteweg-de Vries** (often abbreviated, **KdV**) equation is another important nonlinear partial differential equation and may be one of the most studied nonlinear partial differential equations. Historically, the Korteweg-de Vries equation, named after Korteweg and de Vries, originated from the study of waves in shallow waters. The investigation started with observations made by John Scott Russell[9] during experiments to determine the most efficient hull design for canal boats. He reported [Russell (1845)]:

> I was observing the motion of a boat which was rapidly drawn along a narrow channel by a pair of horses, when the boat suddenly stopped — not so the mass of water in the channel which it had put in motion; it accumulated round the prow of the vessel in a state of violent agitation, then suddenly leaving it behind, rolled forward with great velocity, assuming the form of a large solitary elevation, a rounded, smooth and well-defined heap of water, which continued its course along the channel apparently without change of form or diminution of speed. I followed it on horseback, and overtook it still rolling on at a rate of some eight or nine miles an hour, preserving its original figure some thirty feet long and a foot to a foot and a half in height. Its height gradually diminished, and after a chase of one or two miles I lost it in the windings of the channel. Such, in the month of August 1834, was my first chance interview with that singular and beautiful phenomenon which I have called the "Wave of Translation".

However, the Korteweg-de Vries equation was not formulated using a mathematically rigorous argument until the work of Korteweg and de Vries in 1845 and the true importance of the Korteweg-de Vries equation was not fully realized until the last half of century. This recognition began with a discovery in a numerical simulation carried out by Zabusky[10] and Kruskal[11] in 1965 of a self-enforcing solitary wave that maintains its shape as it propagates with constant velocity. Such a solution is called a soliton solution, see [Zabusky and Kruskal (1965)]. The discovery of soliton solutions of the Korteweg-de Vries equation shed light on many physical phenomena that could not be explained before. Readers are referred to [de Jager (2011)] for the derivation of the Korteweg-de Vries equations.

---

[9]John Scott Russell, Scottish engineer (1808–1882).
[10]Norman J. Zabusky, American physicist (1929–)
[11]Martin D. Kruskal, American mathematician and physicist (1925–2006).

There are several commonly used equivalent formulations of the Korteweg-de Vries equation:

$$u_t + 6u\,u_x + u_{xxx} = 0, \tag{11.32}$$

$$u_t + u\,u_x + k\,u_{xxx} = 0, \tag{11.33}$$

$$u_t - 6u\,u_x + u_{xxx} = 0, \tag{11.34}$$

$$u_t + u\,u_x + u_{xxx} = 0, \tag{11.35}$$

$$u_t + \alpha u\,u_x + \beta u_{xxx} = 0. \tag{11.36}$$

All of these forms are equivalent and may be obtained from each other by rescaling the independent and dependent variables. Suppose $x = A\,X$, $t = B\,T$, and $u = C\,V$ then

$$u_t + 6u\,u_x + u_{xxx} = \frac{C}{B}V_T + \frac{6C^2}{A}V\,V_X + \frac{C}{A^3}V_{XXX} = 0,$$

which is equivalent to

$$V_T + \frac{6B\,C}{A}V\,V_X + \frac{B}{A^3}V_{XXX} = 0.$$

By choosing the constants A, B, and C appropriately, any form of the Korteweg-de Vries equation given in Eqs. (11.33)–(11.36) can be obtained. The details are left for the reader in Exercise 14.

Mathematically, the Korteweg-de Vries equation describes the interaction among the temporal evolution, the nonlinearity, and the dispersion effect (the triple partial derivative term). Solitons are created as a result of this complicated interaction, especially between the nonlinear term and dispersion effects in the medium. A study of the Korteweg-de Vries equation can easily fill several monographs, but the remainder of this section will discuss only the existence of the simplest solitons.

Just as was the case for Burgers' equation, the investigation starts with determining if Eq. (11.32) has a solution of the form given in Eq. (11.1). Setting $\xi = x - ct$ and substituting $u(x,t) = U(\xi)$ into the Korteweg-de Vries equation produce

$$\frac{d^3U}{d\xi^3} + 6U\frac{dU}{d\xi} - c\frac{dU}{d\xi} = 0.$$

Integrating both sides with respect to ξ yields

$$\frac{d^2U}{d\xi^2} + 3U^2 - c\,U = A,$$

where A is an arbitrary constant of integration. To further simplify the problem, assume that $U \to 0$, $U' \to 0$, and $U'' \to 0$ as $\xi \to \infty$ and as $\xi \to -\infty$, then

$$\frac{d^2U}{d\xi^2} + 3U^2 - c\,U = 0.$$

This is a second-order (nonlinear) ordinary differential equation with missing first derivative U'. Multiplying the ordinary differential equation by $U' = dU/d\xi$ and integrating both sides of the resulting equation yield

$$\frac{1}{2}\left(\frac{dU}{d\xi}\right)^2 + U^3 - \frac{1}{2}cU^2 = B,$$

where B is again an arbitrary constant of integration. The assumptions that $U \to 0$ and $U' \to 0$ as $\xi \to \infty$ and $\xi \to -\infty$ imply that $B = 0$, and therefore, after solving for $dU/d\xi$,

$$\frac{dU}{d\xi} = \pm U\sqrt{c - 2U}, \tag{11.37}$$

where the $\pm$ sign should be chosen according to the sign of $dU/d\xi$. In the following, the $\pm$ sign will be dropped since it makes no difference in the integration steps to follow. To solve for U, separate the variables in Eq. (11.37) and integrate,

$$\int \frac{1}{U\sqrt{c - 2U}} dU = \xi + A, \tag{11.38}$$

where A is again a constant. To evaluate the integral on the left side, make a change of variable which, while not obvious, is motivated by identities from hyperbolic trigonometry,

$$U = \frac{c}{2}\operatorname{sech}^2\zeta \implies dU = -c\frac{\sinh\zeta}{\cosh^3\zeta}d\zeta.$$

Performing the substitution in Eq. (11.38) results in

$$\int \frac{1}{U\sqrt{c - 2U}} dU = \int \frac{2}{c\operatorname{sech}^2\zeta\sqrt{c - c\operatorname{sech}^2\zeta}}\frac{-c\sinh\zeta}{\cosh^3\zeta}d\zeta$$

$$= -\frac{2}{\sqrt{c}}\int 1\,d\zeta = -\frac{2}{\sqrt{c}}\zeta = -\frac{2}{\sqrt{c}}\operatorname{sech}^{-1}\sqrt{\frac{2U}{c}}.$$

Thus the equation,

$$-\frac{2}{\sqrt{c}}\operatorname{sech}^{-1}\sqrt{\frac{2U}{c}} = \xi + A$$

is solved for $U = U(\xi)$ to give

$$U(\xi) = \frac{c}{2}\operatorname{sech}^2\left(\frac{\sqrt{c}}{2}\xi + A\right).$$

Therefore, the Korteweg-de Vries equation has the following traveling wave solution:

$$u(x, t) = \frac{c}{2}\operatorname{sech}^2\left(\frac{\sqrt{c}}{2}(x - ct) + A\right).$$

The presence of the term $\sqrt{c}$ and the desire to find a real-valued solution, suggest that $c > 0$. If the constant c is replaced by $4c^2$ for convenience, then

$$u(x,t) = 2c^2 \operatorname{sech}^2(c(x - 4c^2t) + A). \tag{11.39}$$

The solution given by Eq. (11.39) is a soliton solution of the Korteweg-de Vries equation. Note that this soliton solution is positive for all x and t, for each fixed t has a single maximum, and $u(x,t) \to 0$ as $|x| \to \infty$. These features are "particle like". Thus the name "soliton" was used to describe such traveling wave solutions of nonlinear partial differential equations. In addition, all such solutions travel to the right with a speed depending on the amplitude of the solution. A solution with larger amplitude travels faster with a speed equal to twice its amplitude as demonstrated in Fig. 11.7.

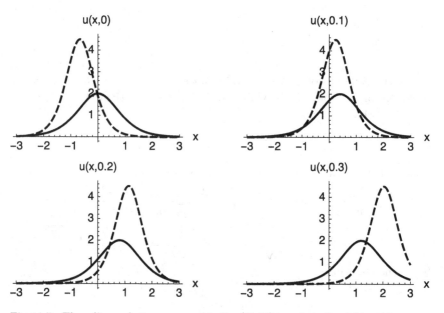

Fig. 11.7   The soliton solution expressed in Eq. (11.39) travels to the right as t increases at a wave speed twice the amplitude on the wave. Two solitons are graphed above. The dashed soliton has $c = 3/2$ and $A = 1$ while the solid curve has $c = 1$ and $A = 0$. It is apparent the dashed soliton is traveling faster to the right than the solid curve.

The soliton solutions given by Eq. (11.39) are not the only traveling wave solutions of the Korteweg-de Vries equation. There are many other solitons and solitary wave solutions of the Korteweg-de Vries equation. For example, if the assumption that the traveling wave solution and its derivatives vanishes at infinity is relaxed, then the function $U(\xi)$ satisfies:

$$\frac{1}{2}\left(\frac{dU}{d\xi}\right)^2 = -U^3 + \frac{c}{2}U^2 + AU + B \equiv F(U) \qquad (11.40)$$

instead of Eq. (11.37), where A and B are arbitrary constants. Equation (11.40) has real solutions if and only if $F(U)$ is positive. Since the function $F(U)$ is a cubic polynomial (with a negative leading coefficient), it may have one or three real zeros. From the properties of the zeros of $F(U)$, the conditions under which Eq. (11.40) has real solutions and the properties of the solutions can be determined. For example, if $F(U)$ has three distinct real zeros, say $U_1 < U_2 < U_3$, then Eq. (11.40) has real bounded solutions if the initial value is between U_2 and U_3 and all such solutions oscillate between U_2 and U_3. In fact, the solution in this case can be represented by elliptic integrals and leads to traveling wave solutions of the Korteweg-de Vries equation, called **cnoidal waves**. For details, see [Drazin and Johnson (1989)]. The soliton solutions given by Eq. (11.39) are called one solitons to indicate they have one peak. There are also two-solitons, three-solitons, and more generally, N-solitons. For a more complete treatment of the topic, see [Drazin and Johnson (1989)] or [Newell (1985)].

One of the most prominent features of soliton solutions is their remarkable stability property. Soliton solutions maintain their form even after two solitons propagating at different speeds collide. The waves seem to simply pass through each other and each keeps its wave profile with only a slight phase shift in wave position (the peak of the wave profile). Consider the initial value problem of the Korteweg-de Vries equation with initial condition,

$$u(x,0) = \frac{9}{2}\operatorname{sech}^2\left(\frac{3}{2}x + 8\right) + \frac{9}{8}\operatorname{sech}^2\left(\frac{3}{4}x\right).$$

If the Korteweg-de Vries equation was linear then the solution to the initial value problem would be

$$u(x,t) = \frac{9}{2}\operatorname{sech}^2\left(\frac{3}{2}(x - 9t) + 8\right) + \frac{9}{8}\operatorname{sech}^2\left(\frac{3}{4}(x - 9t/4)\right).$$

However, due to the nonlinearity of the Korteweg-de Vries equation, the solitons interact in such a way as to produce a phase shift so that for t much larger than the time of interaction,

$$u(x,t) \approx \frac{9}{2}\operatorname{sech}^2\left(\frac{3}{2}(x - 9t) + 8 + \theta_1\right) + \frac{9}{8}\operatorname{sech}^2\left(\frac{3}{4}(x - 9t/4) + \theta_2\right),$$

for some θ_1 and θ_2 [Tao (2009)]. Figure 11.8 shows the interaction of the two solitons over time. The overall shapes of the two solitons appear unchanged after the faster wave passes the slower wave. The shift in phase of the solitons is illustrated in Fig. 11.9. The solid curve is the actual solution of the initial value problem while the dashed curve shows the profile of the sum of the two solitons if there had been no interaction between them. For more details, see [Toda (1989)] or [Drazin and Johnson (1989)].

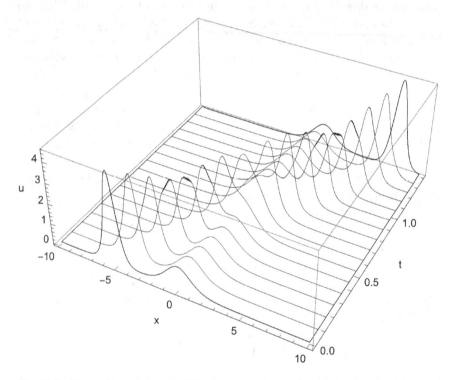

Fig. 11.8   Two solitons interact without constructive or destructive interference. The shapes of the solitons are largely unchanged after the faster soliton passes the slower soliton. The stability of solitons is one of their interesting properties.

11.4   The Nonlinear Schrödinger Equation

This section introduces the nonlinear Schrödinger equation (NLS) also referred to as the Schrödinger wave equation since it describes how the wave function of a particle evolves over time. The wave function of a particle when multiplied by its complex conjugate can be regarded as the probability

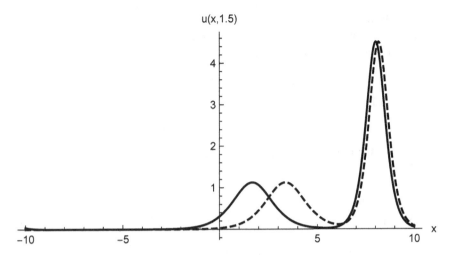

u(x,1.5)

Fig. 11.9   When solitons interact they experience a phase shift. The solid curve depicts the profile of the solution to an initial value problem in which interaction takes place while the dashed curve shows the sum of the solitons had there been no interaction.

density function for the location of a particle at any given time. The Schrödinger equation is the fundamental equation of quantum mechanics, just as Newton's second law of motion is the foundation of classical mechanics. However, along with technological advances, the Schrödinger equation has become even more important due to its connection with the way that signals are propagated in fiber optics, see [Stolen (2008)] and [Agrawal (2013)]. This section will only discuss the existence of soliton solutions to the Schrödinger equation in the one dimensional case.

The standard form of the nonlinear Schrödinger equation in one spatial dimension is

$$i\,\psi_t + \psi_{xx} + \gamma|\psi|^2\psi = 0, \tag{11.41}$$

where $i = \sqrt{-1}$ is the imaginary unit and ψ is a complex-valued function of t and x. To motivate the approach below, assume Eq. (11.41) has a plane wave solution of the form

$$\psi(x,t) = A\,e^{i(kx-\omega t)}, \tag{11.42}$$

where A, k, and ω are constants to be determined. The real or the complex part of $\psi(x,t)$ could have been chosen as the plane wave solution as well. One motivation to look for solutions in the form above is that the method of undetermined coefficients familiar from the study of ordinary differential equation may enable the constants to be found. As mentioned in Sec. 11.1, the constant k in Eq. (11.42) is called the wave number and ω is the frequency. Substituting the plane wave function defined in Eq. (11.42) for ψ in Eq. (11.41) produces the equation

$$i(-i\,\omega) + i^2 k^2 + \gamma A^2 = 0,$$

which can be simplified as

$$\omega = k^2 - \gamma A^2. \tag{11.43}$$

Therefore, for any k, if ω is defined by Eq. (11.43), the function $\psi(x,t)$ is a solution to the nonlinear Schrödinger equation. Equation (11.43) determines the relationship among the velocity of the solution, the wave number, and the amplitude.

A more general solution to the nonlinear Schrödinger equation takes the form

$$\psi(x,t) = f(\xi)e^{i(kx-\omega t)} \tag{11.44}$$

where $\xi = x - c\,t$ and the symbols k, c, and ω are constants. Note that this solution incorporates features of a traveling wave and a plane wave. Substituting Eq. (11.44) into the nonlinear Schrödinger equation of Eq. (11.41) yields

$$0 = -c\,if' + \omega f + f'' + 2i\,kf' - k^2 f + \gamma f^3$$
$$= f'' + i(2k - c)f' + (\omega - k^2)f + \gamma f^3.$$

If f is a real-valued function of ξ, the term in the equation above involving i must vanish, which implies $c = 2k$ and the equation simplifies to

$$f'' + (\omega - k^2)f + \gamma f^3 = 0.$$

Multiplying the equation above by $2f'$ and antidifferentiating both sides yield

$$(f')^2 = A - (\omega - k^2)(f)^2 - \frac{\gamma}{2}(f)^4, \tag{11.45}$$

where A is an arbitrary constant of integration. If the solution $\psi(x,t)$ and its derivative $\psi_x(x,t)$ vanish as $x \to \pm\infty$, then $f \to 0$ and $f' \to 0$ as $x \to \pm\infty$. This implies $A = 0$ and function f satisfies

$$(f')^2 = (k^2 - \omega)(f)^2 - \frac{\gamma}{2}(f)^4.$$

Solving for f' produces

$$\frac{df}{d\xi} = \pm f\sqrt{(k^2 - \omega) - \frac{\gamma}{2}(f)^2}.$$

Separating the variables in this first-order nonlinear ordinary differential equation and integrating both sides imply (taking the negative root),

$$\xi + A = \int \frac{-1}{f\sqrt{(k^2 - \omega) - \frac{\gamma}{2}(f)^2}} \, df$$

where A is an arbitrary constant of integration. The indefinite integral here is similar to the integral seen in Eq. (11.38) during the discussion of the Korteweg-de Vries equation. A similar integration by substitution step can be used here. Making the change of variable,

$$f = \sqrt{\frac{2(k^2 - \omega)}{\gamma}} \operatorname{sech} \zeta \implies df = -\sqrt{\frac{2(k^2 - \omega)}{\gamma}} \operatorname{sech} \zeta \tanh \zeta \, d\zeta$$

enables the indefinite integral to be rewritten as

$$\int \frac{-1}{f\sqrt{(k^2 - \omega) - \frac{\gamma}{2}(f)^2}} \, df = \int \frac{1}{\sqrt{k^2 - \omega}} \, d\zeta = \frac{1}{\sqrt{k^2 - \omega}} \zeta$$

$$= \frac{1}{\sqrt{k^2 - \omega}} \operatorname{sech}^{-1} \left(\sqrt{\frac{\gamma}{2(k^2 - \omega)}} f \right).$$

Hence

$$\frac{1}{\sqrt{k^2 - \omega}} \operatorname{sech}^{-1} \left(\sqrt{\frac{\gamma}{2(k^2 - \omega)}} f \right) = \xi + A$$

or solving for $f(\xi)$,

$$f(\xi) = \sqrt{\frac{2(k^2 - \omega)}{\gamma}} \operatorname{sech}((k^2 - \omega)^{1/2}(\xi + A)). \tag{11.46}$$

Therefore, substituting Eq. (11.46) into Eq. (11.44) yields the following solution of the nonlinear Schrödinger equation,

$$\psi(x, t) = \sqrt{\frac{2(k^2 - \omega)}{\gamma}} e^{i(kx - \omega t)} \operatorname{sech}((k^2 - \omega)^{1/2}(x - 2kt + A)). \tag{11.47}$$

The solution found above is called an **envelope soliton solution** of the nonlinear Schrödinger equation. There are many other interesting solutions of the nonlinear Schrödinger equation. In fact, if the assumption that $\psi(x, t)$ vanishes as $x \to \pm\infty$ is dropped, then the function f may satisfy the more general equation given by Eq. (11.45). In this case, the solution will be represented by elliptic functions of the first kind. The details of this case are beyond the scope of this text. Readers interested in exploring more of this case may consult [Hasegawa and Matsumoto (2012)].

11.5   Further Comments

Nonlinear partial differential equations arise from a great variety of fields ranging from electrostatics, fluid mechanics, and quantum mechanics to signal processing, finance and many more. Not surprisingly, the study of nonlinear partial differential equations has drawn a great amount of attention from many researchers. The material presented in this chapter is merely a brief introduction, which touched only upon a tiny part of this huge area. Only the one-dimensional cases of Burgers' equation, the Korteweg-de Vries equation, and the nonlinear Schrödinger equation are discussed and the discussion is, in large part, concentrated on the existence and basic properties of traveling wave solutions of the partial differential equations. There are many more important nonlinear partial differential equations. One of these equations is the **sine-Gordon equation** which originated from differential geometry and relativistic field theory. This equation, along with several solution techniques, was known even in the nineteenth century. However, it has gained much attention since the 1970s with the discovery that it possesses soliton solutions (the so-called "kink" and "antikink"). Readers will be asked to explore the existence of such solutions in the exercises using the techniques presented in this chapter, see Exercise 12.

Even though there seems to be no general theory for solving nonlinear partial differential equations, there are methods that can be employed to solve certain types of nonlinear partial differential equations. For example, the initial value problems of Burgers' equation can be solved using the Cole-Hopf transformation and certain traveling wave solutions of some nonlinear partial differential equations can be obtained by solving the corresponding ordinary differential equations. The inverse scattering transformation and the Bäcklund[12] transformation stand out among other well-known methods. The inverse scattering method originated from quantum mechanics in the 1930s. The method was applied to solve Korteweg-de Vries equation in 1967, see [Gardener (1967)] and later it was applied to other nonlinear partial differential equations. The inverse scattering transformation represents one of the most important advances in the study of nonlinear partial differential equations. The method of the inverse scattering transformation sometimes is called the nonlinear Fourier transform method since, in some generalized sense, the method is analogous to the method of the Fourier

---

[12]Albert Victor Bäcklund, Swedish mathematician (1845–1922).

transform used to solve linear partial differential equations. The Bäcklund transformation has its origins in differential geometry and may be considered as a transformation among solution surfaces. The basic idea is, through such a transformation, the solutions of a nonlinear partial differential equation can be connected with those of another equation whose solutions can be found and/or better understood. Readers interested in studying these and other methods can find more details in [Ablowitz (1981)] and [Rogers (1982)], or [Debnath (2012)] and the references therein.

11.6 Exercises

(1) Consider the linear partial differential equation,

$$\frac{\partial^2 u}{\partial x^2} + \alpha \frac{\partial u}{\partial t} + \beta \frac{\partial u}{\partial x} = 0.$$

Find all bounded traveling wave solutions of the equation.

(2) The partial differential equation

$$\frac{\partial^2 u}{\partial t^2} - \alpha^2 \frac{\partial^2 u}{\partial x^2} + u = 0$$

is called a (linear) Klein[13]-Gordon[14] equation, which can be considered as a modification of the wave equation. Find all bounded traveling wave solutions of the equation.

(3) Consider the linear partial differential equation,

$$\frac{\partial^3 u}{\partial x^3} + \alpha \frac{\partial u}{\partial t} + \beta \frac{\partial u}{\partial x} = 0.$$

Find all the bounded traveling wave solutions of the equation.

(4) Assume the following equations have plane wave solutions of the form $u(x,t) = A e^{i(kx - \omega t)}$ and find the dispersion relation for each equation.

   (a) The linear Klein-Gordon equation: $u_{tt} + u = \alpha^2 u_{xx}$.

   (b) The beam equation: $u_{tt} + \alpha^2 u_{xxxx} = 0$.

   (c) The linear Korteweg-de Vries equation: $u_t + \alpha u_x + \beta u_{xxx} = 0$.

(5) An example of the bistable equation is

$$u_t - u_{xx} = -u + H(u - 1/2), \qquad (11.48)$$

where $H(z)$ is the Heaviside function. Find the ordinary differential equation satisfied by a traveling wave solution to Eq. (11.48).

---

[13]Oskar Benjamin Klein, Swedish theoretical physicist (1894–1977).
[14]Walter Gordon, German theoretical physicist (1893–1939).

(6) Consider the ordinary differential equation in Eq. (11.13).

    (a) Show that if the equation $U^2 - 2cU - 2A = 0$ has complex roots Eq. (11.13) has no bounded solutions.

    (b) Show that if the equation $U^2 - 2cU - 2A = 0$ has a repeated real root, the only bounded solution to Eq. (11.13) is constant.

(7) Verify graphically if Eq. (11.13) has two distinct equilibrium solutions, say $U_1 < U_2$, that U_1 is stable and U_2 is unstable. Plot several typical integral curves.

(8) Show that if a function of the form given in Eq. (11.15) is a traveling wave solution to Burgers' equation as given in Eq. (11.7), then $c = (U_1 + U_2)/2$.

(9) A slight generalization of Burgers' equation is the following:

$$u_t + (F(u))_x - \nu\, u_{xx} = 0.$$

Assume that $F(u)$ is a smooth function with $F''(u) > 0$ for all u. Derive the ordinary differential equation satisfied by the traveling wave solution $u(x,t) = U(x - ct)$ and discuss the solution properties of the ODE.

(10) Show that Fisher's[15] equation,

$$u_t - u_{xx} = \gamma\, u \left(1 - \frac{u}{K} \right)$$

can be written in the nondimensional form,

$$u_t - u_{xx} = u(1 - u)$$

and derive the ordinary differential equation satisfied by $U(\xi) = U(x - ct)$ if $U(x - ct) = u(x,t)$ is a traveling wave solution of Fisher's equation (in the nondimensional form).

(11) Show that if the initial condition of the inviscid Burgers' equation is $f(x) = e^{-x^2}$, a shock wave forms at $t = \sqrt{e/2}$.

(12) The sine-Gordon equation is another well-known nonlinear partial differential equation:

$$\frac{\partial^2 u}{\partial t^2} - \frac{\partial^2 u}{\partial x^2} + \sin u = 0. \tag{11.49}$$

Let c be any constant and $\gamma = 1/\sqrt{1 - c^2}$. Show that the sine-Gordon equation is invariant under the change of variables:

$$\xi = \gamma(x - ct) \text{ and } \eta = \gamma(t - cx).$$

---

[15]Ronald Aylmer Fisher, English statistician (1890–1962).

(13) Consider the sine-Gordon equation (SGE) shown in Eq. (11.49).

    (a) Assume that the sine-Gordon equation has a traveling wave solution in the form of $u(x,t) = U(x - ct)$ for some constant c and derive the ordinary differential equation for U.

    (b) Assume $U \to 0$ and $U' \to 0$ as $\xi \to \pm\infty$ and solve the ordinary differential equation in the previous part.

(14) Let $x = AX$, $t = BT$, and $u = CV$ and show that the alternative forms of the Korteweg-de Vries equation given in Eqs. (11.33)–(11.36) can be derived from Eq. (11.32) by the appropriate choices of constants A, B, and C.

(15) Show that, if $u(x,t)$ is a solution of the Korteweg-de Vries equation, then

$$w(x,t) = A^2 u(Ax + 6ABt + C, A^3 t + D) - B$$

is also a solution.

(16) Let $c > 0$ and A be any constant. Verify directly that

$$u(x,t) = \frac{c}{2} \operatorname{sech}^2\left(\frac{\sqrt{c}}{2}(x - ct) + A\right)$$

is a solution the Korteweg-de Vries equation (found in Eq. (11.32)) and use technology to graph the function for various values of c as a function of x with different values of t.

(17) Let $f(t)$ be any continuous function on $(-\infty, \infty)$. Show that if $v(x,t)$ is a solution of the equation,

$$v_t + 3(v_x)^2 + v_{xxx} = f(t),$$

then $w(x,t) = v(x,t) - \int_0^t f(s)\, ds$ is a solution to

$$u_t + 3(u_x)^2 + u_{xxx} = 0$$

and that $u = v_x$ and $u = w_x$ are solutions of the Korteweg-de Vries equation (Eq. (11.32)).

(18) The partial differential equation,

$$v_t - 6v^2 v_x + v_{xxx} = 0 \qquad (11.50)$$

is called the modified Korteweg-de Vries equation. Assume that $V \to 0$, $V' \to 0$, and $V'' \to 0$ as $|x| \to \infty$ and follow the procedure of finding the solution to the Korteweg-de Vries equation to show that

$$v(x,t) = -\operatorname{csch}(x - t)$$

is a solution of the equation.

(19) Assume that $V \to A$ (a constant), $V' \to 0$, and $V'' \to 0$ as $|x| \to \infty$ for any fixed t in the modified Korteweg-de Vries equation (see Exercise 18) and show that

$$v(x,t) = A + \frac{2}{e^{-(x+(6A^2-1)t)} - 4A + (4A^2 - 1)e^{x+(6A^2-1)t}}$$

is a solution of the equation.

(20) Show that if $v(x,t)$ is a solution of the modified Korteweg-de Vries equation (Eq. (11.50)), then $u = v^2 + v_x$ is a solution of the Korteweg-de Vries equation (Eq. (11.32)).

(21) The Boussinesq[16] equation is the following:

$$u_{tt} - \alpha^2 u_{xx} - \beta\, u_{xxxx} - \gamma(u^2)_{xx} = 0. \tag{11.51}$$

Assume that $u(x,t) = U(x - ct)$ is a solution of the Boussinesq equation and that $\xi = x - ct$. Find the ordinary differential equation that $U(\xi)$ satisfies.

---

[16] Joseph Valentin Boussinesq, French mathematician (1842–1929).

Chapter 12

Numerical Solutions to PDEs Using Finite Differences

The reader has reached another chapter of optional material intended to give a more complete picture of methods for finding solutions to partial differential equations. In this chapter, a numerical approach to approximating the solutions to partial differential equations will be introduced, the **finite difference method**. A proper study of this method would occupy the space of several textbooks and require a few semesters of study, so this chapter will merely introduce the method and illustrate its use on a sampling of examples. Readers interested in pursuing this topic further will find a wealth of printed and web-based resources. In particular the textbooks [Smith (1985)] and [Strikwerda (2004)] are good starting points. Source code files written in the Java programming language [Schildt (2014)] for the examples presented in this chapter are available for download from the website supporting this textbook. Java was chosen, not because of any special suitability for implementing finite difference methods, but because the language and tools required to develop, compile, debug, and run programs written in it are ubiquitous and generally available for free.

The finite difference method is often said to "solve" a partial differential equation, though in truth it approximates the solution. Thus some attention will be given to measures of accuracy of the algorithms presented here. Often the finite difference method approximation will be compared to an exact solution which can be found using the methods of the previous chapters. Numerical methods are most useful for approximating the solutions to boundary value and initial boundary value problems which have no closed form, analytical solution. Proper discussion of the finite difference method must be conducted using the vocabulary and concepts of **linear algebra**. Readers who have completed an introductory course in linear algebra or numerical analysis may wade into the remainder of this chapter.

Readers needing a brief introduction to the concepts of linear algebra used in this chapter should consult App. B. Finally, the finite difference method is not the only numerical scheme for approximating the solutions to partial differential equations. Another often used method is known as the **finite element method** which, due to its generally more sophisticated mathematical underpinnings and concerns over the space required to introduce it, will not be explored in this textbook.

12.1   Discretization of Derivatives

Finite difference methods often replace the xt-plane with a regularly spaced grid of points as illustrated in Fig. 12.1. This collection of regularly spaced points is often referred to as a **discretization** of the plane. The horizontal spacing of the points is denoted as h while the vertical spacing is denoted as k. Some authors use Δx and Δt in place of h and k respectively. A particular point in the discrete plane will be located by its coordinates $(i\,h, j\,k)$ where i and $j \in \mathbb{Z}$. If u is a real-valued function of (x,t) then the approximation of this function at the point with coordinates $(i\,h, j\,k)$ will be compactly denoted as u_i^j. Higher dimensional spaces can be discretized in a similar manner. For example $\mathbb{R}^3$ can be discretized as points with spacing in the x direction given by h, in the y direction by k, and in the t (vertical) direction by l. Thus a point in the discretization of $\mathbb{R}^3$ would be located at coordinates $(i\,h, j\,k, n\,l)$ where i, j, and $n \in \mathbb{Z}$. A function u of (x, y, t) evaluated at such a point would be noted simply as $u_{i,j}^n$.

The solution of an ordinary or partial differential equation by the finite difference method is a matter of developing formulas to approximate the derivatives or partial derivatives of functions at the discrete grids of points described above. The underlying mechanism which makes these approximations possible is **Taylor's theorem**.

Theorem 12.1. *If function f is $n + 1$ times continuously differentiable on (a, b) and if $a < x_0 < b$ then for any $x \in (a, b)$,*

$$f(x) = f(x_0) + f'(x_0)(x - x_0) + \cdots + \frac{f^{(n)}(x_0)}{n!}(x - x_0)^n$$
$$+ \frac{f^{(n+1)}(z)}{(n + 1)!}(x - x_0)^{n+1} \tag{12.1}$$

where z lies between x_0 and x.

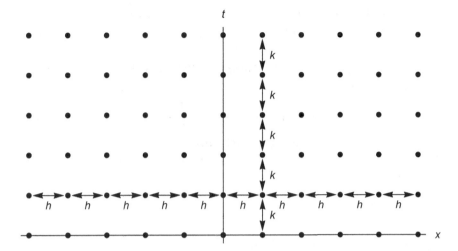

Fig. 12.1 A discretization of the xt-plane as a collection of points with horizontal spacing h and vertical spacing k.

The expression

$$P_n(x) = f(x_0) + f'(x_0)(x - x_0) + \cdots + \frac{f^{(n)}(x_0)}{n!}(x - x_0)^n \qquad (12.2)$$

is called the nth **Taylor polynomial approximation** of $f(x)$ centered at x_0 and

$$R_n(x) = \frac{f^{(n+1)}(z)}{(n+1)!}(x - x_0)^{n+1} \qquad (12.3)$$

is called the nth **Taylor remainder**. Sometimes $R_n(x)$ is called the **truncation error** since it is the difference between the exact value of $f(x)$ and $P_n(x)$. If there exists a constant M_n such that

$$\max_{a \leq z \leq b} \left| \frac{f^{(n+1)}(z)}{(n+1)!} \right| \leq M_n,$$

then the absolute error in approximating $f(x)$ on interval (a, b) by $P_n(x)$ is bounded by $E_n(x) = M_n|x - x_0|^{n+1}$. Proofs of Taylor's theorem can be found in most elementary calculus textbooks, for example [Smith and Minton (2012)].

Suppose function U is twice continuously differentiable on an open interval containing x_0, then applying Taylor's theorem,

$$U(x_0 + h) = U(x_0) + U'(x_0)h + U''(z)\frac{h^2}{2!}$$

where z lies between x_0 and $x_0 + h$. Rearranging terms yields the equation

$$U'(x_0) = \frac{U(x_0 + h) - U(x_0)}{h} - U''(z)\frac{h}{2!}.$$

When $h > 0$ this produces the **forward difference** approximation to U' (x_0) or equivalently,

$$U'(x_0) \approx \frac{U(x_0 + h) - U(x_0)}{h}. \tag{12.4}$$

If $h < 0$, Eq. (12.4) is called the **backward difference** approximation to $U'(x_0)$. The error term $U''(z)h/2!$ is described as being $O(h)$ (pronounced "big oh of h") as $h \to 0$. Since $U(x)$ is assumed to be twice continuously differentiable near x_0, when h is sufficiently small,

$$\left|\frac{U''(z)}{2!}h\right| \leq M|h|,$$

where M is a positive constant. In general a function $f(h)$ is said to be $O(h^n)$ as $h \to 0$ if there exist positive constants M and ϵ such that $|f(h)| \leq M|h|^n$ for $|h| < \epsilon$.

If function U is three times continuously differentiable on (a, b) and $a < x_0 < b$ then

$$U(x_0 + h) = U(x_0) + U'(x_0)h + U''(x_0)\frac{h^2}{2!} + \frac{U'''(z)}{3!}h^3,$$

$$U(x_0 - h) = U(x_0) - U'(x_0)h + U''(x_0)\frac{h^2}{2!} - \frac{U'''(w)}{3!}h^3.$$

If $h > 0$ then $x_0 - h < w < x_0 < z < x_0 + h$. Subtracting these two equations and solving for $U'(x_0)$ result in

$$U'(x_0) = \frac{U(x_0 + h) - U(x_0 - h)}{2h} + \left(\frac{U'''(z) + U'''(w)}{2}\right)\frac{h^2}{3!}.$$

Since $(U'''(z) + U'''(w))/2$ must lie between $U'''(w)$ and $U'''(z)$ and U''' is assumed continuous on the $[w, z] \subset (a, b)$ then by the Intermediate Value Theorem there exists $v \in [w, z]$ such that $U'''(v) = (U'''(z) + U'''(w))/2$ which implies

$$U'(x_0) = \frac{U(x_0 + h) - U(x_0 - h)}{2h} + U'''(v)\frac{h^2}{3!},$$

for some $a < v < b$. The approximation

$$U'(x_0) \approx \frac{U(x_0 + h) - U(x_0 - h)}{2h} \tag{12.5}$$

is known as the **centered difference formula** and has an error term which is $O(h^2)$.

Suppose function U is four times continuously differentiable on (a, b) and $a < x_0 < b$, then

$$U(x_0 + h) = U(x_0) + U'(x_0)h + U''(x_0)\frac{h^2}{2!} + U'''(x_0)\frac{h^3}{3!} + \frac{U^{(4)}(z)}{4!}h^4,$$

$$U(x_0 - h) = U(x_0) - U'(x_0)h + U''(x_0)\frac{h^2}{2!} - U'''(x_0)\frac{h^3}{3!} + \frac{U^{(4)}(w)}{4!}h^4.$$

If $h > 0$ then $x_0 - h < w < x_0 < z < x_0 + h$. Adding the two equations and solving for $U''(x_0)$ produce

$$U''(x_0) = \frac{U(x_0 + h) - 2U(x_0) + U(x_0 - h)}{h^2} + \left(\frac{U^{(4)}(z) + U^{(4)}(w)}{2}\right)\frac{h^2}{12}.$$

Again by the Intermediate Value Theorem there exists $v \in [w, z]$ such that $U^{(4)}(v) = (U^{(4)}(z) + U^{(4)}(w))/2$ which implies

$$U''(x_0) = \frac{U(x_0 + h) - 2U(x_0) + U(x_0 - h)}{h^2} + U^{(4)}(v)\frac{h^2}{12}$$

for some $a < v < b$. The approximation

$$U''(x_0) \approx \frac{U(x_0 + h) - 2U(x_0) + U(x_0 - h)}{h^2} \tag{12.6}$$

is known as the **second derivative midpoint formula** and has an error term which is $O(h^2)$.

Other differencing formulas for the first, second, and higher order derivatives are possible. Some examples and their error terms may be found in [Burden *et al.* (2016)]. Even though these derivative approximation formulas were developed for functions of a single variable, they apply to multivariable functions where the differencing is applied to the variable of interest. For example if u is a function of (x, t) the discrete approximation to the homogeneous heat equation $u_t = u_{xx}$ can be expressed using the forward difference formula in t and the second derivative midpoint formula in x as

$$\frac{u(x, t + k) - u(x, t)}{k} = \frac{u(x + h, t) - 2u(x, t) + u(x - h, t)}{h^2}. \tag{12.7}$$

Other finite difference approximations can be developed depending on the choices of differencing formulas made for u_t and u_{xx}. This topic will be taken up again in the later sections when more specific categories of partial differential equations (parabolic, elliptic, and hyperbolic) are treated. Before that discussion can take place, the issues of discretizing the initial and boundary conditions must be explored.

12.2   Discretization of Initial and Boundary Conditions

Consider an initial boundary value problem with dependent variable u posed on the interval $[0, L]$ with Dirichlet boundary conditions and an initial condition:

$$u(0,t) = a(t) \text{ and } u(L,t) = b(t) \text{ for } t > 0,$$
$$u(x,0) = f(x) \text{ for } 0 < x < L$$

where $a(t)$, $b(t)$, and $f(x)$ are known functions. The interval $[0, L]$ is partitioned into N equally sized subintervals of width $h = L/N$ with end points $x_i = i\,h$ for $i = 0, 1, \ldots, N$. Discretizing the initial condition is straight forward as $u_i^0 = u(i\,h, 0) = f(i\,h)$ for $i = 0, 1, \ldots, N$. The Dirichlet boundary conditions are also easily handled by assigning the known values at the endpoints of the interval, $u_0^j = u((0)h, j\,k) = a(j\,k)$ and $u_N^j = u(N\,h, j\,k) = b(j\,k)$ for $j \in \mathbb{N}$.

Boundary conditions involving a derivative such as the Neumann boundary conditions or Robin boundary conditions, are discretized using a differencing formula. Since the Neumann boundary conditions are a limiting case of the Robin boundary conditions the approximation by differencing formulas will only be described for Robin boundary conditions. Suppose the boundary conditions are:

$$K_0 u_x(0,t) = \alpha(u(0,t) - a(t)),$$
$$K_0 u_x(L,t) = -\beta(u(L,t) - b(t)),$$

where K_0, α, and β are positive constants and $a(t)$ and $b(t)$ are known functions. At $x = 0$ a forward difference formula for the partial derivative with respect to x is employed. For $j \in \mathbb{N}$,

$$K_0 \frac{(u_1^j - u_0^j)}{h} = \alpha(u_0^j - a(j\,k)).$$

At $x = L$ a backward difference formula is used, which when written as

$$K_0 \frac{(u_{N-1}^j - u_N^j)}{h} = \beta(u_N^j - b(j\,k)),$$

eliminates the minus sign from the right-hand side of the boundary condition. The two-point forward and backward difference formulas have $O(h)$ error terms which may be insufficiently accurate for some applications. If two fictitious points[1] at $x = -h$ and $x = L+h$ are introduced, the $O(h^2)$ accurate centered difference formula can be used. In this case the discretized

---

[1] Some authors refer to them as "ghost points".

Robin boundary conditions can be expressed as

$$K_0 \frac{(u_1^j - u_{-1}^j)}{2h} = \alpha(u_0^j - a(j\,k)),$$

$$K_0 \frac{(u_{N+1}^j - u_{N-1}^j)}{2h} = -\beta(u_N^j - b(j\,k)).$$

The values of the dependent variable at the fictitious points can be found in terms of the "real" points used in the difference scheme.

The next section will apply the theory developed so far to the solution of the heat equation with various boundary and initial conditions. The numerical experiments performed will lead to improvements in the finite difference algorithms employed.

12.3   The Heat Equation

The reader may be eager to implement finite difference schemes for approximating the solutions to partial differential equations. In this section several examples borrowed from Chap. 4 will be approximated numerically in order to explore the implementation and behavior of finite difference techniques. This will present the opportunity to improve the basic methods and to understand some of the notions that can be used to describe the methods such as stability, accuracy, and consistency.

Consider developing a numerical solution for Example 4.1. The spatial interval $[0, 1]$ must be subdivided into N intervals of length $h = 1/N$. A forward time step of size $k > 0$ must be chosen and a differencing scheme must be selected for the partial differential equation. The formula in Eq. (12.7) approximates the homogeneous heat equation and is $O(k)$ in t and $O(h^2)$ in x. Thus the finite difference approximation in Eq. (12.8) is classified as having truncation error $O(k) + O(h^2)$. This finite difference scheme can be written in the discrete notation as

$$\frac{1}{k}(u_i^{j+1} - u_i^j) = \frac{1}{h^2}(u_{i+1}^j - 2u_i^j + u_{i-1}^j)$$

$$u_i^{j+1} = ru_{i-1}^j + (1 - 2r)u_i^j + ru_{i+1}^j \tag{12.8}$$

for $i = 1, 2, \ldots, N - 1$, where $r = k/h^2$. Equation (12.8) is an example of an **explicit** finite difference scheme. In general a finite difference method is called an explicit scheme if the value of u_i^{j+1} is a finite sum of u_l^m where $m \leq j$ [Strikwerda (2004), Sec. 1.6]. A finite difference scheme which is not explicit is called **implicit**. For $i = 0$ and $i = N$ the homogeneous Dirichlet boundary conditions provide $u_0^j = u_N^j = 0$ for all $j \in \mathbb{N}$. Note that

Eq. (12.8) gives the value of u_i^{j+1} as a function of three values of u one time step earlier. The following graphic or **stencil** illustrates the dependency of u_i^{j+1} on earlier time values.

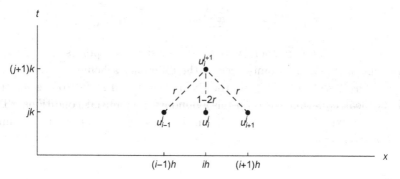

The values of u_i^0 are the initial conditions and are thus known. This enables the solution to be approximated at $t = j\,k$ for $j \in \mathbb{N}$ iteratively.

The explicit finite difference scheme for the heat equation can be expressed in matrix/vector form. If the vector $\mathbf{u}^{(j)} = (u_1^j, u_2^j, \ldots, u_{N-1}^j)^T$ then the system of equations found in Eq. (12.8) may be written as $\mathbf{u}^{(j+1)} = A(r)\,\mathbf{u}^{(j)}$ or

$$
\begin{bmatrix} u_1^{j+1} \\ u_2^{j+1} \\ \vdots \\ u_{N-2}^{j+1} \\ u_{N-1}^{j+1} \end{bmatrix}
=
\begin{bmatrix}
1-2r & r & 0 \cdots 0 & 0 & 0 \\
r & 1-2r & r \cdots 0 & 0 & 0 \\
\vdots & \vdots & \vdots \; \vdots \; \vdots & \vdots & \vdots \\
0 & 0 & 0 \cdots r & 1-2r & r \\
0 & 0 & 0 \cdots 0 & r & 1-2r
\end{bmatrix}
\begin{bmatrix} u_1^j \\ u_2^j \\ \vdots \\ u_{N-2}^j \\ u_{N-1}^j \end{bmatrix} . \quad (12.9)
$$

For example if $h = 1/10$ and $k = 1/1000$ then $r = 1/10$ and the values of the finite difference approximation are shown in Table 12.1. A plot of the infinite series solution and the approximate solution generated by the explicit finite difference method for $t = 0.012$, $t = 0.036$, and $t = 0.060$ are shown in Fig. 12.2. The reader may recognize one drawback to this explicit scheme. Since the forward time step size is $k = 0.001$, one thousand iterations of the explicit formula are required to approximate the solution for $0 < t \leq 1$. To reduce the number of iterations required to reach the $t = 1$ milestone, k could be increased. If $k = 0.004$ only two hundred fifty iterations are required. Figure 12.3 illustrates the comparison of the analytic Fourier series solution to the explicit finite difference approximation at $t = 0.012$, $t = 0.036$, and $t = 0.060$. The numerical solution does not appear to agree as well with the infinite series solution when $r = 0.4$. Finally

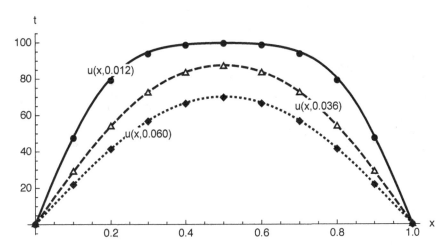

Fig. 12.2 The analytic solution in the form of the Fourier series is plotted along with the approximate solution resulting from the explicit form of the finite difference method. The curves are the graphs of the infinite series solution while the symbols (points, triangles, *etc.*) are the discrete approximations generated by the finite difference technique. The approximate solution is generally in agreement with the analytic solution. The spatial discretization and the forward time step were chosen so that $r = 0.1$.

if $k = 0.006$ and $h = 0.1$, then $r = 0.6$ and the explicit finite difference method does not seem to capture the solution to the homogeneous heat equation with Dirichlet boundary conditions. See Fig. 12.4. The degradation in accuracy of the explicit finite difference scheme as r increases requires further study. This will be explored in more detail in Sec. 12.7.

An alternative approximation to the solution of the heat equation (and parabolic partial differential equations in general) was proposed by Crank[2] and Nicolson[3] [Crank and Nicolson (1947)] and thus is known as the **Crank-Nicolson implicit method**. This method differs from the previously studied explicit method in that the solution is assumed to be found at the time $t = k(j + 1/2)$ rather than at $t = k(j + 1)$.

Suppose f is a twice continuously differentiable function on a neighborhood of $t + k/2$, then expanding f as a Taylor polynomial about $t + k/2$ produces

$$f(s) = f\left(t + \frac{k}{2}\right) + f'\left(t + \frac{k}{2}\right)\left(s - t - \frac{k}{2}\right) + O\left(\left(s - t - \frac{k}{2}\right)^2\right),$$

for s close to $t + k/2$. Replacing s by t and again by $t + k$ yields the two

---

[2]John Crank, British mathematical physicist (1916–2006).
[3]Phyllis Nicolson, British mathematician (1917–1968).

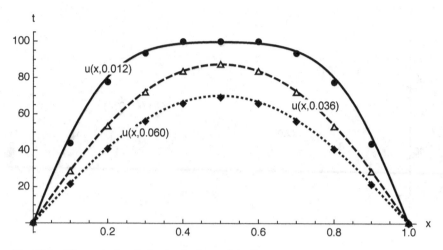

Fig. 12.3   The analytic solution in the form of the Fourier series is plotted along with the approximate solution resulting from the explicit form of the finite difference method. The approximate solution still seems accurate but deviates further from the exact solution when $r = 0.4$. Compare this graph with the one shown in Fig. 12.2.

equations:

$$f(t) = f\left(t + \frac{k}{2}\right) - f'\left(t + \frac{k}{2}\right)\left(\frac{k}{2}\right) + O(k^2),$$

$$f(t + k) = f\left(t + \frac{k}{2}\right) + f'\left(t + \frac{k}{2}\right)\left(\frac{k}{2}\right) + O(k^2).$$

Adding these two equations and solving for $f(t + k/2)$ give the equation

$$f\left(t + \frac{k}{2}\right) = \frac{f(t) + f(t + k)}{2} + O(k^2),$$

which has an $O(k^2)$ truncation error.

Choose f to be $u_{xx}(x, t)$, then

$$\begin{aligned}
u_{xx}(x, t + k/2) &= \frac{u_{xx}(x, t) + u_{xx}(x, t + k)}{2} + O(k^2) \\
&= \frac{u(x + h, t) - 2u(x, t) + u(x - h, t)}{2h^2} \\
&\quad + \frac{u(x + h, t + k) - 2u(x, t + k) + u(x - h, t + k)}{2h^2} \\
&\quad + O(k^2) + O(h^2).
\end{aligned}$$

Thus a formula for $u_{xx}(x, t + k/2)$ has been found by averaging the three-point formulas for the second derivative at times $t = kj$ and $t = k(j + 1)$.

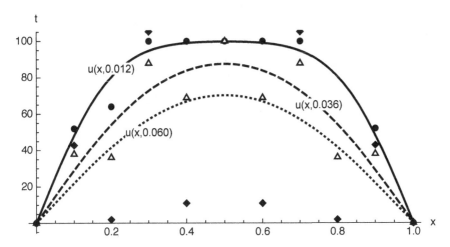

Fig. 12.4 The analytic solution in the form of the Fourier series is plotted along with the approximate solution resulting from the explicit form of the finite difference method. The approximate solution does not agree well with the exact solution when $r = 0.6$.

This formula has $O(k^2) + O(h^2)$ accuracy as h, $k \to 0$. Figure 12.5 demonstrates the improvement in the approximation of the analytic solution provided by the $O(k^2) + O(h^2)$ accuracy of the Crank-Nicolson implicit scheme for the heat equation.

The grid points involved in the Crank-Nicolson implicit method are illustrated below.

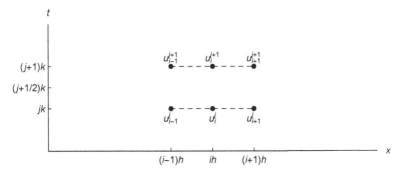

Thus the homogeneous heat equation can be discretized as

$$\frac{u_i^{j+1} - u_i^j}{k} = \frac{u_{i-1}^j - 2u_i^j + u_{i+1}^j}{2h^2} + \frac{u_{i-1}^{j+1} - 2u_i^{j+1} + u_{i+1}^{j+1}}{2h^2}$$

or equivalently as,

$$-ru_{i-1}^{j+1} + 2(1+r)u_i^{j+1} - ru_{i+1}^{j+1} = ru_{i-1}^j + 2(1-r)u_i^j + ru_{i+1}^j \qquad (12.10)$$

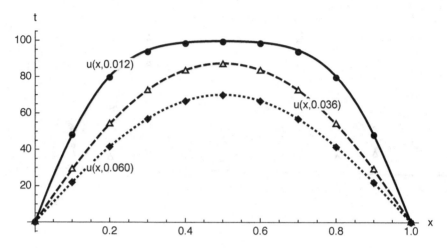

Fig. 12.5   The analytic solution in the form of the Fourier series is plotted along with the approximate solution resulting from the Crank-Nicolson form of the finite difference method. The implicit technique demonstrates greater accuracy than the explicit technique whose results are plotted in Fig. 12.3 when the same values of h and k (and hence r) are used. In this case $r = 0.4$.

for $i = 1, 2, \ldots, N - 1$ and where $r = k/h^2$. Thus relative to the point at $(x, t) = (i\,h, (j + 1/2)k)$, the first derivative in t and the second derivative in x have been replaced with central difference approximations which, in general, are more accurate than forward (or backward) difference approximations. The three values u_{i-1}^{j+1}, u_i^{j+1}, and u_{i+1}^{j+1} are unknown and must be solved for simultaneously and depend on values found on the $t = j\,k$ row (which were either given as initial conditions or were calculated previously).

If the boundary conditions are of homogeneous Dirichlet type, the system of equations can be written in matrix/vector form as

$$A(r)\mathbf{u}^{(j+1)} = A(-r)\mathbf{u}^{(j)} \text{ where} \qquad (12.11)$$

$$A(r) = \begin{bmatrix} 2(1+r) & -r & 0 & \cdots & 0 & 0 & 0 \\ -r & 2(1+r) & -r & \cdots & 0 & 0 & 0 \\ \vdots & \vdots & \vdots & \vdots & \vdots & & \vdots \\ 0 & 0 & 0 & \cdots & -r & 2(1+r) & -r \\ 0 & 0 & 0 & \cdots & 0 & -r & 2(1+r) \end{bmatrix}.$$

A unique solution to Eq. (12.11) exists if and only if $A(r)$ is invertible. In actuality, the inverse of $A(r)$ in Eq. (12.11) always exists for any $r > 0$ since the matrix is **strictly diagonally dominant**. Furthermore the placement of the nonzero entries in this matrix gives it a special structure known as a **tridiagonal matrix** which is easily inverted. A system of equations of the form $A\mathbf{x} = \mathbf{b}$ where A is an $n \times n$ tridiagonal matrix can be solved by a Gaussian elimination algorithm requiring $O(n)$ operations (multiplications and additions) [Golub and Van Loan (1996)]. In this context the "big oh of n" nature of the Gaussian elimination algorithm means that the number of mathematical operations (additions/subtractions or multiplications/divisions) required to compute the solution grows linearly with the number of unknowns n. Even more advantageous is the case that the matrix is **symmetric** and the off-diagonal elements of the matrix are all identical [Hu and O'Connell (1996)]. The performance of the Crank-Nicolson method will be illustrated in the next example. To find solutions to the linear system the **Crout**[4] **matrix decomposition** method will be employed in the Java program examples accompanying this textbook. See [Burden *et al.* (2016)] for a detailed description and pseudocode for the Crout matrix decomposition.

Example 12.1. Consider the following initial boundary value problem for the heat equation:

$$u_t = u_{xx} \text{ for } 0 < x < 1 \text{ and } t > 0,$$

$$u(0,t) = u(1,t) = 0 \text{ for } t > 0,$$

$$u(x,0) = x(1-x)^2 \text{ for } 0 < x < 1.$$

Approximate the solution for $0 \leq t \leq 1/10$ using the Crank-Nicolson method with $h = 1/10$ and $k = 1/100$.

Solution. For these choices of h and k the parameter $r = 1$ which simplifies Eq. (12.11).

$$
\begin{bmatrix}
4 & -1 & 0 & 0 & \cdots & 0 \\
-1 & 4 & -1 & 0 & \cdots & 0 \\
0 & -1 & 4 & -1 & \cdots & 0 \\
\vdots & \vdots & \vdots & \vdots & & \vdots \\
0 & \cdots & 0 & -1 & 4 & -1 \\
0 & \cdots & 0 & 0 & -1 & 4
\end{bmatrix}
\begin{bmatrix}
u_1^{j+1} \\
u_2^{j+1} \\
u_3^{j+1} \\
\vdots \\
u_{N-2}^{j+1} \\
u_{N-1}^{j+1}
\end{bmatrix}
=
\begin{bmatrix}
0 & 1 & 0 & 0 & \cdots & 0 \\
1 & 0 & 1 & 0 & \cdots & 0 \\
0 & 1 & 0 & 1 & \cdots & 0 \\
\vdots & \vdots & \vdots & \vdots & & \vdots \\
0 & \cdots & 0 & 1 & 0 & 1 \\
0 & \cdots & 0 & 0 & 1 & 0
\end{bmatrix}
\begin{bmatrix}
u_1^{j} \\
u_2^{j} \\
u_3^{j} \\
\vdots \\
u_{N-2}^{j} \\
u_{N-1}^{j}
\end{bmatrix}.
$$

If $u_i^0 = (ih)(1 - ih)^2$ with $h = 1/10$ for $i = 1, 2, \ldots, 9$ then Table 12.2 contains the next ten time steps of the solution. The surface plot of the analytic solution is plotted on the same coordinate axes as the numerical

---

[4]Prescott Durand Crout, American mathematician (1907–1984).

approximation in Fig. 12.6. There appears to be good agreement between the analytic (or true) solution and the approximate, numerical solution.

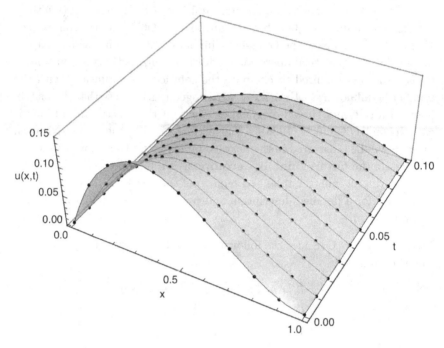

Fig. 12.6   The analytic solution to the initial boundary value problem posed in Example 12.1 is plotted as a surface and the discrete solution found using the implicit Crank-Nicolson technique is plotted as points. There is good agreement between the two solution methods.

The final example of this section will illustrate the treatment of Robin boundary conditions. Readers will be able to adapt this method to Neumann boundary conditions easily. The method used to approximate the boundary conditions is independent of the differencing scheme used to approximate the partial differential equation and thus can be applied to both explicit and implicit techniques. Consider the initial boundary value problem,

$$u_t = u_{xx} \text{ for } 0 < x < 1 \text{ and } t > 0,$$
$$u_x(0, t) = u(0, t) - e^{-t} \text{ for } t > 0,$$
$$-u_x(1, t) = 2u(1, t) - t \text{ for } t > 0,$$
$$u(x, 0) = \frac{1}{3}e^{-2x} \text{ for } 0 < x < 1.$$

Introducing the fictitious points u_{-1}^j, the discretized form of the boundary condition at $x = 0$ can be written as

$$\frac{u_1^j - u_{-1}^j}{2h} = u_0^j - e^{-jk},$$

which implies the values at the fictitious points are

$$u_{-1}^j = -2hu_0^j + u_1^j + 2he^{-jk}.$$

Thus if $i = 0$ in the Crank-Nicolson formula found in Eq. (12.10), this equation can be rewritten as

$$(1+r+hr)u_0^{j+1} - ru_1^{j+1} = (1-r-hr)u_0^j + ru_1^j + hr(e^{-(j+1)k} + e^{-jk}). \quad (12.12)$$

Similarly the boundary condition at $x = 1$ can be approximated using the central difference formula as

$$-\frac{u_{N+1}^j - u_{N-1}^j}{2h} = 2u_N^j - jk.$$

Solving for the fictitious point where $i = N + 1$ yields

$$u_{N+1}^j = u_{N-1}^j - 4hu_N^j + 2hjk.$$

Substituting this expression into Eq. (12.10) with $i = N$ enables the equation to be written as

$$-ru_{N-1}^{j+1} + (1+r+2hr)u_N^{j+1} = ru_{N-1}^j + (1-r-2hr)u_N^j + hr(2j+1)k. \quad (12.13)$$

The linear system of equations in Eq. (12.11) is augmented by two additional equations, one for each boundary condition. Thus the matrix on the left-hand side of Eq. (12.11) has dimensions $(N+1) \times (N+1)$ and resembles:

$$A(r) = \begin{bmatrix} 1+r+hr & -r & 0 & \cdots & 0 & 0 & 0 \\ -r & 2(1+r) & -r & \cdots & 0 & 0 & 0 \\ \vdots & \vdots & \vdots & \vdots & \vdots & & \vdots \\ 0 & 0 & 0 & \cdots & -r & 2(1+r) & -r \\ 0 & 0 & 0 & \cdots & 0 & -r & 1+r+2hr \end{bmatrix}.$$

The matrix on the right-hand side of Eq. (12.11) is likewise modified to become $A(-r)$. If $\mathbf{u}^j = (u_0^j, u_1^j, \ldots, u_N^j)^T$ then the linear system can be expressed as

$$A(r)\,\mathbf{u}^{j+1} = A(-r)\,\mathbf{u}^j + \begin{bmatrix} hr(e^{-(j+1)k} + e^{-jk}) \\ 0 \\ \vdots \\ 0 \\ hr(2j+1)k \end{bmatrix}.$$

Fortunately matrix $A(r)$ is strictly diagonally dominant and tridiagonal with all of its off-diagonal elements equal to one another and thus the linear system can be easily and efficiently solved. Table 12.3 contains the results of the finite difference approximation for the first ten time steps.

The exercises at the end of this chapter contain additional initial boundary value problems which the reader is encouraged to solve numerically.

12.4   The Wave Equation

In this section explicit and implicit finite difference schemes will be presented for approximating the solutions to hyperbolic equations such as the wave equation. The methods explored are well-suited to problems without discontinuities in their initial conditions. If a problem of interest does possess discontinuities in its initial conditions, a better technique would be to numerically approximate the solution along a grid of characteristic curves. However, these problems are more difficult to implement in computer code and thus will not be covered in this textbook.

Consider the initial value problem for the wave equation,

$$u_{tt} = u_{xx} \text{ for } 0 < x < 1 \text{ and } t > 0,$$
$$u(x,0) = f(x) \text{ for } 0 < x < 1,$$
$$u_t(x,0) = g(x) \text{ for } 0 < x < 1.$$

Suppose the interval $[0,1]$ is divided into N equally sized subintervals. The boundary conditions at $x = 0$ and $x = 1$ will be discussed later within the context of specific examples. Using the three-point differencing formula for the second derivative given in Eq. (12.6), the differential equation may be approximated as

$$\frac{u_i^{j+1} - 2u_i^j + u_i^{j-1}}{k^2} = \frac{u_{i+1}^j - 2u_i^j + u_{i-1}^j}{h^2}.$$

The truncation error of this approximation is $O(k^2) + O(h^2)$. The reader should carefully note that the left-hand side of the equation is the approximation to u_{tt} at point $(i\,h, j\,k)$ and the right-hand side is the approximation to u_{xx} at the same point. The time step size is $\Delta t = k$ and the spatial step size is $\Delta x = h = 1/N$. Letting $r = k/h$ and solving for u_i^{j+1} produce the equation,

$$u_i^{j+1} = r^2 u_{i+1}^j + 2(1 - r^2)u_i^j + r^2 u_{i-1}^j - u_i^{j-1}. \tag{12.14}$$

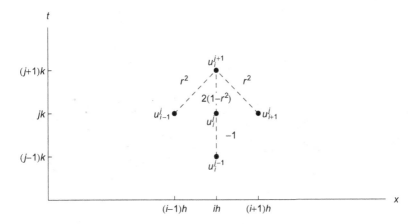

If the initial conditions for the equation are $u(x,0) = f(x)$ and $u_t(x,0) = g(x)$ for $0 < x < 1$ then

$$u_i^0 = f(i\,h) \text{ for } i = 1, 2, \ldots, N - 1. \tag{12.15}$$

Using the centered difference formula for the first derivative with respect to t yields

$$\frac{u_i^1 - u_i^{-1}}{2k} = g(i\,h) \iff u_i^{-1} = u_i^1 - 2kg(i\,h) \tag{12.16}$$

for $i = 1, 2, \ldots, N - 1$. Setting $j = 0$ in Eq. (12.14) and using the expression just derived for u_i^{-1} then

$$u_i^1 = \frac{r^2}{2}f((i + 1)h) + (1 - r^2)f(i\,h) + \frac{r^2}{2}f((i - 1)h) + kg(i\,h) \tag{12.17}$$

for $i = 1, 2, \ldots, N - 1$. Thus the finite difference approximation to the solution to the wave equation may be calculated by using Eq. (12.15) to determine the initial displacement of the wave, Eq. (12.17) to determine the displacement at the first positive time step where $t = k$, and Eq. (12.14) repeatedly to iterate further forward in time.

Example 12.2. Use the explicit finite difference formulas to approximate the solution to the initial boundary value problem in Example 5.1. Use $h = 1/20$ and $k = 1/20$. Compare the finite difference approximation to the infinite series solution at $t = 0, 1/5, 2/5, 3/5, 4/5, 1$.

Solution. The ends of the string Example 5.1 are assumed fixed, thus $u_0^j = u_N^j = 0$ for all j. The initial velocity of the string is zero. Applying Eqs. (12.15) and (12.17) to "bootstrap" the first two time steps and then

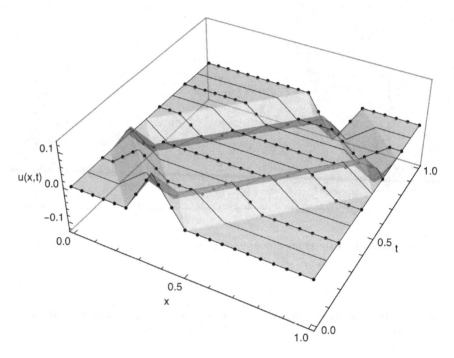

Fig. 12.7   A comparison of the numerically generated explicit finite difference method solution to the wave equation from Example 5.1 to the analytical solution. The numerical solution is in good agreement with the exact solution.

using Eq. (12.14) repeatedly to iterate the remaining time steps generates the displacements shown in Table 12.4. A comparison of the numerical solution and the analytical solution is shown in Fig. 12.7.

There exist several implicit finite difference schemes for approximating the solution to the wave equation. The reader will be asked to develop a Crank-Nicolson implicit method for the wave equation in Exercise 15. Another implicit method will be presented here.

Recall that for a twice continuously differentiable function f,

$$f(x) = \frac{f(x+h) + f(x-h)}{2} + O(h^2)$$

which is equivalent to the following equation,

$$f(x) = \frac{f(x+h) + 2f(x) + f(x-h)}{4} + O(h^2).$$

If $f = u_{xx}(x,t)$ then

$$
\begin{aligned}
u_{xx}(x,t) &= \frac{u_{xx}(x,t-k) + 2u_{xx}(x,t) + u_{xx}(x,t+k)}{4} \\
&= \frac{u(x+h,t-k) - 2u(x,t-k) + u(x-h,t-k)}{4h^2} \\
&\quad + \frac{u(x+h,t) - 2u(x,t) + u(x-h,t)}{2h^2} \\
&\quad + \frac{u(x+h,t+k) - 2u(x,t+k) + u(x-h,t+k)}{4h^2} \\
&\quad + O(h^2) + O(k^2).
\end{aligned}
$$

Therefore the wave equation can be approximated by the differencing scheme,

$$
\frac{u_i^{j+1} - 2u_i^j + u_i^{j-1}}{k^2} = \frac{u_{i+1}^{j+1} - 2u_i^{j+1} + u_{i-1}^{j+1}}{4h^2} + \frac{u_{i+1}^j - 2u_i^j + u_{i-1}^j}{2h^2}
$$

$$
+ \frac{u_{i+1}^{j-1} - 2u_i^{j-1} + u_{i-1}^{j-1}}{4h^2}.
$$

The truncation error in this finite difference approximation is $O(k^2) + O(h^2)$. If both sides of the equation are multiplied by $4k^2$ and terms are rearranged so that the expressions involving u at $t = (j+1)k$ are on one side and all other terms are on the other, then the approximation can be written as

$$
-r^2 u_{i+1}^{j+1} + 2(2 + r^2)u_i^{j+1} - r^2 u_{i-1}^{j+1} \tag{12.18}
$$
$$
= 2r^2 u_{i+1}^j + 4(2 - r^2)u_i^j + 2r^2 u_{i-1}^j + r^2 u_{i+1}^{j-1} - 2(2 + r^2)u_i^{j-1} + r^2 u_{i-1}^{j-1}
$$

for $i = 1, 2, \ldots, N-1$ where $r = k/h$.

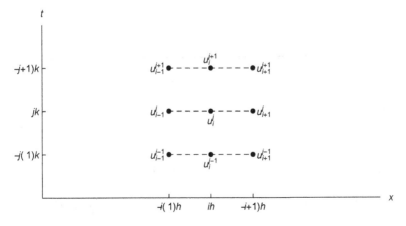

If the boundary conditions are of the homogeneous Dirichlet type (ends of the vibrating string fixed) then this implicit finite difference scheme can be written in matrix/vector form as

$$A(r^2)\mathbf{u}^{j+1} = 2A(-r^2)\mathbf{u}^j - A(r^2)\mathbf{u}^{j-1} \tag{12.19}$$

where $A(r)$ is an $(N-1) \times (N-1)$ matrix:

$$A(r) = \begin{bmatrix} 2(2+r) & -r & 0 & \cdots & 0 & 0 & 0 \\ -r & 2(2+r) & -r & \cdots & 0 & 0 & 0 \\ \vdots & \vdots & \vdots & \vdots & \vdots & \vdots \\ 0 & 0 & 0 & \cdots & -r & 2(2+r) & -r \\ 0 & 0 & 0 & \cdots & 0 & -r & 2(2+r) \end{bmatrix}.$$

Once again, since matrix $A(r^2)$ is tridiagonal and strictly diagonally dominant, the system can be efficiently solved. The initial displacement of the vibrating string determines the values $u_i^0 = u(i\,h, 0)$ for $i = 1, 2, \ldots, N-1$. If $j = 0$ in Eq. (12.19), then

$$A(r^2)\mathbf{u}^1 = 2A(-r^2)\mathbf{u}^0 - A(r^2)\mathbf{u}^{-1} = A(-r^2)\mathbf{u}^0 + kA(r^2) \begin{bmatrix} u_t(h, 0) \\ u_t(2h, 0) \\ \vdots \\ u_t((N-1)h, 0) \end{bmatrix}$$

by Eq. (12.16).

Example 12.3. Use the implicit finite difference scheme described in Eq. (12.18) to approximate the solution to the following initial boundary value problem. Use $N = 10$, $h = 1/10$, and $k = 1/20$ and iterate the solution until $t = 1/2$.

$$u_{tt} = u_{xx} \text{ for } 0 < x < 1 \text{ and } t > 0,$$
$$u(0, t) = u(1, t) = 0 \text{ for } t > 0,$$
$$u(x, 0) = x(1 - x) \text{ for } 0 < x < 1,$$
$$u_t(x, 0) = \sin(5\pi x) \text{ for } 0 < x < 1.$$

Solution. Given the values of h and k the parameter $r = 1/2$ and thus

$$A\left(\frac{1}{4}\right) = \begin{bmatrix} \frac{9}{2} & -\frac{1}{4} & 0 & \cdots & 0 & 0 & 0 \\ -\frac{1}{4} & \frac{9}{2} & -\frac{1}{4} & \cdots & 0 & 0 & 0 \\ \vdots & \vdots & \vdots & & \vdots & \vdots & \vdots \\ 0 & 0 & 0 & \cdots & -\frac{1}{4} & \frac{9}{2} & -\frac{1}{4} \\ 0 & 0 & 0 & \cdots & 0 & -\frac{1}{4} & \frac{9}{2} \end{bmatrix} \text{ and } 2A\left(-\frac{1}{4}\right) = \begin{bmatrix} 7 & \frac{1}{2} & 0 & \cdots & 0 & 0 & 0 \\ \frac{1}{2} & 7 & \frac{1}{2} & \cdots & 0 & 0 & 0 \\ \vdots & \vdots & \vdots & & \vdots & \vdots & \vdots \\ 0 & 0 & 0 & \cdots & \frac{1}{2} & 7 & \frac{1}{2} \\ 0 & 0 & 0 & \cdots & 0 & \frac{1}{2} & 7 \end{bmatrix}.$$

For the sake of brevity denote $A(1/4)$ as simply A and $A(-1/4)$ as simply B. Setting $u_i^0 = (i\,h)(1 - i\,h)$ for $i = 1, 2, \ldots, N - 1$ and solving the tridiagonal system:

$$A\begin{bmatrix} u_1^1 \\ u_2^1 \\ \vdots \\ u_{N-1}^1 \end{bmatrix} = B\begin{bmatrix} h(1-h) \\ 2h(1-2h) \\ \vdots \\ (N-1)h(1-(N-1)h) \end{bmatrix} + kA\begin{bmatrix} \sin(5\pi h) \\ \sin(10\pi h) \\ \vdots \\ \sin(5(N-1)\pi h) \end{bmatrix}$$

produce $\mathbf{u}^1$. Once $\mathbf{u}^0$ and $\mathbf{u}^1$ are calculated, Eq. (12.19) can be used repeatedly to iterate forward in time. Figure 12.8 compares the finite difference approximation to the solution of the initial boundary value problem to the analytical solution. In this example the finite difference approximation is not as accurate as could be desired due to the coarseness of the grid chosen. Table 12.5 contains the numerical results of the implicit finite difference scheme.

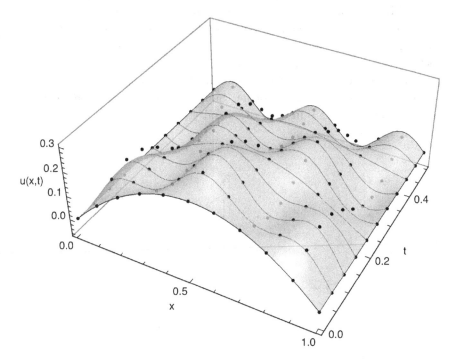

Fig. 12.8 A comparison of the numerically generated implicit finite difference method solution to the wave equation from Example 12.3 to the analytical solution. The accuracy of the numerical solution suffers from the large values chosen for h and k.

12.5   Laplace's and Poisson's Equations

The next family of partial differential equations and boundary value problems to be treated numerically is the family of elliptic equations. This family contains the important examples of Laplace's and Poisson's equations which were treated analytically in Chap. 6. In this section, boundary value problems will only be treated on rectangles though the methods presented here can be extended to disks, sectors, annuli, and connected regions with more general boundaries. For parabolic and hyperbolic equations, both explicit and implicit finite difference schemes were presented. For the implicit methods, the solution results from solving a tridiagonal system of equations, which can be carried out efficiently on computing devices. Elliptic boundary value problems will also have implicit finite difference methods for approximating the solution; however, the linear systems will in general not have tridiagonal form, but instead have a more general banded structure. Solving these linear systems will require much more work. Hence in this section, the finite difference schemes will be presented, but solutions to particular examples will be postponed until Sec. 12.6 when several iterative methods for solving linear systems of equations (as opposed to the direct methods discussed so far) are covered.

The simplest finite difference approximation to the Laplacian operator $\triangle u$ uses five points in the xy-plane. Suppose the plane has been gridded with a spacing of $h > 0$ in both the x and y directions.

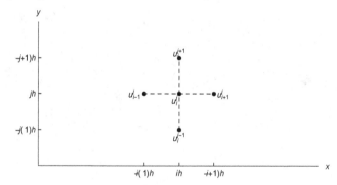

Let $u_i^j = u(i\,h, j\,h)$ for some integers i and j, then

$$
\begin{aligned}
u_{xx} + u_{yy} &= \frac{u(x+h,y) - 2u(x,y) + u(x-h,y)}{h^2} \\
&+ \frac{u(x,y+h) - 2u(x,y) + u(x,y-h)}{h^2} + O(h^2).
\end{aligned}
$$

Hence Poisson's equation can be approximated by the finite difference formula

$$\frac{1}{h^2}(u_{i+1}^j + u_{i-1}^j - 4u_i^j + u_i^{j+1} + u_i^{j-1}) = f(i\,h, j\,k), \qquad (12.20)$$

which has $O(h^2)$ truncation error. The five-point differencing approximation to the Laplacian on the left-hand side of Eq. (12.20) will be denoted as $\Delta_5 u_i^j$. If the domain of Poisson's equation is the rectangle $R = \{(x,y)\,|\,0 < x < N\,h, 0 < y < M\,h\}$ and the boundary conditions are Dirichlet type with $u(x,y) = g(x,y)$ on ∂R then the finite difference approximation can be written in the form of a linear system $A\,\mathbf{u} = \mathbf{b}$. The vector of unknowns $\mathbf{u}$ contains $(M-1)(N-1)$ elements arranged in what is called the **natural ordering** of the interior nodes [Golub and Ortega (1992), Chap. 9]

$$\mathbf{u} = \left(u_1^1, \ldots, u_{N-1}^1, u_1^2, \ldots, u_{N-1}^2, \ldots, u_1^{M-1}, \ldots, u_{N-1}^{M-1}\right)^T.$$

Figure 12.9 illustrates the natural ordering of the interior nodes and the boundary of rectangle R.

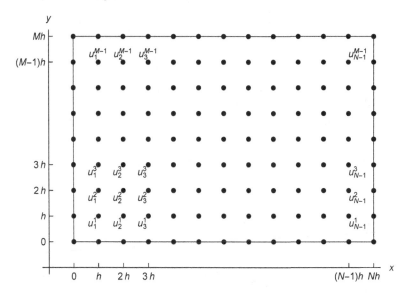

Fig. 12.9 The placement and ordering of the interior points of a rectangular grid discretizing Laplace's equation or Poisson's equation in a rectangle R.

Assuming the Dirichlet boundary condition, the boundary points in the discretization have values given by

$$u_i^0 = g(i\,h, 0), \quad u_i^M = g(i\,h, M\,h), \quad u_0^j = g(0, j\,h), \text{ and } u_N^j = g(N\,h, j)$$

for $i = 1, 2, \ldots, N - 1$ and $j = 1, 2, \ldots, M - 1$.

Matrix A is a square matrix with $(M - 1)^2(N - 1)^2$ entries found by multiplying both sides of Eq. (12.20) by $-h^2$. Matrix A tends to be a sparse matrix having many more zero than nonzero entries. It also has a regular block structure. Let $\hat{A}$ be the $(N - 1) \times (N - 1)$ tridiagonal matrix:

$$
\hat{A} = \begin{bmatrix}
4 & -1 & 0 & \cdots & 0 & 0 & 0 \\
-1 & 4 & -1 & \cdots & 0 & 0 & 0 \\
\vdots & \vdots & \vdots & & \vdots & \vdots & \vdots \\
0 & 0 & 0 & \cdots & -1 & 4 & -1 \\
0 & 0 & 0 & \cdots & 0 & -1 & 4
\end{bmatrix},
$$

let I be the $(N - 1) \times (N - 1)$ identity matrix, and let 0 represent the $(N - 1) \times (N - 1)$ matrix of all zero entries. Matrix A has the block structure,

$$
A = \begin{bmatrix}
\hat{A} & -I & 0 & \cdots & 0 & 0 & 0 \\
-I & \hat{A} & -I & \cdots & 0 & 0 & 0 \\
\vdots & \vdots & \vdots & & \vdots & \vdots & \vdots \\
0 & 0 & 0 & \cdots & -I & \hat{A} & -I \\
0 & 0 & 0 & \cdots & 0 & -I & \hat{A}
\end{bmatrix}.
\tag{12.21}
$$

There are $M - 1$ blocks in each row or column. Note that A is strictly diagonally dominant, thus ensuring there is a unique solution to the linear system $A\mathbf{u} = \mathbf{b}$. Vector $\mathbf{b}$ also has $(M - 1)(N - 1)$ elements. Using the natural ordering of the interior points

$$
\mathbf{b} = -h^2 \left(f_1^1, f_2^1, \ldots, f_{N-1}^1, f_1^2, f_2^2, \ldots, f_{N-1}^2, \ldots, f_1^{M-1}, f_2^{M-1}, \ldots, f_{N-1}^{M-1} \right)^T
$$

$$
+ \left(g_1^0, g_2^0, \ldots, g_{N-1}^0, \underbrace{0, \ldots, 0}_{(N-1)(M-3)}, g_1^M, g_2^M, \ldots, g_{N-1}^M \right)^T
$$

$$
+ \left(g_0^1, \underbrace{0, \ldots, 0}_{N-3}, g_N^1, g_0^2, \underbrace{0, \ldots, 0}_{N-3}, g_N^2, \ldots, g_0^{M-1}, \underbrace{0, \ldots, 0}_{N-3}, g_N^{M-1} \right)^T
$$

where to save space $f_i^j = f(i\,h, j\,h)$ and $g_i^j = g(i\,h, j\,h)$. If M and N are large, solving this system directly can be cumbersome. Many iterative methods employ schemes for numbering the interior points in the discretization of R to reduce the bandwidth of matrix A; however, these techniques will not be used or explored here.

Example 12.4. Derive the linear system of equations for approximating the solution to the following boundary value problem using the 5-point approximation of Eq. (12.20):

$$\triangle u = e^{x-y} \text{ on } R,$$

$$u(x, y) = x\,y \text{ on } \partial R.$$

Let the domain of the boundary value problem be $R = \{(x, y) \,|\, 0 < x < 4,\ 0 < y < 3\}$. Use $h = 1$ for simplicity.

Solution. Use of Eq. (12.20) and the specified boundary condition produce the following system of equations:

$$4u_1^1 - u_2^1 - u_1^2 = -1,$$

$$-u_1^1 + 4u_2^1 - u_3^1 - u_2^2 = -e,$$

$$-u_2^1 + 4u_3^1 - u_3^2 = 4 - e^2,$$

$$-u_1^1 + 4u_1^2 - u_2^2 = 3 - e^{-1},$$

$$-u_2^1 - u_1^2 + 4u_2^2 - u_3^2 = 5,$$

$$-u_3^1 - u_2^2 + 4u_3^2 = 17 - e.$$

Other more accurate approximations to the Laplacian operator can be used to develop linear systems of equations for approximating the solutions to Laplace's and Poisson's equations. Suppose $u(x, y)$ is smooth enough that it can be approximated as a Taylor polynomial in two variables of order five in an open disk centered at (x_0, y_0).

$$u(x, y) = \sum_{n=0}^{5} \left[\sum_{k=0}^{n} \frac{\partial^n u}{\partial x^{n-k} \partial y^k}(x_0, y_0) \frac{(x - x_0)^{n-k}(y - y_0)^k}{(n-k)!k!} \right] + E_6.$$

The expression E_6 denotes error terms of order 6. In Exercise 16 the reader will be asked to show that

$$2h^2(u_{xx}(x_0, y_0) + u_{yy}(x_0, y_0)) \tag{12.22}$$
$$= u(x_0 - h, y_0 - h) + u(x_0 + h, y_0 - h) + u(x_0 + h, y_0 + h)$$
$$+ u(x_0 - h, y_0 + h) - 4u(x_0, y_0)$$
$$- \frac{h^4}{6}(u_{xxxx}(x_0, y_0) + 6u_{xxyy}(x_0, y_0) + u_{yyyy}(x_0, y_0)) + O(h^6),$$

and

$$h^2(u_{xx}(x_0, y_0) + u_{yy}(x_0, y_0)) \tag{12.23}$$
$$= u(x_0 - h, y_0) + u(x_0 + h, y_0) + u(x_0, y_0 - h) + u(x_0, y_0 + h)$$
$$- 4u(x_0, y_0) - \frac{h^4}{12}(u_{xxxx}(x_0, y_0) + u_{yyyy}(x_0, y_0)) + O(h^6).$$

Multiplying both sides of Eq. (12.23) by 4 and adding the result to Eq. (12.22) produce

$$6h^2(u_{xx}(x_0, y_0) + u_{yy}(x_0, y_0)) \tag{12.24}$$

$$= 4u(x_0 - h, y_0) + 4u(x_0 + h, y_0) + 4u(x_0, y_0 - h) + 4u(x_0, y_0 + h)$$

$$+ u(x_0 - h, y_0 - h) + u(x_0 + h, y_0 - h) + u(x_0 + h, y_0 + h)$$

$$+ u(x_0 - h, y_0 + h) - 20u(x_0, y_0)$$

$$- \frac{h^4}{2}(u_{xxxx}(x_0, y_0) + 2u_{xxyy}(x_0, y_0) + u_{yyyy}(x_0, y_0)) + O(h^6).$$

Thus a nine-point $O(h^4)$ finite difference representation of $\triangle u$ is obtained from the following,

$$6h^2 \triangle_9 u_i^j = 4(u_{i-1}^j + u_{i+1}^j + u_i^{j-1} + u_i^{j+1}) - 20u_i^j$$

$$+ (u_{i-1}^{j-1} + u_{i+1}^{j-1} + u_{i-1}^{j+1} + u_{i+1}^{j+1}). \tag{12.25}$$

The truncation error of this approximation can be improved if the $O(h^6)$ terms in Eq. (12.24) are known. If $u(x, y)$ solves Poisson's equation (see Eq. (1.50)) in a neighborhood of (x_0, y_0) then

$$u_{xxxx}(x, y) + u_{xxyy}(x, y) = f_{xx}(x, y),$$

$$u_{xxyy}(x, y) + u_{yyyy}(x, y) = f_{yy}(x, y).$$

Adding these equations together and substituting the sum in Eq. (12.24) generate the following finite difference approximation to Poisson's equation,

$$\triangle_9 u_i^j = f(i\,h, j\,h) + \frac{h^2}{12}\triangle f(i\,h, j\,h). \tag{12.26}$$

This is a finite difference approximation to Poisson's equation with a truncation error of $O(h^4)$. Practical use of Eq. (12.26) requires that function f be known and its Laplacian can be calculated. If f is only represented by a table of values, the three-point differencing formula of Eq. (12.6) for f_{xx} and f_{yy} yields the $O(h^4)$ discrete approximation to Poisson's equation:

$$\triangle_9 u_i^j = \frac{1}{12}(f_{i-1}^j + f_{i+1}^j + 8f_i^j + f_i^{j-1} + f_i^{j+1}). \tag{12.27}$$

Note that to save space, the notation $f_i^j = f(i\,h, j\,h)$ is used. Multiplying both sides of Eq. (12.27) by $-6h^2$ enables the linear system to be written as

$$20u_i^j - 4(u_{i-1}^j + u_{i+1}^j + u_i^{j-1} + u_i^{j+1}) - (u_{i-1}^{j-1} + u_{i+1}^{j-1} + u_{i-1}^{j+1} + u_{i+1}^{j+1})$$

$$= -\frac{h^2}{2}(f_{i-1}^j + f_{i+1}^j + 8f_i^j + f_i^{j-1} + f_i^{j+1}) \tag{12.28}$$

for $i = 1, 2, \ldots, N-1$ and $j = 1, 2, \ldots, M-1$. Other higher accuracy finite difference approximations to Laplace's equation and Poisson's equation can be found in [Rosser (1975)]. When the boundary conditions are of Dirichlet type, the linear system of equations described in Eq. (12.28) can be written in matrix form as $A\mathbf{u} = -(h^2/2)B\mathbf{f} + \mathbf{b}$ where A is an $(M-1)(N-1) \times (M-1)(N-1)$ block banded matrix,

$$
A = \begin{bmatrix}
\hat{A} & -\hat{B} & 0 & \cdots & 0 & 0 & 0 \\
-\hat{B} & \hat{A} & -\hat{B} & \cdots & 0 & 0 & 0 \\
\vdots & \vdots & \vdots & & \vdots & \vdots & \vdots \\
0 & 0 & 0 & \cdots & -\hat{B} & \hat{A} & -\hat{B} \\
0 & 0 & 0 & \cdots & 0 & -\hat{B} & \hat{A}
\end{bmatrix}. \tag{12.29}
$$

Each block denoted by $\hat{A}$, $\hat{B}$, or 0 is an $(N-1) \times (N-1)$ matrix. As before 0 denotes the matrix of all zero entries. Matrices $\hat{A}$ and $\hat{B}$ are tridiagonal matrices with the following structures,

$$
\hat{A} = \begin{bmatrix}
20 & -4 & 0 & \cdots & 0 & 0 & 0 \\
-4 & 20 & -4 & \cdots & 0 & 0 & 0 \\
\vdots & \vdots & \vdots & & \vdots & \vdots & \vdots \\
0 & 0 & 0 & \cdots & -4 & 20 & -4 \\
0 & 0 & 0 & \cdots & 0 & -4 & 20
\end{bmatrix}
\quad \text{and} \quad
\hat{B} = \begin{bmatrix}
4 & 1 & 0 & \cdots & 0 & 0 & 0 \\
1 & 4 & 1 & \cdots & 0 & 0 & 0 \\
\vdots & \vdots & \vdots & & \vdots & \vdots & \vdots \\
0 & 0 & 0 & \cdots & 1 & 4 & 1 \\
0 & 0 & 0 & \cdots & 0 & 1 & 4
\end{bmatrix}.
$$

Matrix B is also $(M-1)(N-1) \times (M-1)(N-1)$ in size with a block banded structure,

$$
B = \begin{bmatrix}
\hat{C} & I & 0 & \cdots & 0 & 0 & 0 \\
I & \hat{C} & I & \cdots & 0 & 0 & 0 \\
\vdots & \vdots & \vdots & & \vdots & \vdots & \vdots \\
0 & 0 & 0 & \cdots & I & \hat{C} & I \\
0 & 0 & 0 & \cdots & 0 & I & \hat{C}
\end{bmatrix}.
$$

Each block denoted by $\hat{C}$, I, or 0 is an $(N-1) \times (N-1)$ matrix. Matrix I is an identity matrix. Matrix $\hat{C}$ is a tridiagonal matrix with the following structure:

$$
\hat{C} = \begin{bmatrix}
8 & 1 & 0 & \cdots & 0 & 0 & 0 \\
1 & 8 & 1 & \cdots & 0 & 0 & 0 \\
\vdots & \vdots & \vdots & & \vdots & \vdots & \vdots \\
0 & 0 & 0 & \cdots & 1 & 8 & 1 \\
0 & 0 & 0 & \cdots & 0 & 1 & 8
\end{bmatrix}.
$$

Finally vector $\mathbf{b}$ contains $(M-1)(N-1)$ elements and will be written as the sum of three vectors, $\mathbf{b} = \mathbf{f}_0 + \mathbf{g}_4 + \mathbf{g}$ where

$$
\mathbf{f}_0 = -\frac{h^2}{2}\left(f_0^1, \underbrace{0,\ldots,0}_{N-3}, f_N^1, f_0^2, \underbrace{0,\ldots,0}_{N-3}, f_N^2, \ldots, f_0^{M-1}, \underbrace{0,\ldots,0}_{N-3}, f_N^{M-1}\right)^T
$$

$$
-\frac{h^2}{2}\left(f_1^0, f_2^0, \ldots, f_{N-1}^0, \underbrace{0,\ldots,0}_{(N-1)(M-3)}, f_1^M, f_2^M, \ldots, f_{N-1}^M\right)^T
$$

has elements which depend on the right-hand side of Poisson's equation.

$$
\mathbf{g}_4 = 4\left(g_1^0, g_2^0, \ldots, g_{N-1}^0, \underbrace{0,\ldots,0}_{(N-1)(M-3)}, g_1^M, g_2^M, \ldots, g_{N-1}^M\right)^T \tag{12.30}
$$

$$
+ 4\left(g_0^1, \underbrace{0,\ldots,0}_{N-3}, g_N^1, g_0^2, \underbrace{0,\ldots,0}_{N-3}, g_N^2, \ldots, g_0^{M-1}, \underbrace{0,\ldots,0}_{N-3}, g_N^{M-1}\right)^T
$$

and finally

$$
\mathbf{g} = \left(g_0^0, g_1^0, \ldots, g_{N-2}^0, \underbrace{0,\ldots,0}_{(N-1)(M-3)}, g_0^M, g_1^M, \ldots, g_{N-2}^M\right)^T \tag{12.31}
$$

$$
+ \left(g_2^0, g_3^0, \ldots, g_N^0, \underbrace{0,\ldots,0}_{(N-1)(M-3)}, g_2^M, g_3^M, \ldots, g_N^M\right)^T
$$

$$
+ \left(g_0^2, \underbrace{0,\ldots,0}_{N-3}, g_N^2, g_0^3, \underbrace{0,\ldots,0}_{N-3}, g_N^3, \ldots, g_0^{M-1}, \underbrace{0,\ldots,0}_{N-3}, g_N^{M-1}, \underbrace{0,\ldots,0}_{N-1}\right)^T
$$

$$
+ \left(\underbrace{0,\ldots,0}_{N-1}, g_0^1, \underbrace{0,\ldots,0}_{N-3}, g_N^1, g_0^2, \underbrace{0,\ldots,0}_{N-3}, g_N^2, \ldots, g_0^{M-2}, \underbrace{0,\ldots,0}_{N-3}, g_N^{M-2}\right)^T.
$$

Example 12.5. Derive the linear system of equations for approximating the solution to the boundary value problem in Example 12.4 using the nine-point approximation to the Laplacian found in Eq. (12.27).

Solution. Using Eq. (12.27) with the specified boundary conditions results in the following system of linear equations:

$$-20u_1^1 + u_2^1 + u_1^2 + 4u_2^2 = 4 + e^{-1} + e,$$

$$u_1^1 - 20u_2^1 + u_3^1 + 4u_1^2 + u_2^2 + 4u_3^2 = 1 + 4e + e^2,$$

$$u_2^1 - 20u_3^1 + 4u_2^2 + u_3^2 = -36 + e + 4e^2 + e^3,$$

$$u_1^1 + 4u_2^1 - 20u_1^2 + u_2^2 = -26 + e^{-2} + e^{-1},$$

$$4u_1^1 + u_2^1 + 4u_3^1 + u_1^2 - 20u_2^2 + u_3^2 = -50 + e^{-1} + e,$$

$$4u_2^1 + u_3^1 + u_2^2 - 20u_3^2 = -104 + 4e + e^2.$$

While the linear systems developed in the previous two examples could be solved by Gaussian elimination, it is far more common to find iterative methods used to solve these types of linear systems. The next section will introduce and explore some iterative techniques for approximating the solution to a linear system of equations.

12.6 Iterative Methods

In this section three methods for approximating the solution to a matrix/vector equation of the form $A\mathbf{u} = \mathbf{b}$ are presented and examples of their use are given. In order of increasing sophistication and computational speed, the methods are the Jacobi[5] method, Gauss-Seidel[6] method, and the successive over-relaxation (SOR) method. All of these iterative methods take a common approach to solving a linear system. If A is an $n \times n$ matrix and $\mathbf{u}$ and $\mathbf{b}$ are column vectors with n components, then let $\mathbf{u}^{(0)}$ be an approximation to $\mathbf{u}$ where $\mathbf{u}$ is the solution of $A\mathbf{u} = \mathbf{b}$. Each iterative method generates a sequence of vectors $\{\mathbf{u}^{(k)}\}_{k=0}^{\infty}$ which converges to $\mathbf{u}$. During the course of this explanation of iterative techniques, several concepts and definitions from linear algebra will be encountered. Readers desiring a more comprehensive introduction to numerical linear algebra may wish to consult [Burden *et al.* (2016), Chap. 7].

12.6.1 *The Jacobi Method*

Consider a linear system of equations expressed in matrix/vector form as $A\mathbf{u} = \mathbf{b}$. Suppose the diagonal entries of matrix A are all nonzero. The

---

[5]Carl Gustav Jacob Jacobi, German mathematician (1804–1851).
[6]Philipp Ludwig von Seidel, German mathematician (1821–1896).

ith equation of this system can be expressed as

$$a_{i1}u_1 + a_{i2}u_2 + \cdots + a_{ii}u_i + \cdots + a_{in}u_n = b_i.$$

Solving this equation for u_i yields

$$u_i = \frac{1}{a_{ii}} \left(b_i - \sum_{j=1, j\neq i}^{n} a_{ij}u_j \right).$$

Thus if $\mathbf{u}^{(k)} = (u_1^{(k)}, u_2^{(k)}, \ldots, u_n^{(k)})^T$ for some $k = 0, 1, \ldots$ is an approximation to the solution of $A\mathbf{u} = \mathbf{b}$ then the next approximation is $\mathbf{u}^{(k+1)}$ where

$$u_i^{(k+1)} = \frac{1}{a_{ii}} \left(b_i - \sum_{j=1, j\neq i}^{n} a_{ij}u_j^{(k)} \right), \tag{12.32}$$

for $i = 1, 2, \ldots, n$. If matrix A is written as the sum of $D + L + U$ where D is a diagonal matrix who entries are the diagonal entries of A, L is a strictly lower triangular matrix (zeros on the diagonal) whose entries are the entries of A strictly below the diagonal, and U is a strictly upper triangular matrix whose entries are the entries of A strictly above the diagonal, then $A\mathbf{u} = (D + L + U)\mathbf{u} = \mathbf{b}$. If the diagonal entries of A are all nonzero then D is invertible and the Jacobi method expressed in Eq. (12.32) may be written in matrix/vector form as

$$\mathbf{u}^{(k+1)} = -D^{-1}(L + U)\mathbf{u}^{(k)} + D^{-1}\mathbf{b} = T_J\mathbf{u}^{(k)} + D^{-1}\mathbf{b} \tag{12.33}$$

where $T_J = -D^{-1}(L + U)$ is called the **Jacobi iteration matrix**. Thus given an initial approximation $\mathbf{u}^{(0)}$ to the solution, an infinite sequence $\{\mathbf{u}^{(k)}\}_{k=0}^{\infty}$ can be calculated.

Any practical iterative method must have a stopping condition to terminate calculation of further entries in the infinite sequence. A stopping condition should indicate that further iteration is unnecessary (either because the solution to the linear system has been found or because the results of further iteration will not appreciably change). Many choices of stopping condition are available and a generally effective choice is that the iteration should cease when the relative change in successive approximations falls below a specified threshold. Thus if the threshold is given by $\epsilon > 0$, calculation of further iterates of the sequence $\{\mathbf{u}^{(k)}\}$ can stop when

$$\frac{\|\mathbf{u}^{(k)} - \mathbf{u}^{(k+1)}\|_2}{\|\mathbf{u}^{(k+1)}\|_2} < \epsilon. \tag{12.34}$$

The subscript of 2 indicates the l_2-norm is used to measure the vectors (for more background on vector and matrix norms, see Sec. B.3). No choice of stopping criterion is perfect and sometimes alterations of the stopping condition are required. For example, the stopping condition given in Eq. (12.34) fails if $\mathbf{u}^{(k+1)} = \mathbf{0}$ for some $k = 0, 1, \ldots$. It may even fail to stop the iteration when $\|\mathbf{u}^{(k+1)}\|_2$ is small. Hence many algorithms will enforce a limit on the number of iterations so that runaway iteration is avoided.

Example 12.6. Use a maximum of five iterations of the Jacobi method to approximate a solution to the following system of linear equations. Let $\epsilon = 10^{-4}$ and $\mathbf{u}^{(0)} = (1, 1, 1, 1)^T$.

$$4u_1 + u_2 - u_3 + u_4 = 6,$$

$$u_1 + 4u_2 - u_3 - u_4 = 6,$$

$$-u_1 - u_2 + 4u_3 + u_4 = 3,$$

$$u_1 - u_2 + u_3 + 3u_4 = -2.$$

For the purposes of comparison the exact solution is $\mathbf{u} = \frac{1}{37}(80, 38, 73, -63)^T$.

Solution. Using Eq. (12.32) the iterates are found in the table below.

k	$u_1^{(k)}$	$u_2^{(k)}$	$u_3^{(k)}$	$u_4^{(k)}$
0	1.0000	1.0000	1.0000	1.0000
1	1.2500	1.7500	1.0000	-1.0000
2	1.5625	1.1875	1.7500	-0.8333
3	1.8490	1.3385	1.6458	-1.3750
4	1.9206	1.1055	1.8906	-1.3854
5	2.0426	1.1462	1.8529	-1.5686

The l_2-norm error in $\mathbf{u}^{(5)}$ is

$$\|\mathbf{u}^{(5)} - \mathbf{u}\|_2 \approx 0.2468$$

In this example the stopping condition expressed in Eq. (12.34) was not achieved in the first five iterations, thus emphasizing the need for practical implementations of the Jacobi method (and other iterative methods) to specify a maximum number of iterations to compute.

12.6.2 The Gauss-Seidel Method

A simple observation leads to an improvement in the Jacobi method. In Eq. (12.32) the ith entry of $\mathbf{u}^{(k+1)}$ is calculated using the entries of $\mathbf{u}^{(k)}$, even though the jth entry for $j = 1, 2, \ldots, i-1$ of $\mathbf{u}^{(k+1)}$ has already been calculated. In general the entries of $\mathbf{u}^{(k+1)}$ are more accurate approximations to the solution of $A\,\mathbf{u} = \mathbf{b}$ than are the entries of $\mathbf{u}^{(k)}$. Use of these more accurate results should reduce the number of iterations required to reach a specified degree of accuracy. Thus the Gauss-Seidel method modifies the Jacobi method by using the already calculated entries of $\mathbf{u}^{(k+1)}$ in place of the corresponding entries of $\mathbf{u}^{(k)}$.

$$u_i^{(k+1)} = \frac{1}{a_{ii}} \left(b_i - \sum_{j=1}^{i-1} a_{ij} u_j^{(k+1)} - \sum_{j=i+1}^{n} a_{ij} u_j^{(k)} \right). \tag{12.35}$$

The same stopping criteria can be used for the Gauss-Seidel method as for the Jacobi method.

Example 12.7. Solve the linear system of Example 12.6 using the Gauss-Seidel method.

Solution. The iterates of the Gauss-Seidel method are tabulated below.

k	$u_1^{(k)}$	$u_2^{(k)}$	$u_3^{(k)}$	$u_4^{(k)}$
0	1.0000	1.0000	1.0000	1.0000
1	1.2500	1.6875	1.2344	−0.9323
2	1.6198	1.1706	1.6807	−1.3766
3	1.9717	1.0831	1.8578	−1.5821
4	2.0892	1.0466	1.9295	−1.6574
5	2.1351	1.0343	1.9567	−1.6858

The error in the last approximation to the solution is

$$\|\mathbf{u}^{(5)} - \mathbf{u}\|_2 \approx 0.0002$$

which is clearly an improvement over the Jacobi method's result after five iterations.

Since the Gauss-Seidel method is no more difficult to carry out than the Jacobi method and usually achieves greater accuracy in fewer iterations, it will now be used to solve Laplace's equation on a rectangle using the 5-point difference approximation to the Laplacian operator. Consider the following boundary value problem:

$$u_{xx} + u_{yy} = 0 \text{ for } 0 < x < 1 \text{ and } 0 < y < 1,$$
$$u(x,0) = e^x \text{ for } 0 < x < 1,$$
$$u(x,1) = \sin(\pi x/2) \text{ for } 0 < x < 1,$$
$$u(0,y) = 1 - y^2 \text{ for } 0 < y < 1,$$
$$u(1,y) = e^{1-y} \text{ for } 0 < y < 1.$$

If $h = 1/10$, then the linear system $A\mathbf{u} = \mathbf{b}$ consists of a 81×81 matrix A and a column vector $\mathbf{b}$ with 81 elements. Setting a numerical tolerance of $\epsilon = 10^{-6}$ in Eq. (12.34) and using the Gauss-Seidel method requires on the order of 100 iterations from a starting approximation of $\mathbf{u}^{(0)} = (1, 1, \ldots, 1)^T$. The solution is plotted in Fig. 12.10.

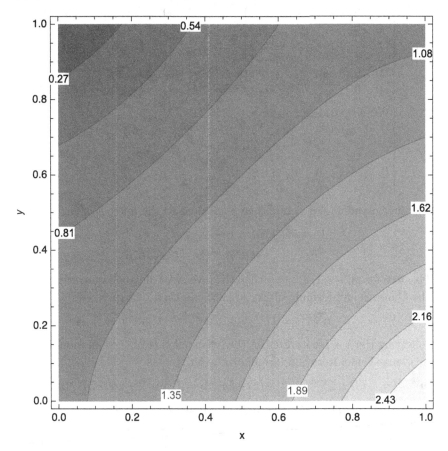

Fig. 12.10 A contour plot of the result of applying the Gauss-Seidel method to the task of solving Laplace's equation on a rectangle. Since nonzero boundary conditions are present on all four sides of the rectangle, the analytical approach to the solution would have required solving four separate boundary value problems (one for each side) and adding the results.

While the Gauss-Seidel method is an improvement over the Jacobi method in terms of reducing the number of iterations required to approximate the solution to a linear system of equations to a given accuracy, there are additional methods which can speed up the convergence of iterative methods. The next section will explore one popular method known as successive over-relaxation. Before entering that discussion the Gauss-Seidel method will be recast in matrix/vector format.

If A is an $n \times n$ matrix, then $A = D + L + U$ where

$$D = \begin{bmatrix} a_{11} & 0 & \cdots & 0 & 0 \\ 0 & a_{22} & \cdots & 0 & 0 \\ \vdots & \vdots & & \vdots & \vdots \\ 0 & 0 & \cdots & a_{n-1,n-1} & 0 \\ 0 & 0 & \cdots & 0 & a_{nn} \end{bmatrix}, \quad L = \begin{bmatrix} 0 & 0 & \cdots & 0 & 0 \\ a_{21} & 0 & \cdots & 0 & 0 \\ \vdots & \vdots & & \vdots & \vdots \\ a_{n-1,1} & a_{n-1,2} & \cdots & 0 & 0 \\ a_{n1} & a_{n2} & \cdots & a_{n,n-1} & 0 \end{bmatrix},$$

$$U = \begin{bmatrix} 0 & a_{12} & \cdots & a_{1,n-1} & a_{1n} \\ 0 & 0 & \cdots & a_{2,n-1} & a_{2n} \\ \vdots & \vdots & & \vdots & \vdots \\ 0 & 0 & \cdots & 0 & a_{n-1,n} \\ 0 & 0 & \cdots & 0 & 0 \end{bmatrix}.$$

The linear system $A\mathbf{u} = \mathbf{b}$ is then equivalent to the linear system

$$(D + L + U)\mathbf{u} = \mathbf{b}$$
$$\mathbf{u} = D^{-1}(\mathbf{b} - L\mathbf{u} - U\mathbf{u}).$$

If $a_{ii} \neq 0$ for all $i = 1, 2, \ldots, n$ then the inverse of diagonal matrix D exists. The Gauss-Seidel formula of Eq. (12.35) can then be expressed as

$$\mathbf{u}^{(k+1)} = D^{-1}(\mathbf{b} - L\mathbf{u}^{(k+1)} - U\mathbf{u}^{(k)}).$$

Collecting the $k + 1$st iterates on the left-hand side of the equation enables the Gauss-Seidel iteration formula to be written as

$$\mathbf{u}^{(k+1)} = -(D+L)^{-1}U\mathbf{u}^{(k)} + (D+L)^{-1}\mathbf{b} = T_{GS}\mathbf{u}^{(k)} + (D+L)^{-1}\mathbf{b} \quad (12.36)$$

where $T_{GS} = -(D+L)^{-1}U$ is called the **Gauss-Seidel iteration matrix**.

The vector $\mathbf{u}^{(k+1)} - \mathbf{u}^{(k)}$ can be pictured geometrically as the vector from $\mathbf{u}^{(k)}$ (the kth approximation) to $\mathbf{u}^{(k+1)}$. In matrix/vector form it can be written as

$$\mathbf{u}^{(k+1)} - \mathbf{u}^{(k)} = D^{-1}(\mathbf{b} - L\mathbf{u}^{(k+1)} - D\mathbf{u}^{(k)} - U\mathbf{u}^{(k)}). \quad (12.37)$$

Thus the Gauss-Seidel method can be thought of as a process of starting with an initial approximation $\mathbf{u}^{(0)}$ of the solution to the linear system $A\mathbf{u} = \mathbf{b}$ and taking a sequence of vector steps given by the expression on the right-hand side of Eq. (12.37) until a sufficiently accurate approximation to the exact solution of the linear system is achieved.

12.6.3   The Successive Over-Relaxation Method

Since $\mathbf{u}^{(k+1)} - \mathbf{u}^{(k)}$ can be thought of as the step to take from approximation $\mathbf{u}^{(k)}$ to the better approximation $\mathbf{u}^{(k+1)}$, a natural improvement to the method is to take a longer step in the same direction. If $\omega > 1$ then $\omega(\mathbf{u}^{(k+1)} - \mathbf{u}^{(k)})$ is a vector in the same direction as $\mathbf{u}^{(k+1)} - \mathbf{u}^{(k)}$ but with greater magnitude. By taking this longer corrective step an algorithm can "accelerate" toward the exact solution. Consider the equation,

$$\mathbf{u}^{(k+1)} = \mathbf{u}^{(k)} + \omega(\mathbf{u}^{(k+1)} - \mathbf{u}^{(k)}). \tag{12.38}$$

When $\omega > 1$ an iterative method based on Eq. (12.38) is called an **over-relaxation method**. Naturally when $0 < \omega < 1$ an iterative method based on Eq. (12.38) would be called an **under-relaxation method**. The **successive over-relaxation method** (abbreviated as SOR) is an accelerated modification of the Gauss-Seidel method created by combining Eqs. (12.37) and (12.38). The SOR technique was initially developed by Young[7] in his Ph.D. thesis [Young (1950)].

$$\mathbf{u}^{(k+1)} = \mathbf{u}^{(k)} + \omega[D^{-1}(\mathbf{b} - L\,\mathbf{u}^{(k+1)} - D\,\mathbf{u}^{(k)} - U\,\mathbf{u}^{(k)})]$$
$$(D + \omega L)\mathbf{u}^{(k+1)} = [(1 - \omega)D - \omega U]\,\mathbf{u}^{(k)} + \omega\mathbf{b} \tag{12.39}$$
$$\mathbf{u}^{(k+1)} = T_{SOR}\mathbf{u}^{(k)} + \omega(D + \omega L)^{-1}\mathbf{b}$$

where $T_{SOR} = (D + \omega L)^{-1}[(1 - \omega)D - \omega U]$ is called the **SOR iteration matrix**. The ith component of $\mathbf{u}^{(k+1)}$ can be written as

$$u_i^{(k+1)} = u_i^{(k)} + \frac{\omega}{a_{ii}}\left(b_i - \sum_{j=1}^{i-1}a_{ij}u_j^{(k+1)} - \sum_{j=i}^{n}a_{ij}u_j^{(k)}\right). \tag{12.40}$$

Note that if $\omega = 1$ then the successive over-relaxation method is just the Gauss-Seidel method.

Example 12.8. Compare the results of the Gauss-Seidel and SOR methods with $\omega = 1.21539$ for approximating the solution to the following linear system of equations. For both methods use $\epsilon = 10^{-4}$ in the stopping criterion.

$$\begin{bmatrix} -4 & 1 & 0 & 0 & 0 \\ 1 & -4 & 1 & 0 & 0 \\ 0 & 1 & -4 & 1 & 0 \\ 0 & 0 & 1 & -4 & 1 \\ 0 & 0 & 0 & 1 & -4 \end{bmatrix} \begin{bmatrix} x_1 \\ x_2 \\ x_3 \\ x_4 \\ x_5 \end{bmatrix} = \begin{bmatrix} 1 \\ 2 \\ 3 \\ 4 \\ 5 \end{bmatrix}.$$

---

[7]David M. Young, Jr., American mathematician (1923–2008).

Solution. In both cases the initial approximation to the solution will be $(0, 0, 0, 0, 0)^T$. The first ten iterations of the Gauss-Seidel method yield a close approximation of the solution to the equation. The exact solution is $\mathbf{x} = (-129/260, -64/65, -75/52, -116/65, -441/260)^T$.

k	$x_1^{(k)}$	$x_2^{(k)}$	$x_3^{(k)}$	$x_4^{(k)}$	$x_5^{(k)}$
0	0.0000	0.0000	0.0000	0.0000	0.0000
1	−0.2500	−0.5625	−0.8906	−1.2227	−1.5557
2	−0.3906	−0.8203	−1.2607	−1.7041	−1.6760
3	−0.4551	−0.9290	−1.4083	−1.7711	−1.6928
4	−0.4822	−0.9726	−1.4359	−1.7822	−1.6955
5	−0.4932	−0.9823	−1.4411	−1.7842	−1.6960
6	−0.4956	−0.9842	−1.4421	−1.7845	−1.6961
7	−0.4960	−0.9845	−1.4423	−1.7846	−1.6961
8	−0.4961	−0.9846	−1.4423	−1.7846	−1.6962
9	−0.4961	−0.9846	−1.4423	−1.7846	−1.6962
10	−0.4962	−0.9846	−1.4423	−1.7846	−1.6962

The SOR method has similar performance though to the four decimal places reported in the table, the 9th and 10th iterates are the same.

k	$x_1^{(k)}$	$x_2^{(k)}$	$x_3^{(k)}$	$x_4^{(k)}$	$x_5^{(k)}$
0	0.0000	0.0000	0.0000	0.0000	0.0000
1	−0.3038	−0.7000	−1.1242	−1.5570	−1.9923
2	−0.4511	−0.9356	−1.4268	−1.9189	−1.6732
3	−0.4910	−0.9889	−1.4878	−1.7625	−1.6944
4	−0.4986	−0.9982	−1.4299	−1.7851	−1.6967
5	−0.4998	−0.9790	−1.4434	−1.7850	−1.6962
6	−0.4937	−0.9854	−1.4424	−1.7846	−1.6961
7	−0.4969	−0.9847	−1.4423	−1.7846	−1.6962
8	−0.4960	−0.9845	−1.4423	−1.7846	−1.6962
9	−0.4962	−0.9846	−1.4423	−1.7846	−1.6962
10	−0.4962	−0.9846	−1.4423	−1.7846	−1.6962

There is no known formula for the optimal value of ω to accelerate the SOR method for a general $n \times n$ matrix A. For certain banded matrices (for example tridiagonal matrices) the optimal value of ω can be determined *a priori* [Ortega (1990)]. As an example, if $\omega \approx 1.52786$ then the SOR method approximates the solution to the boundary value problem explored in the previous section in 25 iterations as opposed to nearly 100 iterations required for the Gauss-Seidel method.

Example 12.9. Use the successive over-relaxation method with $\omega = 1.81621$, the nine-point approximation to the Laplacian operator given in Eq. (12.26), and $h = 1/10$ to approximate the solution to the boundary value problem described in Example 12.4. Use $\epsilon = 10^{-6}$ in the stopping criterion.

Solution. The rectangle over which the solution is sought has dimensions of width 4 and height 3. With Dirichlet boundary conditions and choosing $h = 1/10$ there are

$$\left(\frac{4}{1/10} - 1\right)\left(\frac{3}{1/10} - 1\right) = 1131$$

interior points at which the solution to Poisson's equation must be approximated. The linear system to be solved is of the form $A\mathbf{u} = \mathbf{b}$ where A is a banded tridiagonal matrix of size 1131×1131. Despite the size of the linear system, the SOR technique solves it quickly on most modern computing devices in 92 iterations from a starting approximation of $\mathbf{u}^{(0)} = \mathbf{0}$. A contour plot of the solution is shown in Fig. 12.11.

12.7 Convergence and Stability

As each of the finite difference schemes for the heat, wave, and Poisson's equations was developed, the method was compactly summarized as a linear system of equations in matrix/vector form. The question of convergence and stability of the finite difference schemes is closely tied to the notion of the eigenvalues and eigenvectors of the coefficient matrix present in the linear system.

Lemma 12.1. *Suppose A is an $(n-1) \times (n-1)$ tridiagonal matrix of the form*

$$A = \begin{bmatrix} a & b & 0 & \cdots & 0 & 0 & 0 \\ c & a & b & \cdots & 0 & 0 & 0 \\ \vdots & \vdots & \vdots & & \vdots & \vdots & \vdots \\ 0 & 0 & 0 & \cdots & c & a & b \\ 0 & 0 & 0 & \cdots & 0 & c & a \end{bmatrix}$$

with real or complex entries. If $bc \neq 0$ then the eigenvalues of A are

$$\lambda_j = a + 2b\sqrt{\frac{c}{b}}\cos\frac{j\pi}{n}, \qquad (12.41)$$

534   *A First Course in Partial Differential Equations*

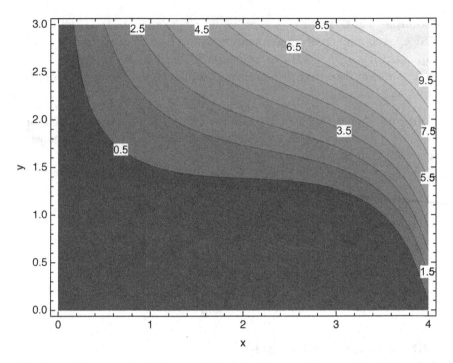

Fig. 12.11   A contour plot of the result of applying the successive over-relaxation method to the task of solving Poisson's equation on a rectangle in Example 12.9.

with corresponding eigenvectors

$$\mathbf{v}_j = \left(\left(\frac{c}{b}\right)^{1/2} \sin\frac{j\pi}{n}, \ldots, \left(\frac{c}{b}\right)^{(n-1)/2} \sin\frac{(n-1)j\pi}{n} \right)^T, \qquad (12.42)$$

for $j = 1, 2, \ldots, n-1$.

Proof. If λ is an eigenvalue with corresponding eigenvector $\mathbf{v}$ then

$$\begin{bmatrix} a & b & 0 & \cdots & 0 & 0 & 0 \\ c & a & b & \cdots & 0 & 0 & 0 \\ \vdots & \vdots & \vdots & & \vdots & \vdots & \vdots \\ 0 & 0 & 0 & \cdots & c & a & b \\ 0 & 0 & 0 & \cdots & 0 & c & a \end{bmatrix} \begin{bmatrix} v_1 \\ v_2 \\ \vdots \\ v_{n-2} \\ v_{n-1} \end{bmatrix} = \lambda \begin{bmatrix} v_1 \\ v_2 \\ \vdots \\ v_{n-2} \\ v_{n-1} \end{bmatrix}.$$

Without loss of generality a zero entry may be prepended and appended to $\mathbf{v}$ so that $\hat{\mathbf{v}} = (0, v_1, v_2, \ldots, v_{n-1}, 0)^T$ and the linear system of equations may be written as

$$cv_{m-1} + av_m + bv_{m+1} = \lambda v_m \iff cv_{m-1} + (a-\lambda)v_m + bv_{m+1} = 0 \quad (12.43)$$

for $m = 1, 2, \ldots, n - 1$. Since $b \neq 0$ and $c \neq 0$ then Eq. (12.43) is an example of a second-order finite difference equation. Such equations can be solved in a manner similar to that used to solve second-order linear differential equations. Assume the solution takes the form $v_m = r^m$ where r is a complex number. Substituting this solution into Eq. (12.43) yields

$$0 = c\, r^{m-1} + (a - \lambda)r^m + b\, r^{m+1}$$

$$0 = b\, r^2 + (a - \lambda)r + c$$

$$r = \frac{\lambda - a \pm ((\lambda - a)^2 - 4bc)^{1/2}}{2b}.$$

Thus there are two possible values for r which will be denoted as r_1 and r_2. If $r_1 \neq r_2$ then by the Principle of Superposition, the general solution to Eq. (12.43) can be expressed as $v_m = \alpha\, r_1^m + \beta\, r_2^m$ for any complex numbers α and β.

Since the prepended entry of the eigenvector $\mathbf{v}$ can be thought of as the initial value of the difference equation, then $v_0 = 0 = \alpha + \beta$, or equivalently $\alpha = -\beta$. Thus α must be nonzero since otherwise, $v_m = 0$ for $m = 1, 2, \ldots, n - 1$ contradicting the assumption that $\mathbf{v}$ is an eigenvector of matrix A. Therefore $v_m = \alpha(r_1^m - r_2^m)$ solves Eq. (12.43). Likewise, since the appended entry of the eigenvector is $v_n = 0$, then $r_1^n = r_2^n$. This implies $(r_1/r_2)^n = 1$ and thus r_1/r_2 must be one of the nth roots of unity which implies

$$\frac{r_1}{r_2} = e^{2j\pi i/n} \tag{12.44}$$

for some $j \in \{1, 2, \ldots, n - 1\}$. The case in which $j = 0$ is not permitted since that would contradict the assumption that $r_1 \neq r_2$. Since r_1 and r_2 are roots of the quadratic equation, then if the quadratic equation can be factored as

$$0 = r^2 + \frac{a - \lambda}{b}r + \frac{c}{b} = (r - r_1)(r - r_2),$$

equating coefficients of r reveals

$$r_1 r_2 = \frac{c}{b} \text{ and } r_1 + r_2 = \frac{\lambda - a}{b}. \tag{12.45}$$

Multiplying both sides of Eq. (12.44) by r_2^2 and using the first part of Eq. (12.45) produce

$$r_1 = \sqrt{\frac{c}{b}}e^{j\pi i/n} \text{ and } r_2 = \sqrt{\frac{c}{b}}e^{-j\pi i/n}.$$

Now that formulas for r_1 and r_2 have been found, the second part of Eq. (12.45) furnishes a formula for the eigenvalue λ.

$$\frac{\lambda - a}{b} = \sqrt{\frac{c}{b}}(e^{j\pi i/n} + e^{-j\pi i/n})$$

$$\lambda - a = b\sqrt{\frac{c}{b}}\left(\cos\frac{j\pi}{n} + i\sin\frac{j\pi}{n} + \cos\frac{j\pi}{n} - i\sin\frac{j\pi}{n}\right)$$

$$\lambda = a + 2b\sqrt{\frac{c}{b}}\cos\frac{j\pi}{n}.$$

Thus it has been established that A has $n - 1$ eigenvalues λ_j of the form given in Eq. (12.41) for $j \in \{1, 2, \ldots, n - 1\}$.

Earlier it was assumed that $r_1 \neq r_2$. This assumption can now be justified using a proof by contradiction argument. If $r_1 = r_2 = (\lambda - a)/(2b)$ then a second solution to Eq. (12.43) is $v_m = m r_1^m$ (see Exercise 21). Once again the Principle of Superposition implies $v_m = (\alpha + \beta m)r_1^m$ solves Eq. (12.43). For $m = 0$ and $m = n$ the following equations hold:

$$\alpha r_1^0 = 0,$$

$$(\alpha + \beta n)r_1^n = 0.$$

The root $r_1 \neq 0$ or else $v_m = 0$ for all $m = 1, 2, \ldots, n - 1$ contradicting the assumption that $\mathbf{v}$ is an eigenvector of A. Hence $\alpha = \beta = 0$ which also implies $\mathbf{v} = \mathbf{0}$, again contradicting the assumption that $\mathbf{v}$ is an eigenvector of matrix A. Thus $r_1 \neq r_2$.

Now that the eigenvalues of A are known the corresponding eigenvectors can be determined. Fix $j \in \{1, 2, \ldots, n - 1\}$ and let $\mathbf{v}_j$ denote the eigenvector corresponding to λ_j. Without loss of generality the coefficients of r_1 and r_2 are chosen to be $\alpha = -i/2$ and $\beta = i/2$ respectively and thus the mth component of $\mathbf{v}_j$ is

$$v_{j,m} = -\frac{i}{2}\left(\frac{c}{b}\right)^{m/2}e^{jm\pi i/n} + \frac{i}{2}\left(\frac{c}{b}\right)^{m/2}e^{-jm\pi i/n}$$

$$= -\frac{i}{2}\left(\frac{c}{b}\right)^{m/2}(\cos\frac{jm\pi}{n} + i\sin\frac{jm\pi}{n} - \cos\frac{jm\pi}{n} + i\sin\frac{jm\pi}{n})$$

$$= \left(\frac{c}{b}\right)^{m/2}\sin\frac{jm\pi}{n}.$$

Consequently the eigenvectors $\mathbf{v}_j$ have the form shown in Eq. (12.42). □

Many of the matrices encountered in the finite difference techniques have a repeating block pattern. The following theorem can be helpful for determining the eigenvalues and eigenvectors of matrices with special block structure.

Theorem 12.2. *Let matrix A be an $NM \times NM$ matrix written in block form as*

$$A = \begin{bmatrix} A_{1,1} & A_{1,2} & \cdots & A_{1,M} \\ A_{2,1} & A_{2,2} & \cdots & A_{2,M} \\ \vdots & \vdots & & \vdots \\ A_{M,1} & A_{M,2} & \cdots & A_{M,M} \end{bmatrix}.$$

Suppose each block $A_{i,j}$ is an $N \times N$ matrix and all the matrices $A_{i,j}$ have a set of N linearly independent eigenvectors $\{\mathbf{v}_1, \mathbf{v}_2, \ldots, \mathbf{v}_N\}$ in common. Then the eigenvalues of matrix A are the eigenvalues of the matrices

$$\Lambda_k = \begin{bmatrix} \lambda_{1,1}^{(k)} & \lambda_{1,2}^{(k)} & \cdots & \lambda_{1,M}^{(k)} \\ \lambda_{2,1}^{(k)} & \lambda_{2,2}^{(k)} & \cdots & \lambda_{2,M}^{(k)} \\ \vdots & \vdots & & \vdots \\ \lambda_{M,1}^{(k)} & \lambda_{M,2}^{(k)} & \cdots & \lambda_{M,M}^{(k)} \end{bmatrix} \quad \text{for } k = 1, 2, \ldots, N,$$

where $\lambda_{i,j}^{(k)}$ is the eigenvalue of $A_{i,j}$ corresponding to the common eigenvector $\mathbf{v}_k$.

Proof. Suppose λ is an eigenvalue of Λ_k. There exists a vector $\boldsymbol{\alpha} = (\alpha_1, \alpha_2, \ldots, \alpha_M)^T \neq \mathbf{0}$ for which $\Lambda_k \boldsymbol{\alpha} = \lambda \boldsymbol{\alpha}$. For each $i = 1, 2, \ldots, M$ and $k \in \{1, 2, \ldots, N\}$,

$$\sum_{j=1}^{M} \lambda_{i,j}^{(k)} \alpha_j = \lambda \alpha_i \implies \sum_{j=1}^{M} \lambda_{i,j}^{(k)} \alpha_j \mathbf{v}_k = \sum_{j=1}^{M} A_{i,j} \alpha_j \mathbf{v}_k = \lambda \alpha_i \mathbf{v}_k.$$

Therefore,

$$A \begin{bmatrix} \alpha_1 \mathbf{v}_k \\ \alpha_2 \mathbf{v}_k \\ \vdots \\ \alpha_M \mathbf{v}_k \end{bmatrix} = \lambda \begin{bmatrix} \alpha_1 \mathbf{v}_k \\ \alpha_2 \mathbf{v}_k \\ \vdots \\ \alpha_M \mathbf{v}_k \end{bmatrix},$$

which implies λ is an eigenvalue of A with corresponding eigenvector $(\alpha_1 \mathbf{v}_k^T, \alpha_2 \mathbf{v}_k^T, \ldots, \alpha_M \mathbf{v}_k^T)^T$. Hence every eigenvalue of Λ_k is an eigenvalue of A for $k = 1, 2, \ldots, N$.

Now suppose A has an eigenvalue μ with corresponding eigenvector $\mathbf{u} = (\mathbf{u}_1^T, \mathbf{u}_2^T, \ldots, \mathbf{u}_M^T)^T$ where for each $i \in \{1, 2, \ldots, M\}$, $\mathbf{u}_i$ is a vector with N components. For each $i = 1, 2, \ldots, M$,

$$\sum_{j=1}^{M} A_{i,j} \mathbf{u}_j = \mu \mathbf{u}_i. \tag{12.46}$$

By assumption $\{\mathbf{v}_1, \mathbf{v}_2, \ldots, \mathbf{v}_N\}$ is a linearly independent set and hence for $i = 1, 2, \ldots, M$ there exist unique scalars $c_{i,k}$ such that $\sum_{k=1}^{N} c_{i,k} \mathbf{v}_k = \mathbf{u}_i$. Therefore Eq. (12.46) can be rewritten as

$$\sum_{j=1}^{M} A_{i,j} \sum_{k=1}^{N} c_{j,k} \mathbf{v}_k = \sum_{j=1}^{M} \sum_{k=1}^{N} c_{j,k} A_{i,j} \mathbf{v}_k = \sum_{j=1}^{M} \sum_{k=1}^{N} c_{j,k} \lambda_{i,j}^{(k)} \mathbf{v}_k = \mu \sum_{k=1}^{N} c_{i,k} \mathbf{v}_k$$

$$\sum_{k=1}^{N} \left(\sum_{j=1}^{M} \lambda_{i,j}^{(k)} c_{j,k} \right) \mathbf{v}_k = \sum_{k=1}^{N} (\mu\, c_{i,k}) \mathbf{v}_k.$$

Thus for $i = 1, 2, \ldots, M$ and $k = 1, 2, \ldots, N$, $\sum_{j=1}^{M} \lambda_{i,j}^{(k)} c_{j,k} = \mu\, c_{i,k}$. Hence for each $k \in \{1, 2, \ldots, N\}$ a system of linear equations can be written in matrix/vector form as

$$\begin{bmatrix} \lambda_{1,1}^{(k)} & \lambda_{1,2}^{(k)} & \cdots & \lambda_{1,M}^{(k)} \\ \lambda_{2,1}^{(k)} & \lambda_{2,2}^{(k)} & \cdots & \lambda_{2,M}^{(k)} \\ \vdots & \vdots & & \vdots \\ \lambda_{M,1}^{(k)} & \lambda_{M,2}^{(k)} & \cdots & \lambda_{M,M}^{(k)} \end{bmatrix} \begin{bmatrix} c_{1,k} \\ c_{2,k} \\ \vdots \\ c_{M,k} \end{bmatrix} = \mu \begin{bmatrix} c_{1,k} \\ c_{2,k} \\ \vdots \\ c_{M,k} \end{bmatrix}.$$

If all of the coefficients $c_{i,k} = 0$ for $i = 1, 2, \ldots, M$ and $k = 1, 2, \ldots, N$ then $\mathbf{u} = \mathbf{0}$ which contradicts the assumption that $\mathbf{u}$ is an eigenvector of A. Hence there exists $k \in \{1, 2, \ldots, N\}$ such that $\mathbf{c} = (c_{1,k}, c_{2,k}, \ldots, c_{M,k})^T$ is an eigenvector of Λ_k with eigenvalue μ. $\square$

With the clever work accomplished of finding a formula for the eigenvalues of special types of tridiagonal and block matrices, attention can return to the notion of stability. In non-technical language, an algorithm (such as a finite difference scheme) is **stable** if small changes in the input data result in proportionately small changes in the output data. If an algorithm is not stable, then it is labeled **unstable**. Adopting the classification defined in [Burden *et al.* (2016), Chap. 1], if e_0 is the error present in the data (either initially or after a given calculation) and e_n is the error after n subsequent calculations, then the growth rate of the error is **linear** if $e_n \propto n\, e_0$ and the growth rate of the error is **exponential** if $e_n \propto \gamma^n e_0$ for some $\gamma > 1$. The symbol $\propto$ indicates proportionality. There are three primary types of error: truncation error, measurement error, and rounding error. Truncation error is due to the use of a finite number of terms taken from a Taylor series to develop the approximations to various derivatives and derivative operators used in the partial differential equations. Measurement error comes from approximations used to set the initial and boundary conditions of an initial boundary value problem and round-off errors result from the machine arithmetic used by computing devices. Even if measurement errors are eliminated, truncation error and rounding error will still be present.

12.7.1 *Stability of Finite Difference Schemes for the Heat Equation*

Consider the explicit finite difference scheme for the heat equation found in Eq. (12.9). The matrix on the left-hand side of Eq. (12.9) (call it $A(r)$) is real, tridiagonal, and symmetric, thus the eigenvalues are all distinct and the eigenvectors form a basis for the vector space $\mathbb{R}^{N-1}$. Suppose after the completion of the calculation of $\mathbf{u}^{(j)}$, the vector can be expressed as $\mathbf{u}^{(j)} = \hat{\mathbf{u}}^{(j)} + \mathbf{e}^{(0)}$ where $\hat{\mathbf{u}}^{(j)}$ is the exact solution of the finite difference equations. The error in the calculation of $\mathbf{u}^{(j)}$ is therefore $\mathbf{e}^{(0)}$. After n additional time steps forward

$$\mathbf{u}^{(j+n)} = (A(r))^n \mathbf{u}^{(j)}$$
$$\hat{\mathbf{u}}^{(j+n)} + \mathbf{e}^{(n)} = (A(r))^n (\hat{\mathbf{u}}^{(j)} + \mathbf{e}^{(0)})$$
$$\mathbf{e}^{(n)} = (A(r))^n \mathbf{e}^{(0)}.$$

Since the eigenvectors of $A(r)$ form a basis for $\mathbb{R}^{N-1}$ then there exist constants $c_1, c_2, \ldots, c_{N-1}$ such that $\mathbf{e}^{(0)} = c_1 \mathbf{v}_1 + c_2 \mathbf{v}_2 + \cdots + c_{N-1} \mathbf{v}_{N-1}$. In this case

$$\mathbf{e}^{(n)} = (A(r))^n \sum_{k=1}^{N-1} c_k \mathbf{v}_k = \sum_{k=1}^{N-1} c_k \lambda_k^n \mathbf{v}_k.$$

Thus the explicit finite difference scheme for the heat equation given in matrix form in Eq. (12.9) is stable if $|\lambda_k| \leq 1$ for $k = 1, 2, \ldots, N-1$. According to Eq. (12.41)

$$\lambda_k = 1 - 2r + 2r \cos \frac{k\pi}{N} = 1 - 4r \sin^2 \frac{k\pi}{2N}$$

and $|\lambda_k| \leq 1$ if $r \leq 1/2$. Thus the explicit finite difference scheme given in Eq. (12.9) is stable if $r \leq 1/2$ and unstable for $r > 1/2$. The encroaching instability of the method could be seen as r was gradually increased in the examples graphed in Figs. 12.2–12.4.

The stability of the explicit finite difference scheme for approximating the solution to the heat equation is dependent on the value of the parameter r, which in turn depends on the step sizes chosen for the discretization in the spatial and temporal directions. This should be compared to the implicit Crank-Nicolson scheme expressed in matrix form in Eq. (12.11), or equivalently

$$\mathbf{u}^{(j+1)} = (A(r))^{-1} A(-r) \, \mathbf{u}^{(j)}.$$

By Lemma 12.1 the eigenvalues of $A(r)$ are

$$\lambda_k = 2(1+r) - 2r \cos \frac{k\pi}{N} = 2 + 4r \sin^2 \frac{k\pi}{2N} \text{ for } k = 1, 2, \ldots, N-1,$$

while the eigenvalues of $A(-r)$ are

$$\mu_k = 2(1-r) + 2r \cos \frac{k\pi}{N} = 2 - 4r \sin^2 \frac{k\pi}{2N} \text{ for } k = 1, 2, \ldots, N-1.$$

Matrices $A(r)$ and $A(-r)$ share the same set of eigenvectors

$$\mathbf{v}_k = \left(\sin \frac{k\pi}{N}, \ldots, \sin \frac{(N-1)k\pi}{N} \right)^T.$$

Furthermore if λ is an eigenvalue of an invertible square matrix A with corresponding eigenvector $\mathbf{v}$ then $1/\lambda$ is an eigenvalue of A^{-1} with the same eigenvector $\mathbf{v}$ (see Exercise 22). It is also the case that if $\mathbf{v}$ is an eigenvector of matrix A corresponding to eigenvalue λ and also an eigenvector of matrix B corresponding to eigenvalue μ then $\mathbf{v}$ is an eigenvector of matrix AB corresponding to eigenvalue $\lambda \mu$ (see Exercise 23). Thus the Crank-Nicolson finite difference scheme expressed in Eq. (12.11) is stable for all $r > 0$ since for $k = 1, 2, \ldots, N-1$ the eigenvalues of $(A(r))^{-1} A(-r)$ all have magnitudes bounded by 1,

$$\left| \frac{\mu_k}{\lambda_k} \right| = \left| \frac{2 - 4r \sin^2 \frac{k\pi}{2N}}{2 + 4r \sin^2 \frac{k\pi}{2N}} \right| \le 1.$$

12.7.2   Stability of Finite Difference Schemes for the Wave Equation

In order that the reader may see a variety of approaches to determining the stability of finite difference schemes, a different approach from that of the previous section will be used to examine the stability of the explicit scheme used to approximate the solution to the wave equation with homogeneous Dirichlet boundary conditions. The $O(k^2) + O(h^2)$ truncation error scheme is found in Eq. (12.14). The analysis performed in this section will rely on the formulas found for the eigenvalues and eigenvectors of simple tridiagonal matrices and on the familiar technique of separation of variables used in earlier chapters to find analytic solutions to initial boundary value problems. Further details and applications of this method can be found in [Golub and Ortega (1992)].

The method described below can be considered as a discrete version of the method of separation of variables. Suppose the solution u_i^j for $i = 1, 2, \ldots, N - 1$ and $j = 1, 2, \ldots$ is a product solution of the form $u_i^j = X_i T_j$. Substituting the product solution into the finite difference scheme of Eq. (12.14) and dividing both sides by $X_i T_j$ produce

$$X_i T_{j+1} = r^2 X_{i+1} T_j + 2(1 - r^2) X_i T_j + r^2 X_{i-1} T_j - X_i T_{j-1}$$

$$\frac{T_{j+1}}{T_j} + \frac{T_{j-1}}{T_j} = r^2 \frac{X_{i+1}}{X_i} + 2(1 - r^2) + r^2 \frac{X_{i-1}}{X_i}.$$

Just as the reader saw when separation of variables was used for partial differential equations, the left-hand side of the equation above depends only on the index j while the right-hand side depends on the index i and therefore the left- and right-hand sides must be constant with respect to i and j. The constant will be denoted as λ. Thus two difference equations are implied:

$$r^2 X_{i-1} + [2(1 - r^2) - \lambda]X_i + r^2 X_{i+1} = 0 \text{ for } i = 1, \ldots, N - 1 \quad (12.47)$$

$$T_{j-1} - \lambda T_j + T_{j+1} = 0 \text{ for } j = 1, 2, \ldots. \quad (12.48)$$

Equation (12.47) is a discrete boundary value problem with boundary conditions $X_0 = X_N = 0$ (assuming homogeneous Dirichlet boundary conditions). This type of equation has been solved in the discussion following Eq. (12.43). Hence the values of λ according to Eq. (12.41) are:

$$\lambda_k = 2(1 - r^2) + 2r^2 \cos \frac{k\pi}{N} = 2 - 4r^2 \sin^2 \frac{k\pi}{2N}$$

and $X_i = \sin(ik\pi/N)$ for $i = 0, 1, \ldots, N$.

Equation (12.48) is an initial value problem instead of a boundary value problem. Assume the initial conditions T_0 and T_1 are known, then

$$T_{j-1} - \lambda_k T_j + T_{j+1} = 0$$

for $j = 1, 2, \ldots.$ If $T_j = s^j$ then

$$0 = s^{j-1} - \lambda_k s^j + s^{j+1} \iff s^2 - \lambda_k s + 1 = 0.$$

Solving this quadratic equation implies s takes on the value

$$s_1 = \frac{1}{2}(\lambda_k - \sqrt{\lambda_k^2 - 4}) \text{ or } s_2 = \frac{1}{2}(\lambda_k + \sqrt{\lambda_k^2 - 4}).$$

By the Principle of Superposition the general solution to Eq. (12.48) is $T_j = \alpha\, s_1^j + \beta\, s_2^j$ where α and β are arbitrary constants. Let α_k and β_k be the solutions to the simultaneous equations

$$T_0 = \alpha_k + \beta_k,$$
$$T_1 = \alpha_k s_1 + \beta_k s_2.$$

The product solution u_i^j can be written as

$$u_i^j = \sum_{k=1}^{N-1} (\alpha_k s_1^j + \beta_k s_2^j) \sin \frac{ik\pi}{N}.$$

The finite difference scheme given Eq. (12.14) is stable whenever u_i^j remains bounded for all j. This condition is met if and only if $|s_1| \leq 1$ and $|s_2| \leq 1$. If $\lambda_k^2 \leq 4$ then s_1 and s_2 are complex conjugates and

$$|s_i|^2 = s_1 s_2 = \frac{1}{4}(\lambda_k^2 - (\lambda_k^2 - 4)) = 1.$$

If $\lambda_k^2 > 4$ then $\lambda_k < -2$ in which case $|s_1| > 1$. A necessary and sufficient condition for the finite difference scheme in Eq. (12.14) to be stable is that $-2 \leq \lambda_k \leq 2$. This inequality is equivalent to having $r = k/h \leq 1$.

12.7.3 *Convergence of Iterative Schemes for Poisson's Equation*

Various iterative methods (Jacobi, Gauss-Seidel, and successive over-relaxation) were used previously to solve the linear systems resulting from finite difference schemes for approximating the solutions to Laplace's and Poisson's equations. In this section the question of under what circumstances an iterative method will converge to the solution will be addressed. Since the Jacobi and Gauss-Seidel methods are so closely related they will be explored together before examining the successive over-relaxation method.

In matrix/vector form the iterative methods can be expressed as the iterative equation $\mathbf{u}^{(k+1)} = T\,\mathbf{u}^{(k)} + \mathbf{b}$. The vector $\mathbf{b}$ encompasses the boundary conditions for a Laplace or Poisson problem and, in the case of a Poisson problem, the nonhomogeneous function on the right-hand side of the equation. The matrix T will be one of T_J (see Eq. (12.33)), T_{GS} (see Eq. (12.36)), or T_{SOR} (see Eq. (12.39)). If $\mathbf{u}^{(k)}$ is a vector with n components then T is an $n \times n$ matrix. Let $\hat{\mathbf{u}}$ be the exact solution of the linear system, then $\hat{\mathbf{u}} = T\,\hat{\mathbf{u}} + \mathbf{b}$. Define the error vector of the kth iterate as $\mathbf{e}^{(k)} = \mathbf{u}^{(k)} - \hat{\mathbf{u}}$, then

$$\hat{\mathbf{u}} + \mathbf{e}^{(k+1)} = T(\hat{\mathbf{u}} + \mathbf{e}^{(k)}) + \mathbf{b}$$

$$\mathbf{e}^{(k+1)} = T\mathbf{e}^{(k)}. \tag{12.49}$$

The expression $\rho(T)$ denotes the spectral radius of matrix T (see Sec. B.4 for the formal definition). If matrix T is diagonalizable, in other words, if the eigenvectors of T form a basis for the vector space $\mathbb{R}^n$, then the proof is relatively straightforward. The proof of the more general case requires the concept of the Jordan[8] canonical form of a matrix and is beyond the scope of this textbook. A convergence test that can be applied to iterative techniques such as the Jacobi and Gauss-Seidel methods is stated next.

Theorem 12.3. *If $\hat{\mathbf{u}} = T\,\hat{\mathbf{u}} + \mathbf{b}$ then the iterates $\mathbf{u}^{(k)} = T\,\mathbf{u}^{(k-1)} + \mathbf{b}$ for $k = 1, 2, \ldots$ converge to $\hat{\mathbf{u}}$ for any choice of $\mathbf{u}^{(0)}$ if and only if $\rho(T) < 1$.*

Proof. Suppose the eigenvectors $\{\mathbf{v}_1, \mathbf{v}_2, \ldots, \mathbf{v}_n\}$ of matrix T form a basis for $\mathbb{R}^n$ and define matrix P as an $n \times n$ matrix whose ith column is the eigenvector $\mathbf{v}_i$. Matrix P is invertible (since its columns are linearly independent) and $P^{-1}TP = B$ is a diagonal matrix for which $b_{ii} = \lambda_i$ the eigenvalue corresponding to eigenvector $\mathbf{v}_i$. Let $\mathbf{u}^{(0)}$ be any vector in $\mathbb{R}^n$ then by Eq. (12.49) the error in the kth approximation to $\hat{\mathbf{u}}$ is

$$\mathbf{e}^{(k)} = T^k(\mathbf{u}^{(0)} - \hat{\mathbf{u}}) = (PBP^{-1})^k(\mathbf{u}^{(0)} - \hat{\mathbf{u}}) = (PB^kP^{-1})(\mathbf{u}^{(0)} - \hat{\mathbf{u}}).$$

If B is a diagonal matrix then $\lim_{k\to\infty} B^k = 0$ (the $n \times n$ zero matrix) if and only if $|b_{ii}| < 1$. Thus $\mathbf{e}^{(k)} \to \mathbf{0}$ if and only if all the eigenvalues of T have magnitude less than 1. $\square$

Before Theorem 12.3 can be applied to an iterative technique, two things about the technique must be established. First, the iterative technique must be written in matrix/vector for $\mathbf{u}^{(k+1)} = T\,\mathbf{u}^{(k)} + \mathbf{b}$ and second, all the eigenvalues of matrix T must be shown to have magnitudes less than 1. For matrices with a very simple structure this could be done by direct calculation of the eigenvalues. However, a different approach will be outlined here, an approach which is based on matrix algebra concepts.

Suppose matrix T has the property that $\|T\|_2 < 1$, then if λ is an eigenvalue of T with corresponding eigenvector $\mathbf{v}$, it is by definition true that $\mathbf{v} \neq \mathbf{0}$ and $T\mathbf{v} = \lambda\mathbf{v}$. This implies

$$\mathbf{u} = \left\| \frac{\mathbf{v}}{\|\mathbf{v}\|_2} \right\|_2 = 1 \text{ and } T\mathbf{u} = T\frac{\mathbf{v}}{\|\mathbf{v}\|_2} = \frac{\lambda\mathbf{v}}{\|\mathbf{v}\|_2} = \lambda\mathbf{u},$$

---

[8]Camille Jordan, French mathematician (1838–1922).

in other words, $\mathbf{u}$ is an eigenvector of norm 1 corresponding to eigenvalue λ. Therefore it is the case that

$$\|T\,\mathbf{u}\|_2 = \|\lambda\,\mathbf{u}\|_2 = |\lambda|\,\|\mathbf{u}\|_2 = |\lambda| < 1.$$

As a consequence, the maximum magnitude of any eigenvalue of matrix T is less than 1. Conversely if T possesses an eigenvalue with magnitude greater than 1, then $\|T\|_2 > 1$. Hence one method for establishing that all the eigenvalues of a matrix lie within the unit circle in the complex plane is to demonstrate that $\|T\|_2 < 1$. For any matrix norm, since $\|T\,\mathbf{x}\| \le \|T\|\,\|\mathbf{x}\|$, the argument above shows that if $\|T\| < 1$, then $\rho(T) < 1$.

Now consider the Jacobi iterative formula applied to a linear system $A\,\mathbf{u} = \mathbf{b}$ as in Eq. (12.33). Specifically the matrices can be expressed in detail as

$$D^{-1} = \begin{bmatrix} a_{11}^{-1} & 0 & \cdots & 0 \\ 0 & a_{22}^{-1} & \cdots & 0 \\ \vdots & \vdots & & \vdots \\ 0 & 0 & \cdots & a_{nn}^{-1} \end{bmatrix} \text{ and } L+U = \begin{bmatrix} 0 & a_{12} & \cdots & a_{1n} \\ a_{21} & 0 & \cdots & a_{2n} \\ \vdots & \vdots & & \vdots \\ a_{n1} & a_{n2} & \cdots & 0 \end{bmatrix}.$$

Multiplying these matrices produces

$$T_J = -D^{-1}(L+U) = -\begin{bmatrix} 0 & a_{12}/a_{11} & \cdots & a_{1n}/a_{11} \\ a_{21}/a_{22} & 0 & \cdots & a_{2n}/a_{22} \\ \vdots & \vdots & & \vdots \\ a_{n1}/a_{nn} & a_{n2}/a_{nn} & \cdots & 0 \end{bmatrix}.$$

Therefore if A is strictly diagonally dominant then for each i, $\sum_{j=1, j\neq i}^{n} |a_{ij}/a_{ii}| < 1$. According to Gershgorin's Theorem (Theorem B.14) all the eigenvalues of T_J lie strictly inside the unit circle, or equivalently $\rho(T_J) < 1$. As a consequence of Theorem 12.3, the Jacobi iterative method converges to the solution of $A\,\mathbf{u} = \mathbf{b}$ for any choice of initial estimate $\mathbf{u}^{(0)}$ when A is strictly diagonally dominant. If A is irreducibly diagonally dominant then for each i, $\sum_{j=1, j\neq i}^{n} |a_{ij}/a_{ii}| \le 1$ with strict inequality holding for at least one i. Gershgorin's Theorem implies $\rho(T_J) \le 1$ which is not sufficiently strong on its own to conclude that the Jacobi method converges. However, suppose T_J has an eigenvalue λ with $|\lambda| = 1$. By the definition of eigenvalue then $T_J - \lambda I$ is a singular matrix and consequently

$$-D(T_J - \lambda I) = \lambda\, D + L + U = \hat{A}$$

is singular. Recall that since $A = D + L + U$ is irreducibly diagonally dominant and $|\lambda| = 1$, then $\hat{A}$ is likewise irreducibly diagonally dominant and hence nonsingular (see Theorem B.20) which is a contradiction. Therefore when A is irreducibly diagonally dominant it is still the case that $\rho(T_J) < 1$ and the following theorem holds.

Theorem 12.4. *If A is an $n \times n$ strictly diagonally dominant or irreducibly diagonally dominant matrix and $\mathbf{b}$ is an $n \times 1$ vector, then the Jacobi iterative method (Eq. (12.33)) converges to the solution of $A\mathbf{u} = \mathbf{b}$ for any choice of initial estimate $\mathbf{u}^{(0)}$.*

Now consider the Gauss-Seidel iterative method in Eq. (12.36) under the assumption that matrix A is irreducibly diagonally or strictly diagonally dominant. Once again, in detail the matrices may be expressed as

$$D + L = \begin{bmatrix} a_{11} & 0 & \cdots & 0 \\ a_{21} & a_{22} & \cdots & 0 \\ \vdots & \vdots & & \vdots \\ a_{n1} & a_{n2} & \cdots & a_{nn} \end{bmatrix} \text{ and } U = \begin{bmatrix} 0 & a_{12} & \cdots & a_{1,n-1} & a_{1n} \\ \vdots & \vdots & & \vdots & \vdots \\ 0 & 0 & \cdots & 0 & a_{n-1,n} \\ 0 & 0 & \cdots & 0 & 0 \end{bmatrix}.$$

Let λ be an eigenvalue of T_{GS} with corresponding eigenvector $\mathbf{v} = (v_1, v_2, \ldots, v_n)^T$.

$$T_{GS}\mathbf{v} = -(D + L)^{-1}U\mathbf{v} = \lambda\mathbf{v} \iff -U\mathbf{v} = \lambda(D + L)\mathbf{v}.$$

Let i be the value of the index such that $|v_i| = \max_{j=1,\ldots,n}|v_j|$, then the ith equation of $-U\mathbf{v} = \lambda(D + L)\mathbf{v}$ is

$$-\sum_{j=i+1}^{n} a_{ij}v_j = \lambda\sum_{j=1}^{i} a_{ij}v_j$$

$$-\sum_{j=i+1}^{n} \frac{a_{ij}v_j}{a_{ii}v_i} = \lambda\left(1 + \sum_{j=1}^{i-1} \frac{a_{ij}v_j}{a_{ii}v_i}\right)$$

$$-\beta = \lambda(1 + \alpha)$$

where $\beta = \sum_{j=i+1}^{n} a_{ij}v_j/(a_{ii}v_i)$ and $\alpha = \sum_{j=1}^{i-1} a_{ij}v_j/(a_{ii}v_i)$. If matrix A is strictly diagonally dominant then $|\alpha| + |\beta| < 1$. Taking absolute values of both sides of the last equation and rearranging terms produce

$$|\lambda| = \frac{|\beta|}{|1 + \alpha|} \leq \frac{|\beta|}{1 - |\alpha|} < 1.$$

If A is irreducibly diagonally dominant then $|\alpha| + |\beta| \leq 1$ which implies $|\lambda| \leq 1$. Again suppose T_{GS} has an eigenvalue λ with $|\lambda| = 1$. The matrix $T_{GS} - \lambda I$ is not invertible which implies

$$-(D + L)(T_{GS} - \lambda I) = \lambda(D + L) + U = \hat{A}$$

is a singular matrix. Reasoning as before in the case of the Jacobi iteration matrix, since $|\lambda| = 1$ and matrix A is irreducibly diagonally dominant, then $\hat{A}$ is irreducibly diagonally dominant and hence invertible. This contradiction implies $\rho(T_{GS}) < 1$ when A is strictly diagonally dominant or irreducibly diagonally dominant. Thus the following theorem holds.

Theorem 12.5. *If A is an $n \times n$ strictly diagonally dominant or irreducibly diagonally dominant matrix and $\mathbf{b}$ is an $n \times 1$ vector, then the Gauss-Seidel iterative method (Eq. (12.36)) converges to the solution of $A\mathbf{u} = \mathbf{b}$ for any choice of initial estimate $\mathbf{u}^{(0)}$.*

The astute reader may have noticed a weakness in the application of Theorem B.14 to the five-point and nine-point finite difference approximations to the Laplacian given in Eqs. (12.21) and (12.29) respectively. These matrices are only weakly diagonally dominant since for both $|a_{ii}| \geq \sum_{j \neq i}^{n} |a_{ij}|$ with equality holding on most (but not all rows). Thus Theorem B.14 cannot guarantee that all eigenvalues of these matrices have magnitudes strictly less than 1, unless it is proved that the matrices are irreducible. However, the work done earlier on the eigenvalues and eigenvectors of tridiagonal and block matrices can be used to establish that all the eigenvalues of matrix A in Eq. (12.21) lie strictly inside the unit circle in the complex plane.

The tridiagonal matrix $\hat{A}$ is one of the blocks comprising matrix A in Eq. (12.21). According to Lemma 12.1, $\hat{A}$ has $N - 1$ distinct eigenvalues $\lambda_i = 4 + 2\cos(i\pi/N)$ for $i = 1, 2, \ldots, N - 1$ with corresponding linearly independent eigenvectors $\mathbf{v}_i \in \mathbb{R}^{N-1}$ given in Eq. (12.42). The remaining $(N - 1) \times (N - 1)$ blocks making up matrix A are either the negative of the identity matrix or the matrix of all zero entries. Since $-I\mathbf{v}_i = (-1)\mathbf{v}_i$ then $\mathbf{v}_i$ is an eigenvector of $-I$ with eigenvalue -1. Likewise $0\,\mathbf{v}_i = (0)\mathbf{v}_i$ which implies $\mathbf{v}_i$ is an eigenvector of the zero matrix with corresponding eigenvalue 0. Consequently all the blocks of matrix A share a common set of linearly independent eigenvectors $\{\mathbf{v}_1, \mathbf{v}_2, \ldots, \mathbf{v}_{N-1}\}$. According to Theorem 12.2 the eigenvalues of matrix A in Eq. (12.21) are the eigenvalues of the matrices

$$
\Lambda_i = \begin{bmatrix}
\lambda_i & -1 & 0 & \cdots & 0 & 0 & 0 \\
-1 & \lambda_i & -1 & \cdots & 0 & 0 & 0 \\
\vdots & \vdots & \vdots & & \vdots & \vdots & \vdots \\
0 & 0 & 0 & \cdots & -1 & \lambda_i & -1 \\
0 & 0 & 0 & \cdots & 0 & -1 & \lambda_i
\end{bmatrix} \quad \text{for } i = 1, 2, \ldots, N - 1.
$$

Using Lemma 12.1, the eigenvalues of the matrices Λ_i are $\lambda_i + 2\cos(j\pi/M)$ for $j = 1, 2, \ldots, M$. Hence (by Theorem 12.2) the eigenvalues of A are

$$\lambda_{i,j} = \lambda_i + 2\cos\frac{j\pi}{M} = 4 + 2\left(\cos\frac{i\pi}{N} + \cos\frac{j\pi}{M}\right),$$

with $i = 1, 2, \ldots, N - 1$ and $j = 1, 2, \ldots, M - 1$.

Proof of the convergence of the Jacobi iteration method for the weakly diagonally dominant matrix A given in Eq. (12.21) for any starting approximation $\mathbf{u}^{(0)}$ can now be given. Recall that $A = D + L + U$ where D is a diagonal matrix whose diagonal entries are the diagonal entries of A and L and U are the strictly lower and upper triangular entries of A respectively. This implies $-D^{-1}A + I = -D^{-1}(L + U)$ and thus λ is an eigenvalue of $-D^{-1}(L + U)$ if and only if λ is an eigenvalue of $-D^{-1}A + I$. Suppose λ is an eigenvalue of $-D^{-1}A + I$, then there is a nonzero vector $\mathbf{v}$ for which

$$(-D^{-1}A + I)\mathbf{v} = \lambda\mathbf{v}$$
$$-\frac{1}{4}A\mathbf{v} + \mathbf{v} = \lambda\mathbf{v}$$
$$A\mathbf{v} = 4(1 - \lambda)\mathbf{v}.$$

This implies $4(1 - \lambda)$ is an eigenvalue of matrix A. Since the eigenvalues of A are known, then for some i and j,

$$4(1 - \lambda) = 4 + 2\left(\cos\frac{i\pi}{N} + \cos\frac{j\pi}{M}\right)$$
$$|\lambda| = \frac{1}{2}\left|\cos\frac{i\pi}{N} + \cos\frac{j\pi}{M}\right| \leq \frac{1}{2}\left(\cos\frac{\pi}{N} + \cos\frac{\pi}{M}\right) < 1.$$

Earlier the matrices in Eqs. (12.21) and (12.29) were described as weakly diagonally dominant. In fact, these matrices are irreducibly diagonally dominant. Thus by Theorems 12.4 and 12.5 the Jacobi and Gauss-Seidel methods (respectively) converge to a solution of Laplace's or Poisson's equation for an arbitrary initial estimate using either the five-point or nine-point approximation to the Laplacian operator. Attention can now be turned to the convergence properties of the successive over-relaxation method. Let matrix $A = D + L + U$ where as before D is the diagonal part of matrix A and L and U are the strictly lower and upper triangular parts of A respectively. The next result, originally due to Kahan,[9] bounds the values of ω which are necessary for the successive over-relaxation method to converge [Kahan (1958)].

---

[9]William Morton Kahan, Canadian mathematician (1933–).

Theorem 12.6. *If $A\mathbf{u} = \mathbf{b}$ is a linear system of equations with $a_{ii} \neq 0$ for $i = 1, 2, \ldots, n$, the successive over-relaxation method converges to the solution $\mathbf{u}$ for an arbitrary initial estimate $\mathbf{u}^{(0)}$ only if $0 < \omega < 2$.*

Proof. Let $\lambda_1, \lambda_2, \ldots, \lambda_n$ be the eigenvalues of matrix T_{SOR}. The determinant of a square matrix is the product of its eigenvalues, thus

$$
\begin{aligned}
\lambda_1 \lambda_2 \cdots \lambda_n &= \det\left((D + \omega L)^{-1}[(1 - \omega)D - \omega U]\right) \\
&= \det(D + \omega L)^{-1} \det\left((1 - \omega)D - \omega U\right) \\
&= \frac{(1 - \omega)^n \det D}{\det D} = (1 - \omega)^n.
\end{aligned}
$$

If $|\lambda| = \max_{1 \leq i \leq n} |\lambda_i|$ then $|\lambda| \geq |\omega - 1|$. Since the iterates of the SOR technique converge for arbitrary starting approximation $\mathbf{u}^{(0)}$ if and only if $|\lambda| < 1$, then convergence occurs only if $|\omega - 1| < 1$ or equivalently only if $0 < \omega < 2$. □

If matrix A possesses another property (positive definiteness), the condition in Theorem 12.6 is also sufficient as stated below [Ortega (1990), p. 201].

Theorem 12.7 (Ostrowski[10]-Reich[11]). *Let $A\mathbf{u} = \mathbf{b}$ be a linear system of equations where A is a positive definite matrix. The successive over-relaxation method converges to the solution $\mathbf{u}$ for an arbitrary initial estimate $\mathbf{u}^{(0)}$ if and only if $0 < \omega < 2$.*

The matrices of the finite difference approximation methods for Laplace's or Poisson's equation are all positive definite and therefore the successive over-relaxation iterative method will converge.

The optimal choice for the successive over-relaxation parameter ω is the one which minimizes the magnitude of the largest magnitude eigenvalue of T_{SOR}. Without proof, the optimal value of ω is

$$
\omega^* = \frac{2}{1 + \sqrt{1 - |\lambda|^2}}
$$

where λ is the eigenvalue of $A^{-1}(L + U)$ with the largest magnitude. The reader may wish to experiment with the SOR technique by examining the number of iterations necessary for the solution to converge to a given accuracy using the optimal choice of ω and other nearby values of ω.

---

[10]Alexander Markowich Ostrowski, Swiss mathematician (1893–1986).
[11]Edgar Reich, American mathematician (1927–2009).

To conclude this chapter and to suggest possible avenues of further study, it should be mentioned that there are other techniques for solving linear systems not explored in this chapter. Some potentially useful methods include the preconditioned conjugate gradient technique and the alternating direction implicit technique.

12.8 Exercises

(1) Let $f(x) = x \sin x$.
   (a) Use the forward difference formula for the first derivative to approximate $f'(1)$ with $h = 1/10$.
   (b) Find a bound for the error in the approximation.

(2) Let $f(x) = x \sin x$.
   (a) Use the centered difference formula for the first derivative to approximate $f'(1)$ with $h = 1/10$.
   (b) Find a bound for the error in the approximation.

(3) Let $f(x) = x \sin x$.
   (a) Use the second derivative midpoint formula to approximate $f''(1)$ with $h = 1/10$.
   (b) Find a bound for the error in the approximation.

(4) Show that the matrix/vector form of the explicit finite difference scheme for the heat equation given in Eq. (12.9) can be rewritten as $\mathbf{u}^{(j)} = (A\ (r))^j \mathbf{u}^{(0)}$ where $(A(r))^j$ denoted the j-fold product of matrix $A(r)$ with itself.

(5) Consider the partial differential equation,

$$u_t = \kappa u_{xx} \text{ for } a < x < b.$$

Show that by an appropriate change of variable that the partial differential equation may be rewritten as

$$u_\tau = u_{\xi\xi} \text{ for } 0 < \xi < 1.$$

(6) Consider the partial differential equation,

$$u_{tt} = c^2 u_{xx} \text{ for } a < x < b.$$

Show that by an appropriate change of variable that the partial differential equation may be rewritten as

$$u_{\tau\tau} = u_{\xi\xi} \text{ for } 0 < \xi < 1.$$

(7) Show that the Neumann boundary conditions of Eq. (1.29) can be discretized using the forward/backward difference formulas as

$$-K_0 \frac{(u_1^t - u_0^t)}{h} = g_1(t),$$

$$K_0 \frac{(u_{N-1}^t - u_N^t)}{h} = g_2(t).$$

(8) Use the explicit finite difference scheme to approximate the solution to the following initial boundary value problem. Use $h = 0.1$ and $k = 0.001$. Iterate the solution method until $t = 0.01$.

$$u_t = u_{xx} + 2u_x \text{ for } 0 < x < 1 \text{ and } t > 0,$$
$$u(0,t) = u(1,t) = 0 \text{ for } t > 0,$$
$$u(x,0) = 100x \sin(2\pi x).$$

(9) Use the explicit finite difference scheme to approximate the solution to the following initial boundary value problem. Use $h = 0.1$ and $k = 0.001$. Iterate the solution method until $t = 0.01$.

$$u_t = u_{xx} + 2u_x \text{ for } 0 < x < 1 \text{ and } t > 0,$$
$$u_x(0,t) = u_x(1,t) = 0 \text{ for } t > 0,$$
$$u(x,0) = \begin{cases} 1 & \text{if } 0 < x < 1/2, \\ 2 & \text{if } 1/2 < x < 1. \end{cases}$$

(10) Use the Crank-Nicolson implicit finite difference scheme to approximate the solution to the following initial boundary value problem. Use $h = 0.1$ and $k = 0.01$. Iterate the solution method until $t = 0.10$.

$$u_t = u_{xx} + 2u \text{ for } 0 < x < 1 \text{ and } t > 0,$$
$$u(0,t) = u(1,t) = 0 \text{ for } t > 0,$$
$$u(x,0) = \sin(\pi x).$$

(11) Use the Crank-Nicolson implicit finite difference scheme to approximate the solution to the following initial boundary value problem. Use $h = 0.1$ and $k = 0.01$. Iterate the solution method until $t = 0.10$.

$$u_t = u_{xx} + 2u \text{ for } 0 < x < 1 \text{ and } t > 0,$$
$$u_x(0,t) = u_x(1,t) = 0 \text{ for } t > 0,$$
$$u(x,0) = \begin{cases} -1 & \text{if } 0 < x < 1/2, \\ 1 & \text{if } 1/2 < x < 1. \end{cases}$$

(12) Use the Crank-Nicolson implicit finite difference scheme to approximate the solution to the following initial boundary value problem. Use $h = 0.1$ and $k = 0.01$. Iterate the solution method until $t = 0.10$.

$$u_t = u_{xx} + 2u_x \text{ for } 0 < x < 1 \text{ and } t > 0,$$
$$u_x(0, t) = u(0, t) \text{ for } t > 0,$$
$$-u_x(1, t) = u(1, t) \text{ for } t > 0,$$
$$u(x, 0) = 1 \text{ for } 0 < x < 1.$$

(13) Use the Crank-Nicolson implicit finite difference scheme to approximate the solution to the following initial boundary value problem. Use $h = 0.1$ and $k = 0.01$. Iterate the solution method until $t = 0.10$.

$$u_t = e^{-x}u_{xx} \text{ for } 0 < x < 1 \text{ and } t > 0,$$
$$u_x(0, t) = u_x(1, t) = 0 \text{ for } t > 0,$$
$$u(x, 0) = 100x^2(1 - x)^2 \text{ for } 0 < x < 1.$$

(14) Use the explicit finite difference scheme with $h = 0.1$ and $k = 0.05$ to approximate the solution to the following initial boundary value problem. Iterate the solution until $t = 0.5$.

$$u_{tt} = u_{xx} \text{ for } 0 < x < 1 \text{ and } t > 0,$$
$$u(0, t) = u(1, t) = 0 \text{ for } t > 0,$$
$$u(x, 0) = 4x(1 - x) \text{ for } 0 < x < 1,$$
$$u_t(x, 0) = \sin(3\pi x) \text{ for } 0 < x < 1.$$

(15) Derive a Crank-Nicolson implicit finite difference approximation to the homogeneous wave equation $u_{tt} = u_{xx}$.

(16) Use the Taylor polynomial expansion for a function of two variables to verify the results shown in Eqs. (12.22) and (12.23).

(17) Use the Jacobi method with $\epsilon = 10^{-4}$ to approximate the solution to the following linear system of equations. Let $\mathbf{u}^{(0)} = \mathbf{0}$.

$$4u_1 + u_2 - u_3 = -7,$$
$$u_1 + 6u_2 - 2u_3 + u_4 - u_5 = 1,$$
$$u_2 + 5u_3 - u_5 + u_6 = 4,$$
$$2u_2 + 5u_4 - u_5 - u_7 - u_8 = 4,$$
$$-u_3 - u_4 + 6u_5 - u_6 - u_8 = 5,$$
$$-u_3 - u_5 + 5u_6 = 1,$$
$$-u_4 + 4u_7 - u_8 = -2,$$
$$-u_4 - u_5 - u_7 + 5u_8 = 8.$$

(18) Solve the linear system in Exercise 17 using the Gauss-Seidel method.

(19) Express the linear system for the nine-point approximation to Poisson's equation in matrix/vector form for the case in which a formula for the function on the right-hand side of Poisson's equation and its Laplacian are known.

(20) Use the five-point approximation to the Laplacian and the Gauss-Seidel method to approximate a solution to the following boundary value problem.

$$\triangle u = 0 \text{ for } 0 < x < 1 \text{ and } 0 < y < 1,$$
$$u(0, y) = e^{1-y} \text{ for } 0 < y < 1,$$
$$u(x, 0) = e^{1-x} \text{ for } 0 < x < 1,$$
$$u_x(1, y) = 0 \text{ for } 0 < y < 1,$$
$$u_y(x, 1) = 0 \text{ for } 0 < x < 1.$$

(21) Show that if $(\lambda - a)^2 = 4bc$ then $x_m = m\, r^m$ solves Eq. (12.43) where $r = (\lambda - a)/(2b)$.

(22) Prove that if λ is an eigenvalue of an invertible square matrix A with corresponding eigenvector $\mathbf{v}$ then $1/\lambda$ is an eigenvalue of A^{-1} with corresponding eigenvector $\mathbf{v}$.

(23) Prove that if $\mathbf{v}$ is an eigenvector of matrix A corresponding to eigenvalue λ and also an eigenvector of matrix B corresponding to eigenvalue μ then $\mathbf{v}$ is an eigenvector of matrix $A\,B$ corresponding to eigenvalue $\lambda\,\mu$.

(24) Determine the values of r for which the implicit finite difference scheme for the wave equation given in matrix form in Eq. (12.19) is stable.

Table 12.1   The numerical approximation of the solution to the heat equation from Example 4.1.

t	$u(0.0, t)$	$u(0.1, t)$	$u(0.2, t)$	$u(0.3, t)$	$u(0.4, t)$	$u(0.5, t)$	$u(0.6, t)$	$u(0.7, t)$	$u(0.8, t)$	$u(0.9, t)$	$u(1.0, t)$
0.000	0.0000	100.0000	100.0000	100.0000	100.0000	100.0000	100.0000	100.0000	100.0000	100.0000	0.0000
0.001	0.0000	90.0000	100.0000	100.0000	100.0000	100.0000	100.0000	100.0000	100.0000	90.0000	0.0000
0.002	0.0000	82.0000	99.0000	100.0000	100.0000	100.0000	100.0000	100.0000	99.0000	82.0000	0.0000
0.003	0.0000	75.5000	97.4000	99.9000	100.0000	100.0000	99.9900	99.9000	97.4000	75.5000	0.0000
0.004	0.0000	70.1400	95.4600	99.6600	99.9900	100.0000	99.9580	99.6600	95.4600	70.1400	0.0000
0.005	0.0000	65.6580	93.3480	99.2730	99.9580	99.9980	99.8935	99.2730	93.3480	65.6580	0.0000
0.006	0.0000	61.8612	91.1715	98.7490	99.8935	99.9900	99.7887	98.7490	91.1715	61.8612	0.0000
0.007	0.0000	58.6061	88.9982	98.1057	99.7887	99.9707	99.6386	98.1057	88.9982	58.6061	0.0000
0.008	0.0000	55.7847	86.8698	97.3633	99.6386	99.9343	99.4406	97.3633	86.8698	55.7847	0.0000
0.009	0.0000	53.3147	84.8106	96.5414	99.4406	99.8752	99.1942	96.5414	84.8106	53.3147	0.0000
0.010	0.0000	51.1329	82.8341	95.6583	99.1942	99.7883	98.9000	95.6583	82.8341	51.1329	0.0000
0.011	0.0000	49.1897	80.9464	94.7294	98.9000	99.6694	98.5599	94.7294	80.9464	49.1897	0.0000
0.012	0.0000	47.4464	79.1490	93.7682	98.5599	99.5155	98.5599	93.7682	79.1490	47.4464	0.0000
⋯	⋯	⋯	⋯	⋯	⋯	⋯	⋯	⋯	⋯	⋯	⋯
0.036	0.0000	28.8251	53.7743	72.2347	83.2386	86.8515	83.2386	72.2347	53.7743	28.8251	0.0000
⋯	⋯	⋯	⋯	⋯	⋯	⋯	⋯	⋯	⋯	⋯	⋯
0.060	0.0000	21.8077	41.3486	56.6866	66.4253	69.7578	66.4253	56.6866	41.3486	21.8077	0.0000

Table 12.2   The numerical approximation of the solution to the heat equation from Example 12.1.

t	$u(0.0,t)$	$u(0.1,t)$	$u(0.2,t)$	$u(0.3,t)$	$u(0.4,t)$	$u(0.5,t)$	$u(0.6,t)$	$u(0.7,t)$	$u(0.8,t)$	$u(0.9,t)$	$u(1.0,t)$
0.00	0.0000	0.0810	0.1280	0.1470	0.1440	0.1250	0.0960	0.0630	0.0320	0.0090	0.0000
0.01	0.0000	0.0577	0.1029	0.1258	0.1282	0.1150	0.0919	0.0646	0.0386	0.0176	0.0000
0.02	0.0000	0.0468	0.0844	0.1072	0.1133	0.1052	0.0874	0.0649	0.0418	0.0201	0.0000
0.03	0.0000	0.0389	0.0713	0.0923	0.1001	0.0958	0.0824	0.0635	0.0424	0.0211	0.0000
0.04	0.0000	0.0331	0.0612	0.0804	0.0889	0.0870	0.0768	0.0608	0.0416	0.0210	0.0000
0.05	0.0000	0.0286	0.0532	0.0706	0.0792	0.0790	0.0710	0.0573	0.0399	0.0204	0.0000
0.06	0.0000	0.0250	0.0467	0.0625	0.0709	0.0716	0.0654	0.0535	0.0376	0.0194	0.0000
0.07	0.0000	0.0220	0.0413	0.0556	0.0637	0.0650	0.0599	0.0495	0.0351	0.0182	0.0000
0.08	0.0000	0.0195	0.0367	0.0497	0.0573	0.0589	0.0547	0.0456	0.0325	0.0169	0.0000
0.09	0.0000	0.0174	0.0328	0.0446	0.0517	0.0534	0.0499	0.0418	0.0300	0.0156	0.0000
0.10	0.0000	0.0156	0.0294	0.0401	0.0467	0.0484	0.0455	0.0382	0.0275	0.0144	0.0000

Table 12.3 The numerical approximation of the solution to the heat equation with Robin boundary conditions.

t	$u(0.0,t)$	$u(0.1,t)$	$u(0.2,t)$	$u(0.3,t)$	$u(0.4,t)$	$u(0.5,t)$	$u(0.6,t)$	$u(0.7,t)$	$u(0.8,t)$	$u(0.9,t)$	$u(1.0,t)$
0.00	0.3333	0.2729	0.2234	0.1829	0.1498	0.1226	0.1004	0.0822	0.0673	0.0551	0.0451
0.01	0.3437	0.2832	0.2324	0.1904	0.1559	0.1276	0.1045	0.0856	0.0701	0.0575	0.0475
0.02	0.3514	0.2921	0.2409	0.1978	0.1622	0.1328	0.1088	0.0891	0.0731	0.0603	0.0506
0.03	0.3580	0.2997	0.2487	0.2050	0.1685	0.1382	0.1133	0.0929	0.0765	0.0636	0.0540
0.04	0.3636	0.3065	0.2558	0.2119	0.1747	0.1436	0.1179	0.0970	0.0801	0.0671	0.0576
0.05	0.3685	0.3126	0.2623	0.2184	0.1807	0.1490	0.1227	0.1012	0.0840	0.0708	0.0615
0.06	0.3729	0.3180	0.2684	0.2245	0.1866	0.1544	0.1275	0.1056	0.0881	0.0748	0.0656
0.07	0.3767	0.3230	0.2739	0.2303	0.1922	0.1597	0.1324	0.1101	0.0924	0.0790	0.0699
0.08	0.3802	0.3275	0.2791	0.2357	0.1976	0.1649	0.1374	0.1147	0.0968	0.0833	0.0742
0.09	0.3832	0.3316	0.2839	0.2409	0.2029	0.1700	0.1423	0.1194	0.1013	0.0878	0.0787
0.10	0.3860	0.3354	0.2884	0.2458	0.2080	0.1751	0.1472	0.1242	0.1059	0.0923	0.0833

Table 12.4   The displacements of the vibrating string described in Example 5.1 approximated numerically using an explicit finite difference method. Only selected displacement times are listed in the table.

x	$t = 0$	$t = 0.2$	$t = 0.4$	$t = 0.6$	$t = 0.8$	$t = 1.0$
0.00	0.000	0.000	0.000	0.000	0.000	0.000
0.05	0.000	0.000	−0.025	0.000	0.000	0.000
0.10	0.000	0.025	−0.025	0.000	0.000	0.000
0.15	0.000	0.050	0.000	−0.025	0.000	0.000
0.20	0.000	0.050	0.000	−0.050	0.000	0.000
0.25	0.000	0.025	0.000	−0.050	0.000	0.000
0.30	0.050	0.000	0.000	−0.025	0.000	0.000
0.35	0.100	0.000	0.000	0.000	−0.025	0.000
0.40	0.100	0.000	0.000	0.000	−0.050	0.000
0.45	0.050	0.000	0.000	0.000	−0.050	0.000
0.50	0.000	0.025	0.000	0.000	−0.025	0.000
0.55	0.000	0.050	0.000	0.000	0.000	−0.050
0.60	0.000	0.050	0.000	0.000	0.000	−0.100
0.65	0.000	0.025	0.000	0.000	0.000	−0.100
0.70	0.000	0.000	0.025	0.000	0.000	−0.050
0.75	0.000	0.000	0.050	0.000	−0.025	0.000
0.80	0.000	0.000	0.050	0.000	−0.050	0.000
0.85	0.000	0.000	0.025	0.000	−0.050	0.000
0.90	0.000	0.000	0.000	0.025	−0.025	0.000
0.95	0.000	0.000	0.000	0.025	0.000	0.000
1.00	0.000	0.000	0.000	0.000	0.000	0.000

Table 12.5 The numerical approximation of the solution to the wave equation in Example 12.3 calculated using an implicit finite difference scheme.

t	u(0.0, t)	u(0.1, t)	u(0.2, t)	u(0.3, t)	u(0.4, t)	u(0.5, t)	u(0.6, t)	u(0.7, t)	u(0.8, t)	u(0.9, t)	u(1.0, t)
0.00	0.0000	0.0900	0.1600	0.2100	0.2400	0.2500	0.2400	0.2100	0.1600	0.0900	0.0000
0.05	0.0000	0.1376	0.1575	0.1575	0.2375	0.2975	0.2375	0.1575	0.1575	0.1376	0.0000
0.10	0.0000	0.1588	0.1501	0.1222	0.2300	0.3178	0.2300	0.1222	0.1501	0.1588	0.0000
0.15	0.0000	0.1425	0.1380	0.1166	0.2175	0.2985	0.2175	0.1166	0.1380	0.1425	0.0000
0.20	0.0000	0.0933	0.1218	0.1376	0.2000	0.2427	0.2000	0.1376	0.1218	0.0933	0.0000
0.25	0.0000	0.0295	0.1025	0.1685	0.1776	0.1673	0.1776	0.1685	0.1025	0.0295	0.0000
0.30	0.0000	-0.0247	0.0815	0.1864	0.1504	0.0960	0.1504	0.1864	0.0815	-0.0247	0.0000
0.35	0.0000	-0.0499	0.0600	0.1728	0.1187	0.0484	0.1187	0.1728	0.0600	-0.0499	0.0000
0.40	0.0000	-0.0395	0.0391	0.1218	0.0830	0.0315	0.0830	0.1218	0.0391	-0.0395	0.0000
0.45	0.0000	-0.0028	0.0191	0.0437	0.0442	0.0372	0.0442	0.0437	0.0191	-0.0028	0.0000
0.50	0.0000	0.0394	-0.0004	-0.0398	0.0037	0.0458	0.0037	-0.0398	-0.0004	0.0394	0.0000

Appendix A

Complex Arithmetic and Calculus

This text on partial differential equations makes liberal use of results from complex analysis, without requiring a course in complex analysis as a prerequisite. Such a course is unnecessary since most of the results used can be readily grasped by the reader. For students who insist on seeing all the complex analysis topics gathered in one place, this appendix serves as a brief introduction or review of topics related to complex arithmetic, algebra, and calculus. In it, complex numbers are defined and some of their properties are listed, complex-valued functions are defined and the topics of limits, continuity, derivatives, and integrals familiar from calculus are summarized, and the complex forms of the exponential, logarithmic, and trigonometric functions are described. Readers interested in a full-fledged course of study of complex analysis may consult any standard textbook on the subject, for instance, [Churchill and Brown (2014)] or [Marsden and Hoffman (1999)].

A.1 Complex Numbers

A **complex number** z can be expressed as $z = a + i\,b$ where a and b are real numbers and $i = \sqrt{-1}$. The quantity i is sometimes called the **imaginary unit**. The quantity a is called the **real part** of z and is denoted as $\mathrm{Re}\,(z) = a$ while b is called the **imaginary part** of z and denoted as $\mathrm{Im}\,(z) = b$. Note that the real and the imaginary parts of complex numbers are both real numbers. The **complex plane** denoted as $\mathbb{C}$ is the set of all complex numbers, $\mathbb{C} = \{x + i\,y \mid x \in \mathbb{R} \text{ and } y \in \mathbb{R}\}$.

Two complex numbers z_1 and z_2 are said to be equal, written as $z_1 = z_2$, if and only if $\mathrm{Re}\,(z_1) = \mathrm{Re}\,(z_2)$ and $\mathrm{Im}\,(z_1) = \mathrm{Im}\,(z_2)$. Arithmetic on the complex plane is similar to arithmetic performed on real numbers. Suppose

$z = a + i\,b$, $w = c + i\,d$, and s is any real number. Addition of complex numbers is defined as

$$z + w = (a + i\,b) + (c + i\,d) = (a + c) + i(b + d).$$

Scalar multiplication of complex numbers by real numbers is carried out as

$$s\,z = s(a + i\,b) = (a\,s) + i(b\,s).$$

If $s = -1$ then $(-1)z = -z$ which is called the additive inverse of z. Multiplication of complex numbers is defined as

$$z\,w = (a + i\,b)(c + i\,d) = (a\,c - b\,d) + i(a\,d + b\,c).$$

If $z \neq 0$ then the multiplicative inverse of z, denoted as z^{-1} or $1/z$, exists and is defined to be

$$z^{-1} = \frac{a}{a^2 + b^2} - \frac{i\,b}{a^2 + b^2}.$$

The operations of addition and multiplication defined above possess the following properties. If v, w, and $z \in \mathbb{C}$, then

(1) $z + w = w + z$ (complex addition is commutative),
(2) $z + (w + v) = (z + w) + v$ (complex addition is associative),
(3) $z + 0 = z$ (additive identity element),
(4) $z + (-z) = 0$ (additive inverse),
(5) $z\,w = w\,z$ (complex multiplication is commutative),
(6) $z(w\,v) = (z\,w)v$ (complex multiplication is associative),
(7) $1\,z = z$ (multiplicative identity element),
(8) $z\,z^{-1} = 1$ if $z \neq 0$ (multiplicative inverse),
(9) $z(w + v) = z\,w + z\,v$ (distributive law).

The set of complex numbers equipped with the defined addition and multiplication forms an algebraic structure called a field or **complex field**. The complex field $\mathbb{C}$ can be identified with $\mathbb{R}^2$ in the sense that the mapping $x + i\,y \mapsto (x, y)$ is a one-to-one correspondence. Just as the Cartesian and polar coordinate systems can be used to locate points in $\mathbb{R}^2$, there is a **polar coordinate representation** for complex numbers. Suppose $z = x + i\,y \in \mathbb{C}$ and $z \neq 0$, then

$$z = \sqrt{x^2 + y^2}(\cos\theta + i\sin\theta)$$

where θ is the angle measured from the positive x-axis to the ray from the origin through the point (x, y). The quantity $\sqrt{x^2 + y^2}$ is called the **norm** of z and is denoted as $|z|$. The angle $\theta \in (-\pi, \pi]$ and is called the **argument** of z and is denoted as $\theta = \arg z$. The polar coordinate representation of z is not unique since

$$z = |z|(\cos(\theta + 2k\pi) + i\sin(\theta + 2k\pi))$$

for all $k \in \mathbb{Z}$.

For every complex number $z = x + iy$ there is an associated complex number called the **complex conjugate** of z. The complex conjugate, denoted as $\bar{z}$, is defined as $\bar{z} = x - iy$. The following result can be verified directly from the definition.

Theorem A.1. *Let w and z be complex numbers, then*

(1) $\overline{z + w} = \bar{z} + \bar{w}$,
(2) $\overline{zw} = \bar{z}\,\bar{w}$,
(3) $\overline{z/w} = \bar{z}/\bar{w}$ *if $w \neq 0$,*
(4) $z\bar{z} = |z|^2$,
(5) $z^{-1} = \bar{z}/|z|^2$ *if $z \neq 0$,*
(6) $z = \bar{z}$ *if and only if $z \in \mathbb{R}$,*
(7) $\operatorname{Re}(z) = (z + \bar{z})/2$ *and* $\operatorname{Im}(z) = (z - \bar{z})/(2i)$,
(8) $\bar{\bar{z}} = z$.

A.2 Complex Functions

Let $D \subset \mathbb{C}$ be a set. A **function** defined on D is a rule of correspondence mapping elements of D to elements of a set $R \subset \mathbb{C}$. The set D is called the **domain** and the set R is called the **range**. The symbols $f : D \to R$ are often used to denote the function, the domain, and the range. It is also common to use $f : D \to \mathbb{C}$ although, in this case, the range is not necessarily the whole complex plane. Since the input and the output of a function are both complex numbers, it is often helpful to decompose the function as

$$f(z) = f(x + iy) = u(x, y) + i\,v(x, y)$$

where u and v are real-valued functions mapping subsets of $\mathbb{R}^2$ into $\mathbb{R}$. Function u is often called the real part of f and v is called the imaginary part of f. In the remainder of this section a few elementary functions of complex numbers will be defined and some of their properties listed.

If $z \in \mathbb{C}$ and n is a positive integer, then the nth power of z is denoted as z^n. The following result shows that finding the nth power of a complex number is straight forward if the polar representation of complex numbers is used.

Theorem A.2 (De Moivre's[12] Formula). *If* $z = |z|(\cos\theta + i\sin\theta)$ *and* $n \in \mathbb{N}$, *then*

$$z^n = |z|^n(\cos(n\theta) + i\sin(n\theta)). \tag{A.1}$$

Proof. Equation (A.1) is true for $n = 1$ since it is merely the polar coordinate representation of z. Suppose Eq. (A.1) holds for some integer $n \geq 1$, then

$$z^{n+1} = z\,z^n = |z|(\cos\theta + i\sin\theta)|z|^n(\cos(n\theta) + i\sin(n\theta))$$

$$= |z|^{n+1}(\cos\theta\cos(n\theta) - \sin\theta\sin(n\theta) + i(\cos\theta\sin(n\theta) + \sin\theta\cos(n\theta)))$$

$$= |z|^{n+1}(\cos((n+1)\theta) + i\sin((n+1)\theta)).$$

Thus by the principle of mathematical induction, Eq. (A.1) holds for all positive integers n. ☐

One of the consequences of De Moivre's Formula is that, for any $z \in \mathbb{C}$, $|z^n| = |z|^n$ and $\arg z^n = n\arg z$. With z^n defined above, polynomial and rational functions of complex numbers can then be defined. Furthermore, exponentiation of z can be extended to rational powers. If z is any complex number and n is a positive integer such that $w^n = z$ for some complex number w then w is called an nth **root** of z, written as $w = \sqrt[n]{z}$ or $w = z^{1/n}$. Note that for most complex numbers z, $w = \sqrt[n]{z}$ is multiple-valued.

Corollary A.1. *If* $z \neq 0$ *is a complex number and* $n \in \mathbb{N}$ *then the* n*th roots of* z *are the* n *complex numbers,*

$$w_k = \sqrt[n]{|z|}\left[\cos\left(\frac{\theta + 2k\pi}{n}\right) + i\sin\left(\frac{\theta + 2k\pi}{n}\right)\right] \tag{A.2}$$

for $k = 0, 1, \ldots, n-1$ *where* $\theta = \arg z$.

Proof. Let w_k be defined as in Eq. (A.2), then

$$w_k^n = |z|(\cos(\theta + 2k\pi) + i\sin(\theta + 2k\pi)) = |z|(\cos\theta + i\sin\theta) = z,$$

and thus w_k is an nth root of z. ☐

The n roots of z, w_k for $k \in \{0, 1, \ldots, n-1\}$ are distinct complex numbers. Let m be any integer, then for any $z \neq 0$, $z^{m/n}$ is defined as

$$z^{m/n} = (z^{1/n})^m.$$

---

[12]Abraham de Moivre, French mathematician (1667–1754).

For most complex numbers $z^{m/n}$ is multi-valued.

Next consider the natural exponential function of complex numbers. It must be defined in such a way that it agrees with the familiar exponential function e^x for $x \in \mathbb{R}$ and it is desirable that it be defined for all $z \in \mathbb{C}$. There are several approaches to do this and the method chosen here is motivated by the Taylor series for e^x. For any $z \in \mathbb{C}$, define

$$e^z = \sum_{n=0}^{\infty} \frac{z^n}{n!}.$$

In particular, for any $y \in \mathbb{R}$,

$$e^{iy} = \sum_{n=0}^{\infty} \frac{(iy)^n}{n!} = \sum_{k=0}^{\infty} (-1)^k \frac{y^{2k}}{(2k)!} + i \sum_{k=0}^{\infty} (-1)^k \frac{y^{2k+1}}{(2k+1)!}$$
$$= \cos y + i \sin y$$

This is the well-known Euler identity. Furthermore, it can be shown that if $z = x + iy$, then

$$e^z = e^x(\cos y + i \sin y)$$

and the following familiar properties hold:

(1) $e^{z+w} = e^z e^w$,
(2) $e^z \neq 0$,
(3) $e^x > 1$ when $x > 0$ and $0 < e^x < 1$ when $x < 0$,
(4) $|e^{x+iy}| = e^x$,
(5) $e^z = 1$ if and only if $z = 2n\pi i$ for some $n \in \mathbb{Z}$,
(6) e^z is periodic and each period has the form $2n\pi i$ for some $n \in \mathbb{N}$.

Armed with Euler's Identity, the relationships between the real-valued trigonometric functions and the complex exponential function can be seen. For example,

$$\cos y = \frac{e^{iy} + e^{-iy}}{2} \quad \text{and} \quad \sin y = \frac{e^{iy} - e^{-iy}}{2i}.$$

These equations lead to a natural way to define the trigonometric functions on the complex numbers. For $z \in \mathbb{C}$,

$$\cos z = \frac{e^{iz} + e^{-iz}}{2} \quad \text{and} \quad \sin z = \frac{e^{iz} - e^{-iz}}{2i}.$$

The remaining trigonometric functions of complex values can be defined as ratios and reciprocals of cosine and sine the same way as these functions are defined for real numbers. The reader can verify that the usual trigonometric identities hold.

Since the complex exponential function is periodic with fundamental period $2\pi i$, it is not invertible unless the domain is restricted to a horizontal "swath" of the complex plane of height $2\pi i$. Thus if the domain of e^z is restricted to the subset $D = \{x + iy \,|\, x \in \mathbb{R}$ and $-\pi < y \leq \pi\}$ the exponential function is a one-to-one and onto mapping (bijection) from D to the range set $R = \mathbb{C} \setminus \{0\}$ (the complex plane with the origin removed). The injective (one-to-one) property is shown by assuming there exist two complex numbers z and w in set D for which $e^z = e^w$. This implies $e^{z-w} = 1$ which in turn means $z - w = 2n\pi i$. If z and w both lie in set D then their imaginary parts are less than $2\pi i$ apart which means $\mathrm{Im}\,(z) = \mathrm{Im}\,(w)$. This in turn implies that $\mathrm{Re}\,(z) = \mathrm{Re}\,(w)$ and, as a result, $z = w$. The surjective (onto) property is shown by demonstrating the equation $e^z = w$ has a solution $z \in D$ for every $w \in R$. Let $w \in R$ and consider the equation $e^z = w$. The complex number z can be written as $z = x + iy$, where x, $y \in \mathbb{R}$. Thus the equation $e^z = w$ becomes

$$e^{x+iy} = e^x e^{iy} = |w| \frac{w}{|w|} = w.$$

Since $|e^{iy}| = 1$ then $e^{iy} = w/|w|$ and thus $e^x = |w|$. Use of the real-valued logarithm function implies $x = \ln|w|$. The equation $e^{iy} = w/|w|$ has a unique solution $y \in (-\pi, \pi]$ which is the argument of w. Thus $z = x + iy \in D$ is the unique solution to the equation. Having established that e^z is a bijection between D and R, the inverse of the complex exponential function can be defined. Let $z \in \mathbb{C} \setminus \{0\}$ then

$$\ln z = \ln|z| + i \arg z \qquad\qquad (A.3)$$

where $\ln|z|$ is the real-valued natural logarithm function, and $\arg z \in (-\pi, \pi]$. In fact, more generally, the logarithm of complex numbers can be defined as

$$\ln z = \ln|z| + i(\arg(z) + 2k\pi)$$

where k is an integer. This results in a multiple-valued function. The restriction of the argument of z results in what is called a **branch** of the logarithm function. The negative part of the real axis is sometimes called a **branch cut**. The function $\ln z$ defined above by restricting $\arg z \in (-\pi, \pi]$ is called the **principal branch** of the logarithm of z. The reader should keep in mind that the restricted interval for $\arg z$ is chosen arbitrarily. Any horizontal band in the complex plane of height $2\pi i$ can be used and therefore the branch cut can be placed anywhere that is convenient.

Now that several important examples of complex-valued functions have been given, attention can be turned to the calculus of complex functions.

Before studying complex differentiation and complex integration in the next two sections, the notion of limit and continuity for complex functions must be defined.

An **open disk** of radius $r > 0$ centered at a complex number z_0 is defined to be the set $\{z \in \mathbb{C} : |z - z_0| < r\}$. An **pierced open disk** of radius $r > 0$ centered at a complex number z_0 is defined to be the set $\{z \in \mathbb{C} : 0 < |z - z_0| < r\}$. Thus the open disk contains z_0 while the pierced open disk does not. An **open set** $O \subset \mathbb{C}$ is a set of complex numbers with the property that for every $z \in O$ there exists an open disk of some radius $r > 0$ centered at z such that the open disk is a subset of O.

Suppose $f : D \subset \mathbb{C} \to \mathbb{C}$ is a function and z_0 is a point such that D contains a pierced open disk centered at z_0, then f has a **limit** L, as z approaches z_0, denoted as

$$\lim_{z \to z_0} f(z) = L \qquad (A.4)$$

provided that for every $\epsilon > 0$ there exists $\delta > 0$ such that for all $0 < |z - z_0| < \delta$ then $|f(z) - L| < \epsilon$. Note that the definition of limit does not require f to be defined at z_0 or, even if $f(z_0)$ is defined, $f(z_0)$ to be the same as L. Suppose f and g are functions for which the following two limits exist:

$$\lim_{z \to z_0} f(z) = L \text{ and } \lim_{z \to z_0} g(z) = M,$$

then the following properties of limits hold:

(1) the limits are unique,
(2) $\lim_{z \to z_0} (f(z) + g(z)) = L + M$,
(3) $\lim_{z \to z_0} (f(z)g(z)) = L M$,
(4) $\lim_{z \to z_0} (f(z)/g(z)) = L/M$ provided $M \neq 0$.

The concept of continuity is built on the concept of the limit. Let $f : D \to \mathbb{C}$, where D is an open set, and let $z_0 \in D$. Function f is **continuous at** z_0 if and only if

$$\lim_{z \to z_0} f(z) = f(z_0).$$

The function is said to be **continuous on** D if and only if f is continuous at each $z \in D$.

A.3 Complex Differentiation

The complex derivative requires more care to define than the real derivative. Lessons learned while defining the derivative of a function of two variables will inform the development here. The reader may wish to review the derivative from multivariable calculus if the definition of the complex derivative given in this section needs further clarification.

Suppose $f : D \subset \mathbb{C} \to \mathbb{C}$ where D is an open set and $z = x + i\,y \in D$. The **derivative** of f at z is defined as

$$\lim_{(h,k)\to(0,0)} \frac{f(x + h + i(y + k)) - f(x + iy)}{h + i\,k}$$

provided the limit exists. If the limit exists, the derivative is denoted as $f'(z)$ and f is said to be **differentiable** at z. A natural question to ask is, given a function f, what are the necessary and sufficient conditions to guarantee the existence of the limit and hence the derivative. Suppose the complex derivative exists, then

$$f(z + h + i\,k) - f(z) = f'(z)(h + i\,k) + \epsilon(h + i\,k)$$

where $\epsilon(h + i\,k)$ is a function with the property that

$$\lim_{(h,k)\to(0,0)} \frac{\epsilon(h + i\,k)}{h + i\,k} = 0. \tag{A.5}$$

Let $f(z) = f(x + i\,y) = u(x,y) + i\,v(x,y)$, $f'(z) = s(x,y) + i\,t(x,y)$, and $\epsilon(h + i\,k) = \alpha(h,k) + i\,\beta(h,k)$, then

$$u(x + h, y + k) + i\,v(x + h, y + k) - u(x,y) - i\,v(x,y)$$
$$= (s(x,y) + i\,t(x,y))(h + i\,k) + \alpha(h,k) + i\beta(h,k)$$
$$= (h\,s(x,y) - k\,t(x,y) + \alpha(h,k)) + i\,(h\,t(x,y) + k\,s(x,y) + \beta(h,k)).$$

Equating the real and imaginary parts on both sides of the previous equation results in the two equations:

$$u(x + h, y + k) - u(x,y) = h\,s(x,y) - k\,t(x,y) + \alpha(h,k), \qquad \text{(A.6)}$$
$$v(x + h, y + k) - v(x,y) = h\,t(x,y) + k\,s(x,y) + \beta(h,k). \qquad \text{(A.7)}$$

Equation (A.5) implies

$$\lim_{(h,k)\to(0,0)} \frac{\alpha(h,k)}{(h^2 + k^2)^{1/2}} = 0 = \lim_{(h,k)\to(0,0)} \frac{\beta(h,k)}{(h^2 + k^2)^{1/2}}.$$

Equations (A.6) and (A.7) imply that functions u and v are differentiable functions (in the sense of real-valued functions of two variables). Setting $k = 0$, dividing both sides of Eq. (A.6) by h, and taking the limit as $h \to 0$ show that $s(x,y) = \partial u/\partial x$. Setting $h = 0$, dividing both sides of Eq. (A.6) by k, and taking the limit as $k \to 0$ show that $t(x,y) = -\partial u/\partial y$.

Likewise a similar manipulation of Eq. (A.7) shows that $s(x, y) = \partial v / \partial y$ and $t(x, y) = \partial v / \partial x$. Hence if f is differentiable at $z = x + i\,y$ then

$$u_x = v_y$$
$$u_y = -v_x. \qquad \text{(A.8)}$$

These are called the **Cauchy-Riemann equations**. To summarize, if f is differentiable at $z = x + i\,y$ then u and v are differentiable in the real variable sense and the partial derivatives of u and v satisfy the Cauchy-Riemann equations of Eq. (A.8). Conversely, given any $z_0 = x_0 + i\,y_0 \in \mathbb{C}$, if u and v are differentiable and their partial derivatives satisfy the Cauchy-Riemann equations at (x_0, y_0), then $f(z) = u + i\,v$ is differentiable at z_0 and, in this case, the derivative of f can be written as

$$f'(z) = f_x(z) = -i\,f_y(z).$$

A complex function which is differentiable at all points in an open region $D \subset \mathbb{C}$ is called an **analytic** function or sometimes a **holomorphic** function on D. If $D = \mathbb{C}$, the function is also called an **entire function**. It can be shown that the elementary complex functions, such as the power function z^n, the exponential function e^z, and the trigonometric functions $\sin z$ and $\cos z$, are all entire functions. Moreover, the usual differentiation formulas and rules developed for functions of real variables have their counterparts for complex valued functions. It can be shown that if $f : D \to \mathbb{C}$ is analytic, then f has continuous derivatives of all orders.

Theorem A.3. *If $f(z)$ is an analytic function on a simply connected domain $D \subset \mathbb{C}$, then $f'(z)$ is also analytic on D.*

This section ends with a result that reveals the close connection between analytic functions and real harmonic functions, the solutions to Laplace's equation.

Theorem A.4. *Let D be an open, connected subset of $\mathbb{C}$. If $f = u + i\,v$ is an analytic function on D and u and v have continuous second partial derivatives, then u and v satisfy Laplace's equation*

$$u_{xx} + u_{yy} = 0 \quad and \quad v_{xx} + v_{yy} = 0.$$

Proof. If f is analytic on D then u and v satisfy the Cauchy-Riemann equations. This implies

$$u_{xx} = (u_x)_x = (v_y)_x = (v_x)_y = (-u_y)_y = -u_{yy}.$$

Function v can be shown to be harmonic in a similar fashion. $\qquad \square$

A.4   Complex Integration

The integral of a complex function is closely related to the notion of the line integral or contour integral from multivariable calculus. Before treating the integral several preliminary definitions are required. If $[a, b]$ is a closed interval on the real number line and if $g : [a, b] \to \mathbb{C}$ is a continuous complex function with derivative g' defined and continuous for all $a < t < b$ then the graph of g in the complex plane is called a **smooth** curve. The derivative of g is defined in the natural way by taking the limit of the difference quotient of g expressed in terms of its real and imaginary parts, that is, if $g(t) = \alpha(t) + i\,\beta(t)$ then $g'(t) = \alpha'(t) + i\,\beta'(t)$. If $g(a) = g(b)$ the graph of g is said to be a **closed curve**. If, in addition, the graph of $g(t)$ does not intersect itself except at the endpoints, the graph is called a **simple closed curve**. If there is a partition $a = t_0 < t_1 < \cdots < t_n = b$ of $[a, b]$ such that g is smooth when restricted to each subinterval $[t_{i-1}, t_i]$ for $i = 1, 2, \ldots, n$ then g is called **piecewise smooth**. The graph of a piecewise smooth function is called a **piecewise smooth curve** often denoted as C.

Let C be a piecewise smooth curve (or path) given by $g : [a, b] \to \mathbb{C}$ and let f be a continuous, complex function defined on a region D containing C, then the path or contour integral of f over C is defined as

$$\int_C f(z)\, dz = \int_a^b f(g(t))g'(t)\, dt. \tag{A.9}$$

The following theorem states some of the properties of the complex integral. In the theorem the piecewise smooth path C can be thought of as a union of non-overlapping piecewise smooth segments $C = C_1 \cup C_2 \cup \cdots \cup C_n$. The initial point of segment C_i is the terminal point of segment C_{i-1}. Since each path has an implied orientation (the direction of the curve as the parameter t increases), the notation $-C$ denotes the path C with the opposite orientation.

Theorem A.5. *Let f and g be continuous complex functions defined on a region D containing the piecewise smooth curve C and let p and q be complex numbers, then*

(1) $\int_C [p\, f(z) + q\, g(z)]\, dz = p \int_C f(z)\, dz + q \int_C g(z)\, dz$,

(2) $\int_C f(z)\, dz = \int_{C_1} f(z)\, dz + \int_{C_2} f(z)\, dz + \cdots + \int_{C_n} f(z)\, dz$,

(3) $\int_{-C} f(z)\, dz = -\int_C f(z)\, dz$.

To describe other important theorems related to the integration of complex functions, the following concepts are necessary. An open set $D \subset \mathbb{C}$ or $\mathbb{R}^2$ is said to be **connected** if any two points in D can be connected with a piecewise smooth path that lies entirely in D. A connected open set is often referred to as a **domain**. A domain D is **simply connected** if every piecewise smooth closed path in D contains only points of D in its interior. Readers must recall from multivariable calculus that, if $M = M(x, y)$ and $N = N(x, y)$ have continuous partial derivatives on a simply connected domain in $\mathbb{R}^2$, then the condition that $\int_C M(x, y)\, dx + N(x, y)\, dy$ is independent of the path C in the domain is equivalent to the condition that the integral along any closed smooth path in the domain is equal to zero. In addition, the latter condition is equivalent to $\partial M / \partial y = \partial N / \partial x$ by Green's theorem. Now consider an analytic function $f(z) = u(x, y) + i\, v(x, y)$ with continuous $f'(z)$ on a simply connected domain $D \subset \mathbb{C}$. Let C be any simple closed smooth path in D described parametrically by $g(t) = x(t) + i\, y(t)$ for $t \in [a, b]$, then by Green's theorem and the Cauchy-Riemann equations,

$$\int_C f(z)\, dz = \int_C (u + i\, v) d(x + i\, y) = \int_C u\, dx - v\, dy + i \int_C v\, dx + u\, dy$$

$$= -\iint_R (v_x + u_y)\, dA + \iint_R (u_x - v_y)\, dA = 0$$

where R is the region enclosed by C. This result is called the **Cauchy integral theorem**. It turns out that the condition u and v have continuous partial derivatives is not necessary since it is automatically true for analytic functions (see Theorem A.3). This leads to the Cauchy-Goursat[13] Theorem, one of the key results in complex analysis.

Theorem A.6 (Cauchy-Goursat). *If $f(z)$ is an analytic function on a simply connected domain $D \subset \mathbb{C}$, then for any simple closed path C in D,*

$$\oint_C f(z)\, dz = 0.$$

The following is another key result related to integrals of complex functions, stated here for completeness.

Theorem A.7 (Cauchy Integral Formula). *If $f(z)$ is an analytic function on a simply connected domain $D \subset \mathbb{C}$ and C is a simple closed path in D, then for any z_0 in the interior of C,*

$$f(z_0) = \frac{1}{2\pi i} \oint_C \frac{f(z)}{z - z_0}\, dz.$$

---

[13]Édouard Jean-Baptiste Goursat, French mathematician (1858–1936).

Both the Cauchy-Goursat integral theorem and Cauchy integral formula hold true for simple closed piecewise smooth curves. The Cauchy integral theorem, Cauchy integral formula, and the residue theorem [Krantz (1999), Sec. 4.4.2] are often used to evaluate the integrals of real functions. The procedure for finding the exact value of the integral of $\int_0^\infty \frac{\sin x}{x}\, dx$ is used to demonstrate the idea.

Let $0 < r < R$ and set

C_1 : the line segment connecting the points $(-R, 0)$ and $(-r, 0)$,

C_2 : the upper semicircle with radius r,

C_3 : the line segment connecting the points $(r, 0)$ and $(R, 0)$,

C_4 : the upper semicircle with radius R.

The path $C = C_1 \cup C_2 \cup C_3 \cup C_4$ is a piecewise smooth, simple closed curve. Let $f(z) = e^{iz}/z$ and note that $f(z)$ is analytic on an open, simply connected region containing C and its interior. Therefore by Theorem A.6,

$$\int_C \frac{e^{iz}}{z}\, dz = \int_{C_1} \frac{e^{iz}}{z}\, dz + \int_{C_2} \frac{e^{iz}}{z}\, dz + \int_{C_3} \frac{e^{iz}}{z}\, dz + \int_{C_4} \frac{e^{iz}}{z}\, dz = 0.$$

Assuming that C is oriented counterclockwise, it can be shown that

$$\lim_{r \to 0} \int_{C_2} \frac{e^{iz}}{z}\, dz = -i\,\pi$$

$$\lim_{R \to \infty} \int_{C_4} \frac{e^{iz}}{z}\, dz = 0.$$

The path integrals along C_1 and C_3 are integrals along portions of the real axis, thus

$$\int_{C_1} \frac{e^{iz}}{z}\, dz + \int_{C_3} \frac{e^{iz}}{z}\, dz = \int_{-R}^{-r} \frac{e^{ix}}{x}\, dx + \int_r^R \frac{e^{ix}}{x}\, dx = 2\int_r^R \frac{\cos x + i \sin x}{x}\, dx.$$

Taking the limit as $r \to 0$ and $R \to \infty$ results in

$$2\int_0^\infty \frac{\cos x}{x}\, dx + 2i \int_0^\infty \frac{\sin x}{x}\, dx = i\,\pi.$$

Equating real and complex parts of the equation implies that

$$\int_0^\infty \frac{\sin x}{x}\, dx = \frac{\pi}{2}.$$

In this textbook the reader will occasionally come across the integral of a complex-valued function of the form $f : [a, b] \subset \mathbb{R} \to \mathbb{C}$. In this case f is integrable on $[a, b]$ provided its real and complex parts are integrable on $[a, b]$. The definite integral of $f(t) = u(t) + i\, v(t)$ can be evaluated as

$$\int_a^b f(t)\, dt = \int_a^b u(t)\, dt + i \int_a^b v(t)\, dt. \tag{A.10}$$

Appendix B

Linear Algebra Primer

In this chapter a brief introduction to matrices, vectors, and linear algebra is given in support of the discussion of finite difference methods for approximating solutions to partial differential equations covered in Chap. 12. The coverage of linear algebra topics will be selective and geared toward the implementation and analysis of the various numerical techniques mentioned in Chap. 12.

B.1 Matrices and Their Properties

A **linear system of equations** with m equations and n unknowns can be written as

$$a_{11}x_1 + a_{12}x_2 + \cdots + a_{1n}x_n = b_1$$
$$a_{21}x_1 + a_{22}x_2 + \cdots + a_{2n}x_n = b_2$$
$$\vdots$$
$$a_{m1}x_1 + a_{m2}x_2 + \cdots + a_{mn}x_n = b_m.$$

The expressions a_{ij} for $i = 1, 2, \ldots, m$ and $j = 1, 2, \ldots, n$ are called the **coefficients** of the linear system and the expressions x_j are called its **unknowns**. The linear system of equations above can be expressed in terms of matrices. A **matrix** is a rectangular array of numbers (real or complex). An $m \times n$ matrix contains mn entries arranged in m rows and n columns. Matrices are often denoted with capital letters such as A, B, C, *etc.* The entries of a matrix associated with a particular linear system are the coefficients of the linear system. For example, a matrix, denoted as A, could be associated with the linear system above and defined as

$$
A = \begin{bmatrix}
a_{11} & a_{12} & \cdots & a_{1n} \\
a_{21} & a_{22} & \cdots & a_{2n} \\
\vdots & \vdots & & \vdots \\
a_{m1} & a_{m2} & \cdots & a_{mn}
\end{bmatrix},
$$

and the linear system of equations can be written as $A\mathbf{x} = \mathbf{b}$ where $\mathbf{x}$ is an $n \times 1$ matrix with entries $x_1, x_2, \ldots, x_n$ and $\mathbf{b}$ is an $m \times 1$ matrix with entries $b_1, b_2, \ldots, b_m$. For convenience, set $(A)_{ij} = a_{ij}$, the entry of matrix A in the ith row and the jth column.

Arithmetic operations for matrices are defined in the natural way. Two matrices A and B are **equal**, denoted as $A = B$, if they have the same number of rows and columns as each other (henceforth referred to as having the same size) and if $(A)_{ij} = (B)_{ij}$ for all i and j. If matrix A and matrix B have the same size, the **sum** of A and B, denoted as $A + B$, is a matrix in which $(A + B)_{ij} = (A)_{ij} + (B)_{ij}$ for all i and j. Matrix addition can be thought of as being carried out component-wise. If c is a scalar (a real or complex number) and A is a matrix, the **scalar product** of c and A is denoted as $c\,A$ and is defined as a matrix for which $(c\,A)_{ij} = c(A)_{ij}$ for all i and j. Matrix subtraction can be defined as $A - B = A + (-1)B$ for matrices of the same size. The **transpose** of an $m \times n$ matrix A is an $n \times m$ matrix, denoted as A^T, where $(A^T)_{ij} = (A)_{ji}$ for all i and j.

The operation of **matrix multiplication** requires more care to define. If A is an $m \times p$ matrix and if B is a $p \times n$ matrix, the **product** is denoted as $A\,B$ and is an $m \times n$ matrix where

$$
(A\,B)_{ij} = \sum_{k=1}^{p} (A)_{ik}(B)_{kj},
$$

for $i = 1, 2, \ldots, m$ and $j = 1, 2, \ldots, n$.

The following theorem presents many of the properties of matrices and matrix operations. Many of these properties are matrix analogues of properties of the real numbers.

Theorem B.1. *Let A, B, and C be matrices of the appropriate size so that the following operations are well-defined and let a and b be scalars, then*

- $A + B = B + A$ *(commutative property of matrix addition)*,
- $A + (B + C) = (A + B) + C$ *(associative property of matrix addition)*,
- $A(BC) = (AB)C$ *(associative property of matrix multiplication)*,
- $A(B + C) = AB + AC$ *(left distributive law)*,

- $(B + C)A = BA + CA$ (*right distributive law*),
- $a(B + C) = aB + aC$,
- $(a + b)C = aC + bC$,
- $a(bC) = (ab)C$,
- $a(BC) = (aB)C = B(aC)$,
- $(AB)^T = B^T A^T$,
- $(aB)^T = aB^T$,
- $(A^T)^T = A$.

Often matrices have special organization or structure. A **square matrix of order** n is a matrix with n rows and n columns. Matrix powers can be defined for square matrices. If A is an $n \times n$ matrix and $k \in \mathbb{N}$ then the kth power of A, denoted as A^k, is defined as

$$A^k = \underbrace{AA \cdots A}_{k \text{ factors}}.$$

A matrix for which all entries are 0 will be called a **zero matrix** and denoted simply as 0. The **diagonal** of a square matrix A of order n is the ordered sequence of entries $(a_{11}, a_{22}, \ldots, a_{nn})$. A **lower triangular matrix** L is a matrix for which $(L)_{ij} = 0$ for each $j = 1, 2, \ldots, n$ for $i = 1, 2, \ldots, j - 1$. An **upper triangular matrix** U is a matrix for which $(U)_{ij} = 0$ for each $j = 1, 2, \ldots, n$ for $i = j+1, j+2, \ldots, n$. A square matrix A for which $A = A^T$ is said to be a **symmetric matrix**. A square matrix of order n is called a **banded matrix** if there exist integers r and s with $1 < r, s < n$ for which $(A)_{ij} = 0$ when $r \leq j - i$ or $s \leq i - j$. The **band width** is $r + s - 1$. Banded matrices of band width 3 are sometimes called **tridiagonal matrices** and have the form

$$A = \begin{bmatrix} a_{11} & a_{12} & 0 & \cdots & 0 & 0 & 0 \\ a_{21} & a_{22} & a_{23} & \cdots & 0 & 0 & 0 \\ \vdots & \vdots & \vdots & & \vdots & & \vdots \\ 0 & 0 & 0 & \cdots a_{n-1,n-2} & a_{n-1,n-1} & a_{n-1,n} \\ 0 & 0 & 0 & \cdots & 0 & a_{n,n-1} & a_{n,n} \end{bmatrix}.$$

Banded matrices of band width 1 are called **diagonal matrices** and may have nonzero entries only on their diagonals. A diagonal matrix with 1's along its diagonal is known as the **identity matrix** and is denoted as I or I_n when the order of the matrix is significant.

Theorem B.2. *If A is an $m \times n$ matrix then $AI_n = A = I_m A$.*

A **column vector** or simply **vector** is an $m \times 1$ matrix. In the interest of saving printed space a vector written out in component form will often be denoted as $\mathbf{x} = (x_1, x_2, \ldots, x_m)^T$. A set of vectors $\{\mathbf{x}_1, \mathbf{x}_2, \ldots, \mathbf{x}_n\}$ is said to be a **linearly independent set** if the equation,

$$c_1\mathbf{x}_1 + c_2\mathbf{x}_2 + \cdots + c_n\mathbf{x}_n = \mathbf{0}$$

implies the scalars $c_1 = c_2 = \cdots = c_n = 0$. When dealing with vectors with n components, a useful set of linearly independent vectors is $\{\mathbf{e}_1, \mathbf{e}_2, \ldots, \mathbf{e}_n\}$ where $\mathbf{e}_i = (0, \ldots, 0, 1, 0, \ldots, 0)^T$ and the 1 appears in the ith position for $1 \leq i \leq n$. Any set of n linearly independent vectors $\{\mathbf{x}_1, \mathbf{x}_2, \ldots, \mathbf{x}_n\}$ in $\mathbb{R}^n$ forms a **basis** for $\mathbb{R}^n$ and hence if $\mathbf{x} \in \mathbb{R}^n$ there exists a unique set of scalars $c_1, c_2, \ldots, c_n$ such that

$$c_1\mathbf{x}_1 + c_2\mathbf{x}_2 + \cdots + c_n\mathbf{x}_n = \mathbf{x}.$$

A related, but stronger property of a set of vectors is known as **orthogonality**. Vectors $\mathbf{v}_1$ and $\mathbf{v}_2$ are orthogonal if $\mathbf{v}_1^T \mathbf{v}_2 = \mathbf{v}_2^T \mathbf{v}_1 = 0$. If $\{\mathbf{v}_1, \mathbf{v}_2, \ldots, \mathbf{v}_n\}$ is a set of nonzero, mutually orthogonal vectors, the set is also linearly independent. To see this, suppose there exist scalars c_1, c_2, $\ldots$, c_n such that

$$c_1\mathbf{v}_1 + c_2\mathbf{v}_2 + \cdots + c_n\mathbf{v}_n = \mathbf{0}.$$

Then for each $i = 1, 2, \ldots, n$,

$$\mathbf{v}_i^T(c_1\mathbf{v}_1 + c_2\mathbf{v}_2 + \cdots + c_n\mathbf{v}_n) = c_1\mathbf{v}_i^T\mathbf{v}_1 + c_2\mathbf{v}_i^T\mathbf{v}_2 + \cdots + c_n\mathbf{v}_i^T\mathbf{v}_n = c_i\|\mathbf{v}_i\|_2^2 = 0.$$

This implies $c_i = 0$ for $i = 1, 2, \ldots, n$ and hence the vectors are linearly independent.

If A and B are $n \times n$ matrices for which $AB = BA = I_n$ then A is said to be an **invertible** matrix and B is the **inverse** of A. In this case matrix B is often denoted as A^{-1}. Likewise A is the inverse of matrix B. If matrix A has no inverse then A is said to be a **singular** matrix.

Theorem B.3. *The inverse of A, if it exists, is unique.*

Proof. Suppose there exist two matrices B and C for which $AB = BA = I$ and $AC = CA = I$, then $AB - AC = A(B - C) = 0$ and thus

$$BA(B - C) = I(B - C) = B - C = 0 \iff B = C.$$

Consequently the inverse of matrix A is unique. $\qquad\square$

Theorem B.4. *If A and B are $n \times n$ invertible matrices and $k \neq 0$ is a scalar, then*

- $(A\,B)^{-1} = B^{-1}A^{-1}$,
- $(A^{-1})^{-1} = A$,
- $(A^m)^{-1} = (A^{-1})^m = A^{-m}$ *for* $m = 0, 1, 2, \ldots$,
- $(k\,A)^{-1} = (1/k)A^{-1}$,
- $(A^T)^{-1} = (A^{-1})^T$.

Invertible matrices play a fundamental role in solving systems of linear equations.

Theorem B.5. *The following statements are equivalent for any square matrix A of order n.*

(1) *Matrix A is invertible.*

(2) *The equation $A\mathbf{x} = \mathbf{b}$ has a unique solution for any column vector $\mathbf{b}$ with n components.*

(3) *The columns of A are linearly independent.*

Proof. Suppose A is invertible and define $\mathbf{x} = A^{-1}\mathbf{b}$, then

$$A\mathbf{x} = A(A^{-1}\mathbf{b}) = (A\,A^{-1})\mathbf{b} = I_n\,\mathbf{b} = \mathbf{b},$$

which establishes the existence of a solution. Now suppose $A\mathbf{x} = \mathbf{b}$ and $A\mathbf{y} = \mathbf{b}$, then

$$\mathbf{x} = (A^{-1}A)\mathbf{x} = A^{-1}(A\mathbf{x}) = A^{-1}\mathbf{b} = A^{-1}(A\mathbf{y}) = (A^{-1}A)\mathbf{y} = \mathbf{y}$$

which establishes the uniqueness of the solution.

Let the columns of matrix A be the vectors $\mathbf{a}_1$, $\mathbf{a}_2$, $\ldots$, $\mathbf{a}_n$ and suppose

$$\begin{aligned} \mathbf{0} &= c_1\mathbf{a}_1 + c_2\mathbf{a}_2 + \cdots + c_n\mathbf{a}_n \\ &= c_1 A\mathbf{e}_1 + c_2 A\mathbf{e}_2 + \cdots + c_n A\mathbf{e}_n \\ &= A(c_1\mathbf{e}_1 + c_2\mathbf{e}_2 + \cdots + c_n\mathbf{e}_n). \end{aligned}$$

If the equation $A\mathbf{x} = \mathbf{b}$ has a unique solution for every vector $\mathbf{b}$, then the equation above has the unique solution

$$c_1\mathbf{e}_1 + c_2\mathbf{e}_2 + \cdots + c_n\mathbf{e}_n = \mathbf{0}.$$

Since the vectors $\mathbf{e}_1$, $\mathbf{e}_2$, $\ldots$, $\mathbf{e}_n$ are linearly independent, then $c_1 = c_2 = \cdots = c_n = 0$. Hence the columns of matrix A are linearly independent.

Now suppose the columns of matrix A are linearly independent. For each canonical basis vector $\mathbf{e}_i$ there exists a unique set of scalars b_{1i}, b_{2i}, $\ldots$, b_{ni} such that

$$b_{1i}\mathbf{a}_1 + b_{2i}\mathbf{a}_2 + \cdots + b_{ni}\mathbf{a}_n = \mathbf{e}_i.$$

Define matrix B as

$$B = \begin{bmatrix} b_{11} & b_{12} & \cdots & b_{1n} \\ b_{21} & b_{22} & \cdots & b_{2n} \\ \vdots & \vdots & & \vdots \\ b_{n1} & b_{n2} & \cdots & b_{nn} \end{bmatrix}$$

where the ith column of B consists of the coefficients of the column representation of e_i. Then $AB = I_n$ and $B^T A^T = I_n$. Suppose $A^T \mathbf{x} = \mathbf{0}$, then $B^T A^T \mathbf{x} = \mathbf{0}$ and consequently $\mathbf{x} = \mathbf{0}$. As a result $A^T \mathbf{x} = \mathbf{0}$ if and only if $\mathbf{x} = \mathbf{0}$. Thus the only linear combination of the columns of A^T equal to the zero vector is the trivial linear combination. This implies the columns of A^T, and hence the rows of A, are linearly independent. Using the argument above there exists an $n \times n$ matrix $\hat{B}$ such that $\hat{B} A = I_n$. Now consider

$$\hat{B} = \hat{B}(AB) = (\hat{B} A) B = B$$

which implies $AB = BA = I_n$ and A is invertible. $\qquad\square$

The reader should note that not all the familiar operations and properties of the real numbers have analogues found among the properties of matrices listed in Theorem B.1. For example there is no commutative property of matrix multiplication, no zero factor property, and no multiplicative inverse for all nonzero matrices. In fact, these properties do not hold in general for matrices.

B.2   Determinant of a Square Matrix

The **determinant** is a number associated with each square matrix and hence the determinant can be thought of as a function mapping the set of square matrices into the scalars (real or complex). In applications and calculations the determinant is often a surrogate quantity giving useful information about properties of the matrix. Computing determinants of 2×2 matrices is trivial and defined as

$$\det \begin{bmatrix} a_{11} & a_{12} \\ a_{21} & a_{22} \end{bmatrix} = a_{11}a_{22} - a_{12}a_{21}.$$

The determinant of a general $n \times n$ matrix is calculated via a recursive procedure making use of quantities known as **cofactors**. If A is an $n \times n$ matrix the ij**th minor**, denoted as M_{ij}, is the determinant of the submatrix which remains after the ith row and jth column of A are deleted. The ijth cofactor is defined as $C_{ij} = (-1)^{i+j} M_{ij}$. The determinant of a general $n \times n$ matrix is then

$$\det A = (A)_{11} C_{11} + (A)_{12} C_{12} + \cdots + (A)_{1n} C_{1n}.$$

Theorem B.6. *The determinant of an $n \times n$ matrix A can be computed by multiplying the entries in any row (column) by their corresponding cofactors and adding the resulting products, that is, for any $i = 1, 2, \ldots, n$ and for any $j = 1, 2, \ldots, n$,*

$$\det A = \sum_{k=1}^{n} (A)_{ik} C_{ik} = \sum_{k=1}^{n} (A)_{kj} C_{kj}.$$

In practice the determinant is most easily calculated by expanding along the row or column containing the most zero entries. The remainder of this section will present some of the properties of determinants.

Theorem B.7. *If A is an $n \times n$ diagonal matrix, lower triangular matrix, or upper triangular matrix,*

$$\det A = \prod_{i=1}^{n} (A)_{ii}.$$

Theorem B.8. *If A is an $n \times n$ matrix and c is a scalar, then $\det(c\,A) = c^n \det A$.*

Theorem B.9. *If A and B are $n \times n$ matrices then $\det(A\,B) = \det(A) \det(B)$.*

Theorem B.10. *Matrix A is invertible if and only if $\det A \neq 0$.*

Theorem B.10 is one of the most fundamentally important results in linear algebra. It can be added as another statement equivalent to the others listed in Theorem B.5.

Corollary B.1. *If A is an invertible matrix, then*

$$\det(A^{-1}) = (\det A)^{-1}.$$

Proof. Suppose A has an inverse, then $A\,A^{-1} = I$ and

$$1 = \det I \text{ (by Theorem B.7)}$$
$$1 = \det(A\,A^{-1}) = \det(A) \det(A^{-1}) \text{ (by Theorem B.9)}$$
$$\det(A^{-1}) = (\det A)^{-1}$$

by Theorem B.10. $\qquad\qquad\qquad\qquad\qquad\qquad\qquad\qquad\qquad\qquad\square$

The notion of the determinant plays a key role in the topic of eigenvalues and eigenvectors coming up in Sec. B.4.

B.3   Vector and Matrix Norms

Two vital concepts used throughout mathematical analysis are the notions of norm (or length) and distance. If $\mathbf{x}$ is a vector of n components, define

$$\|\mathbf{x}\|_2 = \left(x_1^2 + x_2^2 + \cdots + x_n^2\right)^{1/2},$$

which is sometimes called the l_2-**norm** of a vector. Many readers may know the l_2-norm as the Euclidean length of a vector. Another useful norm is the l_∞-**norm** defined as

$$\|\mathbf{x}\|_\infty = \max_{1 \le i \le n} |x_i|.$$

The choice of which vector norm to use is one of convenience since the two norms are equivalent in the sense that

$$\|\mathbf{x}\|_\infty \le \|\mathbf{x}\|_2 \le \sqrt{n}\|\mathbf{x}\|_\infty \tag{B.1}$$

for all $\mathbf{x} \in \mathbb{R}^n$. There are other equivalent norms, but their use is not necessary to establish the results mentioned in this textbook.

The distance between two vectors $\mathbf{u}$ and $\mathbf{v}$ is defined as $\|\mathbf{u} - \mathbf{v}\|_\infty$ (if using the l_∞-norm) or as $\|\mathbf{u} - \mathbf{v}\|_2$ (if using the l_2-norm). Thus the notion of a sequence of vectors $\{\mathbf{u}^{(k)}\}_{k=0}^\infty$ **converging** to a vector $\mathbf{u}$ can be made more precise using the concept of the distance between vectors. The sequence $\mathbf{u}^{(k)} \to \mathbf{u}$ as $k \to \infty$ provided for all $\epsilon > 0$ there exists a natural number N such that

$$\|\mathbf{u}^{(k)} - \mathbf{u}\| < \epsilon \text{ for all } k \ge N.$$

The norm used above can be either the l_2-norm or the l_∞-norm.

Theorem B.11. *The sequence of vectors* $\{\mathbf{u}^{(k)}\}_{k=0}^\infty$ *converges to* $\mathbf{u} \in \mathbb{R}^n$ *with respect to the* l_∞-*norm (or* l_2-*norm) if and only if*

$$\lim_{k \to \infty} u_i^{(k)} = u_i \text{ for each } i = 1, 2, \ldots, n.$$

Proof. Suppose $\mathbf{u}^{(k)} \to \mathbf{u}$ with respect to the l_∞-norm. By definition, given any $\epsilon > 0$ there exists $N \in \mathbb{N}$ such that if $k \ge N$ it is true that

$$\|\mathbf{u}^{(k)} - \mathbf{u}\|_\infty = \max_{1 \le i \le n} |u_i^{(k)} - u_i| < \epsilon.$$

Therefore for each $i = 1, 2, \ldots, n$ it is the case that $|u_i^{(k)} - u_i| < \epsilon$ which implies $u_i^{(k)} \to u_i$ as a sequence of scalars.

Now suppose that for each $i = 1, 2, \ldots, n$ it is true that $u_i^{(k)} \to u_i$ as a sequence of scalars. By definition given $\epsilon > 0$ there exists for each

$i = 1, 2, \ldots, n$ a natural number N_i such that $|u_i^{(k)} - u_i| < \epsilon$ for all $k \geq N_i$. Define $N = \max_{1 \leq i \leq n} N_i$ then if $k \geq N$,

$$\max_{1 \leq i \leq n} |u_i^{(k)} - u_i| = \|\mathbf{u}^{(k)} - \mathbf{u}\|_\infty < \epsilon.$$

This implies that $\mathbf{u}^{(k)} \to \mathbf{u}$ with respect to the l_∞-norm. The case using l_2-norm can be proved similarly. □

The vector norms defined above induce norms on matrices in the following manner. If A is an $m \times n$ matrix and $\mathbf{x} \in \mathbb{R}^n$ with $\|\mathbf{x}\|_\infty = 1$ then the l_∞-norm of A denoted as $\|A\|_\infty$ is defined as

$$\|A\|_\infty = \max_{\|\mathbf{x}\|_\infty = 1} \|A\mathbf{x}\|_\infty.$$

Similarly the l_2-norm of A is defined as

$$\|A\|_2 = \max_{\|\mathbf{x}\|_2 = 1} \|A\mathbf{x}\|_2.$$

Since the geometrical interpretation of norm is length of a vector, the norm of a matrix can be thought of as the maximum factor by which the length of any unit vector is multiplied when the unit vector is multiplied by the matrix. Suppose $\|\mathbf{x}\|_\infty = 1$, then $\|A\mathbf{x}\|_\infty \leq \|A\|_\infty = \|A\|_\infty \|\mathbf{x}\|_\infty$. If $\mathbf{x} \neq \mathbf{0}$, then $\|\mathbf{x}\|_\infty > 0$ and hence $\mathbf{z} = \mathbf{x}/\|\mathbf{x}\|_\infty$ is a unit vector (with respect to the l_∞-norm). Similar to the previous calculation,

$$\|A\mathbf{z}\|_\infty \leq \|A\|_\infty \|\mathbf{z}\|_\infty$$
$$\|\mathbf{x}\|_\infty \|A\mathbf{z}\|_\infty \leq \|\mathbf{x}\|_\infty \|A\|_\infty \|\mathbf{z}\|_\infty$$
$$\|A\|\mathbf{x}\|_\infty \mathbf{z}\|_\infty \leq \|A\|_\infty \|\|\mathbf{x}\|_\infty \mathbf{z}\|_\infty$$
$$\|A\mathbf{x}\|_\infty \leq \|A\|_\infty \|\mathbf{x}\|_\infty.$$

The final inequality is trivially true when $\mathbf{x} = \mathbf{0}$.

Theorem B.12. $\|A\|_\infty = \max\limits_{\mathbf{x} \neq 0} \dfrac{\|A\mathbf{x}\|_\infty}{\|\mathbf{x}\|_\infty}.$

Proof. Let $\mathbf{x} \in \mathbb{R}^n$ with $\mathbf{x} \neq \mathbf{0}$, then the vector $\mathbf{z} = \mathbf{x}/\|\mathbf{x}\|_\infty$ is a unit vector with respect to the l_∞-norm. Hence $\|A\mathbf{z}\|_\infty \leq \|A\|_\infty$ which implies

$$\max_{\mathbf{x} \neq 0} \frac{\|A\mathbf{x}\|_\infty}{\|\mathbf{x}\|_\infty} = \max_{\mathbf{x} \neq 0} \left\| A\left(\frac{\mathbf{x}}{\|\mathbf{x}\|_\infty}\right) \right\|_\infty \leq \|A\|_\infty.$$

Note that

$$\|A\|_\infty = \max_{\|\mathbf{x}\|_\infty = 1} \|A\mathbf{x}\|_\infty = \max_{\|\mathbf{x}\|_\infty = 1} \frac{\|A\mathbf{x}\|_\infty}{\|\mathbf{x}\|_\infty} \leq \max_{\mathbf{x} \neq 0} \frac{\|A\mathbf{x}\|_\infty}{\|\mathbf{x}\|_\infty}.$$

Thus the theorem is proved. □

A similar theorem holds for the l_2-norm of matrices.

B.4   Eigenvalues and Eigenvectors

Matrix multiplication applied to a vector can be thought of as a function mapping a given set of vectors into another set of vectors. The vectors in these two different sets do not necessarily have to have the same number of components, but when the matrix is a square matrix, they will. Thus to each square matrix A of order n there may exist nonzero vectors $\mathbf{v}$ and scalars λ such that $A\mathbf{v} = \lambda\mathbf{v}$. Geometrically this can be imagined as stating that the vector $A\mathbf{v}$ is parallel to vector $\mathbf{v}$ (for the case where λ is a real number, if λ is complex the geometry is more difficult to describe). More formally if $\mathbf{v} \neq \mathbf{0}$ is a vector and λ is a scalar for which $A\mathbf{v} = \lambda\mathbf{v}$ then λ is said to be an **eigenvalue of** A and $\mathbf{v}$ is said to be an **eigenvector of** A corresponding to the eigenvalue λ. The equation $A\mathbf{v} = \lambda\mathbf{v}$ is equivalent to $(A - \lambda I)\mathbf{v} = \mathbf{0}$. According to Theorem B.5 the situation that $\mathbf{v} \neq \mathbf{0}$ solves the equation implies that matrix $A - \lambda I$ is singular. Thus as a consequence of Theorem B.10, $\det(A - \lambda I) = 0$. The expression $\det(A - \lambda I)$ is a polynomial in λ of degree n called the **characteristic polynomial of** A and is sometimes written as $p(\lambda)$. Hence the eigenvalues of A are the roots of the characteristic polynomial of A. As a result, a square matrix of order n will have n eigenvalues counting multiplicity.

The set of eigenvalues of a matrix is known as the **spectrum** of the matrix and is denoted as $\sigma(A)$. The eigenvalue of largest magnitude is called the **spectral radius** of the matrix and will be denoted as

$$\rho(A) = \max_{\lambda \in \sigma(A)} |\lambda|.$$

The following theorem lists some of the properties of the characteristic polynomial and the spectrum of a matrix.

Theorem B.13. *Let A be a square matrix of order n, then*

(1) $p(0) = \det A = \prod_{\lambda \in \sigma(A)} \lambda$,
(2) $\rho(A) \leq \|A\|$, *where* $\|\cdot\|$ *can be any matrix norm.*

Proof. By definition the characteristic polynomial $p(\lambda) = \det(A - \lambda I)$, so $p(0) = \det A$. The expression $p(0)$ is the constant term in the characteristic polynomial and since each eigenvalue is a root of the characteristic polynomial, then

$$p(0) = \prod_{\lambda \in \sigma(A)} \lambda.$$

Suppose λ is an eigenvalue of A, then there exists a nonzero vector $\mathbf{v}$ for which $A\mathbf{v} = \lambda\mathbf{v}$. Without loss of generality assume $\|\mathbf{v}\| = 1$. Thus

$$|\lambda| = |\lambda|\,\|\mathbf{v}\| = \|\lambda\mathbf{v}\| = \|A\mathbf{v}\| \le \|A\|\|\mathbf{v}\| = \|A\|.$$

In turn then $\rho(A) = \max_{\lambda \in \sigma(A)} |\lambda| \le \|A\|$. $\qquad\square$

The following theorem provides some insight as to the locations in the complex plane of the eigenvalues of a matrix.

Theorem B.14 (Gershgorin's[14] Disks). *Let A be an $n \times n$ square matrix, let $r_i = \sum_{j \ne i}^n |a_{ij}|$, and let D_i be the closed disk of radius r_i centered at a_{ii} in the complex plane. Every eigenvalue of A lies within at least one of the D_i.*

Proof. Let λ be an eigenvalue of matrix A with corresponding eigenvector $\mathbf{v} = (v_1, v_2, \ldots, v_n)^T$. Let $i \in \{1, 2, \ldots, n\}$ be the value of the index for which $|v_i| = \max_{j=1,\ldots,n} |v_j|$. Since $\mathbf{v}$ is an eigenvector, $|v_i| > 0$ and $A\mathbf{v} = \lambda\mathbf{v}$. Multiplying the ith row of A component-wise by $\mathbf{v}$ results in

$$\sum_{j=1}^n a_{ij}v_j = \lambda v_i \iff \lambda v_i - a_{ii}v_i = \sum_{j=1,j\ne i}^n a_{ij}v_j.$$

Taking the norm of both sides of the last equation and dividing both sides by $|v_i|$ produce

$$|\lambda - a_{ii}| = \left|\sum_{j=1,j\ne i}^n a_{ij}\frac{v_j}{v_i}\right| \le \sum_{j=1,j\ne i}^n |a_{ij}|\frac{|v_j|}{|v_i|} \le \sum_{j=1,j\ne i}^n |a_{ij}| = r_i.$$

Therefore $\lambda \in D_i$. $\qquad\square$

The disks D_i in the theorem are sometimes called **Gershgorin disks**. Since many of the matrices encountered during the discussion of finite difference method approximations to initial boundary value problems are symmetric matrices, the following theorem is often useful.

Theorem B.15. *If A is symmetric matrix with real entries, the eigenvalues of A are all real.*

---

[14]Semyon Aronovich Gershgorin, Soviet mathematician, (1901–1933).

Proof. Let λ be an eigenvalue of A with corresponding eigenvector $\mathbf{v}$. The notation $\overline{\lambda}$ and $\overline{\mathbf{v}}$ will denote the complex conjugates of λ and $\mathbf{v}$ respectively. Since $A\mathbf{v} = \lambda\mathbf{v}$ then

$$\overline{A\mathbf{v}} = \overline{\lambda}\,\overline{\mathbf{v}}$$
$$A\overline{\mathbf{v}} = \overline{\lambda}\,\overline{\mathbf{v}}$$
$$\mathbf{v}^T A\overline{\mathbf{v}} = \mathbf{v}^T \overline{\lambda}\,\overline{\mathbf{v}}$$
$$(A^T\mathbf{v})^T\overline{\mathbf{v}} = \overline{\lambda}\,\mathbf{v}^T\overline{\mathbf{v}}$$
$$(A\mathbf{v})^T\overline{\mathbf{v}} = \overline{\lambda}\,\|\mathbf{v}\|_2^2$$
$$(\lambda\mathbf{v})^T\overline{\mathbf{v}} = \overline{\lambda}\,\|\mathbf{v}\|_2^2$$
$$\lambda\,\|\mathbf{v}\|_2^2 = \overline{\lambda}\,\|\mathbf{v}\|_2^2.$$

The last equation is equivalent to $(\lambda-\overline{\lambda})\|\mathbf{v}\|_2^2 = 0$. Since $\|\mathbf{v}\|_2^2 > 0$ (because $\mathbf{v}$ is an eigenvector) then $\lambda = \overline{\lambda}$ which implies $\lambda \in \mathbb{R}$. □

As a consequence of the eigenvalues of a real, symmetric matrix being real, the components of the eigenvectors can be chosen to be real as well.

Theorem B.16. *If A is a symmetric matrix, eigenvectors corresponding to different eigenvalues are orthogonal.*

Proof. Let λ_1 and λ_2 with $\lambda_1 \neq \lambda_2$ be eigenvalues of A with corresponding eigenvectors $\mathbf{v}_1$ and $\mathbf{v}_2$ respectively. Then,

$$\lambda_1\mathbf{v}_1^T\mathbf{v}_2 = (A\mathbf{v}_1)^T\mathbf{v}_2 = \mathbf{v}_1^T A^T\mathbf{v}_2 = \mathbf{v}_1^T A\mathbf{v}_2 = \lambda_2\mathbf{v}_1^T\mathbf{v}_2$$

which implies $(\lambda_1 - \lambda_2)\mathbf{v}_1^T\mathbf{v}_2 = 0$ and thus $\mathbf{v}_1^T\mathbf{v}_2 = 0$ which means $\mathbf{v}_1$ and $\mathbf{v}_2$ are orthogonal (and consequently linearly independent). □

In many ways a diagonal matrix is the simplest form of matrix for performing many calculations (*e.g.*, solving a linear system, inverting a nonsingular matrix, finding eigenvalues, *etc.*). Thus a natural question to ask is, under what conditions can a matrix be diagonalized, that is, be replaced with a diagonal matrix preserving the important properties of the original matrix. This leads to two important definitions. First, if A and B are square matrices of the same order, than B is **similar** to A if there exists an invertible matrix P such that $B = P^{-1}AP$. Second, a square matrix A is **diagonalizable** if there is an invertible matrix P such that $D = P^{-1}AP$ is a diagonal matrix. Thus a matrix is diagonalizable if and only if it is similar to some diagonal matrix. The following theorem answers the question as to which matrices are diagonalizable.

Theorem B.17. *A square matrix of order n is diagonalizable if and only if it has n linearly independent eigenvectors.*

Proof. Let A be a diagonalizable matrix of order n. By definition there exists an invertible matrix P such that $D = P^{-1}AP$ is a diagonal matrix. Let the columns of P be the vectors $\mathbf{p}_1, \mathbf{p}_2, \ldots, \mathbf{p}_n$. Since P is invertible, the columns of P are linearly independent (Theorem B.5). Suppose the diagonal entries of D are $\lambda_1, \lambda_2, \ldots, \lambda_n$, then

$$PD = AP,$$

$$[\lambda_1\mathbf{p}_1 \mid \lambda_2\mathbf{p}_2 \mid \cdots \mid \lambda_n\mathbf{p}_n] = [A\,\mathbf{p}_1 \mid A\,\mathbf{p}_2 \mid \cdots \mid A\,\mathbf{p}_n]$$

which implies the eigenvalues of A are $\lambda_1, \lambda_2, \ldots, \lambda_n$ with corresponding linearly independent eigenvectors $\mathbf{p}_1, \mathbf{p}_2, \ldots, \mathbf{p}_n$.

Conversely suppose A has n linearly independent eigenvectors $\mathbf{p}_1, \mathbf{p}_2, \ldots, \mathbf{p}_n$ with corresponding eigenvalues $\lambda_1, \lambda_2, \ldots, \lambda_n$. Define matrix P to be a square matrix whose ith column is the vector $\mathbf{p}_i$. Define matrix D to be the diagonal matrix for which $(D)_{ii} = \lambda_i$.

$$[A\,\mathbf{p}_1 \mid A\,\mathbf{p}_2 \mid \cdots \mid A\,\mathbf{p}_n] = [\lambda_1\mathbf{p}_1 \mid \lambda_2\mathbf{p}_2 \mid \cdots \mid \lambda_n\mathbf{p}_n],$$

$$AP = PD.$$

Since the columns of P are linearly independent, matrix P is invertible and thus $D = P^{-1}AP$. $\qquad\square$

B.5 Diagonally Dominant Matrices

Many of the matrices encountered when developing finite difference methods for approximating solutions to partial differential equations have a special property known as diagonal dominance. A square matrix A of order n is **weakly diagonally dominant** if for each $i = 1, 2, \ldots, n$ the following inequality holds,

$$|(A)_{ii}| \geq \sum_{j=1, j\neq i}^{n} |(A)_{ij}|.$$

The matrix is **strictly diagonally dominant** if

$$|(A)_{ii}| > \sum_{j=1, j\neq i}^{n} |(A)_{ij}|,$$

for each $i = 1, 2, \ldots, n$. One of the benefits of encountering a strictly diagonally dominant matrix is its invertibility.

Theorem B.18. *A strictly diagonally dominant matrix is invertible.*

Proof. This result is established using a proof by contradiction. Suppose A is an $n \times n$ strictly diagonally dominant matrix and there exists a vector $\mathbf{x} = (x_1, x_2, \ldots, x_n)^T \neq \mathbf{0}$ for which $A\mathbf{x} = \mathbf{0}$. Let k be the value of the index for which $|x_k| = \max_{1 \leq j \leq n} |x_j|$. By assumption $A\mathbf{x} = \mathbf{0}$ and thus

$$\sum_{j=1}^{n} (A)_{kj} x_j = 0 \iff (A)_{kk} x_k = -\sum_{j=1, j \neq k}^{n} (A)_{kj} x_j.$$

Taking the absolute value of both sides of the last equation and applying the triangle inequality produce

$$|(A)_{kk}| \, |x_k| \leq \sum_{j=1, j \neq k}^{n} |(A)_{kj}| \, |x_j|.$$

Since $\mathbf{x} \neq \mathbf{0}$ then $|x_k| > 0$ and thus

$$|(A)_{kk}| \leq \sum_{j=1, j \neq k}^{n} |(A)_{kj}| \frac{|x_j|}{|x_k|} \leq \sum_{j=1, j \neq k}^{n} |(A)_{kj}|,$$

which contradicts the assumption that matrix A is strictly diagonally dominant. □

If a weakly diagonally dominant matrix is **irreducible**, then it may be nonsingular as well. An $n \times n$ matrix A is called **reducible** if the set of indices $\{1, 2, \ldots, n\}$ can be partitioned into two nonempty, disjoint subsets $\{i_1, i_2, \ldots, i_\alpha\}$ and $\{j_1, j_2, \ldots, j_\beta\}$ with $\alpha + \beta = n$ and $(A)_{i_\mu j_\nu} = 0$ for all $\mu = 1, 2, \ldots, \alpha$ and $\nu = 1, 2, \ldots, \beta$. If a square matrix is not reducible, then it is irreducible. While it is not immediately apparent from this definition, a square matrix is irreducible if and only if it cannot be placed into upper triangular form by a sequence of row and column permutations [Brualdi and Ryser (1991), Chap. 3]. This definition of irreducibility is difficult to apply when n is large.

In many cases an easier condition to verify is whether the directed graph of matrix A is strongly connected. Let A be an $n \times n$ matrix and $\{P_1, P_2, \ldots, P_n\}$ be n distinct points called **vertices**. If $(A)_{ij} \neq 0$ then the **directed arc** from P_i to P_j is denoted as $\overrightarrow{P_i P_j}$. The **directed graph** of A denoted as $\mathbb{G}(A)$ is the set of directed arcs associated with A. A **directed path** from vertex P_{i_0} to P_{i_m} is a collection of directed arcs

$$\overrightarrow{P_{i_0} P_{i_1}}, \overrightarrow{P_{i_1} P_{i_2}}, \ldots, \overrightarrow{P_{i_{m-1}} P_{i_m}}$$

which connects vertex P_{i_0} to vertex P_{i_m}. In terms of matrix A, a directed path from vertex P_{i_0} to P_{i_m} exists if and only if

$$\prod_{j=0}^{m-1} (A)_{i_j i_{j+1}} \neq 0.$$

Finally a directed graph $\mathbb{G}(A)$ of matrix A is **strongly connected** if for each pair of vertices, for instance P_i and P_j, there is a directed path in $\mathbb{G}(A)$ with initial vertex P_i and terminal vertex P_j. Without proof the following theorem connects irreducible matrices and strongly connected graphs.

Theorem B.19. *An $n \times n$ matrix A is irreducible if and only if its directed graph $\mathbb{G}(A)$ is strongly connected.*

Example B.1. Show that the following matrix is reducible.

$$A = \begin{bmatrix} 0 & 1 & 0 & 0 \\ 1 & 0 & 0 & 0 \\ 0 & 0 & 0 & 1 \\ 0 & 0 & 1 & 0 \end{bmatrix}.$$

Solution. Note that the subsets of indices $\{1,2\}$ and $\{3,4\}$ are each nonempty and the subsets are disjoint. Furthermore $\{1,2\} \cup \{3,4\} = \{1,2,3,4\}$. Since

$$(A)_{13} = (A)_{14} = (A)_{23} = (A)_{24} = 0$$

matrix A is reducible. The directed graph also indicates A is reducible.

Example B.2. Show that the following matrix is irreducible.

$$A = \begin{bmatrix} 0 & 0 & 1 & 0 \\ 0 & 0 & 0 & 1 \\ 0 & 1 & 0 & 0 \\ 1 & 0 & 0 & 0 \end{bmatrix}.$$

Solution. Since the matrix is merely 4×4 then its irreducibility can be demonstrated by applying the definition of irreducibility. The nonempty, disjoint partitions of $\{1,2,3,4\}$ are listed below.

$\{1\}, \{2,3,4\}$	$\{2\}, \{1,3,4\}$	$\{3\}, \{1,2,4\}$	$\{4\}, \{1,2,3\}$	$\{1,2\}, \{3,4\}$
$\{1,3\}, \{2,4\}$	$\{1,4\}, \{2,3\}$	$\{2,3\}, \{1,4\}$	$\{2,4\}, \{1,3\}$	$\{3,4\}, \{1,2\}$
$\{1,2,3\}, \{4\}$	$\{1,2,4\}, \{3\}$	$\{1,3,4\}, \{2\}$	$\{2,3,4\}, \{1\}$	

The reader can check that for each of the 14 partitions of the indices there is at least one instance in which $(A)_{i_\mu j_\nu} \neq 0$. The directed graph provides an easier way to verify that A is irreducible.

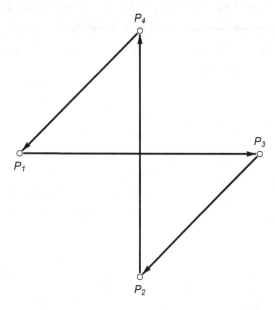

A square matrix A is **irreducibly diagonally dominant** if A is irreducible and

$$|(A)_{ii}| \geq \sum_{\substack{j=1, j \neq i}}^{n} |(A)_{ij}|$$

for $i = 1, 2, \ldots, n$ and if strict inequality holds for at least one value of i. The irreducibly diagonally dominant matrices are a subset of the weakly diagonally dominant matrices. The following theorem is originally due to Olga Taussky[15] [Taussky (1949)].

Theorem B.20. *If an $n \times n$ matrix A is irreducibly diagonally dominant, then A is invertible.*

---

[15]Olga Taussky, Austrian mathematician (1906–1995).

Proof. For the purposes of contradiction, suppose A is singular. This implies there is a vector $\mathbf{x} \neq \mathbf{0}$ such that $A\mathbf{x} = \mathbf{0}$. Since $\mathbf{x} \neq \mathbf{0}$ there exists $i \in \{1, 2, \ldots, n\}$ such that $|x_i| > 0$. Let $m \in \{1, 2, \ldots, n\}$ be the index for which $|x_i|$ is maximal and define vector $\mathbf{z}$ to be the scalar multiple of $\mathbf{x}$,

$$\mathbf{z} = |x_m|^{-1}\mathbf{x}.$$

By construction $|z_i| \leq 1$ for all $i = 1, 2, \ldots, n$. Define S to be the subset of $\{1, 2, \ldots, n\}$ for which $|z_i| = 1$. Note that S is nonempty. Since $A\mathbf{z} = |x_m|^{-1}A\mathbf{x} = \mathbf{0}$, then for any i,

$$0 = \sum_{j=1}^{n} (A)_{ij} z_j \iff -(A)_{ii} z_i = \sum_{j=1, j \neq i}^{n} (A)_{ij} z_j.$$

If $i \in S$ then

$$|-(A)_{ii} z_i| = |(A)_{ii}| = \left| \sum_{j=1, j \neq i}^{n} (A)_{ij} z_j \right| \leq \sum_{j=1, j \neq i}^{n} |(A)_{ij}| \, |z_j| \leq \sum_{j=1, j \neq i}^{n} |(A)_{ij}|.$$

The assumption $|(A)_{ii}| \geq \sum_{j=1, j \neq i}^{n} |(A)_{ij}|$ implies

$$|(A)_{ii}| \geq \sum_{j=1, j \neq i}^{n} |(A)_{ij}| \, |z_j|,$$

and hence the following equation holds for $i \in S$,

$$|(A)_{ii}| = \sum_{j=1, j \neq i}^{n} |(A)_{ij}| \, |z_j| = \sum_{j=1, j \neq i}^{n} |(A)_{ij}|. \tag{B.2}$$

Since the strict inequality

$$|(A)_{kk}| > \sum_{j=1, j \neq k}^{n} |(A)_{kj}|$$

must hold for at least one $k \in \{1, 2, \ldots, n\}$. Hence S is a nonempty, proper subset of $\{1, 2, \ldots, n\}$. Furthermore, for any $i \in S$, if it is the case that $\sum_{j=1, j \neq i}^{n} |(A)_{ij}| = 0$ then by Eq. (B.2), $A_{ij} = 0$ for $j = 1, 2, \ldots, n$. In other words, the ith row of A consists of zero entries. This contradicts the assumption that A is irreducible. Hence there exists $j \neq i$ such that $(A)_{ij} \neq 0$, which implies $|z_j| = 1$ by Eq. (B.2) which implies $j \in S$.

Since matrix A is irreducible, there is a directed path in the graph $\mathbb{G}(A)$ from vertex P_i to any vertex P_s. Suppose the directed path is

$$\overrightarrow{P_i P_{i_1}}, \overrightarrow{P_{i_1} P_{i_2}}, \ldots, \overrightarrow{P_{i_{r-1}} P_{i_r}}, \overrightarrow{P_{i_r} P_s}.$$

which corresponds to the nonzero, off-diagonal entries in A,

$$(A)_{ii_1}, (A)_{i_1 i_2}, \ldots, (A)_{i_{r-1} i_r}, (A)_{i_r s}.$$

Note that by Eq. (B.2), $|z_{i_1}| = 1$ which implies $i_1 \in S$. Once it is established that $i_1 \in S$ then Eq. (B.2) with i replaced by i_1 implies that $|z_{i_2}| = 1$ which implies $i_2 \in S$. In this way it is shown that $s \in S$. Since s can be any integer in $\{1, 2, \ldots, n\}$ then $S = \{1, 2, \ldots, n\}$ which contradicts the statement that S is a proper subset of $\{1, 2, \ldots, n\}$. Hence A is nonsingular. $\square$

B.6   Positive Definite Matrices

The matrices developed for the finite difference method for approximating solutions to Laplace's and Poisson's equations have a property which is useful when trying to establish that the iterative methods for solving linear systems converge. These matrices are called **positive definite**. A real $n \times n$ matrix A is positive definite if A is symmetric and if $\mathbf{x}^T A \mathbf{x} > 0$ for all $\mathbf{x} \neq \mathbf{0}$. For the sake of completeness, there are also concepts of positive semi-definite, negative semi-definite, and negative definite matrices, though they will not play a role in this textbook.

Theorem B.21. *If A is a real $n \times n$ positive definite matrix, then*

(1) *A is nonsingular,*
(2) *$a_{ii} > 0$ for $i = 1, 2, \ldots, n$,*
(3) *$\max_{1 \leq k, j \leq n} |a_{kj}| \leq \max_{1 \leq i \leq n} |a_{ii}|$.*
(4) *$a_{ij}^2 < a_{ii} a_{jj}$ for $i \neq j$.*

Proof. Suppose A is positive definite, then if $\mathbf{x} \neq \mathbf{0}$ and $A\mathbf{x} = \mathbf{0}$ then $\mathbf{x}^T A \mathbf{x} = 0$. This contradicts the assumption that A is positive definite and hence by Theorem B.5 A is nonsingular.

For $i = 1, 2, \ldots, n$ let $\mathbf{e}_i$ be the ith standard basis vector, then $a_{ii} = \mathbf{e}_i^T A \mathbf{e}_i > 0$.

Suppose $k \neq j$ and define $\mathbf{x}$ as

$$x_i = \begin{cases} 0 & \text{if } i \neq j \text{ and } i \neq k, \\ 1 & \text{if } i = j, \\ -1 & \text{if } i = k. \end{cases}$$

Since $\mathbf{x} \neq \mathbf{0}$ then $0 < \mathbf{x}^T A \mathbf{x} = a_{jj} + a_{kk} - a_{jk} - a_{kj}$ which implies $2a_{kj} < a_{jj} + a_{kk}$ since $A = A^T$. Now define $\mathbf{z}$ as the vector with

$$z_i = \begin{cases} 0 & \text{if } i \neq j \text{ and } i \neq k, \\ 1 & \text{if } i = j \text{ or } i = k. \end{cases}$$

Since $\mathbf{z} \neq \mathbf{0}$ then $0 < \mathbf{z}^T A \mathbf{z} = a_{jj} + a_{kk} + a_{jk} + a_{kj}$ which implies $-2a_{kj} < a_{jj} + a_{kk}$ since $A = A^T$. Thus if $k \neq j$,

$$|a_{kj}| < \frac{a_{jj} + a_{kk}}{2} \leq \max_{1 \leq i \leq n} |a_{ii}| \implies \max_{1 \leq k,j \leq n} |a_{kj}| \leq \max_{1 \leq i \leq n} |a_{ii}|.$$

Finally, suppose $i \neq j$ and define $\mathbf{x}$ as

$$x_k = \begin{cases} 0 & \text{if } k \neq j \text{ and } k \neq i, \\ \alpha & \text{if } k = i, \\ 1 & \text{if } k = j, \end{cases}$$

where $\alpha \in \mathbb{R}$. Since $\mathbf{x} \neq \mathbf{0}$ then $0 < \mathbf{x}^T A \mathbf{x} = \alpha^2 a_{ii} + 2\alpha a_{ij} + a_{jj}$. Note this quadratic polynomial in α has no real roots, therefore $4a_{ij}^2 - 4a_{ii}a_{jj} < 0$ or equivalently $a_{ij}^2 < a_{ii}a_{jj}$. $\square$

This appendix concludes with the statement of a theorem which lists other properties of positive definite matrices. This theorem refers to the **leading principal minors** of an $n \times n$ matrix. For a given $n \times n$ matrix A, the leading principal minors are the determinants of the upper-left $k \times k$ sub-matrices of A for $k = 1, 2, \ldots, n$. This is at odds with the definition of a minor given earlier, but given the context, should not confuse the reader.

Theorem B.22. *Let A be an $n \times n$ symmetric matrix. The following statements are equivalent.*

(1) *A is positive definite.*
(2) *All eigenvalues of A are positive.*
(3) *All the leading principal minors of A are positive.*
(4) *There exists a nonsingular square matrix B such that $A = B^T B$.*

Bibliography

Ablowitz, M. J. and Segur H. (1981). *Solitons and the Inverse Scattering Transform*, Society for Industrial and Applied Mathematics, Philadelphia, PA, USA.

Agrawal, G. P. (2013). *Nonlinear Fiber Optics*, 5th edition, Academic Press, Boston, MA, USA.

Al-Gwaiz, M. (2008). *Sturm-Liouville Theory and its Applications*, Springer, New York, NY, USA.

Asmar, N. H. (2016). *Partial Differential Equations with Fourier Series and Boundary Value Problems*, 3rd edition, Dover Publications, Inc., New York, NY, USA.

Atkinson, F. V. (1964). *Discrete and Continuous Boundary Value Problems*, Academic Press, New York, NY, USA.

Awrejcewicz, J. (2012). *Classical Mechanics*, Springer-Verlag, New York, NY, USA.

Bhatia, R. (2005). *Fourier Series*, The Mathematical Association of America, Washington, DC, USA.

Bleecker, D. and Csordas, G. (1996). *Basic Partial Differential Equations*, International Press of Boston, Cambridge, MA, USA.

Bôcher, M. (1906). "Introduction to the Theory of Fourier's Series," *Annals of Mathematics*, (2), **7**, No. 3, pp. 81–152.

Boyce, W. E. and DiPrima, R. C. (2012). *Elementary Differential Equations and Boundary Value Problems*, 10th edition, John Wiley & Sons, Inc., New York, USA.

Brauer, F. and Castillo-Chavez, C. (2014). *Mathematical Models in Population Biology and Epidemiology*, Springer, New York, NY, USA.

Brauer, F. and Nohel, J. A. (1989). *The Qualitative Theory of Ordinary Differential Equations: An Introduction*, Dover Publications, Inc., New York, USA.

Brualdi, R. A. and Ryser, H. J. (1991). *Combinatorial Matrix Theory*, Cambridge University Press, New York, NY, USA.

Burden, R. L., Faires, J. D. and Burden, A. M. (2016). *Numerical Analysis*, 10th edition, Brooks/Cole Publishing Company, Pacific Grove, CA, USA.

Burgers, J. M. (1948). "A Mathematical Model Illustrating the Theory of Turbulence," *Advances in Applied Mechanics*, **1**, pp. 171–199.

Carleson, L. (1966). "On Convergence and Growth of Partial Sums of Fourier Series," *Acta Mathematica*, **116**, No. 1, pp. 135–157.

Churchill R. V. and Brown J. W. (2014). *Complex Variables and Applications*, 9th edition, McGraw-Hill Education, New York, NY, USA.

Coddington, E.A. and Levinson, N. (1955). *Theory of Ordinary Differential Equations*, McGraw-Hill, Boston, MA, USA.

Courant, R. and Hilbert, D. (1989a). *Methods of Mathematical Physics*, Volume I, John Wiley & Sons, Inc., New York, NY, USA.

Courant, R. and Hilbert, D. (1989b). *Methods of Mathematical Physics*, Volume II, John Wiley & Sons, Inc., New York, NY, USA.

Crank, J. and Nicolson, P. (1947). "A Practical Method for Numerical Evaluation of Solutions of Partial Differential Equations of the Heat Conduction Type," *Proceedings of the Cambridge Philosophical Society*, **43**, No. 1, pp. 50–67.

Cushing, J. M. (1994). "The Dynamics of Hierarchical Age-Structured Populations," *Journal of Mathematical Biology*, **32**, pp. 705–729.

Debnath L. (2012). *Nonlinear Partial Differential Equations for Scientists and Engineers*, 3rd edition, Birkhäuser, Boston, MA, USA.

de Jager, E. M. (2011). "On the origin of the Korteweg-de Vries equation," preprint, https://arxiv.org/pdf/math/0602661v2.pdf.

Dirichlet, L. (1829). "On the Convergence of Trigonometric Series Which Serve to Represent an Arbitrary Function between Two Given Limits," *Journal für die reine und angewandte Mathematik*, **4**, pp. 157–169.

Drazin, P. G. and Johnson, R. S. (1989). *Solitons: An Introduction*, 2nd edition, Cambridge University Press, New York, NY, USA.

Evans, L. C. (2010). *Partial Differential Equations*, 2nd edition, American Mathematical Society, Providence, RI, USA.

Flanders, H. (1973). "Differentiation Under the Integral Sign," *American Mathematical Monthly*, **80**, No. 6, pp. 615–627.

Folland, G. B. (1992). *Fourier Series and Its Applications*, Wadsworth & Brooks/Cole, Pacific Grove, CA, USA.

Fulks, W. (1978). *Advanced Calculus*, John Wiley & Sons, New York, NY, USA.

Gardner C. S., Greene, J. M. Kruskal M. D. and Miura R. M. (1967). "Method for Solving the Korteweg-de Vries Equation," *Physical Review Letters*, **19**, pp. 1095–1097.

Garrity, T. A. (2015). *Electricity and Magnetism for Mathematicians: A Guided Path from Maxwell's Equations to Yang-Mills*, Cambridge University Press, Cambridge, UK.

Gelfand, I. M. and Fomin, S. V. (1991). *Calculus of Variations*, Dover Publications, Inc., Mineola, NY, USA.

Golub, G. H. and Ortega, J. M. (1992). *Scientific Computing and Differential Equation: An Introduction to Numerical Methods*, Academic Press, San Diego, CA, USA.

Golub, G. H. and Van Loan, C. F. (1996). *Matrix Computations*, 3rd edition, The Johns Hopkins University Press, Baltimore, MD, USA.

González-Velasco, E. A. (1995). *Fourier Analysis and Boundary Value Problems*, Academic Press, San Diego, CA, USA.

Gradshteyn, I. S. and Ryzhik, I. M. (2007). *Tables of Integrals, Series, and Products*, 7th edition, Academic Press, San Diego, CA, USA.

Griffiths, D. J. (1999). *Introduction to Electrodynamics*, 3rd edition, Prentice-Hall, Inc., Upper Saddle River, NJ, USA.

Guckenheimer, J. and Holmes, P. J. (1983). *Nonlinear Oscillations, Dynamical Systems, and Bifurcations of Vector Fields*, Springer-Verlag, New York, NY, USA.

Haberman, R. (1998). *Mathematical Models: Mechanical Vibrations, Population Dynamics, and Traffic Flow*, Society for Industrial and Applied Mathematics, Philadelphia, PA, USA.

Hale, J.K. (2009). *Ordinary Differential Equations*, Dover Publications, Inc., New York, USA.

Hamilton, M. F. and Blackstock, D. T. (1998). *Nonlinear Acoustics*, Academic Press, San Diego, CA, USA.

Hartman, P. (1982). *Ordinary Differential Equations*, 2nd edition, Birkhäuser, Boston, USA.

Hasegawa, A. and Matsumoto, M. (2012). *Optical Solitons in Fibers*, 3rd revised and enlarged edition, Springer-Verlag, Berlin, Germany.

Hu, G. Y. and O'Connell, R. F. (1996). "Analytical inversion of symmetric tridiagonal matrices". *Journal of Physics A: Mathematical and General*, **29**, pp. 1511–1513.

Kahan, W. M. (1958). "Gauss-Seidel Methods of Solving Large Systems of Linear Equations," Ph.D. Thesis, University of Toronto, Toronto, CA.

Kolmogorov, A. (1927). "Sur la Convergence des Séries de Fonctions Orthogonales," *Mathematische Zeitschrift*, **26**, No. 1, pp. 432–441.

Körner, T. W. (1988). *Fourier Analysis*, Cambridge University Press, Cambridge, UK.

Korteweg, D. J. and de Vries, G. (1948). "On the Change of Form of Long Waves Advancing in a Rectangular Canal, and on a New Type of Long Stationary Waves," *Philosophical Magazine C*, **240**, No. 39, pp. 422–443.

Krantz, S. G. (1999). *Handbook of Complex Variables*, Birkhäuser, Boston, USA.

Lindenstrauss, J., Evans, L. C., Douady, A., Shalev, A. and Pippenger, N. (1994). "Fields Medals and Nevanlinna Prize presented at ICM-94 in Zürich," *Notices of the American Mathematical Society*, **41**, No. 9, pp. 1103–1111.

Marsden, J. E. and Hoffman, M. J. (1999). *Basic Complex Analysis*, 3rd edition, W. H. Freeman and Company, New York, NY, USA.

McOwen, R. C. (2002). *Partial Differential Equations: Methods and Applications*, 2nd edition, Prentice-Hall, Upper Saddle River, NJ, USA.

Mohr, P. J., Newell, D. B., and Taylor, B. N. (2015). "CODATA Recommended Values of the Fundamental Physical Constants: 2014," preprint, National Institute of Standards and Technology, http://arxiv.org/pdf/1507.07956.pdf.

Myint-U, T. and Debnath, L. (2007). *Linear Partial Differential Equations for Scientists and Engineers*, 4th edition, Birkhäuser, Boston, USA.

Newell, A. C. (1985). *Solitons in Mathematics and Physics*, Regional Conference Series in Applied Mathematics, Vol. 48, Society for Applied and Industrial Mathematics, Philadelphia, USA.

Olesen, P. (2003). "An Integrable Version of Burgers Equation in Magnetohydrodynamics," *Physical Review E*, **68**, No. 1, pp. 016307–016312.

Ortega, J. M. (1990). *Numerical Analysis: A Second Course*, Society for Industrial and Applied Mathematics, Philadelphia, PA, USA.

Picone, M. (1910). "Sui Valori Eccezionali di un Parametro da cui Dipende Un'equazione Differenziale Lineare del Secondo Ordine," *Annali della Scuola Normale Superiore di Pisa - Classe di Scienze*, **11**, No. 1, pp. 1–141.

Pinsky, M. A. (1998). *Partial Differential Equations and Boundary-Value Problems with Applications*, 3rd edition, American Mathematical Society, Providence, RI, USA.

Polyanin, A. D. and Manzhirov, A. V. (2008). *Handbook of Integral Equations*, 2nd edition, Chapman & Hall/CRC Press, Boca Raton, FL, USA.

Prüfer, H. (1926). "Neue Herleitung der Sturm-Liouvilleschen Reihenentwicklung stetiger Funktionen," *Mathematische Annalen*, **95**, No. 1, pp. 499–528.

Rauch, J. (2012). *Hyperbolic Partial Differential Equations and Geometric Optics*, American Mathematical Society, Providence, RI, USA.

Rogers, C. and Shadwick, W.F. (1982). *Bäcklund Transformations and Their Applications*, 1st edition, Academic Press, New York, NY, USA.

Rose, N. J. (2003). "On an Eigenvalue Problem Involving Legendre Functions," http://www4.ncsu.edu/~njrose/pdfFiles/Legendre.pdf.

Rosser, J. B. (1975). "Nine-Point Difference Solutions for Poisson's Equation," *Computers & Mathematics with Applications*, **1**, No. 3–4, pp. 351–360.

Rudin, W. (1976). *Principles of Mathematical Analysis*, 3rd edition, McGraw-Hill, New York, NY, USA.

Russell, J. S. (1845). "Report on Waves", Report of the Fourteenth Meeting of the British Association for the Advancement of Science, London.

Schildt, H. (2014). *Java: The Complete Reference*, 9th edition, McGraw-Hill Education, New York, NY, USA.

Smith, G. D. (1985). *Numerical Solution of Partial Differential Equations: Finite Difference Methods*, 3rd edition, Clarendon Press, Oxford, UK.

Smith, R. T. and Minton, R. B. (2012). *Calculus: Early Transcendental Functions*, 4th edition, McGraw-Hill, New York, NY, USA.

Smoller, J. (1994). *Shock Waves and Reaction-Diffusion Equations*, 2nd edition, Springer-Verlag, New York, NY, USA.

Steele, J. M. (2004). *The Cauchy-Schwarz Master Class: An Introduction to the Art of Mathematical Inequalities* (*MAA Problem Books*), Cambridge University Press, Cambridge, UK.

Stolen, R. H. (2008). "The Early Years of Fiber Nonlinear Optics," *IEEE/OSA Journal of Lightwave Technology*, **26**, No. 9, pp. 1021–1031.

Strikwerda, J. C. (2004). *Finite Difference Schemes and Partial Differential Equations*, 2nd edition, Society for Industrial and Applied Mathematics, Philadelphia, PA, USA.

Szegő. G. (1939). *Orthogonal Polynomials*, 4th edition, American Mathematical Society, Providence, RI, USA.

Tao, T. (2009). "Why are solitons stable?" *Bulletin of the American Mathematical Society*, **46**, pp. 1–33.

Taussky, O. (1949). "A Recurring Theorem on Determinants," *The American Mathematical Monthly*, **56**, No. 10, pp. 672–676.

Toda, M. (1989). *Theory of Nonlinear Lattices*, Springer-Verlag, New York, NY, USA.

Trench, W. F. (1978). *Advanced Calculus*, Harper & Row, New York, NY, USA.

Walker, J. S. (1988). *Fourier Series*, Oxford University Press, New York, NY, USA.

Watson, G. N. (1922). *A Treatise on the Theory of Bessel Functions*, 2nd edition, Cambridge University Press, New York, NY, USA.

Weisstein, E. W. (2013). "Bessel Function Zeros," From *MathWorld* — A Wolfram Web Resource. http://mathworld.wolfram.com/BesselFunctionZeros.html.

Whitham, G. B. (1974). *Linear and Nonlinear Waves*, John Wiley & Sons, Inc., New York, NY, USA.

Young Jr., D. M. (1950). "Iterative Methods for Solving Partial Difference Equations of Elliptic Type," Ph.D. Thesis, Harvard University, http://www.ma.utexas.edu/CNA/DMY/david_young_thesis.pdf.

Zettl, A. (2005). *Sturm-Liouville Theory*, American Mathematical Society, Providence, RI, USA.

Zabusky, N. J. and Kruskal, M. D. (1965). "Interaction of "Solitons" in a Collisionless Plasma and the Recurrence of Initial States," *Physical Review Letters*, **15**, No. 6, pp. 240–243.

Index

Printed in the United States
By Bookmasters